FRONTISPIECE. Jacob M. Valentine (right), retired biologist of the U.S. Fish and Wildlife Service, became the first recipient of the L. H. WALKINSHAW CRANE CONSERVATION AWARD in honor of his long-time contribution to the study and conservation of cranes. The award was presented on 11 January 1996 by Scott G. Hereford (left), President of the North American Crane Working Group. Jake has authored approximately 25 papers on cranes and spent more than 30 years conducting research on the Mississippi sandhill crane. His work was a major factor in creation of Mississippi Sandhill Crane National Wildlife Refuge. (Photo by J. W. Hardy.)

Cover Photo: Territorial male Mississippi sandhill crane (*Grus canadensis pulla*) on Mississippi Sandhill Crane National Wildlife Refuge by Ute Bradter.

PROCEEDINGS OF THE SEVENTH NORTH AMERICAN CRANE WORKSHOP

10 – 13 JANUARY 1996
BILOXI, MISSISSIPPI, U.S.A.

Editors
RICHARD P. URBANEK and DALE W. STAHLECKER

Editorial Assistant
VICTORIA J. CRENSHAW

Final Review Editor
THEODORE A. BOOKHOUT

Coordinating Editor
WENDY M. BROWN

Associate Editors
RODERICK C. DREWIEN
DOUGLAS H. JOHNSON
JAMES C. LEWIS
GLENN H. OLSEN
THOMAS V. STEHN
RICHARD P. URBANEK

Published by
NORTH AMERICAN CRANE WORKING GROUP

Publication Co-sponsored by
U.S. FISH AND WILDLIFE SERVICE
THE WILDLIFE SOCIETY

Proceedings of the Seventh North American Crane Workshop

Printed in the United States of America by BookCrafters, Chelsea, Michigan

Available from
International Crane Foundation
E-11376 Shady Lane Road
P.O. Box 447
Baraboo, Wisconsin 53913-0447 USA

$25.00 Postpaid

Proceedings of the North American Crane Workshops:

Lewis, J. C., editor. 1976. Proceedings of the [1975] international crane workshop. Oklahoma State University Publishing and Printing, Stillwater. 355pp.

Lewis, J. C., editor. 1979. Proceedings of the 1978 crane workshop. Colorado State University Printing Service, Fort Collins. 259pp.

Lewis, J. C., editor. 1982. Proceedings of the 1981 crane workshop. National Audubon Society, Tavernier, Fla. 296pp.

Lewis, J. C., editor. 1987. Proceedings of the 1985 crane workshop. Platte River Whooping Crane Maintenance Trust, Grand Island, Nebr. 415pp.

Wood, D. A., editor. 1992. Proceedings of the 1988 North American crane workshop. Florida Game Fresh Water Fish Comm. Nongame Wildl. Program Tech. Rep. 12. 305pp.

Stahlecker, D. W., and R. P. Urbanek, editors. 1992. Proceedings of the sixth North American crane workshop. North American Crane Working Group, Grand Island, Nebr. 179pp.

Urbanek, R. P., and D. W. Stahlecker, editors. 1997. Proceedings of the seventh North American crane workshop. North American Crane Working Group, Grand Island, Nebr. 262pp.

Suggested citation format for articles in the seventh workshop:

Author(s). 1997. Title of paper. Proc. North Am. Crane Workshop 7:000-000.

ISBN 0-9659324-0-0
Library of Congress Catalog Card Number: 97-68410

PREFACE

In 1975 a group of North American crane researchers and managers convened a workshop at the fledgling International Crane Foundation in Baraboo, Wisconsin. At the end they agreed to publish the proceedings of the workshop and to reconvene in 3 years to continue to share information, ideas, and support. This 1975 international crane workshop served as the first North American crane workshop and was followed by workshops in 1978 (Rockport, Texas), 1981 (Grand Teton National Park, Wyoming), 1985 (Grand Island, Nebraska), and 1988 (Kissimmee Prairie, Florida). In these first 5 North American workshops, proceedings were published by the sponsoring agencies or organizations. At the Florida meeting participants formed the North American Crane Working Group (NACWG), an organization of "professional biologists, aviculturists, land managers, and other interested individuals dedicated to the conservation of cranes and their habitats in North America." In 1991 the Sixth North American Crane Workshop was held in Regina, Saskatchewan. This was the first workshop in which the NACWG sponsored the workshop and published the proceedings. Because the quality and commitment of the North American Crane Working Group appear to be substantial and long-lived, the Proceedings of the North American Crane Workshop became a numbered series with Proc. North Am. Crane Workshop 6.

NACWG provides information exchange through a semi-annual newsletter *The Unison Call*, sponsors the North American Crane Workshop every 3–5 years and publishes its proceedings, addresses conservation issues affecting cranes and their habitats, promotes appropriate research on crane conservation and management, and promotes a better understanding and appreciation of cranes and their habitats among the general public. A Board of Directors consisting of 7 members elected to 3-year terms directs the group. Board members in 1997 were Mary Anne Bishop, Wendy Brown, Scott Hereford (CEO), Gary Lingle, Steve Nesbitt, Glenn Olsen, and Richard Urbanek. The NACWG board and membership express their appreciation to the Platte River Whooping Crane Maintenance Trust for contributing office space and equipment, thereby providing an unofficial base of operations for the Working Group during the first decade of its existence.

As the Seventh Workshop opened in Biloxi, Mississippi, a Congressionally-imposed furlough of government employees had just ended, but last minute travel authorizations could not be approved in time for some expected workshop participants to attend. Meanwhile, a blizzard on the East Coast grounded many from the Patuxent contingent and others in the D.C. area. In spite of these difficulties, the Workshop was an outstanding success with 71 crane biologists from 25 states/provinces present. Of 32 scheduled papers, 24 were presented in addition to 4 alternates. A special discussion session on reintroduction techniques for migratory cranes generated 3 more papers. The Proceedings of the Seventh North American Workshop contain 34 of these papers and abstracts of 3 others that will be published elsewhere. The NACWG gratefully acknowledges the U.S. Fish and Wildlife Service; The Wildlife Society, Southeast Section; Chevron Oil Company, Pascagoula Refinery; International Paper Company, Moss Point Mill; and Grand Casino for sponsoring the Workshop and the U.S. Fish and Wildlife Service and The Wildlife Society, Southeast Section, for sponsoring publication of these Proceedings.

Papers submitted for publication in the Proceedings are peer-reviewed according to scientific journal standards. Because the Proceedings are intended to serve as a source of current scientific knowledge and state-of-the-art in crane biology, unpublished as well as published information may be included in the literature cited. We thank each Associate Editor for many hours of paper review and editorial effort. We also thank the following referees:

Jane E. Austin	David H. Ellis	Steven E. Landfried	M. A. Ottinger
Alan J. Bennett	George F. Gee	Julia A. Langenberg	Pamela J. Pietz
Douglas S. Benning	Jeff S. Hatfield	James C. Lewis	Ruth E. Shea
Mary Anne Bishop	Gary L. Ivey	Carroll D. Littlefield	S. Bret Snyder
Theodore A. Bookhout	Susan I. Jarvi	Wayne H. MacAllister	Marilyn G. Spalding
Wendy M. Brown	Douglas H. Johnson	Meenakshi Nagendran	Scott R. Swengel
Felipe Chavez-Ramirez	Joseph D. Kaplan	Stephen A. Nesbitt	Mike R. Szymczak
John C. Clem	Cameron B. Kepler	Glenn H. Olsen	Paul A. Vohs
Roderick C. Drewien	Gary L. Krapu	Kathleen E. O'Malley	Scott R. Winterstein

Theodore A. Bookhout provided valuable editorial advice, assistance, and review. Seney National Wildlife Refuge was instrumental in providing support and facilities to publish these Proceedings. As part of her Wildlife Biology Internship sponsored by the Seney Natural History Association, Vicki Crenshaw organized and typeset the initial draft. Seney Interns Matthew Williams, Stefanie Miklovic, and Nicole Rein proofread final copy. Rebecca Huber and staff at BookCrafters were cooperative in our desire to produce a volume of high quality. Without the efforts of all involved, these Proceedings would not have been possible.

Richard P. Urbanek and Dale W. Stahlecker, Editors
June 1997

Battle Creek, Michigan, 1942

Kissimmee Prairie, Florida, 1988

LAWRENCE H. WALKINSHAW
1904–93

Dr. Lawrence "Larry" Walkinshaw died 16 January 1993 at the age of 88. He was born 25 February 1904 in Calhoun County, Michigan, and graduated from the University of Michigan School of Dentistry in 1929. Though a dentist by training, he was a leading authority on cranes on which he published 61 articles and two books: *The Sandhill Cranes* (1949) and *Cranes of the World* (1973). He described the Canadian sandhill crane (*Grus canadensis rowani*) in 1965. He published on other species as well, most notably Kirtland's warbler, but he was first and foremost a "crane man."

Larry remained an active field ornithologist throughout his life. In 1934, on foot with Clara, his wife and to-be-field companion of 62 years, he recorded some of the first survey data on cranes and other birds in a remote area of the Michigan Upper Peninsula on what is now Seney National Wildlife Refuge, before it was a refuge. After retiring from dentistry, he divided time between homes near sandhill crane habitat in Florida and Michigan. Larry was a master bird observer, world crane lover, determined nest finder, and meticulous data collector. For many years there were few papers on cranes published in North America for which he did not serve as a refereeing authority. The dedication of the proceedings of the First International Crane Workshop to Dr. Walkinshaw appropriately identified him as "The Father of International Studies of Gruiformes." He was the constant mentor to succeeding generations of crane researchers.

One of Larry's abiding concerns was recovery of the whooping crane. He participated in some of the original aerial searches to locate the then unknown breeding grounds of the only remaining migratory population before its eventual discovery in 1954. During his lifetime he saw the species climb back from a perilous low of 22 birds to nearly 250, with captive populations established and efforts underway to reintroduce whooping cranes to former range.

This kind and gentle man, whose contributions to science never flagged, will be missed by those of us who benefited from his counsel and were inspired by his spirit. The Dr. L. H. Walkinshaw Memorial Scholarship Fund has been established in his honor in the Department of Ornithology at Cornell University. The North American Crane Working Group has instituted the L. H. Walkinshaw Crane Conservation Award to recognize the outstanding crane biologists following in his footsteps. His life spanned 10 decades; his legacy will live forever. (Photos courtesy of J. R. Walkinshaw.)

Stephen A. Nesbitt
Gainesville, Florida

Mary Anne Bishop
Cordova, Alaska

Richard P. Urbanek
Seney, Michigan

THOMAS C. TACHA
1951–95

Tom Tacha, age 43, died on 1 May 1995 of a brain aneurysm. At the time he was Professor of Wildlife Sciences at Caesar Kleberg Wildlife Research Institute, Texas A & M University–Kingsville.

Tom was an extremely productive scientist; his accomplishments fill many pages and can only be touched upon here. From 1990 to 1995 he secured more than 1 million dollars in grant money, the most by any professor at Texas A & M–Kingsville. His motto for his graduate students at Southern Illinois University, where he worked from 1983 to 1990, was "work, work, work." As a major professor (I was 1 of his 22 graduate students), he set the ultimate example of dedication and diligence by working long hours, often leaving his office late in the day *after* the graduate students. His lemon-yellow VW Bug was often the only vehicle in the staff parking lot early in the morning and late in the evening. Tom "adopted" graduate students; he spent long hours in the advising process, ensuring that everything from funding to political well-being was satisfied.

Tom earned his B.S. degree from Kansas State University in 1973, M.S. from South Dakota State in 1976, and Ph.D. from Oklahoma State in 1981. He had a deep appreciation for wetlands and their avian inhabitants, and this is reflected in his accomplishments. His Master's work was on Canada geese and his Ph.D. on Mid-continent sandhill cranes. He was actively involved at crane workshops. More of his 71 publications dealt with cranes (30) than any other single species or topic.

Tom loved cranes. As his only "crane student," he took liberties to share his thoughts (and sense of humor) about cranes with me. He would say things like "There are cranes and then there are dickie-birds!" In discussing his heavy involvement with studies of Canada geese, he would say "They aren't cranes but they are perhaps the next best thing."

Tom worked hard to convey the importance of the North Platte River Valley as a staging area for Mid-continent sandhill cranes. When Tom escaped university walls to pay me a visit on our North Platte study area, he was transformed from a rather serious, stern professor into a "kid with a new toy." Speaking of the sandhill cranes, he would say "They may not be the prettiest bird but they sure have a lot of character!" It was appropriate that Tom's ashes be spread in that valley where he studied his favorite species. (Photo by C. R. Paine.)

Martin J. Folk
St. Cloud, Florida

Whooping Crane Coordinator Retires

Dr. James C. Lewis retired from his official duties as U.S. Whooping Crane Coordinator in February 1997. Jim received his B.S. and M.S. degrees from the University of Michigan and Michigan State Universities, respectively, and completed his doctoral dissertation on sandhill cranes at Oklahoma State University. He continued to work with cranes throughout his career and conducted and contributed to many important research and management projects. Jim worked for the U.S. Fish and Wildlife Service for 30 years, beginning in 1967 as Leader of the Oklahoma Cooperative Wildlife Research Unit. Later he became Georgia CWRU Leader, and from 1984 to 1997 he served as the Whooping Crane Coordinator and Recovery Team Leader. He was instrumental in developing international agreements with Canada and 13 U.S. states to protect whooping cranes in migration and on the wintering grounds, worked closely with the Grays Lake and Florida reintroduction projects, and served as the Service's representative to the Avian Power Line Interaction Committee, a consortium of national utility companies which have supported research and developed line markers to reduce bird collisions. He has authored or co-authored numerous book chapters and more than 70 published scientific articles, and he served as editor and co-chairperson of the first 4 Proceedings of the North American Crane Workshop. Jim has been the recipient of many distinguished awards throughout his career, including an honor award from the Whooping Crane Conservation Association. He will no doubt continue to contribute to the conservation of cranes and other wildlife as he enjoys his retirement. (Photo courtesy of U.S. Fish and Wildlife Service.)

Wendy M. Brown
Albuquerque, New Mexico

CONTENTS

HISTORY OF BREEDING PAIRS AND NESTING SITES OF THE MISSISSIPPI SANDHILL CRANE

JACOB M. VALENTINE, JR.,[1] U.S. Fish and Wildlife Service, 1306 Greenbriar Road, Lafayette, LA 70503, USA
SCOTT G. HEREFORD, U.S. Fish and Wildlife Service, Mississippi Sandhill Crane National Wildlife Refuge, 7200 Crane Lane, Gautier, MS 39553, USA

Abstract: Thirty-four Composite Nesting Areas (CNA's) of endangered Mississippi sandhill cranes (*Grus canadensis pulla*) within the current breeding range in Jackson County, Mississippi, were located from 1965 to 1996, primarily by ground searching. Of those 34 CNA's located, 28 were on 1 of the 3 refuge units and 6 were off but adjacent to the refuge's Ocean Springs Unit. Five of the CNA's had 2–3 distinct smaller core nesting areas within. Two CNA's had active nests in more than 20 years, and those nests accounted for 11% of the total. Nine CNA's accounted for 130 (63%) of the nests. Eleven CNA's were used in only 1 year. The first marked cranes nested in 1985; by 1996, 34 marked cranes had nested. The mean distance between different CNA's used by same individual(s) in different years was 2.0 km. The shortest distance between active nests in a year was 0.8 km. The mean distance from release pen to CNA for 32 released cranes was 2.9 km; only 4 cranes used CNA's in different units from their release pens.

PROC. NORTH AM. CRANE WORKSHOP 7:1-9

Key words: *Grus canadensis pulla*, Mississippi, nesting, sandhill crane, territory.

The endangered Mississippi sandhill crane is a non-migratory subspecies found in the wild only on and adjacent to the Mississippi Sandhill Crane National Wildlife Refuge (NWR) in extreme southeastern Mississippi. Earlier studies have described nesting habitats (Valentine and Noble 1970), nesting ecology (Valentine 1982, 1992), captive release efforts (Zwank and Derrickson 1982, Zwank and Dewhurst 1992, Mitchell and Zwank 1987, Wilson 1987, Valentine and Logan 1991), changes in habitat (Smith and Valentine 1987), and status and management (U.S. Fish and Wildlife Service [USFWS] 1991; Valentine 1979, 1981, 1987; Valentine and Noble 1976). This paper is a historical record of the breeding pairs and nest area occupancy during 32 years.

We acknowledge and thank past and present employees of the Mississippi Sandhill Crane NWR, Louisiana State University students, State of Mississippi biologists, Student Conservation Association Resource Assistants, refuge volunteers, interns, Mississippi Gulf Coast Audubon Society members, contract helicopter pilots, and numerous others for participating in nest searching efforts over the years. International Paper Company, DeSoto National Forest, Jackson County, Mississippi, Mississippi Gulf Coast Wastewater Authority, State of Mississippi, N. Jordan, E. Huger, and others allowed nest searchers access to their lands. We gratefully acknowledge J. R. Twiss for preparing the figure.

STUDY AREA AND METHODS

This study was conducted on the 7,770-ha Mississippi Sandhill Crane NWR and adjacent lands in Jackson County east of the Pascagoula River in southeastern Mississippi (Fig. 1) between April 1965 and June 1996. The study area has been described elsewhere (Valentine and Noble 1970, Smith and Valentine 1987). There are 3 units, Gautier, Ocean Springs, and Fontainebleau, each with an enclave of cranes. Although units are only 3.0 km apart, cranes rarely go into the neighboring units during summer, but they may use the same feeding grounds in winter. Cranes also have nested outside the refuge's Ocean Springs Unit on adjacent private and public lands.

Habitat management for breeding sites has included clearing of pine trees by hand, bulldozing, commercial harvest, and prescribed burning. Shallow impoundments were created by raising culvert elevations on roads and with other water-control structures.

Nests were found during 32 annual searches (1965–96) mostly on foot, but also by airplane, helicopter, horse, and radiotelemetry. Nest site locations were plotted on USGS topographic maps. Beginning in the early 1980's, nest sites were permanently marked with a metal stake and tag. Composite Nesting Areas or CNA's (Kuyt 1981) were used to identify the site where cranes habitually nested. Nesting areas were designated only after evidence of an active nest was discovered. Generally, CNA's were designated with unique numbers. Some CNA's consisted of 2–3 smaller nesting areas used by the same pair. A single CNA was used by but 1 pair in a given year.

RESULTS

The crane breeding population was small (Table 1); nesting habitats were widely separated. Three secluded territories, 0.8 km apart, along Ben Williams Swamp (Fig.

[1]Retired.

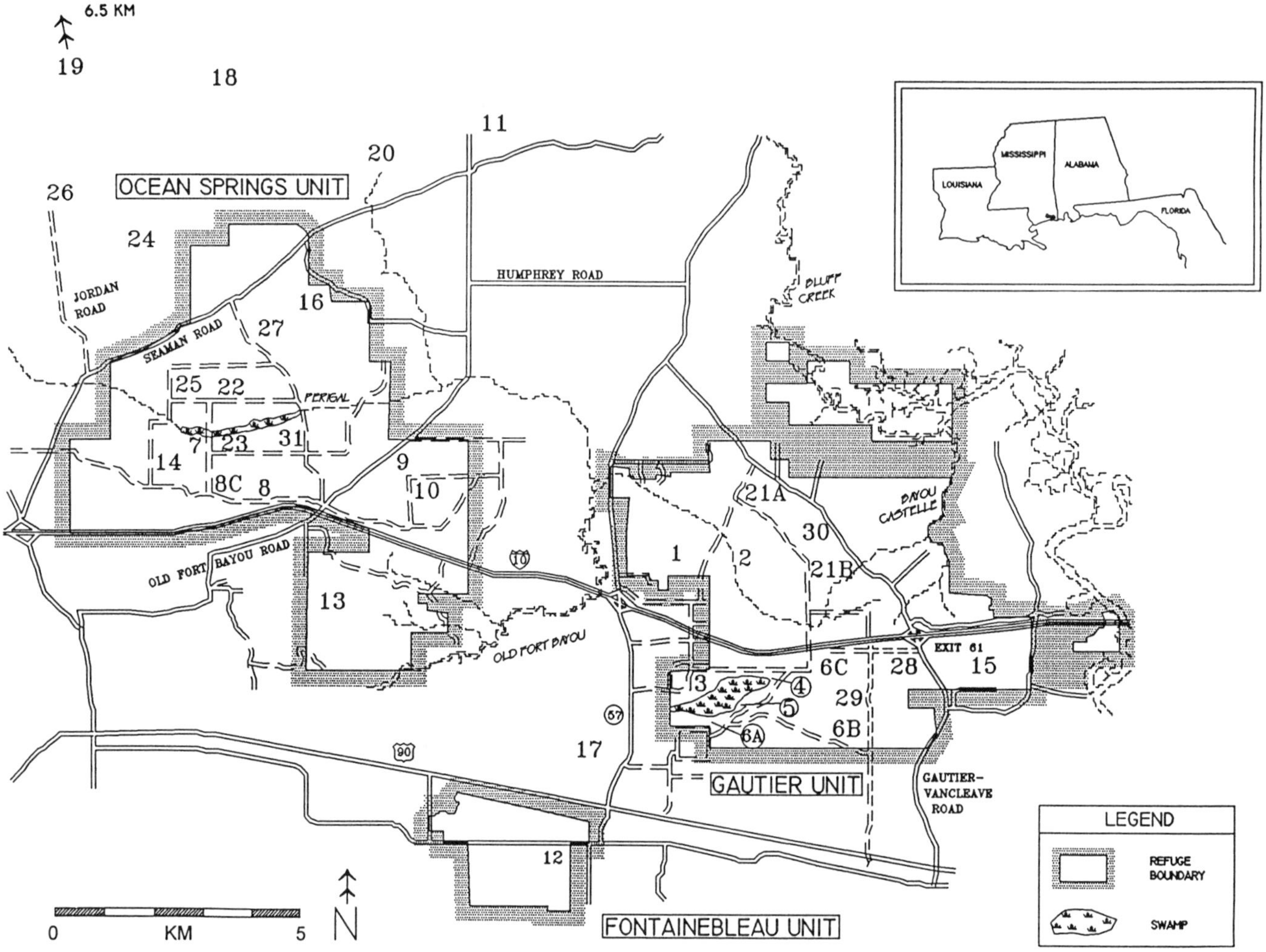

Fig. 1. Locations of Mississippi sandhill crane Composite Nesting Areas (CNA's), Jackson County, Mississippi.

1), were closest. Large open savannas were occupied by only 1 pair at a time. There were 34 CNA's found and used during 1965–96. One of these 34 was described prior to the study period (Walkinshaw 1960). A thirty-fifth CNA was described previously (McIlhenny 1938, Gandy and Turcotte 1979) but not actively used during the study period. Five CNA's included 2 or more separate nesting areas. There were 6 CNA's off the refuge and adjacent to the Ocean Springs Unit. There were 28 CNA's on the refuge: 14 on the Gautier Unit, 13 on the Ocean Springs Unit, and 1 on the Fontainebleau Unit. There were 207 active nests including 20 renests. Among the latter, 9 were used only during 1 or 2 years.

The most frequently occupied CNA had active nests in 21 of the 32 years. Eleven CNA's were used only once. Two CNA's had 44 (11%) of the nests, and 9 CNA's accounted for 130 (63%) of the nests. Five CNA's were not used after 1970. Nesting in 7 CNA's was initiated after 1992.

Crane nesting occurred in the Gautier Unit in all 32 years. Nests were found in 26 years in the Ocean Springs Unit, 10 years in the Fontainebleau Unit, and 16 years in off-refuge CNA's. There were 111 (54%) nests in Gautier, 62 (30%) in Ocean Springs, 10 (5%) in Fontainebleau, and 24 (11%) in off-refuge CNA's.

Marked cranes first nested in 1985. By 1991 the majority of nesting cranes were marked. Six (2 female, 4 male) marked wild-hatched cranes and 28 (9 female, 19 male) marked captive-bred and released cranes nested. The CNA use by certain unmarked cranes was determined from their association with marked cranes and the small population size. Some individual breeding cranes or pairs moved between CNA's (Table 2), nesting in different CNA's in different years and occasionally in the same year. Movement between CNA's was always within a refuge unit, but some movements were across paved public roads, including an interstate highway, and nearly all were across unpaved refuge roads.

Table 1. Occupancy of Mississippi sandhill crane Composite Nesting Areas (CNA's), Jackson County, Mississippi, 1965–96. Number of active nests (first plus renests) is indicated.

	Year																																No. years	No. of
CNA	65	66	67	68	69	70	71	72	73	74	75	76	77	78	79	80	81	82	83	84	85	86	87	88	89	90	91	92	93	94	95	96	occupied	nests
1	1	1		1	1		1	1		1		1	2	1	1	1	2	1	1	1	1	1				1						1	20	22
2		2	1	1	1	1		1	1	1							1															1	10	11
3	1																																1	1
4		1			1		1	1	1	1	1	1	1	1	1	1		2	1				1			1	1		1	1	1	1	21	22
5							1		1	1														1									4	4
6A							1				1	1							1													1	5	5
6B																				1													1	1
6C																									2								1	2
7		1	1		1	1	1																						2				6	7
8		1	1	1				1	1																				1		1	1	8	8
8C																													1				1	1
9			1	1	1																												3	3
10			1																														1	1
11				1	1																												2	2
12				1	1	1	1	1	1	1	1	1	1																				10	10
13					1	1	1	1					1	1			2		1														8	9
14																			1	1	2	1	2	1	2	1	2			1	1		11	15
15											1	1	1		1			1	1	1									2	1	1	1	11	12
16											1						1				1		1		1								5	5
18	1																																1	1
19													1	1							2	1	1	1	2	3							8	12
20														1	1																		2	2
21A																			1		1	1	1		1	2	2	1	1	1	1	1	12	14
21B																							2	1	2	1	1	1	1	1	1	1	10	12
22																				1		1	1			1	1						5	5
23																		1					1	1	1								4	4
24																							1	1		1	1				1	1	6	6
25																									1	1							2	2
26																														1			1	1
27																														1			1	1
28																														1	1	1	3	3
29																																1	1	1
30																																1	1	1
31																																1	1	1

Released cranes tended to use CNA's near their release pens (Table 3).

Gautier Unit

Brown's Trail.— Before 1960 the Brown's Trail savanna was about 1.5 km wide and 3.5 km long. Housing and commercial developments and U.S. I-10 reduced nesting habitat to about 150 ha. There are 2 CNA's in the Brown's Trail area. CNA 1 had active nests in 20 years. Two active nests were found (1965–66) in the 23-ha northeastern savanna (CNA 1A), but the pair abandoned the site after a wildfire coursed through in 1967. No nests were found there until 1990, when an unknown pair nested, but none has nested since.

The original CNA 1A pair may have moved after the fire to the large southwestern "Coke Plant" savanna, CNA 1B, in 1978. In summer 1984, wild female (#627) of the CNA 1B pair was alone with her half-grown colt. Released crane #608 became a surrogate parent that summer and helped raise the juvenile. In 1985, eggs were laid but the nest was abandoned because of disturbance; in December #608 died. In late winter, female #627 mated with wild crane #628 and the 2 nested without success in 1986. Although still paired in 1987, they did not nest. No breeding pairs occupied the territory after 1987 until 1996, although nonbreeding

Table 2. Movement of individual Mississippi sandhill cranes between Composite Nesting Areas (CNA's) in Jackson County, Mississippi, 1965-96.

From		To		Male	Female	Distance between
Year	CNA	Year	CNA	ID	ID	CNA's (km)
1980	4	1981	2	UM[a]	UM	3.0
1981	2	1982	4	UM	UM	3.0
1982	23	1983	14	UM	UM	1.2
1982	4	1983	6A	UM	UM	1.8
1983	6A	1984	6B	UM	UM	3.0
1987	4	1989	6C	UM	634	1.5
1988	24	1989	25		640	1.8
1988	23	1990	24		643	3.3
1988	23	1990	25	644		1.6
1989	16	1991	22	631	UM	2.1
1991	22	1994	27	631	UM	1.3
1993	8C	1994	14	038	031	1.2
1995	28	1996	29		136	1.1
$\bar{x}$						2.0
SD						0.77

[a] Unmarked.

associations of released cranes have occurred there. In 1996 a family (unbanded male, captive-bred #912 female, and chick) appeared on adjacent private lands and presumably the pair nested in this territory.

The adjacent narrow 15-ha CNA 2 (7 nests and 1 renest) in the east savanna, east of the swamp, was occupied concurrently with CNA 1A-B, 1.4-2.2 km apart during 1966-74, except for 1971. This CNA was considered one of the most pristine hydric savannas in the area. After 1974 no nests were found in CNA 2 until 1996, except when the CNA 4 pair nested there in 1981. Five potential nesting openings were hand-cleared and burned several times during 1980-93. A hand-reared pair, #02 male and #933 female, nested in 1996. The Gautier release pen is located at the northern end of the CNA 2 savanna.

Ben Williams Swamp.— The swamp is an area of 104 ha of slash pine (*Pinus elliottii*) and pond cypress (*Taxodium ascendens*) with wet savanna nesting habitat along the edges. During 1965-74, 4 separate nesting areas (Fig. 1) were found along swamp edges. One nest was found in CNA 3 on the north-central edge in 1965; it appeared to have been occupied for several years. After 30 years without occupancy, in 1995 an unknown pair built several start nests in CNA 3 but no eggs were found.

In 1971-74 there were 3 CNA's (4, 5, 6) about 0.8 km apart along the eastern half of the swamp edge. The territory used most (21 years) in the study period was the hydric 6-ha CNA 4 swamp edge along the eastern edge of the Ben Williams Swamp. In a pre-study period, Walkinshaw (1960) found 2 nests in CNA 4 in 1940 and 1 in 1960. The CNA 4 was found again by Valentine in 1966, with nests in 1969 and 1971-80. After their nesting site was burned, CNA 4 pair moved in 1981 3.0 km north to vacant CNA 2 where they nested but destroyed their only egg. They returned to CNA 4 in 1982. After losing a chick in 1982, they nested nearby without success. In 1983 the pair abandoned their non-viable eggs and renested in vacant CNA 6A where they hatched a chick. In 1984 they nested 2.4 km from CNA 4 in a cleared savanna, CNA 6B, without producing young; no nests were found in 1985 or 1986. One of the pair was thought to have been killed by a coyote (*Canis latrans*) in 1985.

A newly formed pair (wild male and captive-bred female #634) nested in CNA 4 in 1987. No nests were found in 1988, but #634 and a wild male nested 1.5 km northeast of CNA 4 in 1989 in CNA 6C. In 1990 female #634 acquired a new mate (captive-bred #855). The pair nested and laid eggs in CNA 4 but deserted. In 1991 the same pair nested, but again deserted the nest. No nests were found in 1992. In 1993 the pair produced a fledgling. Male #855 mated with an unbanded female and nested in 1994, but no chicks were seen. Male #855 disappeared in winter and captive-bred #018 replaced him, nesting with the unbanded female in 1995-96.

A minimum of 3 pairs claiming CNA 4 nested there 22 times during 1966-96. A CNA 4 pair (with changes in spouses) has temporarily nested in nearby vacant territory sites. After CNA 4 burned, the pair nested 1 season in CNA 2 (3.2 km) away. In 1983 they renested in CNA 6A (1.8 km away). After a nest failure in CNA 4 (1984), the pair renested in CNA 6B, cleared savanna 2.4 km away, but returned the next year to CNA 4. The pair nested twice in 1989 in CNA 6C.

The <1-ha CNA 5 was a small opening in the southeastern edge of the swamp, just northwest of the release pen and between CNA's 4 and 6A. Active nests were found in 1971, 1973, and 1974, and none thereafter. The opening has since become overgrown with greenbriar (*Smilax* spp.), and cranes were rarely observed.

There were 5 nests in CNA 6A, an approximately 5-ha, very narrow, hydric area along the south-central edge of the swamp. There were nests in 1971, 1975-76 (CNA 4 pair) but none again until 1996 (probably captive-bred #919 male and #926 female).

The CNA 6B (Fig. 1) is a cleared mesic to xeric savanna north of the Applewhite Dump. There was only 1 nest there, in 1984, when the CNA 4 pair temporarily relocated there.

Also used only 1 year was the narrow 7-ha CNA 6C (Fig. 1), north of the refuge I-10 Crop Unit. The CNA 4 pair nested there in 1989. A renest was not found but was assumed because the pair and a juvenile were seen during the October 1989 census.

Vickers.—The Vickers CNA (15) is located in the southeast part of the unit in a 48-ha mesic savanna (Fig 1). A swamp pond is on the north end, and a narrow marsh runs from the pond southwest through the savanna. Six nests were found in the marsh and 1 in dry wiregrass (*Aristida* spp.) during 1975-84. In 1984 the Vickers pair raised a subadult (#624) and spent the summer and fall 8.0 km north of Vickers. The male appeared to neglect his family and, in winter 1984-85, the female and #624 roosted together in the Bluff Creek Marsh. Telemetry and sightings indicated that #624 and a wild crane (possibly his mother), spent most of March-May 1985 in Vickers. In 1985, 2 large nests were found in the swamp pond, but no eggs were laid.

After a CNA vacancy of 8 years, in 1993 a new pair (released male #922 and wild female) nested through a rare snowfall, then renested in the pond edge and produced a fledgling. The Vickers pair nested again in the pond in 1994, but no chick fledged. In 1995-96, the Vickers pair nested in the swamp edge immediate to the site where in past years 4 nests had been found. During years when no cranes nested, the mesic savanna was hand-cleared and burned several times; the swamp pond was hand-cleared in 1993, just months before the new pair began nesting.

Valentine Savanna.—Two CNA's were defined along the Valentine Road, a pine plantation converted to mesic savannas. Two ponds in wooded drains were created by raising the elevation of culverts in Valentine Road. The 52-ha Valentine North CNA (21A) was discovered in May 1983 when a pair and 2 chicks were seen along Valentine Road. A large nest, apparently used in 1984, contained eggs in 1985. In 1986 a nest with eggs was found 1.0 km south of the 1985 nest. This CNA was the third most frequently used. Between 1983 and 1996, the CNA had active nests every year but 1984 and 1988.

In 1987 a newly formed pair (captive-bred male #623 and wild female) nested near the site of first pair's 1984-85 nests in 21A. No nests were found in 1988, but in 1989 the pair nested in the 5-ha north pond along Valentine Road.

In 1994 the Valentine North pair nested in the savanna about 200 m south of Brown's Trail. The original pair nested 3 times in the savanna. The second pair (#623 and wild female) nested 3 times in the savanna and 3 times in the pond. On 11 November 1994, #623 was killed by a car on the Gautier-Vancleave Road. During the winter the wild female stayed on territory, accepted #883, and the pair nested in the savanna in 1995-96.

The Valentine South CNA (21B) is located farther south along Valentine Road (Fig. 1). In 1987 a pair (captive-bred #636 male and wild female) occupied the small 3.5-ha impoundment south of the Valentine Crop Unit. The CNA has had active nests every year since, totaling 10 nesting attempts. The male #636 died in late 1995. The unbanded female held the territory, mated with captive-bred #214, and nested in 1996. All CNA 21B nests were in the pond along Valentine Road.

Table 3. Mean distance (km) from Mississippi sandhill crane release pens to Composite Nesting Areas (CNA's) used by released cranes in Jackson County, Mississippi, 1980-96.

Pen	n	$\bar{x}$	SD	Min.	Max.
Gautier	8	2.1	1.43	0.5	3.7
Ocean Spring	17	3.6	4.50	0.2	14.0
Ben Williams	7	2.2	1.29	0.4	4.5
Total	32	2.9	3.46	0.2	14.0

Firetower.—Firetower CNA (28), a 1.0-ha wetland cleared of pond cypress in winter of 1993 and adjacent to a 23-ha mesic savanna, is located 1.5 km east of Ben Williams Swamp and 1.6 km northwest of CNA 15 (Fig. 1). Firetower Road was impassable during winter-spring of 1994, which limited human disturbance. In 1994 a nest was found. Both captive-bred #871 male and #136? female were first-time nesters. A pair, presumably the same, nested unsuccessfully in 1995. A new but unknown female nested there in 1996.

Beasley Pond.—The Beasley Pond CNA (29) was used for the first time in 1996 by captive-bred #218 male and #136 female. The nest was in a hydric strand just west of the 1.5-ha Beasley Pond. The female had been mated with #871 in the Firetower CNA 28.

Ricky.—The Ricky CNA (30) was converted back to a 33-ha open savanna in 1989. The mesic savanna borders the Gautier-Vancleave road to the west. A new but unknown banded pair nested in a shallow 0.1-ha pool in 1996, only 0.8 km from the Valentine South CNA (21B).

Fontainebleau Unit

Mary Bourne CNA (12), 130 ha of savanna and pine in the southeast part of the Unit, is located 4.4 km south of CNA 6A, its nearest neighbor. A pair nested each year for 10 years (1968-77). In May 1978 a crane, probably the male, was killed by a vehicle or airplane near the territory. No active nests were found after 1977. A pair was observed in the area until late 1988 after which a single female was seen sporadically until 1993. In late summer 1994 several large areas were cleared, a release pen was built in the autumn, and in March 1995, 12 juveniles were released in the hope of eventually establishing a use area on and adjacent to the Unit for 10-12 cranes, including 2-3 nesting pairs.

Ocean Springs Unit

Perigal Swamp.—The Perigal Swamp area, a region of swamp, wooded creeks, pine plantation, and savanna, was improved for nesting and feeding through bulldozing, timber cutting, hand-clearing, and water control.

The 16-ha CNA 7 is located around the intersection of West Cottonmouth and East Perigal roads. A nest was found in 1966 just west of East Perigal Road next to the swamp (7A); another nest was found in 1967 but none in 1968. In 1969 and 1971, the pair nested about 0.6 km south in a small opening among pines (7B) and returned to nest in CNA 7A in 1970.

Nesting habitats in Perigal Swamp became overgrown with pine trees and shrubs. In the early 1990's, hand-clearing improved nesting habitats on both sides of East Perigal Road. After a 21-year vacancy, a nest was found 50 m east of the road (CNA 7C) in 1993, but the pair abandoned the site. A renest containing 1 egg was found on the west side in CNA 7B. No nests were found in 1994–96.

A pair nested for 2 years (1966–68) in Blue Hole CNA (8A) located north of Glendale Road, 0.6 km west of Old Fort Bayou Road, but left in 1969 when the Interstate Highway 10 right-of-way was cleared. A borrow pit (Blue Hole), excavated for fill, now occupies part of the former nesting site. CNA 7A pair was nearest neighbor (1.8 km apart) to CNA 8A when they nested concurrently during 1966–69. No nests were found in 1969–71, but during 1972–73, 3 nests were found 0.8 km northeast in a small wet marsh (CNA 8B).

In 1993, after a 20-year vacancy, a new pair, a wild male and an unknown banded female, nested in CNA 8B; no nest was found in 1994. After hand-clearing of pines, a nest was found in 1995 in CNA 8A, just north of the Blue Hole. Captive-bred #032 and an unbanded female used the same nest in 1996.

A first-time pair, captive-bred #038 male and #031 female, nested in 1993 at Gwendale CNA (8C), 0.4 km southwest of Bluehole CNA (8A), concurrently with the pair at CNA 8B, in an open, dry pine plantation (Fig. 1). The nest was abandoned after a mock-up TV camera was set up. No nesting has occurred since 1993 after which the pair moved to CNA 14.

The CNA 14 or Utah savanna is the westernmost CNA on the Ocean Springs Unit (Fig. 1). It was the third most used CNA and the most on the Ocean Springs Unit. The 28-ha CNA 14A east of the crop unit was hand-cleared in 1980–81. An old nest was found in 1980. The Cottonmouth (CNA 23) pair moved there to nest in 1983. There were active nests in CNA 14 in all years since except 1991–92 and 1996. Most of the nests in 14A were in a 3-ha core area, circumscribed by a firebreak. The pair moved 0.8 km east/southeast in 1984 and nested in a 28-ha mesic savanna (CNA 14B) restored by bulldozing. The pair nested in CNA 14A during 1985–86 and in 14B in 1987. The male was caught in 1987 and marked as #629. He and an unbanded female were the nesting pair until 1990–91 when captive-bred #642 replaced him. After a 2-year hiatus, captive #038 male and #031 female nested in 1994, after moving from CNA 8C. The 1995 pair was unknown. Although the savanna was frequently used, no fledglings were produced by the Utah pairs.

A nest was found in a 9-ha St. Regis Company clear-cut (Cottonmouth CNA 23) in 1982. The clear-cut became the site of the Ocean Springs release pen and adjacent nesting savanna. The pair moved 1.2 km west in 1983 to Utah CNA (14A). Two captive-bred cranes from the same release cohort, #644 male and #643 female, nested within 100 m of the Ocean Springs pen in 1987. In 1988, a nest was found near the same site. No nests were found near the pen site after 1988, but in 1989 another pair, captive-bred #641 male and a wild female, behaved as though they had a chick near the same area.

Simms Road.—Simms Road area, a 600-ha tract of pine interspersed with savanna openings, is bounded on 3 sides by roads, including Interstate Highway 10 on the south. Two territories, CNA 9 and CNA 10 (Fig. 1), with active nests were found 1.0 km apart in 1967; only 1 (CNA 9) was active in 1968–69. For several nesting seasons, 1 crane, apparently without a mate, was seen.

Eglin Road.—Eglin Road CNA (13), located south of I-10 and Simms Road area, consists of 300 ha of large savannas separated by pine forest. During 1969 both CNA 9 and CNA 13 held nests 2.8 km apart. Nests were found in CNA 13 during 1969–72; then 4 years passed without nests. Nests were found in 1977–78 and a nest and renest in 1981. In spring 1983, CNA 13 was burned by a wildfire, but the pair nested in a small unburned section. Altogether, 8 nests and 1 renest were found in CNA 13 (1969–83). Two cranes (#604 and #619), released from the Gautier pen, and a wild crane spent spring and summer 1986 in CNA 13. A nest start was found, but a permanent pair was not established. A young pair, wild W-1 female and captive-bred #913 male, was observed there in 1991. No more cranes were documented there until 1995 when a savanna bird census team saw a pair. A start nest was found the next spring, possibly from captive-bred #245 male and #W-3 female, the latter hatched at North Valentine, 9 km to the east/northeast.

St. Regis.—St. Regis CNA (16) consists of 2 small wet savannas about 400 m apart, separated by a firebreak. The first nest was found in 1975 in the northwestern savanna (CNA 16A). Two small "start" nests were found in 1976 and

1977. The savanna was burned by a wildfire during the 1977–78 dormant season. The next nest in CNA 16A was found in 1981. In 1985 and 1987, a pair nested in the southeastern savanna (CNA 16B). After the 1987 season a prescribed fire burned the area and no cranes nested in 1988. Mystery Chick (see below) and his mate nested in CNA 16B in 1989; they moved to his natal home (CNA 22) in 1990. No nests were found in CNA 16 after 1989. Costapia CNA (24), located 2.4 km away, was the nearest occupied territory in 1987.

Mystery.—Three wetlands were constructed to collect drain-water from a wastewater treatment facility's spray fields. During 4 of 7 years (1985–87, 1990), a chick or fledgling with its parents were seen, but no nests were found. One chick ("Mystery Chick") was caught in 1986 and radio-tagged (male #631). CNA 22 was not found until 1991 when an active nest was found in Wetland Cell #2. An old nest probably used in 1990 was found nearby.

In 1989 Mystery Chick #631 male and a wild female nested in CNA 16. During the same season, his putative parents nested in CNA 22. However, in 1990, a 14-day-old chick, found north of Wetland Cell #2, was confirmed as the offspring of #631 and the wild female; the chick later fledged. In 1991 their nest was discovered in Wetland Cell #2. In 1992 the pond was dry and the surrounding lands had been burned and no nests were found. The Mystery Chick pair eluded searchers until 1994 when they nested at "Sawdust Savanna" CNA 27, 1.4 km northwest of Wetland Cell #2.

Doubletree.—Doubletree CNA (25) was found when an unbanded male and captive-bred #640 female nested in 1989 in a 27-ha reclaimed savanna located west of Wet Cell #3. Later the same season the male was replaced by #644, a release cohort-mate of #640, and that pair nested in 1990. No nests were found after 1990.

Sawdust.—The Sawdust CNA (27) is a 26-ha mesic treeless savanna cleared in the late 1980's and an adjacent (south) 42-ha hydric savanna with slash pines. Crane #631 and an unbanded mate nested there in 1994 on the south edge of the open savanna; the pair was observed with a fledgling well into the spring 1995. No nest was found in 1995–96, but a start nest was found in 1996 in the southern part of the hydric section.

Sullivan.—A pair with twins was observed in the Greenpond Crop Unit in 1996. The Sullivan CNA (31), formerly Central Greenpond Savanna, was a 22-ha mesic savanna with a 1.5-ha pond along the north edge. A pair was flushed from the area during the 1996 nesting season, but no nest was found. Later, an unbanded male, a banded but unknown female, and 2 young were observed repeatedly at the Greenpond Crop Unit, 0.8 km east of the middle of the savanna. The pair returned to that savanna when flushed.

Off-refuge

Fort Bayou Church.—Fort Bayou CNA (11) was a small wet savanna. An old nest was found in 1967 and active nests in 1968–69. After 1969 human disturbance caused the cranes to abandon the area. Several years later the savanna was plowed and destroyed as nesting habitat.

Bear Pond.—This historical CNA (17) has not been used for nesting since 1963, before the start of this study. It was a 15-ha wet savanna (Fig. 1) northwest of the intersections of Highways 57 and 90. Nests were found in 1938 (McIlhenny 1938), 1939, 1941, and 1963 (Gandy and Turcotte 1970). Valentine flushed a single crane from the south edge in the 1960's. No cranes have been noted there since.

Dummyline.—The Dummyline CNA (18) was a small (<1 ha) opening in the flatwoods just south of the upper drain of the west Costapia River (Fig. 1). This CNA was discovered when a chick was seen by a St. Regis Company employee in 1965. It is 5.5 km north of the Ocean Springs Unit, 5 km south-southeast of CNA 19, and 3 km north-northeast of CNA 26. There was no known nesting after 1965, although a pair was seen in the area from a helicopter in 1967. A set of tracks was seen in 1974 along John Smith Road just east of the CNA. Without management, the small opening has long since become overgrown with pines and gallberry (*Ilex* spp).

Weber.—The Weber CNA (19), on International Paper Co. and DeSoto National Forest lands, is 10 km north of nearest neighbor (Costapia CNA 24). The International Paper nesting sites are in 2 separated ponded areas (CNA 19A); the DeSoto site (CNA 19B) is wet savanna. An active nest was found during 1977–78 in CNA 19A. No nests were found in the next 6 years (1979–84).

In 1985 International Paper disked and furrowed CNA 19A, but despite habitat loss, a wild pair nested in 1985 in the pond. Considering the nesting hiatus, we assume the pair was new. In 1986 the pair moved 0.6 km to DeSoto National Forest (19B) but returned to CNA 19A where they produced eggs in 1987–90 (including 4 renests) in the south pond but no chicks fledged. The pair disappeared in the winter of 1990–91, except for a sighting on the January 1993 census at the pond.

Mallette.—The Mallette CNA (20), in Section 16, owned by the State of Mississippi, is 2.8 km southwest of Fort Bayou CNA 11. The nesting habitat, a narrow seep bog situated between low pine-covered hills, was kept relatively open by wildfires. Cranes frequented the area in the late 1960's but no nests were found until 1978–79, then none was seen after. In April 1982 a crane carcass, slightly scavenged, was found; cause of death not determined.

In 1982 released #602 and a wild mate frequented the area. Two small start nests were found in 1983. In August 1983, #602 was shot and killed adjacent to the territory. The school section was leased in 1990 to a hunting club, fenced, and closed.

Costapia.—Soon after female crane #640 was released, she began an association with a wild crane that lasted from early 1985 to spring 1987 in Costapia CNA (24). CNA 24 is a 50-ha savanna and dense swamp, an eastern part now leased to USFWS by Mississippi Gulf Coast Regional Waste Water Authority and a western part owned by Jackson County. In spring 1987 a new pair, female #640 and wild #632 male, nested in a pine plantation west of a treatment lagoon and south of the gas line right-of-way. In 1988, #632 was replaced by an unbanded male. The 1988 nest was in a swamp 400 m west of the sewage lagoons, in an area owned by Jackson County. No eggs were laid in 1989, and in 1990, #640 was replaced by #643. The pair nested in 1991, but no chick was seen. Crane #643 died in September 1992 and was replaced by an unbanded female, and that pair nested in 1995-96.

Jordan.—Jordan CNA (26), a 0.5-ha former fish pond vegetated with Gulf Coast spike rush (*Eleocharis cellulosa*), is located northwest of the refuge (Fig. 1). The pond has been a crane roost site for many years. A nest was found in 1994, but a chick did not survive. In 1995 the same nest was flooded before eggs were laid. Jordan's farm is managed under a "Partners for Wildlife" agreement.

DISCUSSION

We inferred that a territory with interrupted use lost the original pair and that later occupancy was by a new pair. Territories may be inherited by the surviving mate and a new spouse and passed on from 1 generation to the next. Walkinshaw (1960) found 2 nests in CNA 4 in 1940 and 1 in 1960. During 53 years pairs have nested there occasionally, certainly not the same individuals and probably not of the same lineage. Nine CNA's were recolonized with active nests after at least 5 years vacancy. Seven of the recolonizations were with different pairs. The reuse of CNA 19 probably involved at least 1 member of the original pair. The unbanded female in the reuse of CNA 15 may have been the same from the original pair. CNA's 7 and 8 were reused after being vacant for more than 20 years. Five CNA's (2, 6A, 7, 8, 15) were recolonized after recent hand-clearing. Three new CNA's (27, 28, 30) were first colonized after recent hand- or machine-clearing of savannas.

By 1996, 13 former CNA's had been vacant for at least the past 5 years, 10 of those for more than 10 years. A fourteenth, pre-study CNA 17, was vacant throughout the study period. The initial cause of abandonment was in many cases the death of one of the spouses. The nesting habitats of 4 CNA's (off refuge) were destroyed; 9 other refuge CNA's, with habitat maintenance, were in good condition. Habitat maintenance on 3 refuge CNA's (5, 9, 10) was difficult due to their location in very wet areas. Crane use of CNA's 9 and 10 has been very limited over the last 25 years. Six CNA's were on the fringe of the nesting range and/or refuge where the pairs may have been killed or where disturbance was greater. Roads running through or on edges may invite trespass and shooting. Off-refuge CNA 24 was used recently but was rapidly deteriorating without habitat restoration. CNA 19, not used the last 6 years because of the disappearance of the nesting pair, was in good condition, but unused, probably because the distance from the Ocean Springs Unit limited recolonization. However, start nests were found in CNA's 3 and 18 in 1995-96. CNA 12 was not used after 1977, but reuse looked possible with the building of an acclimation pen and captive releases beginning in 1995.

At least 7 females (#627 in CNA 1B, #634 in CNA 4, #640 in CNA 24, and unmarked females in CNA's 4, 14, 21A, 21B, 24) retained their territories after the male spouse died. We know that males held territory after their mate's death in CNA 4 (#855) and CNA 24 (unmarked), and unmarked males probably held onto CNA's 4 and 24.

When a population is expanding, newly formed pairs usually occupy vacant territories. During the 1960's to 1980's, the Mississippi sandhill crane nesting population was stable. The nesting population did not increase until released cranes began breeding. Among 12 new territories established after 1984, 11 were taken by pairs with at least 1 released crane spouse.

Released birds tended to nest within a few kilometers of their release sites. The exceptions to the small distance between release pen and subsequent CNA were movements by 4 cranes released at Ocean Springs that later nested in the Gautier Unit. This was probably due to better habitat conditions in the Gautier Unit; habitat restoration on the Ocean Springs with wetter soil conditions and thicker undergrowth has been more difficult. The number and rate of release of cranes in the Ocean Springs Unit will need to be adjusted to fit the success of habitat restoration there. As the cranes released in the much smaller Fontainebleau Unit age, increased movement by these cranes to locate suitable nesting areas in other refuge units may occur.

Pens will need to be built to facilitate colonization of areas not presently occupied. With the large release cohorts of December 1989-92 coming into breeding age, preparing new nesting sites and improving vacant sites will have high priority.

LITERATURE CITED

GANDY, B. E., and W. H. TURCOTTE. 1970. Catalog of Mississippi bird records. Vol. 1. Mississippi Game and Fish Comm., Jackson. 105pp.

KUYT, E. 1981. Population status, nest site fidelity, and breeding habitat of whooping cranes. Pages 119-129 *in* J. C. Lewis and H. Masatomi, eds. Crane research around the world. Int. Crane Found., Baraboo, Wis.

MCILHENNY, E. A. 1938. Florida crane, a resident of Mississippi. Auk 55:598-602.

MITCHELL, L. C., and P. J. ZWANK. 1987. Comparison of release methods for parent-reared Mississippi sandhill cranes. Pages 399-409 *in* G. W. Archibald and R. F. Pasquier, eds. Proc. 1983 int. crane workshop. Int. Crane Found., Baraboo, Wis.

SMITH, E. B., and J. M. VALENTINE. 1987. Habitat changes within the Mississippi sandhill crane's range in Jackson County, Mississippi (1942–1984). Pages 341-354 *in* J. C. Lewis, ed. Proc. 1985 crane workshop. Platte River Whooping Crane Maintenance Trust, Grand Island, Nebr.

U.S. FISH AND WILDLIFE SERVICE. 1991. Mississippi sandhill crane recovery plan (third revision). U.S. Fish Wild. Serv., Atlanta, Ga. 42pp.

VALENTINE, J. M., JR. 1979. The Mississippi sandhill crane–a status update. Pages 135-138 *in* J. C. Lewis, ed. Proc. 1978 crane workshop. Colorado State Univ. Printing Serv., Fort Collins.

______. 1981. The Mississippi sandhill crane, 1980. Pages 167-174 *in* J. C. Lewis and H. Masatomi, eds. Crane research around the world. Int. Crane Found., Baraboo, Wis.

______. 1982. Breeding ecology of the Mississippi sandhill crane in Jackson County, Mississippi. Pages 63-72 *in* J. C. Lewis, ed. Proc. 1981 crane workshop. Natl. Audubon Soc., Tavernier, Fla.

______. 1987. Management of the Mississippi sandhill crane. Pages 387-397 *in* G. W. Archibald and R. F. Pasquier, eds. Proc. 1983 int. crane workshop. Int. Crane Found., Baraboo, Wis.

______. 1992. Renesting of Mississippi sandhill cranes in Jackson County, Mississippi, 1965–1989. Pages 173-176 *in* D. A. Wood, ed. Proc. 1988 North Am. crane workshop. Florida Game Fresh Water Fish Comm. Nongame Wildl. Program Tech. Rep. 12.

______, and T. J. LOGAN. 1991. The captive release program for the Mississippi sandhill crane. Pages 415-421 *in* J. Harris, ed., Proc. 1987 int. crane workshop. Int. Crane Found., Baraboo, Wis.

______, and R. E. NOBLE. 1970. A colony of sandhill cranes in Mississippi. J. Wildl. Manage. 34:761-768.

______, and ______. 1976. The Mississippi sandhill crane–endangered or doomed? Pages 343-346 *in* J. C. Lewis, ed. Proc. [1975] int. crane workshop. Oklahoma State Univ. Publ. Printing, Stillwater.

WALKINSHAW, L. H. 1960. Some Mississippi crane notes. Migrant 31:42-43.

WILSON, C. D. 1987. A comparison of habitat-use by Mississippi sandhill cranes released on different locations on a refuge. M.S. Thesis, Louisiana State Univ., Baton Rouge. 89pp.

ZWANK, P. J., and S. R. DERRICKSON. 1982. Gentle release of captive parent-reared sandhill cranes into the wild. Pages 112-116 *in* J. C. Lewis, ed. Proc. 1981 crane workshop. Natl. Audubon Soc., Tavernier, Fla.

______, and D. A. DEWHURST. 1992. Dispersal of pen-reared Mississippi sandhill cranes. Pages 263-270 *in* D. A. Wood, ed. Proc. 1988 North Am. crane workshop. Florida Game Fresh Water Fish Comm. Nongame Wildl. Program Tech. Rep. 12.

MONOGAMY AND PRODUCTIVITY IN SANDHILL CRANES

STEPHEN A. NESBITT, Florida Game and Fresh Water Fish Commission, Wildlife Research Laboratory, 4005 South Main Street, Gainesville, FL 32601, USA
THOMAS C. TACHA,[1] Caesar Kleberg Wildlife Research Institute, Campus Box 218, Texas A&M University-Kingsville, Kingsville, TX 78363, USA

Abstract: Marked pairs of Florida (*Grus canadensis pratensis*) and greater sandhill cranes (*G. c. tabida*) were monitored for 426 pair-years. Among pair members, females averaged 3.7 ± 3.6 (SD) years older than males. Pair longevity was not correlated with pair productivity ($P = 0.279$) for pairs that were together ≥3 years. Forty-four percent of 72 Florida sandhill crane pair bonds were broken during the study, 67.7% due to death and 32.3% to divorce. Among both Florida and greater sandhill cranes, the incidence of divorce was related to a failure to reproduce; 52.6% of pairs that failed to produce young divorced, while only 10% of pairs that divorced had a history of successful reproduction. The development of a monogamous mating strategy is tied to the need to secure and maintain exclusive territories in order to reproduce successfully.

PROC. NORTH AM. CRANE WORKSHOP 7:10-13

Key words: cranes, *Grus canadensis*, mating systems, monogamy, perennial monogamy, productivity, sandhill cranes.

Successful reproduction in sandhill cranes involves extended biparental care of young, maintaining comparatively large exclusive territories during the nesting season, and close physiological synchrony over a sometimes extended (5-6-month) nesting season. These requirements are best met with a perennially monogamous mating system. Perennial monogamy is defined as "monogamy in which mates are pair-bonded throughout their lives" (Barrows 1995). This has come to be loosely interpreted as pairing for life.

Pair bonds among sandhill cranes are maintained year-round, and no differences in pair bond characteristics have been noticed between subspecies or populations (Tacha et al. 1992). Sandhill cranes can reproduce at ≤3 years of age, but most do not succeed until later (Tacha et al. 1992), and 25% of some populations never succeed in reproducing (Nesbitt 1992). Though aspects of the development and duration of pair bonds are known, the mechanism through which mate selection is made and a pair bond initiated is not known.

Development of the initial pair bond in cranes is a protracted process that involves species-specific displays, mutual association, and concurrent behaviors (Bishop 1984, Tacha 1988). Three to 6 short-term (≤2 weeks) preliminary diadic relationships are formed and broken before a bond develops that results in a breeding attempt (Nesbitt and Wenner 1987).

Pairs of individually marked sandhill cranes from 3 populations were monitored for several years as part of other studies (Nesbitt 1988, Nesbitt and Wenner 1987, Tacha 1988, Tacha et al. 1989). We looked at the fate and reproductive success of these pairs as a function of pair longevity to see if pair-bond duration was correlated with reproductive success.

Support was provided by the Florida Game and Fresh Water Fish Commission and the Caesar Kleberg Wildlife Research Institute at Texas A&M University-Kingsville. Research funding was provided in part by the U.S. Fish and Wildlife Service (USFWS) under Section 6 of the Endangered Species Act (PL 93-205). We are indebted to R. D. Bjork, G. C. Iverson, D. C. Martin, S. T. Schwikert, A. S. Wenner, and K. S. Williams for assistance with data collection, and to J. M. Black and P. J. Pietz for several helpful comments on earlier drafts of this manuscript.

METHODS

We studied birds from 3 populations: eastern greater sandhill cranes; the mid-continent population of greater, Canadian (*G. c. rowani*), and lesser sandhill cranes (*G. c. canadensis*) (all migratory subspecies); and Florida sandhill cranes (a nonmigratory subspecies) (see Tacha et al. 1994 for details of these populations). Data for analysis were from eastern greater and Florida sandhill crane populations. Cranes were captured with the use of rocket nets (Tacha 1988) or oral tranquilizers (Nesbitt 1984). Each bird was banded with a USFWS leg band and age was determined (Nesbitt 1987). Birds that were ≥3 years old were designated adults, subadults were 1-3 years old, and juveniles <1 year old. They also were individually marked with colored plastic leg bands, numbered neck collars, or wing streamers for subsequent identification in the field (Tacha 1988, Nesbitt et al. 1992). Some types and sizes of color markers have been shown to cause mortality or changes in social status or behavior (Ramakka 1979, Hoffman 1985, Bennett 1992); therefore, we used only markers that were not known to affect behavior. Only individually marked birds that were paired with another marked individual were included in the pair-bond duration, reproductive success, and comparative

---

[1]Deceased (see page v).

size assessments. Pairs with the minimum ages of both members known were included in the age comparison of pair members. To avoid biasing results, we included only breeding-age pairs observed for more than 3 years in the calculation of pair productivity. Productivity refers to raising young that survive to the age when they would leave the company of their parents, about 10 months of age (Tacha et al. 1992).

RESULTS

Seventy-two pairs of breeding-age Florida and 35 of eastern greater sandhill cranes were observed in Florida for a total of 426 crane-years from 1972 through 1992. In addition, subadults were studied, some from hatching into adulthood. A pair composed of birds with no previous nesting experience usually (5 of 7 times) pioneered a new territory together. These territories were generally smaller than average and were usually initiated between the boundaries of existing actively defended territories. Among experienced adults, new pairs often evolved from trios that developed during the post-nesting season and persisted for several weeks. Trios most often consisted of the resident, territorial male and female and a third bird, usually (14 of 18 times) another female. The resident female was sometimes supplanted by the interloping bird. In the cases in which the trio consisted of 2 males and a female, the males were without a territory and could have been vying as much for the female's territory as for her. Aggression that could result in significant injury or death often occurred within these trios.

Pair Characteristics-Age and Size

Among eastern greater and Florida sandhill cranes, females were older than males in 24 of 32 known-age pairs. This difference was significant ($\chi^2 = 8.0$, $P < 0.01$). The disparity in age ranged from 1 to 17 years. When females were older (24 pairs) they averaged 3.7 ± 3.6 (SD) years older than their mates, and when males were older (8 pairs) they averaged 1.9 ± 1.5 (SD) years older than their mates.

We ranked the pair members by size relative to that population's mean for total mass, bill length, and leg length (Nesbitt et al. 1992). Pairs that consisted of birds larger than the population average were, for the life of the pair bond, 3.8 times more reproductively successful (1.6 young produced per pair ± 1.87 [SD], $n = 25$) than pairs with one or both members smaller than average (0.42 young per pair ± 0.61 [SD], $n = 19$). This difference was significant ($\chi^2 = 9.70$, $P < 0.01$).

Productivity and Pair Longevity

The mean minimum pair bond duration among eastern

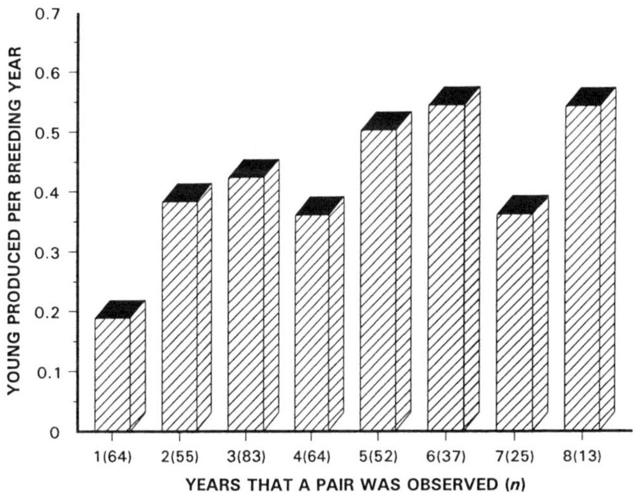

Fig. 1. Productivity of sandhill crane pairs and years that a pair was observed, 1977-92.

greater sandhill cranes was 5.3 ± 1.7 (SD) years, and for Florida sandhill cranes it was 4.0 ± 1.8 (SD) years. Pairs composed of 2 first-time nesters ($n = 13$) did not successfully reproduce during their initial breeding attempt. Recently formed pairs, irrespective of their age or past nesting experience (including those that did not stay together ≥3 years), nesting for the first time were significantly ($\chi^2 = 9.64$, $P < 0.01$) less productive (0.17, $n = 59$ breeding pair-years) than those that nested together for ≥3 years (0.44, $n = 274$ breeding pair-years). We compared reproductive success and pair longevity using linear regression. Though there was an apparent increase in productivity between the first year of observation (for some of these pairs [64%] this was also their first year of nesting together) and subsequent years (Fig. 1), there was no significant increase in productivity with time ($r^2 = 0.083$, $P = 0.279$) for pairs that remained together ≥3 years. When we separated the data for the migratory population ($n = 136$ pair-years), the effect of years of pair longevity on productivity was not significant ($r^2 = 0.02$, $P = 0.748$). For the nonmigratory population ($n = 257$ pair-years), there was again no positive effect on reproductive success with pair longevity ($r^2 = 0.206$, $P = 0.258$).

Pair Fidelity and Territory Retention

During the study, 44.4% ($n = 32$) of the Florida sandhill crane pair bonds we monitored were broken at a rate of 5.5% per year. The cause of break-up was known in 31 cases; death of a pair member accounted for 21 (67.7%) of the broken bonds, and divorce for 10 (32.3%). Divorce was most frequently seen in pairs that had failed to produce young. However, 1 divorce (10%) occurred in a pair with a history of successful reproduction. Among both the migratory and nonmigratory population, 10 of 19 (52.6%) pairs that did not produce young divorced. Pairs that had never produced young were 5 times more likely to divorce than pairs that had successfully reproduced in the past 3 or 4 years. Divorce or "mate deserting" (Wickler and Seibt 1983) has been recognized as a consequence of reproductive failure in black-legged kittiwakes (*Rissa tridactyla*) (Coulson and Thomas 1980, 1985). In 26 cases of crane mate death or divorce, males retained the territory 92.9% and females 69.2% of the time. Females retained the territory following divorce in only 1 of 7 cases. The male retained the territory 100% of the time following death of his mate, while the rate for females was 66%.

DISCUSSION

The 3 preconditions for the development of monogamy (Wittenberger 1979:323) are met in sandhill cranes. First, monogamous pairing should be a benefit to the female in a way not otherwise attainable. Paired female sandhill cranes had access to exclusive territory as a member of a perennial pair to a greater extent than they could without a mate. Second, females must be able to identify unpaired or weakly paired males when searching for a mate. Among sandhill cranes, paired birds are always in association; a female member of the pair responded aggressively to a potentially challenging female. Weakened pair bonds (i.e., those pairs without young) were perhaps the most likely to support development of trios. Preconditions that allowed development of trios (reduced aggression from the resident female) would be discernible to an unpaired female or a female interested in forming a pair bond. The third precondition for the development of monogamy, that "neither sex be prone to desert mate following fertilization" was present because extended, biparental care (9 to 10 months) is necessary for sandhill cranes to successfully reproduce.

Retaining the territory (even one with initially poor reproductive potential) puts the pair (or territory-retaining bird) in a position to improve the quality of the territory as changes (e.g., death or divorce) occur among neighboring pairs. Improvement in territory quality as time on nesting territory increases is known for other species (Rowley 1983). Attrition among neighboring crane pairs is apt to occur with enough frequency so that waiting for the opportunity to expand the size or quality of the territory or acquire a vacated neighboring territory would be a valid option for enhancing reproduction. A bird with a territory can usually secure another mate, but, without a territory, the bird must initiate the process of first securing a breeding territory. The advantage given by greater size in securing and maintaining a territory was not studied. However, a territorial advantage with size could explain why pairs composed of larger-than-average individuals were more reproductively successful than pairs with one or both members smaller than the average.

Females were less likely to retain the territory after loss of mates. They sometimes adopt an unrelated chick, perhaps in order to attain a breeding opportunity with a territorial male (Nesbitt and Wenner 1987). They were also more likely to risk death or injury to actively compete with other females (trios) for the opportunity to breed with a territorial male. The relative age disparity we found between females and males, females being older than their mates, on average, may have been due to females maturing at a slower rate than males (Nesbitt 1992, Tacha et al. 1989). There were slightly more females than males in the Florida sandhill crane population; the composition of post-nesting season flocks was 49.3% males and 50.7% females (S. A. Nesbitt, unpubl. data). Neither this difference, nor the difference in age of first breeding (2 or 3 years for males, 3 or 4 years for females [Nesbitt 1992, Tacha et al. 1989]), seems great enough to account for the age disparity we observed among pairs. It is possible that males, because of their role as primary territory defender, are more likely to die at an earlier age, but this has not been studied. There may be other factors that could account for a disproportionate number of older females in the population and explain the age disparity we observed among pair members.

Formation of new pairs following death or divorce occurred at all times of year, frequently within a few days. The change in pair membership was often so subtle that, had the birds not been individually color-marked, there would have been no change in pair membership detected.

CONCLUSIONS

Pair bonds endured for multiple nesting seasons only as long as the pair had been reproductively successful at least once. It appears that once a pair has successfully reproduced, their reproductive success is not significantly enhanced through time, but more study is needed to fully test this. Our study did not follow any cranes for their maximum potential breeding life ($\pm$20 years), but during the period we observed them, half of the pairs that failed to reproduce switched mates. Plasticity of pair membership may be a reason that productivity was not correlated with time together. The members remated and could keep doing so until reproductive success was attained or a bird left the breeding population. Reproductive success may be more closely correlated with the relative amount and value of available resources (e.g., food or escape cover) associated with each pair's respective nesting territory. Changes in pair membership occurred throughout the population and with such frequency that it is perhaps best to use the phrase "pairing for life" with some caution. Because pair membership can change unnoticeably, it is possible that a territory would appear to be occupied consistently for years when actually pair membership had changed several times.

LITERATURE CITED

BARROWS, E. M. 1995. Animal behavior desk reference. CRC Press, Boca Raton, Fla. 671pp.

BENNETT, A. J. 1992. Sandhill crane mortality resulting from neckbands. Pages 203-206 *in* D. A. Wood, ed. Proc. 1988 crane workshop. Florida Game Fresh Water Fish Comm. Nongame Wildl. Program Tech. Rep. 12.

BISHOP, M. A. 1984. The dynamics of subadult flocks of whooping cranes wintering in Texas, 1978-79 through 1982-83. M.S. Thesis, Texas A&M Univ., College Station. 128pp.

COULSON, J. C., and C. THOMAS. 1980. A study of the factors influencing the duration of the pair bond in the kittiwake gull *Rissa tridactyla*. Proc. Int. Ornithol. Congr. 17:823-833.

______, and ______. 1985. Differences in the breeding performance of individual kittiwake gulls, *Rissa tridactyla* (L.). Pages 489-503 *in* R. M. Sibly and R. H. Smith, eds. Behavioural ecology: ecological consequences of adaptive behaviour-25th Symp. Br. Ecol. Soc. Blackwell Scientfic, Oxford, U.K.

HOFFMAN, R. H. 1985. An evaluation of banding sandhill cranes with colored leg bands. North Am. Bird Bander 10:46-49.

NESBITT, S. A. 1984. Effects of an oral tranquilizer on survival of sandhill cranes. Wildl. Soc. Bull. 12:387-788.

______. 1987. A technique for aging sandhill cranes using wing molt: preliminary finding. Pages 224-229 *in* J. C. Lewis, ed. Proc. 1985 crane workshop. Platte River Whooping Crane Maintenance Trust, Grand Island, Nebr.

______. 1988. Nesting, renesting, and manipulating nesting of Florida sandhill cranes. J. Wildl. Manage. 52:758-763.

______. 1992. First reproductive success and individual productivity in sandhill cranes. J. Wildl. Manage. 56:573-577.

______, and A. S. WENNER. 1987. Pair formation and mate fidelity in sandhill cranes. Pages 117-122 *in* J. C. Lewis, ed. Proc. 1985 crane workshop. Platte River Whooping Crane Maintenance Trust, Grand Island, Nebr.

______, C. T. MOORE, and K. S. WILLIAMS. 1992. Gender prediction from body measurements of two subspecies of sandhill cranes. Proc. North Am. Crane Workshop 6:38-42.

RAMAKKA, J. M. 1979. Capturing sandhill cranes in Texas and Nebraska. Pages 75-80 *in* J. C. Lewis, ed. Proc. 1978 crane workshop. Colorado State Univ. Printing Serv., Fort Collins.

ROWLEY, I. 1983. Re-mating in birds. Pages 331-336 *in* P. P. G. Bateson, ed. Mate choice. Cambridge Univ. Press, New York, N.Y.

TACHA, T. C. 1988. Social organization of sandhill cranes from mid-continental North America. Wildl. Monogr. 99:1-37.

______, D. C. HALEY, and P. A. VOHS. 1989. Age of sexual maturity of sandhill cranes from mid-continental North America. J. Wildl. Manage. 53:43-46.

______, S. A. NESBITT, and P. A. VOHS. 1992. Sandhill crane. A. Poole, P. Stettenheim, and F. Gill, eds. The birds of North America. No. 31. Acad. Nat. Sci., Philadelphia, Pa., and Am. Ornithol. Union, Washington, D.C. 24pp.

______, ______, and ______. 1994. Sandhill crane. Pages 77-94 *in* T. C. Tacha and C. E. Braun, eds. Migratory shore and upland game bird management in North America. Int. Assoc. Fish Wildl. Agencies, Washington, D.C.

WITTENBERGER, J. F. 1979. The evolution of mating systems in birds and mammals. Pages 271-349 *in* P. Marler and J. G. Vandenbergh, eds. Handbook of behavioral neurobiology. Plenum Press, New York, N.Y.

WICKLER, W., and U. SEIBT. 1983. Monogamy: an ambiguous concept. Pages 33-51 *in* P. P. G. Bateson, ed. Mate choice. Cambridge Univ. Press, New York, N.Y.

MORTALITY OF RADIO-EQUIPPED SANDHILL CRANE COLTS AT MALHEUR NATIONAL WILDLIFE REFUGE, OREGON

GARY L. IVEY, Malheur National Wildlife Refuge, HC 72 Box 245, Princeton, OR 97721, USA
ERIC J. SCHEUERING, Malheur National Wildlife Refuge, HC 72 Box 245, Princeton, OR 97721, USA

Abstract: We radio-equipped 142 greater sandhill crane (*Grus canadensis tabida*) colts at Malheur National Wildlife Refuge, Oregon, during 1991–95 to document prefledging mortality factors. Fates of 19 colts were undetermined. A total of 23 colts survived to fledge (19% of known fates). Evidence collected from the carcasses and death sites indicated that predators were responsible for the largest number of colt deaths (64), followed by unknown causes (13), parasitic gapeworms (*Cyastoma spp.*) (8), drowning (7), intraspecific aggression (5), study-related mortality (1), roadkill (1), and hay swather (1). Of 64 colts killed by predators, 26 were lost to mink (*Mustela vison*), 10 to great horned owls (*Bubo virginianus*), 9 to golden eagles (*Aquila chrysaetos*), 7 to unidentified predators, 5 to coyotes (*Canis latrans*), 5 to unidentified raptors, 1 to a northern harrier (*Circus cyaneus*), and 1 to a raccoon (*Procyon lotor)*. Because mink were the most significant predator, we initiated an experimental mink control program in 1993. Losses from mink in 1993 were reduced to 9%, compared with 36% and 27% in 1991 and 1992, respectively. We concluded that an integrated approach involving habitat management and predator and parasite control could be used to enhance crane productivity.

PROC. NORTH AM. CRANE WORKSHOP 7:14-17

Key words: greater sandhill crane, *Grus canadensis tabida*, mortality, Oregon, parasites, predators, radiotelemetry.

Malheur National Wildlife Refuge (Malheur) supports the largest local breeding flock of the Central Valley Population of greater sandhill cranes (Littlefield 1976). This population has had the second lowest recruitment of the 5 populations of the subspecies (Drewien et al. 1995). Cranes breeding at Malheur declined from a peak of 236 pairs in 1975 to 168 in 1989, primarily because of predator impacts on productivity (Littlefield 1995*a*).

A radiotelemetry study at Malheur in 1983 and 1984 showed predators were severely limiting crane colt survival, with coyotes being the most important predator, taking 52% of the young (Littlefield and Lindstedt 1992). Based on these and other data, a predator control program was initiated in 1986 and continued through 1993. Coyotes, raccoons, and common ravens (*Corvus corax*) were controlled in the most important crane breeding areas on up to 35,000 ha of the 76,000-ha refuge during the cranes' breeding seasons. In 1993, mink were added to the control program on 12,000 ha of the refuge. With the exception of 1988, a drought year, crane productivity improved during the first 4 years of the control program, but was lower than expected. Consequently, in 1991 we initiated this second radiotelemetry study to determine causes of crane colt mortality. This paper presents study results.

We wish to thank the many volunteers and refuge staff who helped monitor cranes and gather data during the study, including J. K. Davison, L. Grall, G. A. Green, S. A. Gall, D. Hollis, E. P. Kelchlin, P. Kerhulas, T. Mahan, K. McIntyre, E. M. Nowak, S. J. Rule, S. A. Smith, R. T. Vetter, R. E. White, and J. Yacabucci. G. P. Keister and C. E. Trainer of the Oregon Department of Fish and Wildlife allowed us to use the state's assigned radio frequencies for our transmitters. We also thank the staff of the National Wildlife Health Center for conducting necropsies of cranes from the study, with special thanks to L. E. Hayes-Creekmore. Finally, we thank D. H. Johnson, an anonymous reviewer, and especially C. D. Littlefield for constructive editorial comments on an earlier draft of this paper.

STUDY AREA

Malheur, comprising 75,119 ha, lies within the northern Great Basin of southeastern Oregon. The area is characterized by a semi-arid climate, shrub-steppe rangelands, north-south oriented fault block mountain ranges, and internal drainage basins. The refuge is an oasis within a dry environment that contains freshwater marshes, flood-irrigated meadows, and 2 large lakes. Most of the refuge's nesting cranes rely on the irrigated meadows and marshes for breeding habitats.

METHODS

Sandhill crane nests were located annually on the refuge and eggs were floated to estimate hatching date (Westerskov 1950). We revisited nests shortly after predicted hatching dates to assess nest fate and attach transmitters to colts. Older colts were captured by hand and radio-equipped as they were encountered on their territories throughout the brooding period. We attached transmitters to colts from 1 day to about 9 weeks of age.

In 1991, we used a transmitter similar to that described by Mauser and Jarvis (1991), with some modifications. The transmitter weighed 4.1 g (<5% of mean weight of newly hatched colts), and had a life expectancy of 120 days. It consisted of an SM1 transmitter (AVM Instrument Co., Livermore, Calif.) coupled to a mercuric oxide battery. The 15-cm antenna extended back at a 45° angle from the transmitter. After a few trials, we shortened the wire to 12 mm. A length of stainless steel wire, formed into an anchor, protruded 18 mm from the front of the transmitter and was incorporated into the potting material (dental acrylic) . We used monofilament nylon sutures and cyanoacrylate glue instead of dental floss to secure transmitters to colts.

From 1992 through 1994 we used Holohil Systems Ltd. (Woodlawn, Ont., Canada) transmitters which were modified for sandhill crane colts. The transmitter weighed 3.8 g and had a life expectancy of 120 days. It was powered by a mercuric oxide battery, and the 22-cm antenna extended horizontally from the base. Two suture tubes were in the base of the transmitter.

In 1995 we used transmitters designed by Advanced Telemetry Systems, Inc. (Isanti, Minn.). The transmitter weighed 3.5 g and had a 90-day life expectancy. It was powered by a silver oxide battery with the antenna extended horizontally 18 cm from the transmitter base. It also had 2 basal suture tubes.

Transmitters were attached to the colts' backs, between the wings, with a sterile suture. Beginning in 1992, chromic gut sutures were used as the only transmitter attachment material. Attachment site and sutures were sterilized with a dilute iodine solution. Two single sutures were placed just beneath the skin and tied off with several square knots.

We monitored colts daily and recorded their locations, movements, and habitats, and noted if they were accompanied by parents. We tried to obtain a visual sighting on colts at least once each week. If parents were not observed at the location of a radio signal, we moved in closer to confirm whether the colt was still alive. After colts were about 3 weeks old, they were recaptured by hand and transmitters were removed from their backs and taped to plastic tarsal bands.

Mortality causes were sometimes difficult to determine, but carcass condition and evidence at the site were used to identify causative factors. Intact colt carcasses were shipped to the National Wildlife Health Research Center in Madison, Wisconsin, for necropsy. Although predation was not directly observed, predator identification was possible in many cases by examination of the carcass and death site for evidence such as tooth marks, scats, and tracks.

RESULTS

Between 1991 and 1995, 142 crane colts were transmitter-equipped. A total of 23 colts (19% of known fates) survived to fledge, while fates of 19 colts were undetermined. Predators were responsible for the most losses (64), followed by unknown causes (13), parasitic gapeworms (8), drowning (7), intraspecific aggression (5), study-related mortality (1), roadkill (1), and hay swather (1). Fates of colts are summarized in Table 1.

At least 26 of the 64 colts lost to predators were killed by mink. Mink victims were generally easy to identify; they exhibited characteristic bites at the back of the neck and skull, and were rarely consumed. Three carcasses were excavated from mink dens. Great horned owls were responsible for loss of 10 colts. Decapitations, talon wounds, and in one case a feather at the death scene identified the predator. Golden eagles were responsible for the loss of 9 colts. These carcasses were stripped of flesh and had sheared bones. Transmitters from 2 birds were found at an eagle feeding perch. One colt was recovered from a northern harrier nest, and 5 others were found with talon wounds or nearby scats, indicating predation by raptors. Coyotes were responsible for 5 deaths, which were determined by tooth marks, tracks, and scats. Large tooth marks on the vertebrae of 1 colt, the aquatic location, and flattened vegetation at the site suggested a raccoon as the culprit. An additional 7 colts were lost to unidentified predators.

Seven colts drowned. Drowning is a natural mortality factor, as drowned colts have been found on the refuge in the past (C. D. Littlefield, Wildlife Consultant, pers. commun.; G. L. Ivey and E. J. Scheuering, unpubl. data); however, the disturbance from transmitter attachment may have resulted in premature departure from the nest platform. In 1991, 1 colt apparently died from complications resulting from transmitter attachment. The National Wildlife Health Center diagnosed 8 colt losses from parasitic gapeworm infestations.

Evidence of brood strife, or intraspecific aggression, was found in 5 cases. These colts exhibited stab wounds, broken wings, or bruises. Based on the evidence, birds were determined to have died from injuries caused either by siblings, parents, or perhaps neighboring adults. One colt was found injured about 400 m from its parents and larger sibling the day before it was found dead at the nest platform.

One colt was killed by a vehicle while crossing a highway. Another colt, about 10 days from flight stage, moved off the refuge to a private hay meadow and was killed by a hay swather when the field was mowed on 13 July 1993.

DISCUSSION

Many factors including habitat changes and human activities since European settlement have influenced the distribution and abundance of predator species at Malheur (Littlefield 1995*b*). Artificially created wetlands have increased densities of cranes and other wetland birds, subse-

Table 1. Fates of transmitter-equipped greater sandhill crane colts at Malheur National Wildlife Refuge, Oregon, 1991–95.

Fate	1991	1992	1993	1994	1995	Total	% of known fate
Fledged	1	5	5	10	2	23	19
Unknown causes of death	1	2	4	5	1	13	11
Gapeworm parasites	0	6	2	0	0	8	7
Drowning	2	0	5	0	0	7	6
Intraspecific aggression	1	2	0	0	2	5	4
Study related	1	0	0	0	0	1	1
Roadkill	0	1	0	0	0	1	1
Hay swather	0	0	1	0	0	1	1
Predator	9	14	16	7	18	64	52
Mink	5	8	3	5	5	26	21
Coyote	0	2	3	0	0	5	4
Raccoon	1	0	0	0	0	1	1
Great horned owl	1	2	2	1	4	10	8
Golden eagle	0	1	4	0	4	9	7
Northern harrier	1	0	0	0	0	1	1
Unidentified raptor	1	0	0	1	3	5	4
Unidentified predator	0	1	4	0	2	7	6
Known fate subtotal	15	30	33	22	23	123	
Fate unknown	6	2	5	4	2	19	
Total	21	32	38	26	25	142	

quently increasing the prey base for predators (Littlefield and Thompson 1979, 1987). We believe that higher prey densities combined with the elimination of wolves (*Canis lupus*), livestock and agricultural practices, management for optimal big game populations, introductions of exotic game birds, and road-killed wildlife provided by automobiles have resulted in high local populations of certain predators.

Littlefield and Lindstedt (1992) noted that coyotes killed 52% of colts during their study at Malheur. Their study was conducted at a time when no coyote control had been practiced during the previous 9 years (1972–81). Many individual coyotes had become resident on the refuge and were experienced in finding crane colts. Coyotes killed only 4% of the colts during this study. After the 8-year coyote control program ended in 1993, no colt losses to coyotes were documented in 1994 or 1995. Coyotes were relatively common on the refuge during these 2 years, but abundant small mammal populations in 1994 likely provided an alternate prey base. However, in 1995 small mammal populations crashed while coyote numbers remained high (G. L. Ivey and E. J. Scheuering, unpubl. data). We believe the primary reason for low losses to coyotes during 1994 and 1995 was the coyotes' inexperience in finding young cranes.

Our study indicated that mink had replaced coyotes as the principal predator of colts. Mink were responsible for 21% of known colt mortalities. Based on high losses from mink in 1991 and 1992, an experimental mink control program was initiated in 1993. Losses from mink in 1993 were reduced to 9%, compared with 36% and 27% in 1991 and 1992, respectively. In 1994 and 1995, years in which there was no mink control, mink were responsible for 23% and 22% of colt losses, respectively.

Golden eagles and great horned owls accounted for 20% of known colt mortality. Colt losses to raptors varied annually during the study. The lowest rates occurred during the years following good water conditions, which apparently resulted in higher rodent populations (G. L. Ivey and E. J. Scheuering, unpubl. data). Raptor kill rates were lowest in 1994 (9%), a year with peak rodent populations, and were highest in 1995 (49%), after rodent populations "crashed" (G. L. Ivey and E. J. Scheuering, unpubl. data).

Colts whose deaths were attributed to gapeworm infestations were also emaciated. This condition could have been caused directly by the parasite, but six of the colts died in 1992, an extremely dry year. It is possible that drought conditions either provided a more favorable environment for gapeworms, or reduced food availability, which resulted in colt emaciation and exacerbated parasite effects.

MANAGEMENT IMPLICATIONS

The goal of this study was to collect additional data for

application in increasing crane productivity by reducing colt mortality at Malheur. The coyote control program did have positive effects on productivity for the first few years; however, it became evident that predator mortality was compensatory: as coyote impacts were reduced, mink and raptor impacts increased.

Managers considering predator control to enhance productivity of cranes should consider the following:

1. The local composition of predator communities and the potential for compensatory predation should be appraised. The proportion of young cranes consumed by specific predator species depends on several factors, including alternate prey availability, predator population densities, and predator behavior. In some instances, individual predators may become efficient at finding young cranes, and removing these individuals may temporarily provide satisfactory results.
2. Some problem predator species may be reduced through habitat management or elimination of denning or nesting sites. Perhaps eliminating or modifying artificial mink denning sites (e.g., dikes or nesting islands) would reduce local mink numbers.
3. Colt deaths resulting from parasites could possibly be reduced by developing effective medication methods. Such a program would probably be utilized only when gapeworms were having a severe impact on threatened crane subspecies or local subpopulations.

We conclude that an integrated approach involving habitat management in conjunction with predator control and possibly parasite control may be necessary to optimize crane productivity on managed areas.

LITERATURE CITED

DREWIEN, R. C., W. M. BROWN, and W. L. KENDALL. 1995. Recruitment in Rocky Mountain greater sandhill cranes and comparison with other crane populations. J. Wildl. Manage. 59:339-356.

LITTLEFIELD, C. D. 1976. Productivity of greater sandhill cranes on Malheur National Wildlife Refuge, Oregon. Pages 86-92 *in* J. C. Lewis, ed. Proc. [1975] int. crane workshop. Oklahoma State Univ. Publ. Printing, Stillwater.

______. 1995*a*. Demographics of a declining flock of greater sandhill cranes in Oregon. Wilson Bull. 107: 667-674.

______. 1995*b*. Sandhill crane nesting habitat, egg predators, and predator history on Malheur National Wildlife Refuge, Oregon. Northwest. Nat. 76:137-143.

______, and S. M. LINDSTEDT. 1992. Survival of juvenile greater sandhill cranes on Malheur National Wildlife Refuge, Oregon. Pages 21-31 *in* D. A. Wood, ed. Proc. 1988 crane workshop. Florida Game Fresh Water Fish Comm. Nongame Wildl. Program Tech. Rep. 12.

______, and S. P. THOMPSON. 1979. Distribution and status of the Central Valley Population of greater sandhill cranes. Pages 113-120 *in* J. C. Lewis, ed. Proc. 1978 crane workshop. Colorado State Univ. Printing Serv., Fort Collins.

______, and ______. 1987. Greater sandhill cranes and common ravens on Malheur National Wildlife Refuge, Oregon: their history and relationship. Pages 156-165 *in* J.C. Lewis, ed. Proc. 1985 crane workshop. Platte River Whooping Crane Maintenance Trust, Grand Island, Nebr.

MAUSER, D. M., and R. L. JARVIS. 1991. Attaching radio transmitters to 1-day-old mallard ducklings. J. Wildl. Manage. 55:488-491.

WESTERSKOV, K. 1950. Methods for determining the age of game bird eggs. J. Wildl. Manage. 14:56-67.

SURVIVAL AND HABITAT USE OF GREATER SANDHILL CRANE COLTS ON MODOC NATIONAL WILDLIFE REFUGE, CALIFORNIA

KEVIN J. DESROBERTS,[1] U.S. Fish and Wildlife Service, Modoc National Wildlife Refuge, P.O. Box 1610, Alturas, CA 96101, USA

Abstract: Radiotelemetry was used to monitor 13 (1990) and 14 (1992) greater sandhill crane (*Grus canadensis tabida*) colts on Modoc National Wildlife Refuge, California, to determine causes of mortality, brood habitat utilization, and effects of habitat management on colt survival. Colt survival rates were 0.36 (1990) and 0.54 (1992). Coyotes (*Canis latrans*) killed 3 colts and mink (*Mustela vison*) killed 4; 1 colt died because of a bacterial infection (*Staphylococcus aureous*). Broods used 6 different habitat types and 79% used more than 1 type. Irrigated meadows (74%), cultivated uplands (53%), and marsh (42%) were used by most broods. Habitat management practices included hay harvesting (15 July–30 August) and fall grazing (September–December) and did not directly cause mortality, but hay harvest may have increased vulnerability to predation.

PROC. NORTH AM. CRANE WORKSHOP 7:18-23

Key words: California, colts, habitat, mortality, predation, radiotelemetry, sandhill crane, survival.

Modoc National Wildlife Refuge (NWR), located near Alturas in northeast California, has historically been an important nesting and staging area for members of the Central Valley Population of greater sandhill cranes. The Central Valley Population was listed as a threatened species by the state of California in 1983 (Schlorff 1988).

Sandhill cranes begin arriving at Modoc NWR during mid- to late February. Most pairs establish nesting territories by mid-March and initiate nesting by mid-April. Crane use of the refuge has been monitored to some extent since the mid-1970's. From 1980 to present, special emphasis has been placed on crane habitat management, especially for the breeding population. Breeding pairs increased from 17 in 1976 to 36 in 1992 (Modoc NWR, unpubl. data).

A radiotelemetry study of greater sandhill crane colts was conducted on Modoc NWR during 1990 and 1992. Low nesting success (39%) caused by flooding in May 1991 produced too few colts to monitor (Modoc NWR, unpubl. data). The objectives of the study were (1) to determine causes of crane colt mortality, (2) to determine brood habitat utilization, and (3) to evaluate the effects of refuge management practices on colt survival.

Modoc NWR Manager E. C. Bloom procured funding and Assistant Manager D. Hardt developed the project proposal and acquired transmitters. E. C. Bloom, D. Hardt, R. L. Ryno, B. Storm, and A. Kahlen aided in capture, transmitter attachment, and monitoring of colts. Advice on various aspects of this project was graciously provided by C. D. Littlefield, G. L. Ivey, and M. Nagendran. The National Wildlife Health Center (NWHC), Madison, Wisconsin, performed 1 necropsy. E. C. Bloom, C. D. Littlefield, G. L. Ivey, M. Nagendran, and D. H. Johnson reviewed earlier drafts of this manuscript.

STUDY AREA

Modoc NWR (2,543 ha) is located at an elevation of 1,329 m in northeastern California. Climatic conditions consist of cold, wet winters and cool, dry summers. Annual precipitation averages 30.5 cm. The refuge contains irrigated meadows, a natural flood plain, marsh communities, riparian areas, grasslands, cultivated uplands, and uplands dominated by big sagebrush (*Artemisia tridentata*) and western juniper (*Juniperus occidentalis*).

Hay was removed from several wet meadow units during late July and early August during both years of research. Irrigation was terminated between 1 and 15 July to facilitate drying of meadows for haying to occur between 15 July through 30 August.

MATERIALS AND METHODS

I used a 60× spotting scope and 10×40 binoculars to locate cranes on territories during March and April. I conducted nest searches on foot in areas where cranes were observed on territory from April to June. I estimated the incubation stage of each nest using the flotation method (Westerskov 1950). Crane nests were monitored closely within a few days of the expected hatching date to determine precisely when colts hatched.

1990

Crane colts were captured by hand from previously located nests and opportunistically in other territories. Most colts were located and captured within 24–48 hours after hatching. Each colt weighing >110 g was equipped with a

[1]Present address: Ruby Lake National Wildlife Refuge, HC 60, Box 860, Ruby Valley, NV 89833-9802, USA.

collar-mounted transmitter (SM1-H Mortality) weighing 5.5–6.0 g. Colts were recaptured after 4–5 weeks, the collar-mounted transmitter was removed, and a solar-powered leg-band transmitter (Solar Assisted H Module with Mortality) weighing 58–60 g was placed above the tibial-tarsal joint. A size 9, lock-on federal band was placed on the opposite leg. Both collar-mounted and solar-powered leg-band transmitters were identical to those described by Littlefield and Lindstedt (1992) and constructed by AVM Instrument Co., Livermore, Calif. (use of trade names does not imply endorsement by the U.S. Fish and Wildlife Service).

Crane colts with transmitters were monitored daily throughout the 70-day period prior to fledging and 3–5 times weekly from fledging to migration. Locations of colts, types of habitats occupied, and, when applicable, causes of mortality were recorded. Crane colts with transmitters that failed were located when possible and the transmitter removed. When mortality was suspected, but there was no transmitter signal, the area where the colt(s) was last seen was searched on foot and scanned with the receiver. If the colt(s) could not be found the parents were identified by territory presence and/or color mark identification and monitored until it was certain that the colt(s) was no longer present.

1992

Crane colts were again captured by hand from previously located nests and opportunistically in other territories. Due to problems associated with the 1990 transmitters, I used a PD-2T (Holohil Systems, Ltd., Ont., Canada) transmitter weighing 4 g in 1992. Transmitters were sutured onto the back of colts less than 28 days old. The down feathers were trimmed and transmitters were attached with Dermal 000 suture to the skin on the middle of the back with the antenna pointing down. Two suture tubes located on the transmitters allowed secure attachment and prevented transmitters from twisting. After 28 days each colt was recaptured, the sutures were cut, and the transmitter was reattached with duct tape to a plastic leg-band placed above the tibial-tarsal joint. When colts reached 50 days they were captured again and banded above the tibial-tarsal joint with a white color band (2.5 cm wide) and a red numbered band (9 cm wide) on 1 leg and a federal and 2 color bands (2.5 cm wide) on the opposite leg. All color bands were plastic, and each color sequence was unique for each colt. The transmitter was attached to the red numbered band with duct tape.

Crane colts were monitored daily throughout the 70-day period before fledging. Locations of colts, habitat types occupied, and, when applicable, causes of mortality were recorded. I monitored fledged colts with operating transmitters 3–5 times weekly until they migrated. Fledged colts lacking functional transmitters were identified by color band combination and monitored by visual observation until migration.

During 1990 and 1992, habitat use was recorded whenever colts were contacted by radio or observed. Sampling biases prevented statistical analysis of habitat use data. Instead, I calculated use based on the percentage of habitat types used by broods. The same variable was used for each habitat type regardless of how many times broods were observed in each type.

RESULTS

Crane Colt Fates in 1990

Transmitters were attached to 13 crane colts from 12 broods in 1990 (Table 1). Collar-mounted transmitters were attached to 8 colts ranging in age from 1 to 8 days; 7 disappeared before large enough for leg-band transmitters. Cause of mortality could be determined for only 2 of 7 colts with collar-mounted transmitters; transmitters failed and/or antennas detached on the other 5. One colt was apparently killed by a coyote when 2 days old. The transmitter had been removed and the colt was entirely consumed. The second colt died of starvation at 33 days because the neck collar had constricted the esophagus and halted the passage of food items.

In the 5 cases of transmitter failure, mortality was verified by conducting a ground search and monitoring the adults, who often stayed in the area where mortality occurred for 2–5 days. The absence of colts verified mortality occurred, but cause of death could not be determined. One malfunctioning collar-mounted transmitter was removed from a 25-day-old colt. This colt was never seen again and apparently died before fledging. Thus 6 of 7 (85%) colts marked (not including the colt that was starved by its neck collar) died during their first 5 weeks. Only 1 marked colt was known to have survived beyond 5 weeks; this colt and its sibling later fledged.

Because of the high mortality rate, a new sample was required for the attachment of the solar-powered transmitters. In addition to the surviving colt, transmitters were placed on 5 other colts, ranging in age from 32 to 45 days. The age of colts that had hatched from located nests was determined from hatching dates. Size and weight were used to estimate age of colts that had hatched from unknown nests.

Two colts from the new sample died before fledging. One colt died at 55 days and was located about 24 hours after mortality had occurred. The carcass was severely emaciated (1,200 g) and the right foot was distended. NWHC necropsy

Table 1. Suspected fates of radiotagged greater sandhill crane colts, Modoc NWR, California, 1990 and 1992.

	Age (days)		
Colt no.	Transmitter attachment	At death	Suspected fate
90-01	1	21	Unknown
90-02	2	19	Unknown
90-03	2	2	Coyote predation
90-04	1		Fledged
90-05	1	11	Unknown
90-06	6		Unknown[a]
90-07	7	33	Starvation
90-08	2	28	Unknown
90-09	32		Fledged
90-10	42	55	Infection[b]
90-11	45		Fledged
90-12	45		Fledged
90-13	35	62	Coyote predation
92-01	NA	<8	Unknown
92-02	4		Unknown[c]
92-03	8		Unknown[c]
92-04	23	41	Mink predation
92-05	23	75	Mink predation
92-06	25		Fledged
92-07	25		Fledged
92-08	28		Fledged
92-09	28		Fledged
92-10	27	69	Coyote predation
92-11	35	50	Mink predation
92-12	35		Fledged
92-13	35	60	Mink predation
92-14	35		Fledged
92-15	35		Fledged

[a] Transmitter removed at 25 days because of failure.
[b] *Staphylococcus* infection.
[c] Transmitter detached.

found that a bacterial infection (*Staphylococcus aureous*) was present in the right foot and 1 digit was dislocated; the emaciation was secondary to the infection. The second colt was killed by a coyote when 65 days old. The carcass was found in very dense cattail (*Typha* sp.) and rushes (*Juncus* sp.) 1.5 m tall in 18 cm of water. The vegetation surrounding the carcass was matted. No scat was found, but the carcass condition was typical of coyote predation.

Crane Colt Fates in 1992

Most crane pairs nested about 2 weeks earlier than normal due to mild spring conditions in 1992. Early hatching dates and the late transmitter arrival prevented their attachment to day-old colts. Three of the broods (6 colts) were monitored visually from hatching to capture. One colt was lost during its first week to an unknown cause, but is included in the results. Transmitters were attached to 14 crane colts from 9 broods ranging in age from 4 to 35 days (Table 1), but most colts (86%) were between 22 and 35 days old. Transmitters became detached from 2 colts and their fate was not determined. Seven of 13 (54%) monitored colts survived to fledging.

Four colts were killed by mink. Two colts (41 and 60 days old) killed by mink were found in dens. One den was located within 5 m of the Pit River, the other within 40 m of a canal. The third colt (50 days old) was found in a dry slough in an irrigated meadow that had been recently drained to permit hay harvest. The colt had been dead for 2 days and most flesh had been removed from the carcass. Mink scat was present at several locations in the slough within 1 m of the carcass. It appeared that a mink had killed the colt in the meadow and dragged it into the slough. The fourth colt was 75 days old and had fledged at least 6 days before being preyed upon. The colt was located within 12 hours following predation, and a mink was observed leaving the carcass. This colt was found within 50 m of the area where its sibling had been previously killed by a mink.

One colt was 69 days old when killed by a coyote. The primaries were almost completely developed and the colt probably would have fledged within 2-3 days. The carcass was located within 30 m of a pond in an irrigated pasture on adjacent private land. The carcass condition was typical of coyote predation.

Brood Habitat Utilization

Brood habitat used consisted of 6 types: (1) irrigated meadows consisting of a variety of rushes, sedges (*Carex* sp.), and grasses that were flood irrigated from 1 April through 15 July, hayed during August, and grazed during autumn (some units were grazed September-December); (2) agricultural fields consisting of cultivated uplands planted with winter wheat, barley, or winter rye; (3) irrigated pasture (refuge and private land) consisting of grasses and sedges which are subject to spring and/or summer cattle grazing; (4) marsh consisting of bulrush (*Scirpus* sp.), cattail (*Typha latifolia*), and rushes; (5) floodplain, consisting of willow (*Salix* sp.), reed canary grass (*Phalaris arundinacea*), bulrush, rushes and sedges; and (6) grassland consisting of western wheatgrass (*Agropyron smithii*), crested wheatgrass (*A. cristatum*), intermediate wheatgrass (*A. intermedium*), and Great Basin wild rye (*Elymus cinereus*).

During 1990, colts from 12 nest sites were monitored. Six nests were located in irrigated meadows and 4 in marsh habitat; 2 nests were not located before hatching. All 6 broods that hatched in irrigated meadows used these areas as brood habitat. In addition, 1 brood also used irrigated pasture, 1 also used agricultural fields, and 1 also used marsh habitat. The 4 broods that hatched from marsh nest sites also used these areas as brood habitat. In addition, 1 also used floodplain, 1 also used grasslands, 1 also used agricultural fields, and 1 also used irrigated pasture. Two broods hatched in unknown habitat types; 1 used irrigated meadow and the second a combination of marsh, irrigated meadow, and agricultural fields as brood habitat.

During 1992, colts from 7 nest sites were monitored. Four nests were located in irrigated meadows and 1 in marsh habitat; 2 nests were not located before hatching. All broods that hatched at irrigated meadow nest sites used this habitat for brooding. In addition, 1 brood also used marsh and agricultural fields, 1 also used irrigated pasture, 1 also used agricultural fields, and 1 also used floodplain. The brood that hatched at the marsh nest site used marsh, irrigated meadow, agricultural fields, and grasslands. Two broods hatched in unknown habitat types: 1 used agricultural fields, floodplain, and marsh; the other used agricultural fields as brood habitat.

Habitat Management

During 1990, 4 broods (4 colts) were using irrigated meadow habitat before it was mowed. Most colts were 8-9 weeks old and highly mobile when hay harvesting began. Mowing machinery did not cause any direct mortality, but the colts were displaced. Displacement of 1 colt may have caused its death. Its parents moved the colt and its sibling about 800 m north into marsh habitat with dense stands of rushes and cattail. Within 48 hours the monitored colt (62 days old) was killed by a coyote. The dense vegetation may have limited visibility for the cranes and provided easy concealment for the coyote, increasing the colt's vulnerability.

During 1992, 3 broods (4 colts, 7-9 weeks old) used irrigated meadows and adjacent habitats before meadows were mowed. Temporary displacement occurred, but no mortality occurred during mowing, and all 3 broods continued to use the meadows after hay removal. Two colts were killed after hay harvest: 1 by a coyote and one by a mink. Both colts were killed in adjacent habitat that had also been used before haying of meadow habitat. Therefore, displacement did not appear to increase exposure to predation.

DISCUSSION

Crane Colt Survival

Capture of crane colts and attachment of transmitters caused temporary disturbance and stress to both colts and adult cranes. However, the adults and colts returned to areas they had used prior to disturbance within 4-12 hours. No colts were injured during capture and handling. With exception of the colt that starved, no other colts were known to have died as a result of transmitter attachment or disturbance.

However, causes of crane colt mortality in 1990 could not be determined in most instances because of transmitter problems during the first 5 weeks of the study. The close monitoring of adults only helped to determine when mortality occurred. In 1990, mortality rates appeared to be very high (86%) during the first 5 weeks (Table 1), which was similar to data reported by Littlefield and Lindstedt (1992). Colts which survived through this critical period subsequently had a survival rate of about 71%. Two colts were known to have been killed by a coyote, 1 died from a *Staphylococcus* infection, and 4 died of unknown causes. In contrast, during 1992 the mortality rate (46%) was lower and mortality primarily occurred when colts were between 41 and 75 days old (Table 1). Four colts were killed by mink, 1 by a coyote, and 1 died of an unknown cause.

The increase in survival during 1992 may be attributed to the difference in age distribution of the colts monitored and small sample size. Thirty-eight percent of colts were monitored within 1-7 days following hatching in 1992, compared to 55% in 1990.

In the past, mink were not considered a significant predator of sandhill crane colts (Littlefield and Lindstedt 1992). The incidence of mink predation on Modoc NWR is comparable to recent documentation at Malheur NWR (Ivey and Scheuering 1997). Mink are very efficient and opportunistic predators. The importance of prey items varies with prey availability and season, with high use of specific prey items at certain times of the year (Linscombe et al. 1982). The high incidence of mink predation on crane colts during 1992 was probably influenced by the drought that prematurely dried irrigated meadows, marshes, and ponds. These conditions probably reduced many prey species normally taken by mink, which may have increased the predation rate on crane colts.

Coyotes have been documented in the past as a major predator of waterfowl and sandhill crane eggs and young (Littlefield and Lindstedt 1992; Modoc NWR, unpubl. data). Coyotes have been controlled on Modoc NWR since 1975 to increase waterfowl and crane production. In addition, the refuge is bordered by livestock and sheep ranches where predator control is annually practiced, usually by Animal Damage Control (U.S. Department of Agriculture) personnel. The low incidence of coyote predation on crane colts during this study may be attributed to these control efforts.

Brood Habitat Utilization

During 1990 and 1992, a total of 19 broods used 6 different habitat types. Fifteen broods (79%) used more than 1 type. Fourteen broods (74%) used irrigated meadows, 10 (53%) used cultivated uplands, 8 (42%) used marsh, 3 (16%) used irrigated pasture, 3 (16%) used floodplain, and 2 (11%) used grassland.

With an abundance of invertebrates and small vertebrates, wet meadows provide the best foraging habitat for crane colts (Schlorff 1988), and were most frequently used by pairs with young.

Severe drought conditions were prevalent during 1992, and water was insufficient to properly irrigate meadows and maintain marsh habitat on the refuge. Many areas dried prematurely. Some crane clutches originally laid in nests over water hatched on dry land. Drought conditions may have forced pairs to use other habitats of poorer quality to provide food for their young. Following hatching, pairs with broods often move their young into areas of suitable habitat with abundant invertebrate populations (Walkinshaw 1973). The impact of the 1992 drought was also apparent in the use of cultivated uplands. Seventy-one percent of the broods foraged in cultivated uplands during 1992 compared with 42% in 1990.

In most (91%) cases, cultivated uplands were used in addition to other habitat types and were primarily used from just before fledging until migration. Only 1 brood was observed foraging exclusively in cultivated uplands. Agricultural fields contained standing wheat, barley, and rye, which are very important to juvenile cranes from post-fledging to migration (Walkinshaw 1973, Littlefield 1986). On Modoc NWR, these fields are also used extensively by adult cranes during autumn migration.

Habitat Management

Refuge management practices had no known direct effect on the survival of crane colts using irrigated meadows during this study. Since meadows were mowed after most colts could escape mowing machinery, colts were able to move to adjacent habitat. Broods often returned to meadows following haying. However, drying meadows probably reduced food availability, and displacement of broods during haying probably increased vulnerability to predation, dependent upon colt age and the habitat type into which they moved.

RESEARCH AND MANAGEMENT IMPLICATIONS

Crane Colt Survival and Habitat Use

Capturing, attaching transmitters, banding, and, to a lesser extent, monitoring resulted in disturbance to both colts and adults. Colts were captured by hand and some had to be chased to be captured, but none were injured. The adults were disturbed but remained nearby while colts were handled. Colts promptly rejoined their parents when released. Handling caused temporary stress and displacement, but did not result in known mortality or cause permanent habitat abandonment.

Some adults with broods reacted negatively to me in my vehicle during monitoring; they would often sit down in an apparent attempt to hide. However, since I was present for only a few minutes at a time, these reactions did not appear to have significant impacts on colt survival or habitat use.

Radio Transmitters

There were several problems with the collar-mounted radio transmitters in 1990. The elastic collar (2.54 cm in diameter) did not expand enough to allow passage of large food items. This was first noted when a collar was removed from a colt (25 days old) after the transmitter had failed. There were 2 red-winged blackbird (*Agelaius phoeniceus*) chicks and 1 frog (*Rana* sp.) in the colt's esophagus above the neck collar, and it probably would have starved had the collar not been removed. Another colt did starve as a result of collar restrictions similar to those described above. Though 4 colt disappearances were unexplained (Table 1), starvation due to the elastic collar could have been responsible for 1 of those deaths, the colt that disappeared when 28 days old. The other 3 colts disappeared before they were old enough to be starved by the collar.

The location of the antenna, ventrally on the colt, limited the signal range, especially in dense cover. In addition, antennas were in constant contact with vegetation, and 4 antennas broke off.

The solar powered leg-band mounted transmitters functioned satisfactorily. Attachment did not appear to cause discomfort or limit mobility. However, when 1 colt (55 days old) was recovered after dying from a *Staphylococcus* infection, it was apparent that the tibial-tarsal joint was swollen on the leg with the transmitter. Whether the swelling was caused by the infection or the transmitter is unknown, but it seems possible that the transmitter weight (60 g) might cause irritation to the tibial-tarsal joint.

Reception ranged from 1,600 to 3,200 m. However, cloudy days and cool nights decreased battery power, which subsequently reduced reception distance to as low as 800 m. When cranes returned as juveniles in 1991, 2 of 4 still had functional transmitters.

The transmitters used on small colts in 1992 were much improved. Transmitters sutured on the back of colts did not cause any observed discomfort or mortality, and reception (800 m) was more than adequate. However, some of the sutures did become detached as a result of growth and poor knot tying. When transmitters were attached to leg bands with duct tape, they remained attached but did not inhibit movement, and the attachment location increased reception distance from 800 m to 1,200 m.

Habitat Management

During this study, irrigated meadows were the most used habitat type. In the past, irrigation of most meadows was terminated 1-15 July to permit hay harvest between 15 July and 30 August. The majority of crane nests hatch on Modoc NWR between mid-May and early June. Most colts were only 5 weeks old when irrigation was terminated and 7-8 weeks old when mowing began. Termination of irrigation likely resulted in a decline in populations of both invertebrates and vertebrates during a period when crane colts were most dependent on these food sources. In addition, removing vegetation before colts have fledged increases vulnerability to predation (Littlefield and Lindstedt 1992), especially by coyotes. Irrigation should be sustained at least through 1 August and hay mowing not permitted before 15 August. This delay will maintain important brood habitat and should allow the majority of colts to fledge prior to hay removal.

LITERATURE CITED

IVEY, G. L., and E. J. SCHEUERING. 1997. Mortality of radio-equipped sandhill crane colts at Malheur National Wildlife Refuge, Oregon. Proc. North Am. Crane Workshop 7:14-17.

LINSCOMBE, A., N. KINLER, and R. J. AULERICH. 1982. Mink (*Mustela vison*). Pages 634-635 *in* J.A. Chapman and G.A. Feldhamer, eds. The wild mammals of North America. Johns Hopkins Univ. Press, Baltimore, Md.

LITTLEFIELD, C. D. 1986. Autumn sandhill crane habitat in southeast Oregon. Wilson Bull. 98:131-137.

______. 1989. Status of greater sandhill crane breeding populations in California, 1988. Unpubl. Rep., Wildl. Manage. Div., California Dep. Fish Game, Sacramento. 40pp.

______, and S. M. LINDSTEDT. 1992. Survival of juvenile greater sandhill cranes at Malheur National Wildlife Refuge, Oregon. Pages 21-31 *in* D. A. Wood, ed. Proc. 1988 North Am. crane workshop. Florida Game Fresh Water Fish Comm. Nongame Wildl. Program Tech. Rep. 12.

SCHLORFF, R. W. 1988. Five-year status report-greater sandhill cranes. California Dep. Fish Game, Sacramento. 9pp.

WALKINSHAW, L.H. 1973. Cranes of the world. Winchester Press, New York, N.Y. 370pp.

WESTERSKOV, K. 1950. Methods for determining the age of game bird eggs. J. Wildl. Manage. 14:56-67.

PAIR FORMATION BY COLOR-MARKED WHOOPING CRANES ON THE WINTERING GROUNDS

THOMAS V. STEHN, U.S. Fish and Wildlife Service, Aransas National Wildlife Refuge, P.O. Box 100, Austwell, TX 77950, USA

Abstract: Observations were made of color-marked whooping cranes (*Grus americana*) at Aransas during the 1977–94 winters. Subadult cranes wintered near their juvenile home ranges in flocks averaging 3.12 birds. One and 2 winters prior to nesting, subadults at Aransas tended to form duos apart from other subadults. Frequencies of association 1 year prior to nesting averaged 65.4%. However, 27.7% of pair bonds (n = 18) formed during spring migration or on the breeding grounds without any observed prior association at Aransas. Sixty-eight banded cranes formed pairs. Pair members averaged 1.06 years difference in age. Winter territories of adults tended to be established near the juvenile home range of the male. Information on typical subadult behavior and pair formation at Aransas can be used to compare with pairing behavior in newly established whooping crane flocks.

PROC. NORTH AM. CRANE WORKSHOP 7:24-28

Key words: Aransas, associations, behavior, *Grus americana*, pair formation, subadult, territory establishment, whooping crane.

The only remaining natural flock of whooping cranes is still very much endangered. This Aransas-Wood Buffalo flock has recovered from a low of 15 birds in 1941 (U.S. Fish and Wildlife Service 1994) to reach 133 birds in the spring of 1995 (T. Stehn, unpubl. data). Efforts to establish a second flock of whooping cranes at Grays Lake National Wildlife Refuge in Idaho by using sandhill cranes as foster parents failed when no adult birds formed breeding pairs (U.S. Fish and Wildlife Service 1994). Efforts to establish a non-migratory flock in Florida by using captive-raised whooping cranes soft-released into the wild began in 1993. Pair formation in Florida occurred in 1995 when 2 3-year-olds built a nest (Nesbitt et al. 1997). It is, however, too early to evaluate the success of pair formation and breeding within this experimental flock. Quantification of pair formation behaviors in the natural Aransas-Wood Buffalo flock could be used to ascertain if behavior is normal in any newly established flock.

The formation of subadult flocks at Aransas National Wildlife Refuge (ANWR) is believed to provide conditions under which potential mates meet and select a suitable partner. Frequent associations of subadults observed over 1–3 winter seasons resulted in 6 pair bonds (Bishop 1984). Blankinship (1976) and Stehn (1992*a*,*b*) have observed widowed whooping cranes pairing with new mates at Aransas. This paper describes typical subadult behavior, pair formation, and territory establishment on the wintering grounds for birds color-banded during 1977–88.

I thank ANWR managers J. B. Giezentanner and the late E. F. Johnson for supporting this research. Excellent piloting by J. Kosler, J. Miller, G. Shore, T. Taylor, J. Winship, and the late R. Tanner have enabled much of these data to be collected. I am very appreciative of Canadian Wildlife Service (CWS) biologists J. P. Goossen, B. W. Johns, and E. Kuyt for color-marking cranes and sharing their observations to better understand the Aransas-Wood Buffalo population. I thank R. A. Stehn for statistical analysis, and D. H. Johnson and R. A. Stehn for editing the manuscript.

STUDY AREA

The whooping crane winter range is situated in 8,175 ha of salt marsh on the Texas coast northeast of Rockport (Stehn and Johnson 1987). During the 1994–95 winter, cranes wintered on ANWR (n = 48), West St. Charles Bay (8), San Jose Island (25), Matagorda Island (37), and Welder Flats (14) (T. Stehn, unpubl. data). The entire crane wintering area is commonly referred to as Aransas. Wintering whooping cranes primarily use salt marsh and open bay habitats but sometimes fly to adjacent upland habitat for food and fresh water (U.S. Fish and Wildlife Service 1994).

METHODS

One hundred thirty-five whooping cranes were color-banded from 1977 to 1988 as pre-fledged chicks in Wood Buffalo National Park (WBNP) (Kuyt 1979; Kuyt and Goossen 1987; E. Kuyt, CWS, Edmonton, Alta., unpubl. data). Nomenclature of banded birds follows that of Stehn and Johnson (1987).

Bishop (1984) studied color-marked cranes during the winters of 1980–81 through 1982–83 by using extensive ground observations. Color-marked birds were recorded during weekly aerial census flights during the winters 1977–78 to 1981–82 by S. E. Labuda and 1982–83 to 1994–95 by T. V. Stehn (U.S. Fish and Wildlife Service, Austwell, Tex., unpubl. data). From 1986–87 through 1994–95, I attempted to locate and identify bands on most birds encountered during aerial counts. Bands were identified by flying at 80 knots past the cranes at an altitude of approximately 20 m. I also observed color-banded cranes from boats and vehicles from 1982–83 through 1994–95.

Data collected for each color-banded bird sighting included date, location, and identification of associated cranes. The sex of most marked whooping cranes was determined by behavioral observations (Bishop 1984; T. Stehn, unpubl. data), chromosome analysis of blood components (Kuyt and Goossen 1987), vocalizations (Carlson 1991), and comparison of weights and measurements of pre-fledged chicks during banding operations (B. W. Johns, CWS, Saskatoon, Sask., pers. commun.). Cranes were designated as adults after having made a nesting attempt at WBNP (B. W. Johns and E. Kuyt, CWS, unpubl. data).

Distances from the center of juvenile home ranges to the center of adult territories were measured on a U.S. Geological Survey 1:100,000-scale metric topographic map of San Antonio Bay, Texas. Adult winter territories were also classified as being or not being in the same general area as their juvenile home ranges. General areas were categorized as San Jose Island, Welder Flats, north and south halves of ANWR, and north and south halves of Matagorda Island. For example, a juvenile from the north end of the refuge that established an adult territory on the south end of the refuge (up to 16 km and 8+ territories away), was not considered in the same general area even though it was still on the refuge.

Frequencies of association (FOA) (Hawkins and Klimstra 1970) were calculated each winter between a particular crane and its future mate:

$$\mathrm{FOA} = \frac{\mathrm{C}}{\mathrm{A} + \mathrm{B} - \mathrm{C}} \times 100\%$$

where A = total number of sightings of Bird A, B = total number of sightings of Bird B, and C = total number of times birds A and B were sighted together.

For each color-banded bird, I recorded the sex, age at first nesting, winter territory, winter juvenile home range, winter subadult home range prior to nesting, FOA's with its future mate, and average subadult group size in marsh and bay habitats. Flock size data on uplands (areas that were usually prescribed burns) were excluded since subadults often join adult pairs or families in these areas. Average flock size was also calculated the first winter after nesting to determine if new pairs had established winter territories. I excluded data on 6 widowed birds that had re-paired,1 pair of birds that wintered apart between their first and second nesting attempts, and 2 marked cranes of unknown identity that had lost their color bands and had only aluminum bands remaining. All comparisons of sample means were made with a standard t-test statistic with a pooled variance estimate and 2-tailed probability set at $P < 0.05$.

RESULTS

Banded Cranes

Of the 135 chicks banded in WBNP, 119 survived their first fall migration. Sixty-eight color-marked cranes formed pairs and nested in or near WBNP; 33 (48.5%) were males and 35 (51.5%) were females. The approximately equal sex ratio supports Kuyt and Goossen (1987), who found an even sex ratio in 22 pre-flight chicks. Forty-two banded cranes paired with banded mates (21 pairs), and 26 paired with unbanded mates. Thirty-five of the 68 cranes were alive in November 1995. Pair members averaged 1.06 years apart in age (n = 17, SE = 0.22, range 0–3), with no noticeable trend for either males or females to be older.

One other crane (GwG-BwB, 1988) may have nested in 1992, but it was not positively identified. This crane failed to arrive at Aransas in the fall of 1992 and was excluded from data analysis. As of November 1995, crane r-r (1988) was the only banded crane not yet known to have bred, although this 7-year-old male had an FOA of 100% with an unbanded crane during the 1994–95 winter (n = 17 observations) and was presumably paired.

Six cranes re-paired with subadults after the loss of a mate. In December 1990, 1 widowed male crane re-paired with a younger adult female who had nested once unsuccessfully. The first mate of that female re-paired with a subadult. This is the only known case of a whooping crane pair breaking up after 1 nesting attempt.

Group Size

The average size of subadult flocks containing at least 1 color-banded crane was 3.12 birds (n = 189, SE = 0.09). Each individual sample unit represented average flock size for all sightings of 1 color-banded bird during 1 winter. Flock sizes of marked subadult males ($\bar{x}$ = 3.01, n = 80, SE = 0.14) and marked subadult females ($\bar{x}$ = 3.20, n = 109, SE = 0.12) were not significantly different (t = 1.055, df = 187, P = 0.29).

Flock size was smaller for cranes during the winters 1 and 2 years prior to nesting, averaging 2.87 (n = 56, SE = 0.16) and 2.84 (n = 49, SE = 0.15) birds, respectively. Group size for winters 1 and 2 were not significantly different (t = 0.148, df = 103, P = 0.88). Average group sizes for the winters 3–6 years prior to nesting were 3.39 (n = 42, SE = 0.21), 3.70 (n = 24, SE = 0.29), 3.14 (n = 15, SE = 0.23), and 3.81 (n = 3, SE = 1.04), respectively. There was a highly significant difference (t = 3.30, df = 187, P = 0.001) between average group size for winters 1 and 2 ($\bar{x}$ = 2.86, n = 105, SE = 0.11) versus winters 3–6 ($\bar{x}$ = 3.45,

n = 84, SE = 0.14).

Subadult flocks were largest on ANWR ($\bar{x}$ = 3.50, n = 102, SE = 0.12), next largest on Matagorda Island ($\bar{x}$ = 2.40, n = 32, SE = 0.10), and smallest on San Jose ($\bar{x}$ = 2.04, n = 11, SE = 0.20) and Welder Flats ($\bar{x}$ = 2.27, n = 15, SE = 0.14). Refuge flocks were significantly larger (t = 6.76, df = 158, $P < 0.001$) than those on the other wintering areas. Flock size on Matagorda Island did not differ significantly from the combined values for San Jose and Welder Flats (t = 1.35, df = 56, P = 0.183).

One-year-old birds were found in flocks averaging 3.26 birds (n = 48, SE = 0.18, range 1.00–7.12). There was no significant difference between flock size for 1-year-old versus all older subadults (t = 0.871, df = 187, P = 0.385), or between 1-year-old males and females (t = -0.682, df = 46, P = 0.498).

Adult pairs their first winter after nesting were sometimes found with subadult cranes. Group size the first winter after nesting averaged 2.32 (n = 51, SE = 0.09) compared with average subadult flock size prior to nesting of 3.12. The flock size the first post-breeding winter was significantly smaller (t = -4.397, df = 238, $P < 0.001$) than average flock size of 3.12 for all subadults, and from the flock size of 2.87 (n = 56, SE = 0.17) for the first winter prior to nesting (t = -2.853, df = 105, P = 0.005).

Frequencies of Association

FOA's were calculated for 18 pairs with both birds color-banded. Five additional pairs formed that included widowed birds. Since widowed birds can re-pair quickly (Stehn 1992*a*), and a long association prior to re-pairing may not occur, data for these pairs were excluded. In no instances did siblings (juveniles from different years raised by the same parents) pair. There were no "twin" chicks brought to Aransas during the years cranes were color-marked, partly because of the removal of 1 egg from most of the nests for management purposes.

FOA's 1 year prior to nesting averaged 65.4% (n = 18, SE = 10.77). Ten pairs (55.6%) had FOA's of 100% ($\bar{x}$ = 22 observations per pair, range 8–34) throughout the winter prior to nesting and were presumably paired. Some of these pairs exhibited territorial behavior at Aransas the winter, or part of the winter, prior to nesting (Bishop 1984). Five pairs were not observed in association with their future mates in subadult flocks at Aransas at any time prior to nesting, although in 2 cases associations may have occurred due to proximity of subadult winter ranges. Based on groupings observed prior to the spring migration immediately prior to their first nesting attempt, pair formation for all 5 pairs must have occurred in migration or on the nesting grounds.

During the winter 2 years prior to nesting, FOA's averaged 40.2% (n = 16, SE = 11.37), with members within 3 pairs associating 100% of the time and members of 9 pairs that were never observed together. FOA's were only 15.6% 3 winters prior to nesting (n = 15, SE = 8.52), 17.7% (n = 7, SE = 13.23) 4 winters prior to nesting, and only 4.2% (n = 6, SE = 4.17) 5 winters prior to nesting.

Three pairs associated as subadults every year prior to pair formation. Pair G-YbY and YbY-YbY were the same age and had juvenile winter home ranges on adjacent ANWR territories.

Distances

Adult males established winter territories an average of 5.6 km from their juvenile winter home range (n = 33, SE = 1.08), whereas adult females averaged a distance of 11.9 km (n = 35, SE = 1.74). Distances for males were significantly less (t = -3.04, df = 66, P = 0.003) than female distances.

Adult winter territories were located within the same general area of juvenile home ranges for 25 of the 33 (75.8%) banded males. For those 25 males, 2 of the adult territories were established on the juvenile home range, 18 were on adjacent areas, 3 were 1–2 territories away, and 2 were 3+ territories away. For females, 10 of 35 (28.6%) adult females established territories in the general area of juvenile home ranges. However, for 6 of the 10 females, the male juvenile territory was also located nearby and presumably influenced territory selection. Thus, there were only 4 cases (11.4%) where the adult territory was established in the general area of the female juvenile home range but not in the general area of the male juvenile home range.

Distance between juvenile home ranges of color-marked cranes and their future mates averaged 12.0 km (n = 17, SE = 2.35). For 11 of 16 (68.8%) pairs, juvenile home ranges were from different areas. None of these pairs associated together at all the third winter prior to nesting, and only 3 associated the second winter prior to nesting. In 4 of 16 pairs (25.0%), juvenile home ranges were from the same general area and long associations occurred for at least 3 winters prior to nesting. Third winter associations averaged 58.4% (n = 4). One pair had juvenile home ranges from the same general area but only associated together for 2 winters prior to nesting, with no association the third winter prior to nesting.

DISCUSSION

Group Size

Subadults wintered in flocks averaging 3.12 birds during

1978–94. Bishop and Blankinship (1982) documented larger winter flocks during 1978–80 that averaged between 4.4 and 5.6 cranes forming on the Aransas NWR to exploit unusual food concentrations. They suggested that fluctuations in flock size appeared to be influenced by seasonal availability of food in the marsh.

Subadult flock size was usually larger in areas with a greater number of adult territories. During 1994–95, Aransas NWR had 17 adult territories, St. Charles Bay 4, Matagorda Island 14, San Jose Island 8, and Welder Flats 3. During the past 10 years, 51.1% of winter territories have been at Aransas/St. Charles (T. Stehn, unpubl. data). Subadult flocks were significantly larger on the refuge (3.50), than on Matagorda Island (2.40), San Jose Island (2.04), and Welder Flats (2.27), where only 26.7%, 14.0% and 8.3% of the adult winter territories have been located, respectively, over the past 10 years.

Subadults usually winter in the general area of their juvenile home range (Bishop 1984; T. Stehn, unpubl. data). Thus, the number of juveniles successfully raised in a wintering area will strongly influence how many subadults subsequently winter in that area. This in turn appears to influence subadult flock size, with larger subadult flocks forming when more subadults are in an area. Thus, subadult flocks were largest on the Aransas NWR compared with other wintering areas with fewer winter territories.

Some subadults spend large portions of a particular winter(s) as singles. I postulated that 1-year-old subadults might return to their juvenile home range and winter in the marsh adjacent to their parent's territory rather than join nearby subadult flocks. One-year-olds, especially females, would be expected to be at the bottom of the social structure in subadult flocks and thus might choose to be solitary. This turned out not to be true. One-year-olds actually tended to winter in slightly larger subadult groups (3.26 cranes) than subadults of all ages (3.12). Thus, subadults that are solitary may be choosing to winter in a particular area whether or not other subadults are wintering nearby.

Flock size was smaller for cranes both the first (2.87) and second (2.84) winters prior to nesting. This occurred because strong associations began to form as birds approached breeding age. I found that duos occasionally established winter territories prior to nesting and would not tolerate the presence of subadult flocks, substantiating observations previously made by Bishop (1984). Bishop and Blankinship (1982) also noted that pair bonding influenced subadult flock size.

Flock size was significantly smaller (2.33) the first winter after nesting compared with the winter prior to nesting (2.87). As territories became more firmly established, adult pairs rarely joined subadult flocks in marsh and bay habitats. However, some new pairs continued to tolerate the presence of subadults at Aransas even after nesting. For example, crane pair RwR-O and BwB-r/w took 3–4 winters to establish a territory on the refuge (T. Stehn, unpubl. data). The first winter after nesting, the pair remained in 1 area on ANWR, sometimes with subadults (2.65 group size), and were chased by neighboring pairs. The second and third winters after nesting, they occasionally spent time with subadults, but were never observed defending the marsh they frequented (group sizes 2.19 and 2.24, respectively). Finally, in their fourth winter after first nesting (group size 2.0), they vigorously defended a territory in the same area. They did not successfully bring a juvenile to Aransas until their fifth year nesting.

Frequencies of Association

Although the majority of subadults formed strong associations at least 1 winter prior to nesting, some did not. Five of 18 pairs (27.8%) with both adults color-banded appeared to form pair bonds in a short period of time during the spring migration and/or in WBNP. These 5 pairs were never observed associating at Aransas prior to nesting. This is contrary to pairing behavior described by Bishop (1984), in which she observed formation of pair bonds after long periods of association.

In a majority of cases (68.8%), color-marked cranes selected a mate that did not have a juvenile home range nearby and usually had little association with its future mate 2–3 winters prior to nesting. Pairs were never observed between siblings, but I do not know if there is a behavioral mechanism to prevent such pairing.

Distances

Most adult territories established by color-marked males were close to their juvenile home ranges. This supports the hypothesis that an adult territory is normally established as close as possible to the male's juvenile home range (Stehn and Johnson 1987).

MANAGEMENT IMPLICATIONS

Current management efforts to recover the whooping crane include establishing a non-migratory flock in Florida. Chicks are hatched in captivity where they associate with other juveniles, then small groups of juveniles or subadults are soft-released in pens in Florida (Nesbitt et al. 1997). Some aspects of behavioral development set in these soft-release pen groups may be analogous to effects associated with juvenile home ranges at Aransas. Since pair bonds at

ANWR generally form after long associations within subadult flocks, it seems important that released cranes be provided conditions under which subadult flocks can form and the normal composition of social groups is represented.

Effects of releasing same age cohorts on future mate selection are not known. Nesbitt and Carpenter (1993) did not find any sandhill crane (*Grus canadensis*) pairs formed between members of release cohorts. The first nesting attempt by whooping cranes in Florida was by 2 3-year-olds from different release cohorts, but 2 other potential pairs may have formed between birds from the same release cohort (Nesbitt et al. 1997). Since siblings have never been known to pair at Aransas, this could imply that same-age cohorts socialized together in new flocks will not form pair bonds. It is probable that adults will establish territories in Florida near the release pens, so release pens should be near suitable nesting habitat. Nesbitt et al. (1997) suggested that releasing several small groups of birds each year rather than a single large group may foster pair formation between birds released in that year. Additional release pens, such as the satellite pens used by Nesbitt et al. (1997), should be added to help establish nesting areas throughout the available habitat.

LITERATURE CITED

BISHOP, M. A. 1984. The dynamics of subadult flocks of whooping cranes wintering in Texas, 1978-79 through 1982-83. M.S. Thesis, Texas A&M Univ., College Station. 128pp.

______, and D. R. BLANKINSHIP. 1982. Dynamics of subadult flocks of whooping cranes at Aransas National Wildlife Refuge, Texas, 1981. Pages 180-189 *in* J. C. Lewis, ed. Proc. 1981 crane workshop. Natl. Audubon Soc., Tavernier, Fla.

BLANKINSHIP, D. R. 1976. Studies of whooping cranes on the wintering grounds. Pages 197-206 *in* J.C. Lewis, ed. Proc. [1975] int. crane workshop. Oklahoma State Univ. Publ. Printing, Stillwater.

CARLSON, G. 1991. The feasibility of individual identification and sex determination of whooping cranes (*Grus americana*) by analysis of vocalizations. M.S. Thesis, Idaho State Univ., Pocatello. 60pp.

HAWKINS, R. E., and W. D. KLIMSTRA. 1970. A preliminary study of the social organization of white-tailed deer. J. Wildl. Manage. 31:124-131.

KUYT, E. 1979. Banding of juvenile whooping cranes and discovery of the summer habitat used by nonbreeders. Pages 109-111 *in* J. C. Lewis, ed. Proc. 1978 crane workshop. Colorado State Univ. Printing Serv., Fort Collins.

______, and J. P. GOOSSEN. 1987. Survival, age composition, sex ratio, and age at first breeding of whooping cranes in Wood Buffalo National Park, Canada. Pages 230-244 *in* J. C. Lewis, ed. Proc. 1985 crane workshop. Platte River Whooping Crane Maintenance Trust, Grand Island, Nebr.

NESBITT, S. A., and J. W. CARPENTER. 1993. Survival and movements of greater sandhill cranes experimentally released in Florida. J. Wildl. Manage. 57:673-679.

______, M. J. FOLK, M. G. SPALDING, J. A. SCHMIDT, S. T. SCHWIKERT, J. M. NICOLICH, M. WELLINGTON, J. C. LEWIS, and T. H. LOGAN. 1997. An experimental release of whooping cranes in Florida-the first three years. Proc. North Am. Crane Workshop 7:79-85.

STEHN, T. V. 1992*a*. Re-pairing of whooping cranes at Aransas National Wildlife Refuge. Pages 185-187 *in* D. A. Wood, ed. Proc. 1988 North Am. crane workshop. Florida Game Fresh Water Fish Comm. Nongame Wildl. Program Tech. Rep. 12.

______, 1992*b*. Unusual movements and behaviors of color-banded whooping cranes during winter. Proc. North Am. Crane Workshop 6:95-101.

______, and E. F. JOHNSON 1987. Distribution of winter territories of whooping cranes on the Texas coast. Pages 180-195 *in* J. C. Lewis, ed. Proc. 1985 crane workshop. Platte River Whooping Crane Maintenance Trust, Grand Island, Nebr.

U.S. FISH AND WILDLIFE SERVICE. 1994. Whooping crane recovery plan. U.S. Fish Wildl. Serv., Albuquerque, N.M. 92pp.

VOCALIZATIONS OF SANDHILL CRANES

STEPHEN A. NESBITT, Florida Game and Fresh Water Fish Commission, Wildlife Research Laboratory, 4005 South Main Street, Gainesville, FL 32601, USA
RICHARD A. BRADLEY, The Ohio State University, 1465 Mt. Vernon Avenue, Marion, OH 43302-5695, USA

Abstract: The structure and context of 20 vocalizations given by adult or juvenile sandhill cranes (*Grus canadensis*) are described. Purr calls were given under various circumstances (e.g., pre-flight, feeding, aggression) and communicated information over a short distance. Flight calls, guard calls, and unison calls were full-voiced calls that carried information over greater distances. Calls given in flight were distinguishable from other calls. Guard calls were given in several different circumstances and served multiple functions; other calls were given in very specific contexts. Beyond the basic communication of information to other cranes, vocalizations may function in mate selection and in synchronizing physiological condition of pair members.

PROC. NORTH AM. CRANE WORKSHOP 7:29-35

Key words: cranes, *Grus canadensis*, duetting, vocalizations, voice.

In many species, a specific vocalization is definitive of a particular behavior or display. An understanding of calls and their contextual relationships is often useful in defining social status and interpreting the behavior of individuals (Smith 1963). Sandhill cranes have a varied vocal repertoire, and many of their calls have been described, but with little or no contextual information; others have never been discussed. Archibald (1976) mentioned a number of calls but dealt in depth only with the *Unison* and *Guard Calls*. Walkinshaw (1949, 1973) identified several calls phonetically but provided little interpretation. Voss (1976) described and discussed the ontogeny of vocalizations in juvenile sandhill cranes and also provided some information on adult vocalizations. However, the majority of Voss' study involved captive individuals and consequently the contextual reference for many calls was incomplete. We recorded and described the context associated with the audible vocal repertoire of free-living sandhill cranes in Florida. Our purpose was a better understanding of the interaction of voice and behavior in sandhill cranes.

We are indebted to H. Rothman, Department of Speech, University of Florida, for the use of the Voice Ident Analyzer; P. Bernstein, B. Barbour and J. Hardy reviewed earlier drafts of this manuscript and offered many helpful suggestions.

METHODS

Recordings were made of both the migratory greater sandhill crane (*G. c. tabida*) and the non-migratory Florida sandhill crane (*G. c. pratensis*) in Florida under various circumstances between 1975 and 1985. Both subspecies associate freely during the winter and no attempt was made to differentiate between them since their voices are indistinguishable. Recordings were made with a Uher 4000 Report IC reel-to-reel recorder operating at 19 cm/sec and a Dan Gibson parabolic microphone or with a Technics RS-646DS cassette recorder and a Sennheiser ME 88 microphone. Most of the recordings were of free-ranging wild birds, though a few were of captured individuals. A Voice Ident UTI Audio Spectro Analyzer was used to evaluate vocalizations and generate sonographs. Original tapes and a composite tape used to produce the figures were deposited at the Bioacoustics Laboratory, Florida Museum of Natural History, Gainesville, Florida.

RESULTS

Purrs

Adult cranes produced a graded series of calls that are best described as *Purrs*. They were of low amplitude and communicated information within a flock or family group over short distances. The purr sound was produced as 4 to 12 muffled click-like sounds covering a wide frequency spectrum delivered at a rate of 12 to 20 notes per second. *Feeding Purrs* (Fig. 1A) were given almost constantly as birds fed. They had a relatively low fundamental frequency and were delivered more slowly than the slightly higher-pitched (rising inflection) *Flight Intention Purrs* (Fig. 1B). The *Flight Intention Purr* (Fig. 1B) consisted of 2 parts, the second half at a higher frequency than the first and with rising inflection. These calls were given by birds in flight intention posture (Voss 1976, Tacha 1984) and preceded unforced flight; they were not given if birds were startled, rushed to fly, or were following others to take off. A *Flight Intention Purr* often gradually changed into a full-voiced flight call. The highest-pitched, most rapidly delivered purrs were *Flight Purrs* (Fig. 1C) given by birds in flight.

The *Growl Purr* (Fig. 1E) was an aggressive signal given as a part of the *Low Bow* display (Voss 1976, Nesbitt and Archibald 1981). This low-frequency, low-volume call carried only a few meters. Voss (1976) identified 2 additional

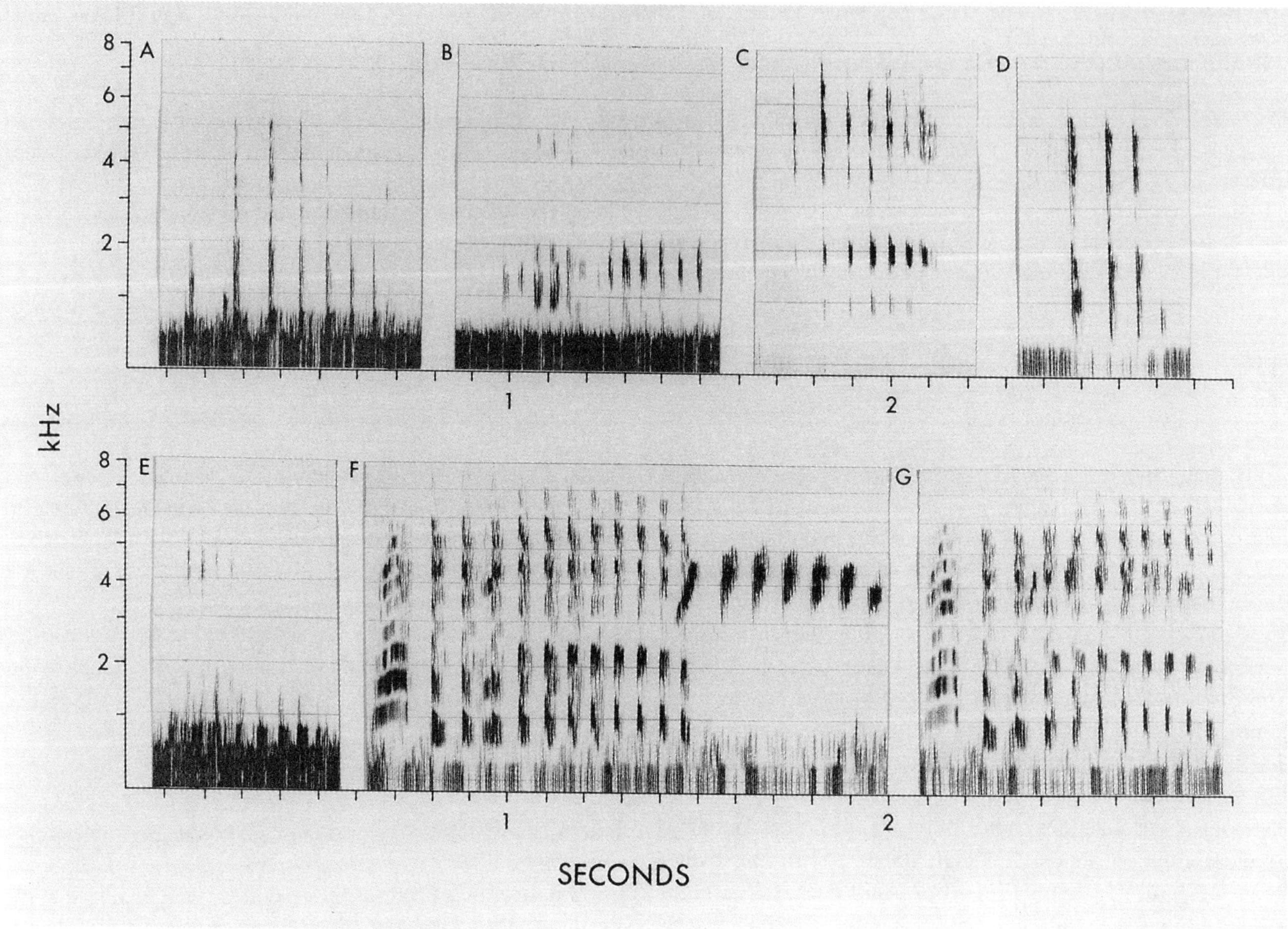

Fig. 1. ***Purr Calls*** **and** ***Guard Calls*** **of sandhill cranes recorded in Florida, 1975–85: (A)** ***Feeding Purr*****–artifact @ 1 KHz; (B)** ***Flight Intention Purr*****–artifact @ 1 KHz; (C) Flight Purr; (D)** ***Perrt Call*****; (E)** ***Growl Purr*****; (F)** ***Guard Calls*****, male and female calling together with juvenile voice following; (G)** ***Guard Calls*****, male, female, and juvenile calling simultaneously.**

vocalizations that we did not record; these calls could be considered with the purr calls. The *Pre-copulation Call* was "a regular series of purr-like calls, gradually swelling in volume," and the *Nesting Call* was "a low moan-like call" given at or near the nest site while birds were building the nest, incubating, or brooding young.

Flight Calls, Guard Calls, and Unison Calls

The loud, full-voiced calls (*Flight Call, Guard Call, Unison Call*) of the sandhill crane have been described as a rattling, bugling, or trumpeting. They consisted of 7 to 20 notes delivered at a rate of about 15 notes per second. The fundamental frequency was about 0.73 KHz for males and 1.0 KHz for females (Weekley 1985). The individual notes were usually brief, often less than 0.05 seconds; occasionally the introductory or terminal note may last 0.1 or 0.2 seconds. Under optimum conditions these calls, especially a male's, carried for 2–3 km.

The predominant vocalization given by cranes on the ground were 2 duet calls, typically given in unison, by pairs or family groups. The *Guard Call* was given when "the bird is afraid but not afraid enough to flee" (Archibald 1976) or as part of aggressive displays. It was given in unison by a pair, and juveniles (when present), during "Directed Walk Threats" (Nesbitt and Archibald 1981, Tacha 1988). The male call began with 1 sustained note (0.2 sec) that was followed by 6–8 shorter notes, somewhat lower in pitch (Figs. 1F, 1G). The female call typically did not have the sustained first note but consisted of a series of 8–10 evenly-spaced short notes. A sequence usually began with the male, joined 0.2–0.4 seconds later by the female, then in another brief interval by the juvenile. The vocalization was often repeated many times a minute with little variation, the

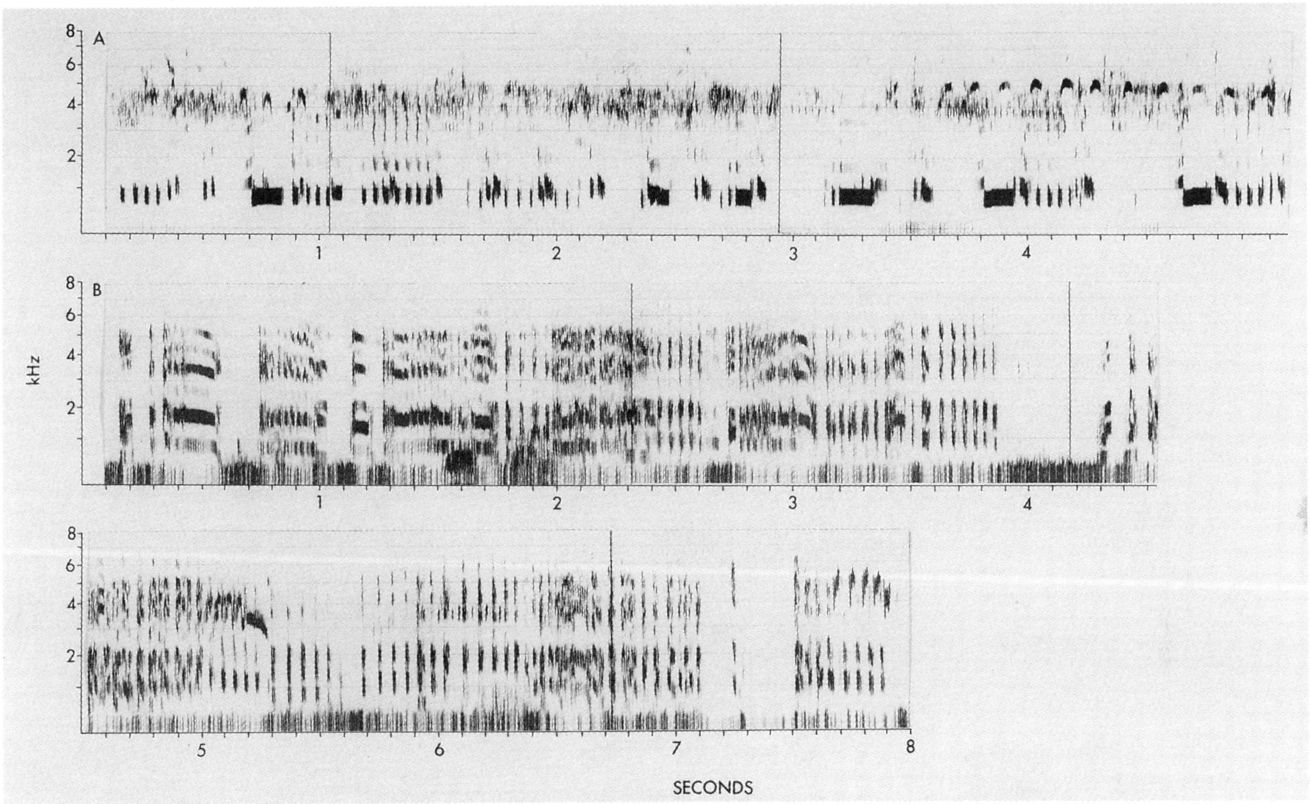

Fig. 2. *Unison Calls* and *Bill Sparring Calls* of sandhill cranes recorded in Florida, 1975–85: (A) *Unison Call*, female part above male part below; (B) *Bill Sparring Call*, in 2 parts.

male, female, and juvenile overlapping their calling. A pair usually alternated their individual parts, giving the vocalization a pronounced antiphonal quality. Detection of a mammalian predator, such as a bobcat (*Lynx rufus*) or dog, usually elicited vociferous *Guard Calling* and vaulting and often included approaching the predator by a flock or family group, as if mobbing. *Guard Calls* were also given in flight by a pair or by individuals when flying near a nesting territory.

The *Unison Call* (Fig. 2A), a duet call characteristic of a bonded pair, was sexually differentiated by voice and posture (Archibald 1976). The male's part was lower pitched and more legato than the staccato, higher-pitched female part (Fig. 2A). The call was initiated by either member of the pair. In agonistic encounters with other cranes, the *Unison Call* became a threat display and a triumph ceremony which followed the successful domination of another individual or family group. Juveniles ≥8 months old participated in the *Unison Call* during familial agonistic encounters. When excited, as during aggressive encounters or migration departure, *Unison Calls* were sometimes given in flight. Single birds separated by some distance from their mates (e.g., during incubation or territorial defense) occasionally gave their respective part of the *Guard Call* or *Unison Call*.

Cranes called more frequently in flight than they did on the ground. *Flight Calls* (Fig. 3A) were given when birds flew to and from feeding, loafing, or roosting locations, particularly as flocks took flight and perhaps served to avoid collisions during the confusion of group departures. *Flight Calls* were quite variable in inflection, either rising or falling in pitch, but were more rapidly delivered and higher in basal frequency than ground calls. These latter features made *Flight Calls* distinguishable from ground calls.

Limited-Context Calls

Perrt Calls (Fig. 1D) were often given when birds landed and took flight. The call consisted of 4 notes rising in pitch, separated by unequal intervals. *Alarm Perrts* (Fig. 4A), higher pitched and of shorter duration, were given to alert others in the flock. "Bill Sparring" (Nesbitt and Archibald 1981) or "Bill Duel" (Tacha 1988), a ritualized aggressive display, had an associated vocalization, *Bill Sparring Call* (Fig. 2B), that consisted of a loud series of rapid slurred notes covering a wide frequency range lasting up to 5

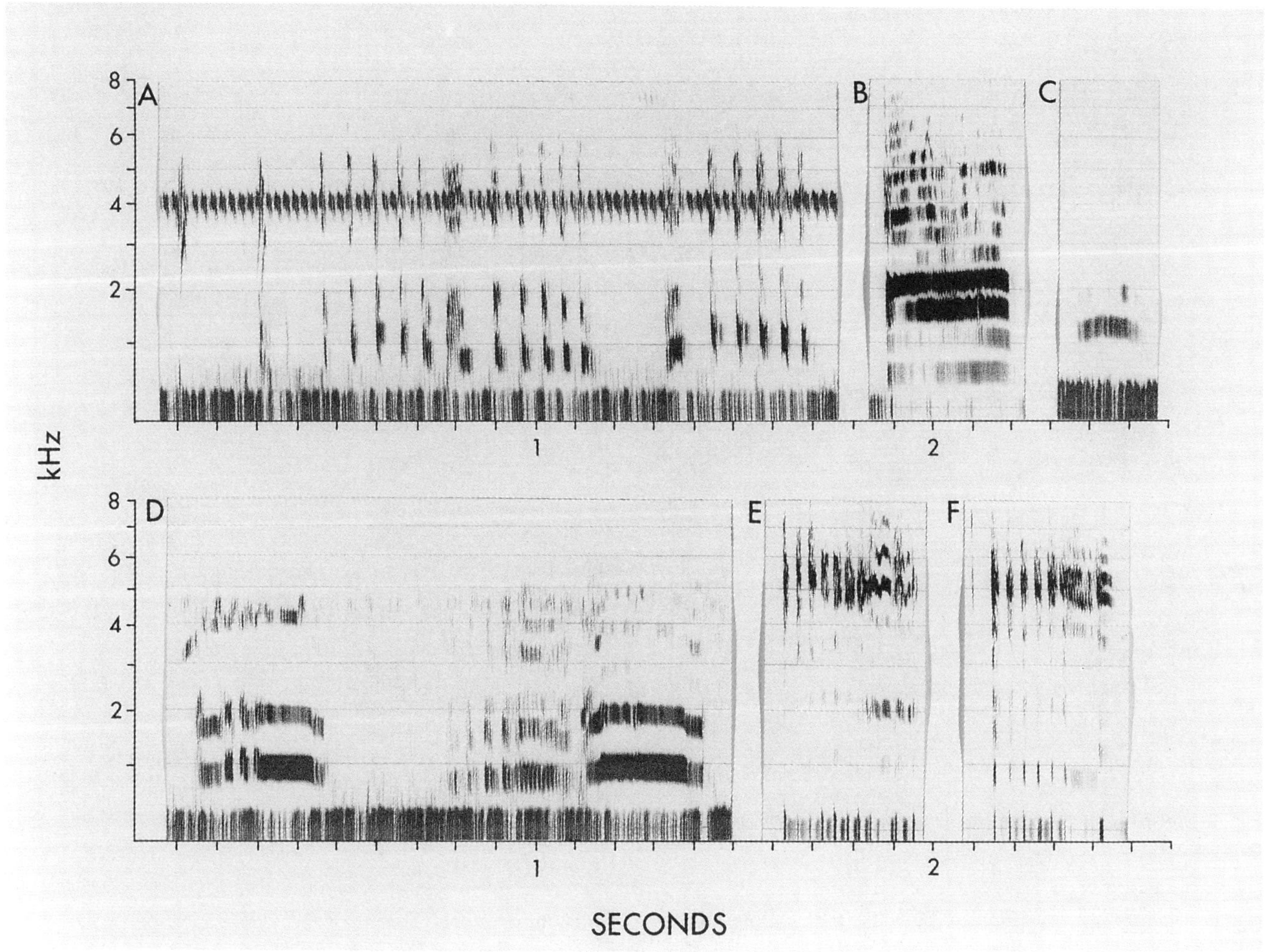

Fig. 3. Calls of sandhill cranes recorded in Florida, 1975–85: (A) *Flight Calls*–artifact @ 4 KHz; (B) *Honk*; (C) *Moan*; (D) *Yelp* "location call"–3 examples with artifact below 1 KHz; (E) and (F) cranes migrating.

seconds. "Bill Sparring" most commonly occurs among cranes in large wintering flocks. In Florida cranes, "Bill Sparring" was most frequent between birds that had not associated previously. This display seems to establish social order. The *Bill Sparring Call* was rarely heard from a flock with a well established dominance hierarchy.

During actual attacks and sometimes during "Bill Sparring" and dancing episodes, cranes gave a loud *Hiss* (Fig. 4B). This broad-band frequency sound was produced by the rapid exhalation of air seemingly without involving the syrinx. A shrill *Tremolo* (no example) with a descending inflection was given by a bird in great distress, as when a wild crane was captured.

Goose-like *Honks* (Fig. 3B) and *Moans* (Fig. 3C) were often part of a *Flight Call* sequence. A *Moan* lacked the 2-syllable yodel-like quality of a *Honk*, otherwise they were similar. These 2 rapidly delivered calls were intermediate between the *Flight Purr* and the full voiced *Flight Call*. *Honks* and *Moans* were used frequently by cranes during migration or when departing from feeding or roosting areas. Perhaps they served to encourage others to join the flock or to maintain contact within a large flock of birds in flight.

A *Yelp* (Fig. 3D) or "location call" (Voss 1976) was given to locate other individuals or flocks. *Yelps* consisted of 3 or more high pitched, broken notes, followed by a sustained note. *Yelps* were given from the ground in answer to those given in flight, especially during migration when yelps may be used to maintain contact between flocks.

Juvenile Voice

The vocal repertoire of young sandhill cranes consisted

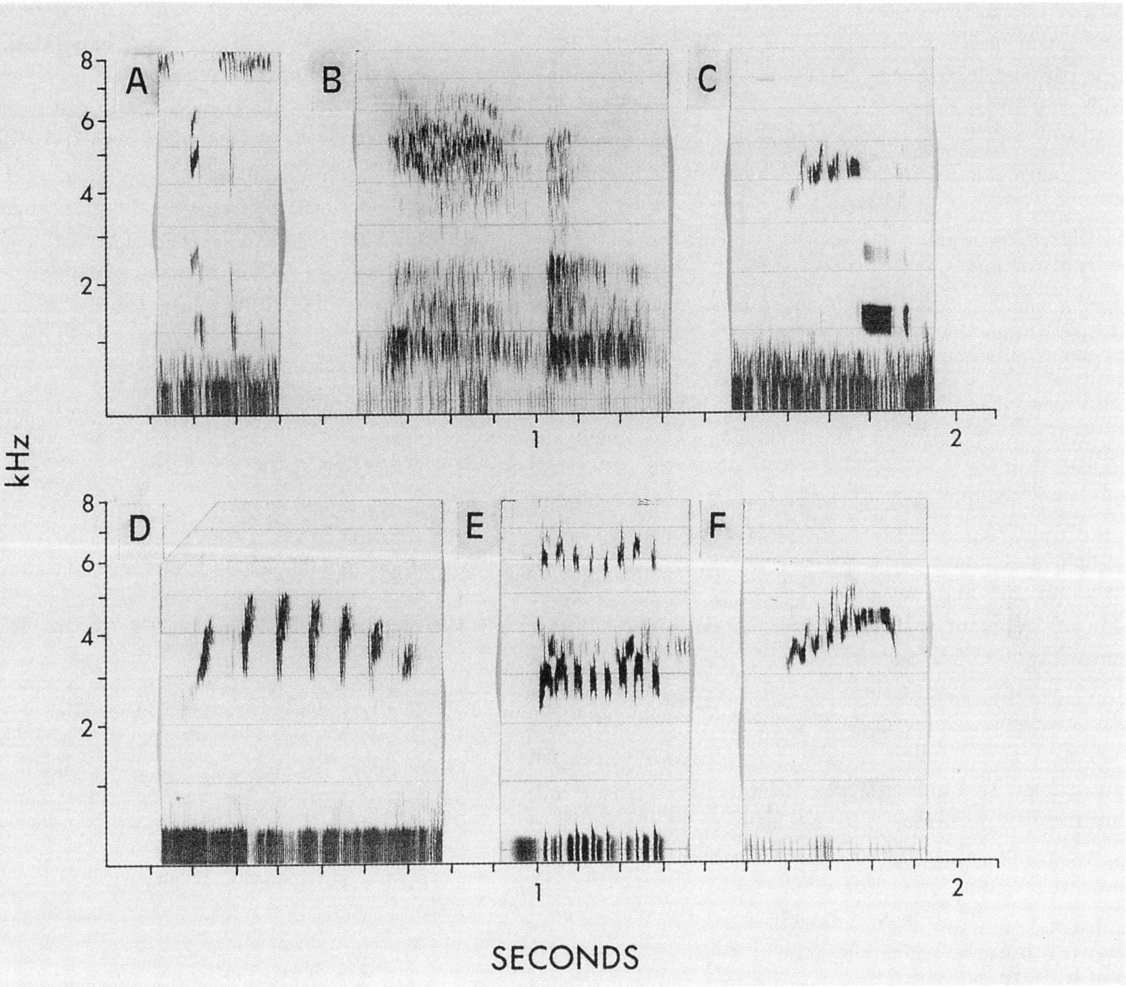

Fig. 4. Limited context calls and chick calls of sandhill cranes recorded in Florida, 1975–85: (A) *Alarm Perrt*-artifact @ 1 KHz; (B) *Hiss*; (C) Juvenile voice change-artifact @ 1 KHz; (D) *Juvenile Flight Call*-artifact @ 1 KHz; (E) *Juvenile Purr Call*-artifact @ 1 KHz; (F) *Juvenile Stress Call*.

of 6–8 distinct calls (Voss 1976). The typical juvenile voice was a whistle or high-pitched trill.

The *Juvenile Guard Call* (Fig. 1G) and *Juvenile Flight Calls* (Fig. 4D) varied little with context, except for inflection, over a frequency range of 3–5 KHz. *Juvenile Flight Calls* were more rapid but otherwise similar in quality to the other juvenile calls. The *Juvenile Stress Call* had a rising inflection (Fig. 4F) given in stressful situations as when young became separated from their parents. *Food-begging Calls* (Archibald 1976) were loud emphatic calls similar to *Stress Calls* but given repeatedly to "stimulate the adults to offer food to the chick."

Juvenile Purr Calls (Fig. 4E) were also less variable than those of the adults, and were given in a variety of contexts (e.g., feeding, contact). Between 9 and 10 months of age, the juvenile voice in the Florida sandhill crane changes to the adult voice (Nesbitt 1975). A recording of a bird in flight made 2 December 1977 (Fig. 4C) was of a juvenile undergoing voice change. The call begins with a typical juvenile whistle but ends with a distinctly adult sound, similar to a *Yelp*.

DISCUSSION

Unique vocal characteristics may be an important aspect of individual and sexual recognition among sandhill cranes. There was a considerable range of variation among the voices of individual cranes irrespective of gender. Sexual dimorphism was apparent in the vocal repertoire; the mean basal frequency of the male sandhill crane voice during *Guard*

Calling (0.58 KHz) was lower in frequency by 33% than the female's (0.87 KHz [n = 57]) (Weekley 1985). A comparison of average weight and bill length of male and female Florida and greater sandhill cranes (n = 114) shows that males (n = 64) were 13.6% heavier and bills were 6.8% longer than females (n = 50) (Nesbitt et al. 1992), so differences in voice basal frequency were not a function of size differences alone.

Individual and sexual differences in voice among cranes may have functions additional to recognition. During *Unison Calling* the male's lower pitch gives the call a better sound transmission quality while the higher pitched, more pulsed signal of the female increases the receiver's ability to locate the sound's source (Wiley and Richards 1982). Thus the combination of the 2 voices functions better to announce that a territory is occupied than either part alone. There are other features of the full voice that facilitate transmission, detection, and localization of the signal. The mean frequency of the ground calls falls within the optimum frequency range (0.05-1.0 KHz) for transmission of a signal generated near ground level (Wiley and Richardson 1982). The posture assumed by the pair during *Unison Calling* and to some extent *Guard Calling*, male's head held vertically and female's at a 45° angle, is also advantageous for signal transmission. The times of day when most territorial announcements were made, sunrise and sunset, are the times of least atmospheric disturbance when sound will carry farthest (Waser and Waser 1977, Henwood and Fabrick 1979). The prolonged call duration and repetition of the call several times with little variation also add to transmission success (Waser and Waser 1977). By making long-distance threats more effective, these calls serve to reduce the necessity for direct agonistic interactions with other territorial conspecifics, reducing the frequency of unnecessary and possible injurious agonistic encounters.

Detection by predators or potential competitors is best avoided if some information is transmitted within the pair or family structure over just a few meters. This information is passed with *Purrs* or, in the case of juvenile cranes, a high pitched *trill* or *whistle*. These signals do not have enough volume to carry well and are pitched above 2 KHz, making them subject to attenuation from vegetation and reducing detection and source location (Wiley and Richards 1982) over more than a few meters.

Beyond the simple transfer of information, some calling in sandhill cranes may function to alter the physiological condition of sender and receiver alike. Vocalizing of pair members can bring about sexual readiness in a mate (Lehrman and Friedman 1969) and promote reproductive synchrony. Such calling may play a part in initiating egg production. In ring doves (*Streptopelia risoria*) the female courtship vocalization altered a female's neuroendocrine state, preparing it for ovulation (Cheng 1986, 1991). There is evidence that *Unison* and perhaps *Guard Calling* in sandhill cranes may induce and sustain reproductive synchrony. Tacha (1988) demonstrated that increased gonad size and blood hormone levels in sandhill cranes correlated with increased frequency of *Unison Calling*. The amplitude of the male's voice at 3 m, a typical separation for calling partners, is 78 db (Archibald 1976) and a female's own voice would be even louder. Frequent calling at this sound intensity could have a hormonal stimulating capacity that would benefit a pair trying to achieve and maintain reproductive synchrony for copulation and egg laying.

LITERATURE CITED

ARCHIBALD, G. W. 1976. The unison call of cranes as a useful taxonomic tool. Ph.D. Thesis, Cornell Univ., Ithaca, N.Y. 167pp.

CHENG, M. F. 1986. Female cooing promotes ovarian development in ring doves. Physiol. Behav. 37:371-374.

_____. 1992. For whom does the female dove coo? A case for the role of vocal self-stimulation. Anim. Behav. 43:1035-1044.

HENWOOD, K., and A. FABRICK. 1979. A quantitive analysis of the dawn chorus: temporal solution for communicatory optimization. Am. Nat. 114:260-274.

LEHRMAN, D. S., and N. FRIEDMAN. 1969. Auditory stimulation of ovarian activity in the ring dove (*Streptopelia risoria*). Anim. Behav. 17:494-497.

NESBITT, S. A. 1975. Voice maturity in Florida sandhill cranes. Fla. Field Nat. 3:19.

_____, and G. W. ARCHIBALD. 1981. The agonistic repertoire of sandhill cranes. Wilson Bull. 93:99-103.

_____, C. T. MOORE, and K. S. WILLIAMS. 1992. Gender prediction from body measurements of two subspecies of sandhill cranes. Proc. North Am. Crane Workshop 6:38-42.

SMITH, W. J. 1963. Vocal communication of information in birds. Am. Nat. 97:117-125.

TACHA, T. C. 1984. Preflight behavior of sandhill cranes. Wilson Bull. 96:471-477.

_____. 1988. Social organization of sandhill cranes from midcontinental North America. Wildl. Monogr. 99:1-37.

VOSS, K. S. 1976. Behavior of the greater sandhill crane. M.A. Thesis, Univ. Wisconsin, Madison. 144pp.

WALKINSHAW, L. H. 1949. The sandhill cranes. Cranbrook Inst. Sci. Bull. 29. 202pp.

_____. 1973. Cranes of the world. Winchester Press, New York, N.Y. 370pp.

WASER, P. N., and M. S. WASER. 1977. Experimental studies of primate vocalization specializations for long distance propagation. Z. Tierpsychol. 43:239-263.

WEEKLEY, F. L. 1985. Individual and regional variation in calls of the greater sandhill crane. M.S. Thesis, Univ. Wisconsin, Stevens Point. 67pp.

WILEY, R. H., and D. G. RICHARDS. 1982. Adaptation for

acoustic communication in birds: sound transmission and signal detection. Pages 131-181 *in* D. E. Kroodsma and E. H. Miller, eds. Acoustic communication in birds. Vol. 1. Academic Press, New York, N.Y.

MICROSATELLITE DNA LOCI FOR GENETIC STUDIES OF CRANES

TRAVIS C. GLENN,[1] **Laboratory of Molecular Systematics, MRC 534, Smithsonian Institution, Washington, DC 20560, USA and Department of Zoology, University of Maryland, College Park, MD 20742, USA**
RYAN S. OJERIO, Laboratory of Molecular Systematics, MRC 534, Smithsonian Institution, Washington, DC 20560, USA
WOLFGANG STEPHAN, Department of Zoology, University of Maryland, College Park, MD 20742, USA
MICHAEL J. BRAUN, Laboratory of Molecular Systematics, MRC 534, Smithsonian Institution, Washington, DC 20560, USA

Abstract: Microsatellites are short tandem arrays of simple DNA sequences (e.g., $[AC]_n$, where $n \geq 10$), which are often highly polymorphic among individuals. These repetitive elements are rapidly becoming the molecular genetic marker of choice for genetic mapping, parentage analyses, and fundamental population genetics. We have isolated more than 50 microsatellite loci from the whooping crane (*Grus americana*). The microsatellites cloned from whooping cranes are unusually short and infrequent in comparison to other birds and especially so in comparison to mammals or reptiles. However, at least 13 of the 20 primer pairs developed amplify polymorphic loci. Paternity for 2 individuals sharing the same mother and 5 potential fathers was established by investigating 6 polymorphic loci. Additionally, the primers developed yield amplicons of the expected size from other crane species with greater than 90% success. Therefore, the markers we have developed will be useful for addressing questions about the captive management, population structure, and phylogeography of all cranes.

PROC. NORTH AM. CRANE WORKSHOP 7:36-45

Key words: DNA, endangered species, genetic variability, *Grus americana*, hypervariable markers, microsatellite, whooping crane.

Managers of captive and endangered species are increasingly aware of the deleterious effects associated with inbreeding and the overall loss of genetic diversity (Ralls and Ballou 1979, Mirande et al. 1991). The use of molecular techniques to study the genetics of endangered species provides useful information for managers attempting to maintain genetic diversity in small isolated populations (Schoenwald-Cox et al. 1983). Information on the genetic structure of wild populations allows managers to mimic that structure in captive breeding programs and reintroduction efforts. Knowledge of the genetic distinctiveness and phylogenetic structure of populations, subspecies, and species promotes informed decisions about the relative importance of those entities when deciding how to allocate scarce resources to conserve threatened or endangered groups (May 1990, Geist 1992).

The crane family, Gruidae, consists of 15 living species (Walkinshaw 1973), 7 of which are endangered or have endangered forms (U.S. Fish and Wildlife Service 1994). It is likely that habitat destruction will continue to threaten many of these birds. Captive propagation from small founding populations has been a crucial component for the rescue of the whooping crane and Mississippi sandhill crane (*Grus canadensis pulla*) from the brink of extinction (Doughty 1989, Gee et al. 1992). Assessment and conservation of the genetic diversity of these and other cranes will play an important role in their successful management and preservation.

Many molecular genetic techniques have been used to assay genetic variation within and among cranes. Isozymes have been used to study genetic variability and phylogenetic relationships in wild and captive populations of cranes (Dessauer et al. 1992). Jarvi et al. (1995) used serological techniques to determine Major Histocompatibility Complex (MHC) genotypes of Florida sandhill cranes (*G. c. pratensis*). Similar serological studies have been completed for the Mississippi sandhill and whooping cranes (S. Jarvi, pers. commun.). DNA fingerprinting has been used to study genetic variation and determine paternity of captive whooping cranes (Longmire et al. 1992). Several studies have used DNA-DNA hybridization to investigate the phylogeny and evolution of cranes (Ingold et al. 1989; Krajewski 1989, 1990). Most recently, Krajewski and Fetzer (1994) sequenced most of the mitochondrial DNA (mtDNA) cytochrome b gene from 1 individual of all crane species for reconstruction of their phylogeny. Several additional studies assaying mtDNA variation within and among species and subspecies have been completed (C. Krajewski, pers. commun.).

The previous genetic analyses have provided much useful information, but additional data are needed to answer remaining questions. For instance, the techniques used previously are limited in their ability to provide precise estimates for several measures of genetic diversity (e.g., allelic diversity, heterozygosity, relatedness, and population differentiation [Lewin 1989]) that would be valuable for management of captive and wild populations of cranes. Furthermore, techniques that avoid handling or invasive sampling would be desirable for endangered cranes. Thus, it would be advantageous to have and use molecular genetic

---

[1]Present address: Department of Biological Sciences, University of South Carolina, Columbia, SC 29208, USA.

markers that (1) provide discrete, single locus data, (2) are highly variable (with many polymorphic loci and many alleles per locus), (3) are present throughout the genome in large numbers, (4) can be screened relatively quickly and inexpensively, and (5) can make use of small and non-traditional tissue samples (such as feathers, droppings, and museum specimens). Microsatellite loci have all these characteristics (Ellegren 1992).

Microsatellites are tandem repeats of simple DNA sequence (1–6 base pairs [bp], e.g., ACAC...AC = $[AC]_n$ where n usually ranges from 10 to 40 [Tautz 1993, Charlesworth et al. 1994]). These small repetitive elements occur more-or-less randomly throughout the genomes of most eukaryotes (Tautz and Renz 1984, Stallings et al. 1991). The total number of unique microsatellite motifs possible is quite small, because apparently distinct motifs are identical, e.g., a repeat of AC is identical to a repeat of CA and, by examining the complementary DNA strand, GT or TG (Fig. 1, Table 1). Microsatellites are related to minisatellites and satellite DNA in that all are tandem arrays of repeated DNA sequence. The size of the repeat unit increases from microsatellites to minisatellites. The number of repeat units usually increases from microsatellite to minisatellite to satellite DNA (Tautz 1993).

In the human genome, microsatellites composed of AC repeats are the most common and are estimated to occur every 30,000 bp (30 kb) in euchromatic regions (Stallings et al. 1991). When all repeat motifs are considered, a microsatellite of some type occurs about every 6 kb (Beckman and Weber 1992). Thus, the microsatellites themselves are a type of interspersed repetitive DNA. Any particular microsatellite can be identified by using the unique single-copy DNA that flanks it. Specific microsatellite loci may be amplified quickly and easily by using the Polymerase Chain Reaction (PCR) with primers that are complementary to the unique single-copy DNA flanking the repetitive array (Fig. 1). More than 8,000 such loci have been identified and characterized in the human genome (Hudson et al. 1995).

Microsatellites tend to mutate by the gain or loss of repeat units at rates of 10^{-5} to 10^{-2} mutations per generation (Di Rienzo et al. 1994). These rates are many orders of magnitude higher than the mutation rate of single-copy DNA, which is about 10^{-9} mutations per year per bp (Li and Graur 1991:69-71). The high mutation rate at microsatellite loci is believed to be primarily due to slipped-strand DNA synthesis, resulting in deletions or additions of repeat units (Levinson and Gutman 1987, Strand et al. 1993). The high mutation rate generates many alleles, which results in a high degree of polymorphism at many loci in large populations. If PCR products (amplicons) for a particular microsatellite locus are amplified from chromosomes carrying alleles with different copy numbers of repeats, the amplicons will differ in size (due to the difference in the number of repeat units). The size differences among amplicons (alleles) can then be detected easily by using standard electrophoretic techniques for DNA (Weber and May 1989, Litt and Luty 1989, Tautz 1989, Ziegle et al. 1992, Morin and Smith 1995).

```
    Flanking DNA          Microsatellite          Flanking DNA

5'-GATCCGTCGATGTAAGACACACACACACACACACACTCCGAATCGGCTGATC-3'
3'-CTAGGCAGCTACATTCTGTGTGTGTGTGTGTGTGTGAGGCTTAGCCGACTAG-5'

      Primer a                                  Primer b
5'-TCCGTCGATGTAA->                       <-GCTTAGCCGACTA-5'
```

Fig. 1. The sequence of a typical AC microsatellite locus with a pair of oligonucleotide primers that could be used for PCR amplification.

Table 1. List of all unique mono- , di-, and tri-nucleotide repeat motifs. The motifs which make the most useful molecular markers are in bold. The most abundant repeat motifs in vertebrates are underlined.

Mono[a]	Di	Tri
A	**AC**	**AAC**
C	**AG**	**AAG**
	AT[b]	**AAT**
	CG[b]	**ACC**
		ACG
		ACT
		AGC
		AGG
		ATC
		CCG

[a] Alleles of mononucleotide repeats are not easily separated.

[b] The dinucleotide repeats AT and CG are self-complementary and are therefore difficult to isolate.

We are thankful for the assistance of D. Albright, R. Bavis, R. Brumfield, C. Huddleston, T. Parsons, J. Sakamoto, G. Sattler, D. Waller, and K. Winker during various phases of this research. We acknowledge the National Cancer Institute for allocation of computing time and staff support at the Frederick Biomedical Supercomputer Center of the Frederick Cancer Research and Development Center. This research has been supported by the Smithsonian's Laboratory of Molecular Systematics (LMS) and NSF grants DEB-9321604 (to W.S. and T.G.) and BIR-9300225 (to the Smithsonian's Research Training Program–for R.O.).

METHODS

When we first considered developing microsatellites from

whooping cranes during the summer of 1992, very little was known about microsatellites in birds. The only publication that identified avain microsatellites (Ellegren 1991) described only 2 loci. Our goal was to identify about 50 microsatellite loci from whooping cranes. Our first priority was to estimate the frequency of microsatellites in birds to assess the magnitude of this task. We began by searching all sequences in the Genbank database entered before sequences specifically isolated to contain microsatellites reached large numbers. The vast majority of avian sequences in Genbank at that time were from domestic chicken (*Gallus gallus*).

Samples and DNA Isolation

With the cooperation of personnel of the U.S. Fish and Wildlife Service, the National Biological Service, the International Crane Foundation (ICF), the Canadian Wildlife Service, the Louisiana State University Museum of Natural Science, and S. Jarvi and H. Dessauer, we obtained samples from almost all captive and many wild whooping cranes. Samples from all known crane species were generously provided by the ICF (blood) or C. Krajewski (DNA). Additionally, with the cooperation of many museums, we have obtained samples of skin (~ 3 cm^2) and, in some cases, a few contour feathers, from more than 100 whooping cranes collected prior to their population bottleneck from throughout the original breeding and migratory range. These samples represent about one-third of all whooping crane specimens in museums (cf. Hahn 1954) and most of the specimens with age and locality data. The museum samples form the basis of continuing research into the amount of genetic variation actually lost during the whooping crane bottleneck.

Prior to 1995, genomic DNA was usually isolated from blood and tissue by standard phenol/chloroform/isoamyl alcohol (PCI) extraction and ethanol precipitation (Sambrook et al. 1989:9.17-9.19). Because bird blood is nucleated, the blood was treated as "tissue," except that it was not homogenized. During 1995, genomic DNA was usually isolated from blood by using guanidine thiocyanate and silica (cf. Boom et al. 1991). The exact protocol used can be downloaded from the LMS Internet server (e.g., with Netscape) by using the URL: ftp://onyx.si.edu/protocols/TSilicaDNA.rtf. The advantages and disadvantages of the silica technique are enumerated in the protocol. Genomic DNA was isolated from museum skins and feathers using a protocol modified from Pääbo et al. (1988). The major modification includes use of a digestion buffer containing 1 mM Ca^{2+} and no EDTA (see Sambrook et al. 1989:B.16). This protocol can also be obtained from the LMS Internet server under the filename MuseumDNA.rtf.

Isolation of Microsatellite Loci

We used small insert plasmid libraries to isolate microsatellites from whooping and sandhill cranes (cf. Hughes and Queller 1993, Ostrander et al. 1992). Briefly, this approach involves digesting total genomic DNA from the species of interest into small fragments with a restriction enzyme. Fragments which were sufficiently long to contain whole microsatellite loci, but still short enough to be conveniently sequenced (300–700 bp), were separated from all other fragments via size selection following agarose gel electrophoresis. Following elution from the agarose, the fragments were then ligated into a plasmid vector and used to transform bacteria. Each resulting bacterial colony contained a random fragment of crane DNA which was screened for the presence of microsatellites by hybridization with radioactively labeled oligonucleotides representing 1 or more simple sequence repeat motifs. Plasmid DNA from the positive colonies was isolated and sequenced. Oligonucleotide primers complementary to the unique sequences flanking the microsatellite were designed and synthesized (Fig. 1). The primers were then used in PCR-based assays as described above. In this paper, we refer to the collection of genomic DNA fragments ligated into plasmids and the resulting transformed colonies as a library. Subsequent collections of plasmids with crane DNA that were isolated from many bacterial colonies are called sub-libraries. Explicit (step-by-step) protocols used to obtain and score microsatellites may be downloaded from the LMS Internet server under the filename MsatManV6.rtf.

Random Libraries.—In random libraries, the microsatellite loci should be detected at a frequency similar to their frequency in the genome (number of colonies screened $\times$ average insert size per colony / number of colonies with microsatellites $\approx$ average number of bp between each microsatellite locus). Many different ligations were used to provide clones for the unenriched libraries.

Enriched Library.—Several library enrichment strategies have been devised which greatly increase the frequency of colonies with microsatellites among those plated and hence reduce the total number of colonies which must be screened (Ostrander et al. 1992, Kandpal et al. 1994, Pulido and Duyk 1994). We used a method derived from Ostrander et al. (1992) which uses the random library we had already made (see also Kunkel et al. 1987, Pulido and Duyk 1994:2.2.1). Briefly, the ligated plasmid vector + crane DNA was transformed into a strain of bacteria that was deficient for uracil DNA glycosylase (*ung*$^-$) and dUTPase (*dut*$^-$). These bacteria incorporated uracil in their DNA at a fraction of the positions (about 0.3%) normally occupied by thymine (Kunkel 1985). Single-stranded plasmid DNA was then recovered from a large number of these bacterial colonies. This sub-library contains many copies of many different pieces of crane DNA, though each was now present in multiple copies. The single-stranded sub-library was subjected to second strand synthesis primed by an oligonucleotide composed of a simple sequence repeat motif. The primer selects only those plasmids which contain complementary repeats, and the newly synthesized (complementary) strands contain no uracil. The resulting double-stranded plasmids transform normal (*ung*$^+$) bacteria at a rate much higher than the remaining single-stranded plasmids which are destroyed by the bacteria because they contain uracil. Thus, sub-libraries were enriched for several simple repeats.

RESULTS

Genbank Searches

Birds belong to the Genbank group "Other Vertebrates." Although microsatellites were detected in Other Vertebrates about as often as in the groups Primates and Other Mammals, microsatellite motifs were detected in birds much less often than in the remainder of the Other Vertebrates. Microsatellites were detected about half as often in birds as in Primates or Other Mammals (Fig. 2). This result implied that avian microsatellites probably occurred in high enough frequency to proceed with isolation by using standard techniques (cf. Hughes and Queller 1993).

Random Libraries

Our initial experiments estimated a frequency of AC microsatellites in only about 0.06% of all clones from the whooping cranes. If an average insert size of 500 bp is assumed, this frequency would imply that AC microsatellites occur about once every 800 kb. Thus, we estimated that we could isolate our goal of 50 microsatellite loci by screening about 100,000 colonies (about 10 times more than required for mammals). More than 100,000 colonies containing random pieces of whooping crane DNA were screened from 1992 to 1994. This effort, however, yielded only enough clones for 11 primer pairs to be designed.

We also screened a sample of about 10,000 colonies from sandhill cranes and red pandas (*Ailurus fulgens*) to ensure that the low frequency of microsatellites observed in whooping cranes was not an artifact of library preparation. The sandhill crane library yielded positive clones at about the same frequency as the whooping crane libraries. Microsatellites are much more common in the red panda genome with an estimated frequency of AC microsatellites of about 0.42% of all clones. Again, if an average insert size of 500 bp is assumed, this frequency would imply that AC microsatellites occur about once every 100 kb. Although we detected AC repeats in red pandas about half as frequently as we expected (based on the frequency detected in dogs [Ostrander et al. 1992, 1993]), the difference between the pandas and the cranes is similar to the difference detected between other mammals and birds (e.g., Crooijmans et al. 1993).

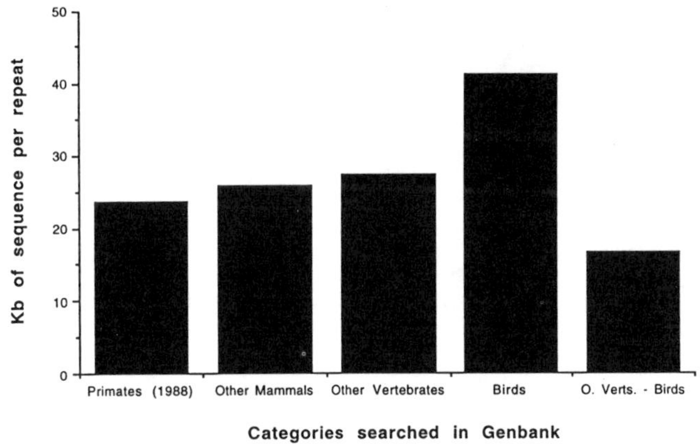

Fig. 2. The frequency of microsatellites in sequences deposited in the Genbank database prior to large-scale isolation of microsatellites (see Devereux 1989 for Genbank categories). The Primate category includes sequences deposited prior to September 1988 and had not been updated through May 1993 (release 79.0). All other categories include sequences deposited prior to October 1992 (release 73.1). "Other Mammals" excludes Primates and Rodents. Birds are included in the "Other Vertebrates" category.

Subsequent screenings of additional colonies revealed that scaling up reduced efficiency of detection, thus there was a limit of how many colonies could be efficiently screened at once with our equipment. Although more than 100,000 colonies were plated, it is difficult to know how many were actually screened with our procedure. Thus, it was clear that this approach would not effectively yield the desired number of microsatellite loci. Therefore, we sought to obtain additional microsatellite loci by enrichment.

Enriched Sub-Library

About 150,000 colonies from the whooping crane library were subjected to enrichment. From the crane sub-library, 4,058 colonies were replated for screening by hybridization. Following colony hybridization, 439 (about 11%) of the colonies were picked as putative positives. This represents about a 180-fold enrichment relative to the random library.

Sequencing was attempted for a total of 180 clones from the enriched crane sub-library. Of those sequences, microsatellites were detected in a total of 60 unique clones.

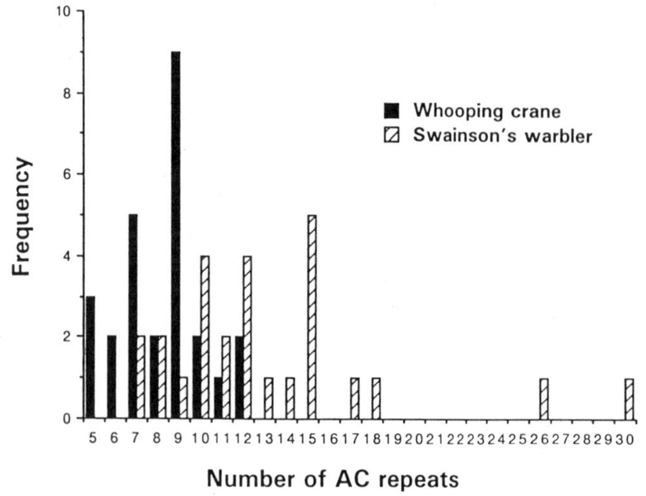

Fig. 3. Comparison of the distribution of the copy number of repeats in microsatellite clones from whooping crane and Swainson's warbler (*Limnothlypis swainsonii*) libraries enriched with use of $(TG)_{15}$. Warbler data are courtesy of K. Winker, G. Graves, and M. J. Braun, Smithsonian Institution.

An accurate determination of microsatellite length was made for 45 clones. The distribution of repeat lengths in the crane clones was dominated by microsatellites with less than 10 repeats, whereas all other libraries we have constructed and enriched from other species (avian, mammalian, and reptilian) are dominated by sequences with ≥10 repeats (e.g., Fig. 3). From these new crane sequences, 26 additional PCR primer pairs were synthesized (24 for AC repeats and 2 for AGG repeats).

Amplification of Microsatellites

PCR amplification conditions have been optimized for 20 of the 37 primer pairs synthesized to date. In an initial survey of cross-species utility, 4 primer pairs were tested on 4 *Grus*, 1 *Bugeranus*, and 1 *Anthropoides* species. Because almost all samples yielded amplicons of the expected size, DNA from at least 1 individual from each of the 15 crane species was subsequently tested for amplification by using 8 primer pairs (Table 2). Amplification was achieved in more than 90% of the species/primer pair combinations with the family Gruidae. However, only 3 of the 8 primer pairs supported amplification in other gruiform birds. Two of these 3 also supported amplification from birds of other orders and American alligators (*Alligator mississippiensis*) (Table 2). Whether these amplicons represent orthologous microsatellite loci in the other taxa remains to be determined, but the fact that they fall in the expected size range encourages optimism. Some loci were screened on acrylamide gels, which allows detection of heterozygotes and thus polymorphism. Polymorphic loci have been detected in red-crowned (*G. japonensis*), sarus (*G. antigone*), sandhill, and Siberian (*G. leucogeranus*) cranes.

The 20 optimized loci were also tested for polymorphism within a small sample of whooping cranes. Thirteen of these loci are polymorphic within a sample of 4 living individuals (Table 3). Genotypes were subsequently obtained for a sample of 16 whooping cranes for 6 of the polymorphic loci. This group contained a known breeding pair with 7 offspring and 2 captive-bred offspring with the same mother and a group of 5 potential fathers. Mendelian inheritance was observed within the known family for all 6 loci (Table 4). Paternity was unambiguously determined for both captive-bred offspring (Table 4). For example, whooping crane 74-001 is the only male not excluded as the potential sire of 85-001 by simply assessing genotypes at locus AGG-2. Initial tests on 4 loci have also shown that successful amplification of DNA isolated from 10 museum specimens could be achieved for about 50% of the specimen/locus pairs.

DISCUSSION

The overall frequency of microsatellites in cranes appears to be similar to that of most other birds (e.g., Croojimans et al. 1993; T. C. Glenn, unpubl. data). Because the frequency is low relative to most mammals, randomly searching small insert plasmid libraries in bacterial colonies is not recommended. Use of enrichment strategies (e.g., Ostrander et al. 1992, Kandpal et al. 1994) or large insert libraries (Margolis et al. 1993, Litt and Browne 1994, Rothuizen and van Raak 1994) is highly recommended when microsatellites are isolated from cranes or other birds.

The average length of the microsatellite loci detected in whooping cranes was quite short relative to other birds (Fig. 3, Crooijmans et al. 1993), mammals, or other vertebrates (e.g., Weber 1990; Serikawa et al. 1992; Ostrander et al. 1993; FitzSimmons et al. 1995; T. C. Glenn, unpubl. data). It is not known if this length bias is shared among other cranes, or is peculiar to whooping cranes. It is possible, though unlikely, that the bottleneck experienced by whooping cranes could have caused the observed bias. Characterization of loci from whooping cranes and additional loci isolated from non-bottlenecked crane species will be needed to elucidate this issue.

Advantages of Microsatellites

Once PCR primers have been designed for particular loci, amplicons are easily synthesized from multiple individuals. When the repeat count in the cloned allele is ≥10, the

Table 2. Ability of PCR primers developed from whooping cranes to amplify other species.

Group	Species[a]	Locus AA-1	AC-1	AT/AC-1	AAC-1	ACC-1	ATC-1	AAGC-1	SH-AG-1[b]
Gruidae	Whooping	+	+	+	+	+	+	+	+
Gruidae	Red-crowned	+	+	+	+	+	+	+	+
Gruidae	Black-necked	+	+	+	+	+	-	+	+
Gruidae	Common	+	+	+	+	+	+	+	+
Gruidae	Hooded	+	+	+	+	+	+	+	+
Gruidae	Sarus	+	-	+	+	+	+	+	+
Gruidae	Brolga	+	+	+	+	+	+	+	+
Gruidae	White-naped	+	+	+	+	+	+	+	+
Gruidae	Sandhill	+	+	+	+/-	+	-	+	+
Gruidae	Wattled	+	+/-	+	+/-	+	+	+	+
Gruidae	Siberian	+	+	+	+	+	-	+	+
Gruidae	Blue	+	+	+	+	+	+	+	+
Gruidae	Demoiselle	+	+	+	+	+	+	+	+
Gruidae	Gray crowned	+	+	+	-	+	-	+	+
Gruidae	Black crowned	+	+	+	-	+	?	+	+
Gruiformes	Trumpeter	+	-	+	-	+	?	-	-
Gruiformes	Rail	+	-	-	-	+	-	-	-
Aves	Brant	-	-	-	-	+/-	-	-	-
Aves	Chicken	+	-	-	-	+	+/-	-	-
Aves	Manakin	+	+	-	-	-	-	-	-
Aves	Cowbird	+	+	-	+	-	?	-	-
Vertebrate	Alligator	+	-	-	-	+	-	-	-
Vertebrate	Raccoon	-	-	-	-	-	+	-	-
	Negative control	-	-	-	-	-	-	-	-
	Expected size	+	+	+	+	+	-	+	+
	Glycerol[c]	-	-	-	+	+	-	-	-

[a] Scientific names of species not mentioned in text: black-necked (*Grus nigricollis*), common (*G. grus*), hooded (*G. monacha*), brolga (*G. rubicunda*), white-naped (*G. vipio*), wattled (*Bugeranus carunculatus*), blue (*Anthropoides paradisea*), demoiselle (*A. virgo*), gray crowned (*Balearica regulorum*), and black crowned (*Balearica pavonina*) cranes; gray-winged trumpeter (*Psophia crepitans*); gray-necked wood rail (*Aramides cajanea*); black brant (*Branta bernicla nigricans*); white-collared manakin (*Manacus candei*); brown-headed cowbird (*Molothrus ater*); raccoon (*Procyon lotor*).

[b] Developed from sandhill crane genomic DNA.

[c] Glycerol composed 10% of PCR reactions designated +.

proportion of polymorphic microsatellite loci exceeds 90% for most taxa (e.g., Weber 1990). Microsatellite loci have been shown to be highly polymorphic, with many alleles per locus, even in organisms with very low allozyme variation (Hughes and Queller 1993, Taylor et al. 1994, Glenn et al. 1996). Because thousands of microsatellites occur within the genomes of most complex organisms, they represent a rich source of genetic markers. Additionally, because the mutation rate is high, each locus is likely to have many alleles, and populations with little gene flow will differentiate quickly. However, the mutation rate is low enough for alleles to be relatively stable within small numbers of generations, allowing for genealogical analysis. Thus, microsatellites are becoming the molecular genetic marker of choice for ecological, human, medical, and population genetics (e.g., Bruford and Wayne 1993, Kuhl and Caskey 1993, Weissenbach 1993, Slatkin 1995).

Microsatellites have several additional benefits. First, the amount of material needed for analysis via PCR is much less than that required for protein studies or Southern blot analysis (i.e., DNA fingerprinting or MHC Restriction Fragment Length Polymorphism [RFLP]), an important consideration for work with endangered species, where destructive sampling and excessive handling are not desirable. For example, the same amount of DNA used for 1 DNA fingerprint can be used to amplify at least 100 microsatellite loci. Second, microsatellites may be amplified from degraded DNA, such as DNA recovered from museum specimens (Ellegren 1991). This was our primary interest in developing these loci for whooping cranes. Other non-traditional sources of tissue have

Table 3. Comparison of the number of polymorphisms detected in whooping cranes by allozymes and microsatellites.

	Allozymes[a]	Microsatellites
Individuals surveyed	34	4
Loci surveyed	24	22
Polymorphic loci	5	13
Polymorphic alleles[b]	11	27

[a] From Dessauer et al. (1992).
[b] Alleles detected at polymorphic loci.

also been used for PCR based assays (e.g., droppings [Kohn et al. 1995] or feathers [Leeton et al. 1993]). Third, once primer sets for polymorphic loci are available, large numbers of individuals can be analyzed quickly at many loci. Microsatellites are particularly amenable to automation (Ziegle et al. 1992).

A final useful feature of microsatellites is that primer pairs developed from 1 species can often be used to amplify orthologous polymorphic loci in other closely related taxa. Many early attempts at trans-species amplification of microsatellites focused on rather distantly related taxa (e.g., Moore et al. 1991, Schlötterer et al. 1991, Stallings et al. 1991) and had varying levels of success. More recent studies have focused on species within taxonomic families (e.g., Breen et al. 1994, Hanotte et al. 1994, Pepin et al. 1995) and have been highly successful. Our results indicate many of the loci we have developed will be of use in studying other cranes (Table 2). Similarly, it appears that genetic analyses of many threatened and endangered species can make use of primers developed from other species (e.g., Morin et al. 1993, Gotelli et al. 1994). Recent studies also suggest that a small percentage of microsatellite loci are conserved across even very distantly related taxonomic groups (e.g., AA-1, Table 2; FitzSimmons et al. 1995; Stallings 1995; Glenn et al. 1996). Through time, it may be possible to accumulate enough such conserved loci that development of new primers will not be necessary for most studies using microsatellites.

Disadvantages of Microsatellites

Although microsatellites occur in all eukaryotes (Hamada et al. 1982, Tautz and Renz 1984), their distribution, overall frequency, and average length vary widely (Hamada et al. 1982, Langercrantz et al. 1993). Thus, it has been much more difficult to obtain microsatellites from whooping cranes than other taxa, even other birds. In addition to the difficulty of isolating microsatellite loci, the proportion of primers that give unsatisfactory results can vary from taxon to taxon, with whooping cranes among the worst we have studied (T. C. Glenn, unpubl. data). Thus, the difficulty of isolating additional loci and developing subsequent primers for cranes should not be underestimated. However, the primer pairs we have developed usually amplify DNA fragments of the correct size and are often polymorphic in related taxa.

As with other techniques, problems such as null alleles (Pemberton et al. 1995), anomalous signficance of linkage (Knowles et al. 1992), and inconsistency between studies due to small variations in technique (Knowles et al. 1992) can affect microsatellites. These problems, however, have been less likely to affect microsatellites than DNA fingerprinting (Jin and Chakraborty 1995). Additionally, by use of captive cranes with known pedigrees, many of these problems can be detected and dealt with appropriately.

Table 4. Genotypes of captive whooping crane family members at 6 microsatellite loci.

			Locus[b]					
ID no.	Status[a]	Sex	AC-2	AC-5	AC-15	AAT-1	ACC-1	AGG-2
71-001	Dam1	F	ab	ac	bb	bb	ab	–
83-003	Fam1	M	ab	ab	bb	bb	bb	ab
84-003	Fam1	M	aa	ac	bb	–	aa	bb
88-022	Fam1	M	bb	ac	bb	ab	bb	bb
90-018	Fam1	M	ab	cc	bb	bb	aa	bc
90-026	Fam1	F	aa	ac	bb	bb	aa	bc
91-006	Fam1	F	ab	bc	bb	ab	ab	bc
64-001	Sire1,?	M	ab	bc	ab	ab	ab	bc
74-001	Sire?,2	M	ab	aa	ab	–	bb	ac
13-7	Sire?	M	ab	aa	ab	bb	bb	bc
68-001	Sire?,3	M	ab	ab	ab	–	ab	bc
13-17	Sire?	M	bb	ab	aa	ab	bb	bb
85-002	Fam3	F	ab	ac	ab	ab	ab	bc
85-001	Fam2	F	bb	ac	ab	ab	bb	aa
13-16	Dam2,3	F	bb	cc	aa	bb	bb	ab

[a] Numbers refer to the family group. Fam designates offspring of a particular mating pair (e.g., individuals designated Fam1 are offspring of 71-001 and 64-001). Sperm from males 64-001, 74-001, 13-7, 68-001, and 13-17 was used to artificially inseminate 13-16 in 1985. Thus any of those males could be the sire of 85-002 and 85-001. The status of "Sire2" and "Sire3" are determined from the genotypes in this table.
[b] – designates genotypes that could not be unambiguously determined from the first attempt at genotyping.

MANAGEMENT IMPLICATIONS

We recognize that ecological and demographic factors are the predominant proximate causes of endangerment (Lande 1988, Caro and Laurenson 1994, Caughley 1994). However, the negative effects of decreased genetic variability and inbreeding have been documented for several vertebrate species (Ralls et al. 1979, Soule 1979, O'Brien et al. 1985, Wildt et al. 1987, Harvey and Read 1988, Leberg 1990), and we believe the risk of genetic defects will often be additive to other causes of mortality. Thus, small inbred populations experience increased risks of extinction. If a lack of genetic diversity in a population increases its chances of extinction from epidemic diseases or other causes (Harvey and Read 1988), thereby erasing all previous conservation efforts, it is prudent for managers to attempt to preserve genetic diversity when managing species or ecosystems.

Without reserves large enough to maintain viable populations for many vertebrate species, managers have looked for alternatives to increase the effective population size of isolated populations. Methods of increasing gene flow and managing genetic diversity include the use of habitat corridors, translocation, and captive propagation (Foose et al. 1986, Simberloff and Cox 1987, Conant 1988). However, without detailed and accurate genetic information, such practices are difficult to implement and monitor wisely. Microsatellite loci can often provide just such information. The microsatellite loci we have developed for cranes can be used to provide answers to such questions as: who are the parents of a given individual; how related are 2 individuals within a population; how differentiated are populations, subspecies, and species; and what fraction of the original genetic diversity remains in highly endangered subspecies and species. Cranes can be better managed by knowing the answers to such questions.

LITERATURE CITED

BECKMAN, J. S., and J. L. WEBER. 1992. Survey of human and rat microsatellites. Genomics 12:627-631.

BOOM, R., C. J. A. SOL, M. M. M. SALIMANS, C. L. JANSEN, P. M. E. WERTHEIM-VAN DILLEN, and J. VAN DER NOORDAA. 1990. Rapid and simple methods for purification of nucleic acids. J. Clin. Microbiol. 28:495-503.

BREEN, M., P. DOWNS, Z. IRVIN, and K. BELL. 1994. Intrageneric amplification of horse microsatellite markers with emphasis on the Przewalski's horse *(E. przewalskii*). Anim. Genet. 25:401-405.

BRUFORD, M. W., and R. K. WAYNE. 1993. Microsatellites and their application to population genetic studies. Curr. Op. Genet. Dev. 3:939-943.

CARO, T. M., and M. K. LAURENSON. 1994. Ecological and genetic factors in conservation: a cautionary tale. Science 263:485-486.

CAUGHLEY, G. 1994. Directions in conservation biology. J. Anim. Ecol. 63:215-244.

CHARLESWORTH, B., P. SNIEGOWSKI, and W. STEPHAN. 1994. The evolutionary dynamics of repetitive DNA in eukaryotes. Nature 371:215-220.

CONANT, S. 1988. Saving endangered species by translocation. Bioscience 38:254-257.

CROOIJMANS, R. P. M. A., A. J. A. VAN KAMPEN, J. J. VAN DER POEL, and M. A. M. GROENEN. 1993. Highly polymorphic microsatellite markers in poultry. Anim. Genet. 24:441-443.

DESSAUER, H. C., G. F. GEE, and J. S. ROGERS. 1992. Allozyme evidence for crane systematics and polymorphisms within populations of sandhill, sarus, Siberian, and whooping cranes. Mol. Phylogenet. Evol. 1:279-288.

DI RIENZO, A., A. C. PETERSON, J. C. GARZA, A. M. VALDES, M. SLATKIN, and N. B. FREIMER. 1994. Mutational processes of simple-sequence repeat loci in human populations. Proc. Nat. Acad. Sci., USA 91:3166-3170.

DOUGHTY, R. W. 1990. Return of the whooping crane. Univ. Texas Press, Austin. 182pp.

ELLEGREN, H. 1991. DNA typing of museum birds. Nature 354:113.

______. 1992. Polymerase-chain-reaction (PCR) analysis of microsatellites - a new approach to studies of genetic relationships in birds. Auk 109:886-895.

FITZSIMMONS, N. N., C. MORITZ, and S. S. MOORE. 1995. Conservation and dynamics of microsatellite loci over 300 million years of marine turtle evolution. Mol. Biol. Evol. 12:432-440.

FOOSE, T. J., R. LANDE, N. R. FLESNESS, G. RABB, and B. READ. 1986. Propagation plans. Zoo Biol. 5:139-146.

GEE, G. F., H. C. DESSAUER, J. LONGMIRE, W. E. BRILES, and R. C. SIMON. 1992. The study of relatedness and genetic diversity in cranes. Pages 225-230 *in* D. A. Wood, ed. Proc. 1988 North Am. crane workshop. Florida Game Fresh Water Fish Comm. Nongame Wildl. Program Tech. Rep. 12.

GEIST, V. 1992. Endangered species and the law. Nature 357:274-276.

GLENN, T. C., W. STEPHAN, H. C. DESSAUER, and M. J. BRAUN. 1996. Allelic diversity in alligator microsatellite loci is negatively correlated with GC content of flanking sequences and evolutionary conservation of PCR amplifiability. Mol. Biol. Evol. 13:1151-1154.

GOTTELLI, D., C. SILLERO-ZUBIRI, G. D. APPLEBAUM, M. S. ROY, D. J. GIRMAN, J. GARCIA-MORENO, E. A. OSTRANDER and R. K. WAYNE. 1994. Molecular genetics of the most endangered canid: the Ethiopian wolf *Canis simensis*. Mol. Ecol. 3:301-312.

HAHN, P. 1954. Where is that vanished bird? An index to the known specimens of the extinct and near extinct North American species. Royal Ontario Mus., Univ. Toronto, Toronto, Ont. 347pp.

HAMADA, H., M. G. PETRINO, and T. KAKUNAGA. 1982. A novel repeated element with Z-DNA-forming potential is widely found in evolutionarily diverse eukaryotic genomes. Proc. Natl. Acad. Sci., USA 79:6465-6469.

HANOTTE, O., C. ZANON, A. PUGH, C. GREIG, A. DIXON, and T. BURKE. 1994. Isolation and characterization of microsatelllite loci in a passerine bird: the reed bunting *Emberiza schoeniclus*. Mol. Ecol. 3:529-530.

HARVEY, P. H., and A. F. READ. 1988. When incest is not best.

Nature 336:514-515.

HUDSON, T. J., et al. 1995. An STS-based map of the human genome. Science 270:1945-1954.

HUGHES, C. R., and D. C. QUELLER. 1993. Detection of highly polymorphic microsatellite loci in a species with little allozyme polymorphism. Mol. Ecol. 2:131-137.

INGOLD, J. L., J. C. VAUGHN, S. I. GUTTMAN, and L. R. MAXSON. 1989. Phylogeny of the cranes (Aves: Gruidae) as deduced from DNA-DNA hybridization and albumin micro-complement fixation analyses. Auk 106:595-602.

JARVI, S. I., G. F. GEE, M. M. MILLER, and W. E. BRILES. 1995. A complex alloantigen system in Florida sandhill cranes, *Grus canadensis pratensis*: evidence for the major histocompatibility (B) system. J. Hered. 86:348-353.

JIN, L., and R. CHAKRABORTY. 1995. Population structure, stepwise mutations, heterozygote deficiency and their implications in DNA forensics. Heredity 74:274-285.

KANDPAL, R. P., G. KANDPAL, and S. M. WEISSMAN. 1994. Construction of libraries enriched for sequence repeats and jumping clones, and hybridization selection for region-specific markers. Proc. Natl. Acad. Sci., USA 91:88-92.

KNOWLES, J. A., V. J. VIELAND, and T. C. GILLIAM. 1992. Perils of gene mapping with microsatellite markers. Am. J. Hum. Genet. 51:905-909.

KOHN, M., F. KNAUER, A. STOFFELLA, W. SCHRÖDER and S. PÄÄBO. 1995. Conservation genetics of the European brown bear - a study using excremental PCR of nuclear and mitochondrial sequences. Mol. Ecol. 4:95-103.

KRAJEWSKI, C. 1989. Phylogenetic relationships among cranes (Gruiformes: Gruidae) based on DNA hybridization. Auk 106:603-618.

______. 1990. Relative rates of single-copy DNA evolution in cranes. Mol. Biol. Evol. 7:65-73.

______. and J. W. FETZER, JR. 1994. Phylogeny of cranes (Gruiformes: Gruidae) based on cytochrome-b DNA sequences. Auk 111:351-365.

KUHL, D. P. A., and C. T. CASKEY. 1993. Tinucleotide repeats and genome variation. Curr. Op. Genet. Dev. 3:404-407.

KUNKEL, T. A. 1985. Rapid and efficient site-specific mutagenesis without phenotypic selection. Proc. Natl. Acad. Sci., USA 82:488-492.

______, J. D. ROBERTS, and R. A. ZAKOUR. 1987. Rapid and efficient site-specific mutagenesis without phenotypic selection. Methods Enzymol. 154:367-382.

LANDE, R. L. 1988. Genetics and demography in biological conservation. Science 241:1455-1460.

LANGERCRANTZ, U., H. ELLEGREN, and L. ANDERSSON. 1993. The abundance of various polymorphic microsatellite motifs differs between plants and vertebrates. Nucleic Acids Res. 21:1111-1115.

LEBERG, P. L. 1990. Influences of genetic variability on population growth: implications for conservation. J. Fish Biol. 37(A):397-401.

LEETON, P., L. CHRISTIDIS, and M. WESTERMAN. 1993. Feathers from museum bird skins-a good source of DNA for phylogenetic studies. Condor 95:465-466.

LEVINSON, G., and G. A. GUTMAN. 1987. Slipped-strand mispairing: a major mechanism for DNA sequence evolution. Mol. Biol. Evol. 4:203-221.

LEWIN, R. 1989. Limits to DNA fingerprinting. Science 243:1549-1551.

LI, W. H., and D. GRAUR. 1991. Fundamentals of molecular evolution. Sinauer Assoc. Inc., Sunderland, Mass. 284pp.

LITT, M., and D. BROWNE. 1994. Characterization of (CA)n microsatellite repeats from large-insert clones. Unit 2.4.1-2.4.6 *in* N. C. Dracopoli, ed. Current protocols in human genetics. John Wiley and Sons, Inc., New York, N.Y.

______, and J. A. LUTY. 1989. A hypervariable microsatellite revealed by in vitro amplification of a dinucleotide repeat within the cardiac muscle actin gene. Am. J. Human Genet. 44:397-401.

LONGMIRE, J. L., G. F. GEE, C. L. HARDEKOPF, and G. A. MARK. 1992. Establishing paternity in whooping cranes (*Grus americana*) by DNA analysis. Auk 109:522-529.

MARGOLIS, R. L., S. H. LI, and C. A. ROSS. 1993. A rapid method for sequencing trinucleotide repeats. Nucleic Acids Res. 21:4983-4984.

MAY, R. M. 1990. Taxonomy as destiny. Nature 347:129-130.

MIRANDE, C., R. LACY, and U. SEAL, eds. 1991. Whooping crane (*Grus americana*) conservation viability assessment. Captive Breeding Specialist Group, Apple Valley, Minn. 119pp.

MOORE, S. S., L. L. SARGEANT, T. J. KING, J. S. MATTICK, M. GEORGES, and D. J. S. HETZEL. 1991. The conservation of dinucleotide microsatellites among mammalian genomes allows the use of heterologous PCR primer pairs in closely related species. Genomics 10:654-660.

MORIN, P. A., and D. G. SMITH. 1995. Nonradioactive detection of hypervariable simple sequence repeats in short polyacrylamide gels. Biotechniques 19:223-227.

______, J. WALLIS, J. J. MOORE, R. CHAKRABORTY, and D. S. WOODRUFF. 1993. Non-invasive sampling and DNA amplifaction for paternity exclusion, community structure, and phylogeography in wild chimpanzees. Primates 34:347-356.

O'BRIEN, S. J., M. E. ROELKE, L. MARKER, A. NEWMAN, C. A. WINKLER, D. MELTSER, L. COLLY, J. F. EVERMANN, M. BUSH, and D. E. WILDT. 1985. Genetic basis for species vulnerability in the cheetah. Science 227:1428-1434.

OSTRANDER, E. A., P. M. JONG, J. RINE, and G. DUYK. 1992. Construction of small-insert genomic DNA libraries highly enriched for microsatellite repeat sequences. Proc. Natl. Acad. Sci., USA 89:3419-3423.

______, G. F. SPRAGUE, JR., and J. RINE. 1993. Identification and characterization of dinucleotide repeat (CA)n markers for genetic mapping in dog. Genomics 16:207-213.

PÄÄBO, S., J. A. GIFFORD, and A. C. WILSON. 1988. Mitochondrial DNA sequences from a 7000-year old brain. Nucleic Acids Res. 16:9775-9787.

PEMBERTON, J. M., J. SLATE, D. R. BANCROFT, and J. A. BARRETT. 1995. Nonamplifying alleles at microsatellite loci: a caution for parentage and population studies. Mol. Ecol. 4:249-252.

PEPIN, L., Y. AMIGUES, A. LEPINGLE, J. BERTHIER, A.

BENSAID, and D. VAIMAN. 1995. Sequence conservation of microsatellites between *Bos taurus* (cattle), *Capra hircus* (goat) and related species: examples of use in parentage testing and phylogeny analysis. Heredity 74:53-61.

PULIDO, J. C., and G. M. DUYK. 1994. Construction of small-insert libraries enriched for short tandem repeat sequences by marker selection. Unit 2.2.1-2.2.33 *in* N. C. Dracopoli, ed. Current protocols in human genetics. John Wiley and Sons, Inc., New York, N.Y.

RALLS, K., K. BRUGGER, and J. BALLOU. 1979. Inbreeding and juvenile mortality in small populations of ungulates. Science 206:1101-1103.

ROTHUIZEN, J., and M. VAN RAAK. 1994. Rapid PCR-based characterization of sequences flanking microsatellites in large-insert libraries. Nucleic Acids Res. 22:5512-5513.

SAMBROOK J., E. F. FRITSCH, and T. MANIATIS. 1989. Molecular cloning: a laboratory manual. Second ed. Cold Spring Harbor Laboratory Press, Cold Spring Harbor, N.Y.

SCHLÖTTERER, C., B. AMOS, and D. TAUTZ. 1991. Conservation of polymorphic simple sequence loci in cetacean species. Nature 354:63-65.

SCHONEWALD-COX, C. M., S. M. CHAMBERS, B. MACBRYDE, and L. THOMAS, eds. 1983. Genetics and conservation: a reference for managing wild animal and plant populations. Bejamin/Cummings, Menlo Park, Calif. 722pp.

SERIKAWA, T., T. KURAMOTO, P. HILBERT, M. MORI, J. YAMADA, C. J. DUBAY, K. LINDPAINTER, D. GANTEN, J. GUENET, G. M. LATHROP, and J. S. BECKMANN. 1992. Rat gene mapping using PCR-analyzed microsatellites. Genetics 131:701-721.

SIMBERLOFF, D., and J. COX. 1987. Consequences and costs of conservation corridors. Conserv. Biol. 1:63-71.

SLATKIN, M. 1995. A measure of population subdivision based on microsatellite allele frequencies. Genetics 139: 457-462.

SOULE, M. E. 1979. Heterozygosity and developmental stability: another look. Evolution 33:396-401.

STALLINGS, R. L. 1995. Conservation and evolution of (CT)n/(GA)n microsatellite sequences at orthologous positions in diverse mammalian genomes. Genomics 25:107-113.

______, A. F. FORD, D. NELSON, D. C. TORNEY, C. E. HILDERBRAND, and R.K. MOYZIS. 1991. Evolution and distribution of (GT)n repetitive sequences in mammalian genomes. Genomics 10:807-815.

STRAND, M. K., T. A. PROLLA, R. M. LISKAY, and T. D. PETES. 1993. Destabilization of tracts of simple repetitive DNA in yeast by mutations affecting DNA mismatch repair. Nature 365:274-276.

TAUTZ, D. 1989. Hypervariability of simple sequences as a general source for polymorphic DNA markers. Nucleic Acids Res. 17:6463-6471.

______. 1993. Notes on the definition and nomenclature of tandemly repetitive DNA sequences. Pages 21-28 *in* S. D. J. Pena, R. Chakraborty, J. T. Epplen, and A. J. Jeffreys, eds. DNA fingerprinting: state of the science. Basel, Boston, Mass.

______. and M. Renz. 1984. Simple sequences are ubiquitous repetitive components of eukaryotic genomes. Nucleic Acids Res. 12:4127-4138.

TAYLOR, A. C., W. B. SHERWIN, and R. K. WAYNE. 1994. Genetic variation of microsatellite loci in a bottlenecked species: the northern hairy-nosed wombat *Lasiorhinus krefftii*. Mol. Ecol. 3:277-290.

U.S. FISH AND WILDLIFE SERVICE. 1994. Endangered and threatened wildlife and plants. U.S. Federal Register 50 CFR 17.11 and 17.12, Washington, D.C. 42pp.

WALKINSHAW, L. H. 1973. Cranes of the world. Winchester Press, New York, N.Y. 370pp.

WEBER, J. L. 1990. Informativeness of human (dC-dA)n·(dG-dT)n polymorphisms. Genomics 7:524-530.

______, and P. E. May. 1989. Abundant class of human DNA polymorphisms which can be typed using the polymerase chain reaction. Am. J. Hum. Genet. 44:388-396.

WEISSENBACH. J. 1993. Microsatellite polymorphisms and the genetic linkage map of the human genome. Curr. Op. Genet. Dev. 3: 414-417.

WILDT, D. E. , M. BUSH, K. L. GOODROWE, C. PACKER, A. E. PUSEY, J. L. BROWN, P. JOSLIN, and S. J. O'BRIEN. 1987. Reproductive and genetic consequences of founding isolated lion populations. Nature 329:328-330.

ZIEGLE, J. S., Y. SU, K. P. CORCORAN, L. NIE, P. E. MAYRAND, L. B. HOFF, L. J. MCBRIDE, M. N. KRONICK, and S. R. DIEHL. 1992. Application of automated DNA sizing technology for genotyping microsatellite loci. Genomics 14:1026-1031.

RECENT ADVANCES IN SCHEDULING STRATEGIES AND PRACTICAL TECHNIQUES IN CRANE ARTIFICIAL INSEMINATION

SCOTT R. SWENGEL, International Crane Foundation, P.O. Box 447, Baraboo, WI 53913, USA
MONICA L. TUITE,[1] International Crane Foundation, P.O. Box 447, Baraboo, WI 53913, USA

Abstract: We analyzed fertility patterns of 339 eggs laid 1985–93 at the International Crane Foundation by cranes whose sole means of fertility was artificial insemination (AI). Ten of 15 crane species were studied. Fertility rates increased significantly ($P < 0.05$) with (1) higher sperm density, (2) greater sperm motility, and (3) 2 vs. 1 vs. 0 inseminations 4–7 days before oviposition. Semen sample size was inversely related to sperm density and had no strong relationship with egg fertility. Inseminations less than 2.5–3.0 days before an egg is laid are probably too late to contribute to fertilizing the egg but may contribute substantially to fertilizing the next egg. We also describe practical knowledge gained during the past 10 years that apply to certain species or situations. Wattled cranes (*Bugeranus carunculatus*) have a long fertile period post-AI (up to 16 days), but may require more careful methods and special AI scheduling strategies. We also describe techniques for AI just after oviposition that are especially beneficial for Siberian cranes (*Grus leucogeranus*), and methods for maximizing sample delivery to the female.

PROC. NORTH AM. CRANE WORKSHOP 7:46-55

Key words: artificial insemination, *Bugeranus*, captive breeding, cranes, Gruidae, *Grus*, sperm, crane propagation, egg fertility.

At a captive breeding center, the managers' skill in prioritizing resources to maintain the health and propagation of the species is crucial. Artificial insemination (AI) is 1 means of species propagation that the International Crane Foundation (ICF) in Baraboo, Wisconsin, strategically utilizes. Natural copulation is preferred, but AI is necessary for the propagation of some cranes at ICF. Various physical disabilities, behavioral problems, and desired genetic outcomes, as described by Gee and Temple (1978), Gee and Sexton (1980), Russman (1980), and Gee (1983), necessitate this procedure.

Artificial insemination is risky for both staff and cranes. Errors in technique and timing can result in injuries or infertility. Although AI, with proper conditioning and management of cranes, often produces higher fertility rates than natural matings (Gee and Mirande 1996), it can be detrimental to breeding programs without adequate knowledge and preparation. For example, a few of our whooping cranes delay the onset of egg laying or stop laying eggs if we inseminate them more than twice per week. Our Siberian cranes do not behave in this manner, and to inseminate them less frequently than 3 times per week would result in lower egg fertility. Sources describing AI procedures include Archibald (1974), Gee and Temple (1978), Russman (1980), and Gee and Mirande (1996). This study focuses on when to inseminate females to maximize egg fertility and on the effects of semen variables on fertility rates. Russman (1987) determined that the optimal time to inseminate is 2–4 days before egg laying and just after oviposition. An ultrasonographic study of egg formation in cranes (*Grus* spp.) found that approximately 56–60+ hours elapse from ovulation to oviposition (Putnam 1983). Putnam estimated that the last chance to fertilize crane eggs is approximately 2.5 days before oviposition, and recommended that 1 insemination be scheduled 2.5 days before the earliest expected date of the female's next egg, based on past laying history.

We analyzed the effects of the following variables on crane egg fertility rates: (1) sperm density, (2) sperm motility, (3) semen volume, and (4) timing of insemination(s). This information might help managers improve their AI scheduling. Practical knowledge acquired by the ICF staff regarding AI technique will also be discussed, with the hope that other crane breeders may apply this knowledge.

We thank the ICF staff who performed the AI, recorded data, and otherwise helped with the AI program. G. F. Gee, C. M. Mirande, G. H. Olsen, D. W. Stahlecker, and an anonymous reviewer made suggestions that improved the manuscript. J. S. Hatfield provided helpful statistical advice.

STUDY AREA AND METHODS

AI and egg records from 1985 to 1993 of captive cranes at ICF in southern Wisconsin (43°29'–43°32'N, 89°45'W) were incorporated into a data base. Inseminations that contained sperm and occurred 2–38 days prior to oviposition were considered "viable" and included in this study. We analyzed fertility effects of the last 2 inseminations only; including more inseminations exponentially increases the complexity of analysis, for no or questionable gain. Only cases in which AI was the sole means of egg fertilization were included; eggs whose fertility could not be determined

---

[1]Present address: 7733 Hillcrest Ave., Middleton, WI 53562, USA.

(11/350, or 3.1% of eggs) were excluded.

Variables included in the data base were (1) egg sequence number, (2) year laid, (3) number of days between the last insemination and oviposition, (4) number of days between the second to last insemination and oviposition, (5) number of hours between the last inseminations and ovipositions which occurred 2–3 days later, (6) number of hours between the next to last inseminations and ovipositions which occurred 2–3 days later, (7) semen volume, (8) sperm density of semen sample, (9) sperm motility of sample, and (10) egg fertility. Ten of the world's 15 crane species were studied (Table 1).

SYSTAT software (version 4.0, 1988) was used for statistical analysis. We compared mean fertility rates (±1 SD) among different groups of semen sample volume, sperm density, and motility, and for different times of insemination vs. oviposition date. We tested for significant differences in fertility rates ($P < 0.05$) using multiple linear regression, Student's *t*-test, Pearson's product-moment correlation matrix (Pearson correlation), Spearman rank correlation, and Chi-square goodness-of-fit tests. Because Spearman and Pearson tests yielded the same result in all but 2 analyses, we present only the more conservative Spearman correlation results except the times where the 2 disagreed.

Sperm Density

Sperm "quality" or density is a letter-graded scale (Russman 1980) in which "A" indicates the highest sperm density in a semen sample observed in the microscope field at 400×. "D" represents the lowest density, and "F" represents no sperm. We scored "A" as 4, "B" as 3, "C" as 2, and "D" as 1. We excluded all "F" samples.

Percentages of fertile eggs were calculated for each grade category (1–4) of sperm density. We did this separately for the last, and next to last, insemination prior to oviposition. *T*-tests were used to test for significant differences in fertility. Percent of fertile eggs was then calculated for combinations of sperm densities present in the last 2 inseminations prior to oviposition. We also analyzed the effect of 0 vs. 1 vs. 2 "A" or "B" graded samples in the last 2 inseminations by using Chi-square goodness-of-fit tests. A Spearman correlation determined the relationship between sperm density and resulting fertility.

Sperm Motility

Motility is the percentage of sperm exhibiting forward movement in a semen sample. Motility is rated from 0 to 95, with 0 meaning no sperm were moving forward. A Spearman correlation analyzed the relationship between sperm motility and resulting fertility. Percent of fertile eggs was calculated for the following scenarios: (1) sperm in both semen samples from the last 2 inseminations prior to oviposition were >50% motile, (2) 1 of the last 2 samples had >50% motility, (3) neither of the samples had motility >50%. We compared fertility rates of these groups by using *t*-tests.

Table 1. Eggs analyzed and fertility rates for 10 crane species (*n* = 339) at the International Crane Foundation, 1985–93.

Species	No. of eggs	% of total	% fertile
Grus leucogeranus (Siberian)	84	24.8	65.5
G. canadensis (sandhill)	61	18.0	77.7
G. japonensis (red-crowned)	54	15.9	38.9
G. antigone (sarus)	42	12.4	31.0
G. vipio (white-naped)	32	9.4	28.1
G. americana (whooping)	23	6.8	65.2
Bugeranus carunculatus (wattled)	13	3.8	30.8
G. monacha (hooded)	12	3.5	66.7
G. rubicunda (brolga)	12	3.5	50.0
Anthropoides paradisea (blue)	6	1.8	100.0

Semen Volume

Semen sample volumes are recorded at ICF on both male and female AI records. Semen volumes ranged from <0.01 ml to 1.1 ml. Those recorded as <0.01 ml were entered in the database as 0.005 ml. Some semen samples were split between 2 females. Before 1989 ICF rarely split samples between 2 females if the sample was ≤0.05 ml. Since then we have increasingly split samples <0.05 ml in AI on Siberian cranes. Semen volumes entered in the database were the amounts inseminated into the females.

Spearman and Pearson correlations were performed to determine the relationship between sample volume and resulting fertility. Pairwise comparisons of fertility rates (using *t*-tests) were performed between cases in which both semen volumes of the last 2 inseminations prior to oviposition were small (≤0.01 ml), when both semen volumes were medium-sized (>0.01 and <0.10 ml), and when both were large (≥0.10 ml). The same was done for cases in which neither of the last 2 inseminations contained >0.025 ml of semen compared to cases in which both samples were at least 0.025 ml.

We used 2 types of analyses to assess the relationship of sperm density to sample volume. First, we used a *t*-test to compare mean sperm density of small (≤0.01 ml), medium (>0.01 ml and <0.10 ml), and large (≥0.10 ml) semen samples. This was done separately for the last and for the penultimate insemination prior to oviposition. Second, we

Table 2. Fertility (±SD) rate vs. sperm density of the last ($n=334$) and second to last ($n=292$) inseminations before oviposition for cranes at the International Crane Foundation, 1985-93.

	Last insemination		Second to last insemination	
Density	n	Fertility rate[a]	n	Fertility rate[a]
A	79	0.722 ± 0.451	66	0.758 ± 0.432
B	133	0.590 ± 0.494	118	0.602 ± 0.492
C	93	0.380 ± 0.487	88	0.420 ± 0.496
D	29	0.410 ± 0.501[b]	20	0.350 ± 0.489[c]

[a] Last insemination fertility rates were only statistically compared with one another, as were second to last insemination rates.

[b] Not significantly different ($P < 0.05$) from B or C; all other densities were significantly different from those below them.

[c] Not significantly different from C; all other densities were significantly different from those below them.

used Spearman and Pearson correlations to test whether sample volume and density covaried with each other for each of the last and next to last inseminations.

Semen Sample: Multivariate Analysis

A multiple regression (MGLH model) analyzed fertility effects of the last insemination before oviposition; egg fertility was the dependent variable, and sperm density, motility, and semen volume were independent variables.

Timing of Inseminations

In an attempt to minimize disturbance during the breeding season, ICF staff examines the birds once in the morning during servicing, and once in afternoon. Staff is not generally present at the time of oviposition. Therefore, to calculate the estimated number of hours between inseminations, and those ovipositions which occurred 2-3 days later, we estimated ranges of times for these ovipositions. This could help pinpoint the moment when it becomes too late to fertilize the next egg.

Percent of fertile eggs was calculated for each day (2 to >15) prior to oviposition in which the female was last inseminated. To search for combinations of last 2 insemination times that resulted in the highest fertility rates, we examined multiple scenarios. Egg fertility was compared for combinations in which the number of days between the last 2 inseminations was 1-2 vs. >2 days apart. Several more combinations which, in general, included both the greatest number of cases and the highest percentage of fertility were analyzed. We compared cases in which both inseminations occurred either 2-3 days, 4-7 days, or 8+ days prior to oviposition with a *t*-test. We used Chi-square tests to compare fertility rates when 0 vs. 1 vs. 2 inseminations were done 4-7 days prior to oviposition.

RESULTS

We analyzed a total of 339 eggs. Overall egg fertility was 54.3% (0.543 ± 0.499) (Table 1).

Sperm Density

"A" samples in the last 2 inseminations resulted in significantly higher egg fertility than "B", "C", or "D" samples (Table 2). Egg fertility of "B" samples was significantly higher than that of "C" samples. No significant fertility difference existed between "B" and "D", or between "C" and "D", samples.

Chi-square tests detected significantly higher fertility rates for cases in which both of the last 2 inseminations prior to oviposition contained "A" or "B" sperm density than when only 1 of the last 2 samples ($P < 0.001$), or neither of the 2 last insemination samples ($P < 0.001$), consisted of "A" or "B" density (Table 3). There was no significant difference between instances in which 1 of the 2 last insemination samples consisted of "A" or "B" density and cases in which neither of the samples contained "A" or "B" sperm density ($P > 0.1$).

Spearman rank correlation showed a significant positive relationship between egg fertility and the sperm density of both the last ($R = 0.257$, $n = 334$, $P < 0.001$) and second to last ($R = 0.270$, $n = 292$, $P < 0.001$) inseminations.

Sperm Motility

Spearman rank correlation showed a significant positive relationship between egg fertility and the sperm motility of both the last ($R = 0.263$, $n = 332$, $P < 0.001$) and second to last ($R = 0.381$, $n = 296$, $P < 0.001$) inseminations. The

Table 3. Fertility of eggs (±SD) at the International Crane Foundation, 1985-93, when both, 1, or neither of the last 2 inseminations had high (A or B) sperm density.

Last insemination	Second to last insemination	n	Fertility rate
A or B	A or B	146	0.719 ± 0.459
A or B	C or D	44	0.500 ± 0.506
C or D	A or B	41	0.488 ± 0.506
C or D	C or D	63	0.349 ± 0.481

more samples with high motility the female received, the higher the egg fertility. Significantly higher egg fertility was found for 2 vs. 1, 2 vs. 0, and 1 vs. 0 semen samples containing sperm motility >50% (Table 4).

Semen Volume

Spearman and Pearson correlations detected an inverse relationship between sample volume and egg fertility. This pattern was significant ($P < 0.05$) for the second to last insemination. For the last insemination, Spearman correlation revealed a marginally significant result ($R = -0.106$, $n = 339$, $P = 0.05$), whereas Pearson correlation showed no significant result.

Small- and medium-sized semen samples produced higher fertility than large samples. The fertility rate for eggs in which both of the last 2 semen volumes inseminated contained ≤0.01 ml (small volume) was 66.7% ($n = 27$); the fertility rate when both samples were >0.01 ml and <0.1 ml (medium volume) was 62.7% ($n = 133$); and the fertility rate when each of the last 2 samples inseminated contained ≥0.10 ml (large volume) was 40% ($n = 35$). Medium-sized samples produced significantly higher fertility than large samples (*t*-test, df = 166, $P = 0.017$) and small samples had nearly significantly higher fertility than large samples (*t*-test, df = 54, $P = 0.055$). In a separate test of smaller vs. larger semen volumes, there was no significant fertility difference (*t*-test, df = 202, $P = 0.234$) between cases in which neither of the last 2 sample volumes was >0.025 ml (66.7% fertility, $n = 39$) and cases in which both samples were ≥0.025 ml (55.9% fertility, $n = 152$).

The mean sperm density of the last insemination did not vary significantly among the 3 semen volume groups ≤0.01 ml, >0.01 and <0.10 ml, and ≥0.10 ml (Table 5). In the second to last insemination, however, medium volume semen samples >0.01 ml and <0.1 ml had significantly higher sperm density than either small or large semen samples (Table 5).

Table 4. Fertility of eggs (± SD) at the International Crane Foundation, 1985–93, when both, 1, or neither of the last 2 inseminations had relatively high (>50%) sperm motility.

>50% motility	*n*[a]	Fertility rate[b]
2	60	0.800 ± 0.403
1	89	0.629 ± 0.486
0	190	0.421 ± 0.495

[a] Includes some cases in which only 1 motility was recorded in the 2 inseminations, or where there was only 1 insemination, not 2.

[b] Fertility rates all significantly different from each other ($P < 0.05$).

Table 5. Sperm density (± SD) of small (≤0.01 ml), medium (>0.01 and <0.10 ml), and large (≥0.10 ml) crane semen volumes at the International Crane Foundation, 1985–93, in instances where both of the last 2 inseminations were small (n=21), both were medium (n=118), or both were large (n=35). Density is measured on a scale of 1 (low) to 4 (high).

Semen volume (ml)[a]	Last insemination		Second to last insemination	
	n	Sperm density	*n*	Sperm density
Small	21	2.476 ± 1.209	21	2.500 ± 1.140
Medium	116	2.879 ± 0.876	118	2.958 ± 0.891
Large	35	2.886 ± 0.631	35	2.571 ± 0.655

[a] Small (≤0.01), medium (>0.01 and <0.10), and large (≥0.10).

The negative correlation between semen volume and egg fertility is consistent with a slightly negative relationship between semen volume and sperm density: this pattern was statistically significant for both the last 2 inseminations in the Pearson correlation ($P < 0.05$), but was not significant in the Spearman correlations.

Semen Sample: Multivariate Analysis

Initially, the timing variable in which the last insemination occurred >3 days prior to oviposition was included in the multiple regression along with the 3 semen variables. The timing variable had a strong non-linear relationship with fertility and was thus excluded when the final multiple regressions were run. Among semen variables, sperm density ($R = 0.168$, $t = 2.835$, $P = 0.005$) and motility ($R = 0.187$, $t = 3.150$, $P = 0.002$) were nearly equally important to egg fertility. Semen volume was not significantly related to egg fertility ($R = 0.025$, $t = 0.466$, $P = 0.641$). The ANOVA statistics were $n = 332$, df = 3,328, $F = 10.854$, and $P < 0.001$.

Timing of Inseminations

Attempts to determine the critical point (the number of hours) after which the egg cannot be fertilized by an insemination done 2–3 days before egg laying were of limited success due to the small sample of cases in which there was no prior insemination that could have contributed to egg fertility. Our limited results suggest a threshold near 2.5–3 days pre-oviposition after which AI is unlikely to fertilize the egg. There were 5 cases in which a lone insemination was done between 30.75 and 54.75 hours (approximately 2 days) prior to oviposition. All resulted in infertility. Sample sperm densities included 2 "A", 2 "B", and 1 "C". Likewise, none

Table 6. Fertility rate (±SD) of eggs at the International Crane Foundation, 1985–93, vs. time from last insemination to oviposition ($n = 336$).

Days since last insemination	n	Fertility rate
>15	20	0.050 ± 0.224
15	4	0.000 ± 0.000
14	1	0
13	1	1
12	3	0.333 ± 0.577
11	2	0.500 ± 0.707
10	1	0
9	9	0.333 ± 0.500
8	11	0.545 ± 0.522
7	8	0.750 ± 0.463
6	13	0.692 ± 0.480
5	24	0.500 ± 0.511
4	48	0.750 ± 0.438
3	102	0.578 ± 0.496
2	89	0.528 ± 0.502

of the 8 eggs in which a lone insemination was done approximately 3 days (49.25–72.75 hours) prior to oviposition was fertile. Sample sperm densities included 2 "A", 2 "B", 3 "C", and 1 "D".

For fertile cases in which there were 2 inseminations done 31.5–81.5 hours prior to oviposition, a third insemination existed (which had been done 5–11 days prior to oviposition) in 6 of these 7 fertile cases. Sperm densities of the 2–3 inseminations included 4 "A", 14 "B", and 2 "C". In those infertile cases in which there were 2 inseminations done between 29.75 and 78.75 hours prior to oviposition, a third insemination existed (which had been done 5–11 days prior to oviposition) in 7 of these 12 infertile cases. Sperm densities of the 2–3 inseminations included 1 "A", 8 "B", 18 "C", and 4 "D".

The smallest time period between the last insemination and oviposition and the next to last insemination and oviposition in which a fertile egg was produced, was approximately 2 and 3 days, respectively. Such cases, where both inseminations were ≤3 days before a fertile egg was laid, would usually have had a third prior insemination that could have fertilized the egg. Our study examined the contributions of a maximum of 2 inseminations. The longest time period in this study between the last 2 inseminations and oviposition of a fertile egg, was 16 and 18 days, respectively (wattled crane).

Tables 6 and 7 provide the groundwork for further analysis of insemination timing vs. fertility rates. For the last insemination, the highest percentage of fertility (where $n > 1$ egg) was 75% and occurred when the female was last inseminated 4 ($n = 48$) and 7 ($n = 8$) days prior to oviposition (Table 6). In cases in which the last 2 inseminations occurred within 2 days of one another, 85.7% of 21 cases resulted in fertility when females were inseminated 5–6 days, and again 4 days, prior to oviposition (Table 7). When the last 2 inseminations occurred >2 days apart from one another, 71% of 21 cases resulted in fertility when females were inseminated >6 days, and again 4 days prior to oviposition (Table 7).

T-tests that compared differences in timing and resultant egg fertility showed no significant difference between cranes inseminated last 2–3 days vs. 4–7 days prior to oviposition ($P = 0.064$). There was significantly higher egg fertility, however, in those cranes inseminated 4–7 days vs. those inseminated 8+ days prior to oviposition, when only the last insemination was counted (Table 8).

When both of the last inseminations occurred 2–3 days, 4–7 days, or ≥8 days prior to oviposition, resultant egg fertility was significantly higher in those cranes inseminated 4–7 days prior to oviposition than 2–3 or ≥8 days before egg laying (Table 9).

Chi-square analysis further confirmed the fertilizing power of insemination(s) done 4–7 days prior to oviposition. Results show a significant increase in egg fertility with 2 inseminations in this time period (84.6% fertile, $n = 39$) compared with 1 (61.2%, $n = 178$) ($\chi^2 = 7.62$, df = 1, $P < 0.01$) or 0 (34.0%, $n = 97$) ($\chi^2 = 25.833$, df = 1, $P < 0.001$) inseminations. Similarly, 1 insemination done 4–7 days prior to oviposition yielded significantly higher fertility rates than if neither insemination occurred within this time period ($\chi^2 = 15.515$, df = 1, $P < 0.001$).

DISCUSSION

Semen Sample Characteristics

Our study found a strong relationship between sperm density and egg fertility. This is consistent with the findings of Gee and Temple (1978), who reported that a high percentage of living, active spermatozoa, and a high sperm concentration give excellent fertility rates after insemination. As sperm density improved, fertility improved.

Sperm density and motility had significantly greater effects on egg fertility than semen volume. Although one would expect larger semen volumes to contain more sperm, thus yielding higher egg fertility rates, our data do not support this. Although Hereford (1987) found a significant correlation between mean semen volume and "A" density of samples, our results (Pearson and Spearman rank correlations) indicate that as volume increases, sperm density decreases. These conflicting findings could be caused by

Table 7. Egg fertility rate (±SD) at the International Crane Foundation, 1985–93, vs. days to oviposition when the last 2 inseminations occurred within a 1–2-day period ($n = 132$) and when they occurred >2 days apart ($n = 144$).

	Last 2 inseminations 1–2 days apart			Last 2 inseminations >2 days apart		
Last insemination	Second to last insemination	*n*	Fertility rate	Second to last insemination	*n*	Fertility rate
9 days	10–11 days	4	0.500 ± 0.577	>11 days	5	0.200 ± 0.447
8 days	9–10 days	2	0.500 ± 0.707	>10 days	6	0.500 ± 0.548
7 days	8–9 days	3	0.667 ± 0.577	> 9 days	1	1
6 days	7–8 days	4	0.750 ± 0.500	> 8 days	7	0.714 ± 0.488
5 days	6–7 days	8	0.625 ± 0.518	> 7 days	13	0.385 ± 0.500
4 days	5–6 days	21	0.857 ± 0.359	> 6 days	21	0.714 ± 0.463
3 days	4–5 days	51	0.627 ± 0.488	> 5 days	45	0.600 ± 0.495
2 days	3–4 days	39	0.561 ± 0.502	> 4 days	46	0.540 ± 0.504

differences in the males used in the 2 studies, as well as the differences between our focus on the full range of sperm densities, and Hereford's (1987) on statistical patterns of "A" density samples. Our result may be explained, in part, by the presence of lymph, urates, or other contaminants, which are harmful to sperm (Smyth 1968, Lake 1971, Gee and Temple 1978) and more commonly found in larger sample volumes. We therefore suggest that once a sample as small as ≤ 0.01 ml is obtained from males that tend to give contaminated samples, attempts to obtain additional semen may be detrimental to the fertilizing capacity of the sample. To guard against this, use a second receptacle to attempt additional collection from a male if the original sample is at risk of further contamination. Furthermore, samples ≤ 0.05 ml may have sufficient sperm density to split in half and inseminate into 2 females. ICF has used this strategy successfully with Siberian cranes when female egg layers outnumber semen donors. Tiny semen samples (<0.01 ml) extended to 10–20× their volume by using semen extender had equal or greater fertilizing power as large (0.1 ml or larger) samples that were minimally extended. This suggests that fertility is not reduced when samples are extended to many times their volume.

Timing of Inseminations

Our limited results on lone inseminations close to the ovulation boundary of 2.5 days suggest that 1 or 2 inseminations less than 3 days before the egg is laid have only a small chance of contributing to the egg's fertility. Most fertile cases in which insemination(s) were done 2–3 days prior to oviposition would have had a second or third prior insemination that could have fertilized the egg.

Gee and Temple (1978) found when cranes were inseminated a few hours after oviposition, the fertility of subsequent eggs was greater than when birds were inseminated 2–3 days prior to oviposition. Likewise, we describe lower fertility rates when insemination(s) occurred 2–3 and 8+ days prior to oviposition. Our results indicate the optimal time to inseminate in order to fertilize the next egg is 4–7 days prior to oviposition, with 2 inseminations resulting in higher fertility than 1. In a similar finding, Gee and Temple (1978) found as the frequency of inseminations increased from 0 to 4 times in a 10-day period, the percentage of fertility increased as well. We believe it is very important to tailor insemination times to individual birds. It is helpful, however, to have a weekly schedule plan for your flock. Historically, semen collection and deposition was done at ICF and at Patuxent Wildlife Research Center every third day (Archibald 1974). Archibald and Viess (1979) subsequently reported an increase in frequency of inseminations to 3 times weekly at ICF. Gee (1983) also increased the recommended insemination frequency to 3 times per week to achieve the best fertility rate. In our study, 2 inseminations 4–7 days prior to oviposition translates into a schedule of 3 times per week for most crane species, with some adjustments for when cranes are expected to lay their next egg.

Although sperm host glands were discovered in birds 50 years ago (Van Drimmelen 1946) and have been confirmed in several species (e.g., Bakst and Bird 1987, Birkhead and

Table 8. Comparison of egg fertility rates (±SD) at the International Crane Foundation, 1985–93, during 3 time periods for the last insemination prior to oviposition.

Days before oviposition	*n*	Fertility rate
2–3	195	0.559 ± 0.498
4–7	92	0.674 ± 0.471
8–15	52	0.250 ± 0.437[a]

[a] Significantly different ($P < 0.05$) from other rates.

Table 9. Comparison of egg fertility rates (±SD) at the International Crane Foundation, 1985–93, when both of the last 2 inseminations occurred within selected time periods prior to oviposition.

Days before oviposition	*n*	Fertility rate
2–3	22	0.364 ± 0.492
4–7	39	0.846 ± 0.366[a]
8–15	35	0.229 ± 0.426

[a] Significantly different ($P < 0.05$) from other rates.

Hunter 1990), they have not yet been found in cranes (Russman 1987). But the lengthy fertile periods we observed (up to 16 days post-insemination) demonstrate sperm storage is occurring in cranes. Most females with the longest periods between AI and a fertile egg were inseminated infrequently. If they were inseminated frequently with high quality semen to fill up the putative sperm host glands, then left without any AI for some time, higher fertility rates ≥8 days post-insemination than we found might be detected (G. Gee, pers. commun.).

We summarize our analysis by stating that each bird is unique. While this is obvious, keep this in mind when using these study results to plan strategies. Our results may be skewed for several reasons: (1) only 10 of the 15 species were studied, (2) study species are not equally represented and therefore results are weighted more by species in which egg numbers are higher, (3) subjective variations between record keepers regarding sperm density and motility are likely, and (4) many more variables other than those analyzed in this study, such as weather, individual crane maturity, and changing pen arrangements, probably affected fertility rates.

Special AI Strategies and Techniques

Meticulous record keeping is crucial in enabling the crane propagator to detect quality patterns of semen samples obtained from the males, as well as egg laying patterns in females. Without such records it becomes much more difficult to strategically plan and execute artificial insemination effectively. Essential data on semen collections include date, time, temperature, sky conditions, person stroking crane, person teasing (operator) cloaca, person collecting semen, response of crane to stroking (struggled, relaxed, relaxed and raised tail, or purred), semen volume, apparent degree of semen contamination, amount of extension with semen extender, relative semen density, percent sperm motility, comments on any unusual crane behavior post-collection, and female inseminated. For females being inseminated, the same data should be collected, plus whether the insemination was placed in the oviduct or cloacally, the name of the male donor, dates of all subsequent eggs, and egg fertility. Optional data to collect include absolute sperm density and live/dead sperm ratios (both collected in lab after AI), distance between the points of the female's pubic bone (used to estimate when female will lay eggs), and appearance of blood spots on the cloaca (may indicate ovulation has occurred or will soon occur). Pubic bone spread is valuable in timing AI (or terminating/delaying it when it is not needed) if normals for the female have been previously established.

Wattled Cranes.—The following pertains especially to wild-caught or parent-reared wattled cranes, but the recommendations with regard to capture apply to any wild-caught crane. Less restrictive strategies may be used for hand-reared cranes.

1. Quick capture, by using 3 people, is essential. If the male becomes upset during a poorly executed capture, he is much less likely to provide a sample of acceptable quality. One month before egg-laying begins, prepare a capture area in the pen corner with the most acute fence angle. Visual barriers like reed mats or tennis wind netting (J. A. Cissel Co., P.O. Box 2025, Lakewood, NJ 08701, USA) attached to the fence, plus a soft material added to it, aids in safe captures. Used Christmas trees (with decorations removed) tied to the fence from the corner to 3 m out on each fence line work well. Dense trees retaining many needles are best.
2. Use the same team in their respective roles, once a semen sample has been successfully obtained by a particular team. Team choice can be based on an examination of past records. Vocalizations simulating crane copulatory calls may be executed by the "stroker" (handler) to enhance stimulation of the male. Because most of the sample will emerge when the "teaser" (operator) squeezes the cloaca the first time, the collection begins then. It may be necessary for the teaser to squeeze the everted cloaca twice before attempted collection in order to obtain very small samples (<0.02 ml). All team members should learn to perform each others' roles whenever possible, to increase chances for success in the event of a particular team member's absence.
3. Several of our wattled crane semen samples have been very small (0.005–0.01 ml) and contained acceptable sperm concentration and motility. Thus, no matter how small or discolored the sample appears, dilute the semen with semen extender and inseminate it immediately into the desired female.
4. Make a concerted effort to create a sense of security for

your pairs, especially for females during the breeding season. "Happy cranes lay eggs." We experimented during winter/spring 1994–95 by giving 1 particularly difficult pair daily rewards (approximately 5 newborn mice or 5 8-cm smelt, and a small handful of corn) several days before and after inseminations and before anticipated oviposition. Although we have no scientific evidence that this contributes to higher fertility rates, we believe it increased the chances for optimal insemination scheduling. The first season we tried this, the female appeared to lay eggs more reliably during the desired 15-day window after AI; she also laid 11 eggs that year instead of the usual 7. The additional protein provided by the smelt and newborn mice could increase the female crane's protein reserves for forming eggs. Furthermore, these tactics seem to calm the cranes and increase their comfort with our presence. We have observed these wild-caught cranes approach within 1.5–3 m of us (whereas previously they had fled) and dance upon provision of rewards. This positive association with humans also helps offset the presumably negative experience of being handled for AI.

5. Timing of insemination is difficult with wattled cranes. Our pairs usually lay 1-egg clutches anywhere from 10 to 50 days apart. However, 1 pair lays most eggs 12–18 days apart, and certain eggs (i.e., the fifth and seventh eggs) are usually clutched with the prior eggs and laid 3–6 days later. We do not recommend performing AI 3 times per week, as we do with certain other species. Minimal disturbance with wild caught birds is preferable.

We had 9/24 (38%) fertile wattled eggs during 1992–95 and of those, successful inseminations or pairs of insemination days prior to oviposition were 5, 6, 6, 6 + 13, 9, 9 + 11, 12, 12, and 16 + 18. Unsuccessful combinations included 2 + 29, 4, 6, 8, 9 + 16, 11, 15, 17, 17 + 29, 18, 19, 19 + 33, 20, 27 + 29, and 38.

Inseminations 4–16 days before oviposition resulted in 60% (9/15) fertility, while all 9 inseminations >16 days beforehand resulted in infertility. We aim for inseminating the female 6–8 days after oviposition so that most eggs will be laid 4–12 days later. Six of the 9 fertile eggs appear to have been fertilized by just 1 sample, because the next to last insemination in 6 cases was >38 days before oviposition, more than twice our longest confirmed fertile period. The other 3 fertile eggs may have been helped by their second to last inseminations 9–18 days pre-oviposition.

In our experience, however, wild-caught wattled cranes artificially inseminated more than once between 2 eggs cease laying eggs for 30–40 days and become unpredictable in their egg-laying pattern after that. Because predictability is the key to successful AI on these birds, we analyze the crane's past egg-laying record and try for 1 insemination between eggs. Sometimes the female clutches her eggs; when this happens we either have a chance at 2 fertile eggs or the second egg comes too late for fertility. It is best, however, to adhere to the most probable interval between eggs of 12–18 days for our pair.

Siberian Cranes.—Quick capture, once again, is vital. Use of 1 person (the assigned stroker) to capture the bird is recommended for birds who behave aggressively toward 1 individual but may be intimidated and become upset in the presence of >1 person. This type of crane usually advances toward the stroker to attack, at which time the stroker should capture the crane and immediately initiate stroking. Other persons must stay out of sight and remain quiet until the capture. After capture, the rest of the team should assume their respective roles and proceed.

Individual Siberian cranes consistently begin laying eggs the same time each year, so for most pairs we begin AI 1 week before that date to cover the first 2 eggs. Siberian cranes also have predictable laying patterns, particularly for the first 3–4 eggs of the season; 16/21 (76%) of the second eggs were laid 3 days after the first eggs. The other 5/21 were laid either 2 or 4 days after the first eggs. While this degree of predictability is good, the interval may frequently be too small to allow an insemination between the eggs to fertilize the next egg. Because it is necessary to inseminate at least 56–60+ hours prior to oviposition in order to fertilize the egg before shell membrane formation (Putnam 1983), it is essential to inseminate the female immediately after the first egg is laid (vs. found). On the anticipated lay date, we observe Siberian cranes more often during daylight hours to inseminate immediately after oviposition. Despite these efforts, it is still possible for the second egg to be infertile, particularly if the first egg is laid during the night, but not found until morning. A frequent consequence, however, of insemination just after the first egg is "found" is fertilization of the third egg. This is also a consequence of the short second-to-third egg laying interval in Siberian cranes.

Different females have individual patterns that differ from others of their species. Keeping a chart of the time intervals between eggs facilitates identification of patterns within the flock that provide the basis for selected insemination strategies (Table 10). For example, ICF routinely inseminates Siberian cranes 3 times per week in order to fertilize the next two eggs laid. Insemination times are tailored to individual birds, however, such that if a female were to lay on an alternate ("off") day, we inseminate her the same day, a few hours after she has laid.

AI After Oviposition.—There are techniques we use for cranes that exhibit one or both of the following behaviors,

Table 10. Examples of egg laying interval (in days) records kept for females at the International Crane Foundation. Grouping females by species (e.g., Siberian cranes here) assists in recognition of species-specific egg laying trends. In 1991, Ramsar's second egg was laid 3 days after the first egg, her third egg was laid 5 days after the second egg, etc. Ranjit's 12-day interval before her third egg in 1995 may have been caused by excessive AI.

Name (no.)	Year	Total no. of eggs	Season length	First lay date 1	Days between eggs 2	3	4	5	6	7	8	9	Last lay date
Ramsar	1991	3	9	2 May	3	5							10 May
6-14	1992	6	23	4 April	3	3	5	5	6				26 April
	1993	5	17	27 March	3	5	4	4					12 April
	1994	8	33	28 March	3	3	5	5	5	6			29 April
	1995	5	20	5 April	2	5	5	8					25 April
Ranjit	1992	6	21	10 April	3	3	4	5	5				30 April
6-22	1993	6	19	2 April	3	3	3	6	3				20 April
	1994	9	29	20 March	3	3	4	3	4	3	5	3	17 April
	1995	9	41	26 March	4	12	3	3	5	3	5	6	6 May

which frequently occur when we remove an egg from the cranes' nest (in an effort to keep the pair in egg production) and subsequently attempt to collect a semen sample to perform AI: (1) extreme aggression because we have taken the egg, and (2) significant fecal elimination during AI because cranes store up feces during incubation to avoid soiling the nest. These undesirable behaviors usually result in either a very poor sample or no sample at all. We recommend that personnel wait at least 3 hours after egg removal before collecting sperm and/or inseminating. This amount of time usually increases the possibility that birds will have "regrouped" and be more receptive to AI. This also enables incubating birds time to eliminate fecal matter and urates. Frequently, both birds have fecal retention. One crane may need to be restrained or locked indoors while the other is being handled for AI. Usually females become very aggressive when incubating. Take advantage of aggression to effect a quick capture.

One exception, in which AI should be done immediately, is when cranes will be keeping the first egg of the clutch and you hope to fertilize the next egg. The longer the cranes have the egg, the more difficult it will be to get a clean semen sample from the male. In Siberian cranes the chances are poor, but the best hope is to try immediately, reasoning that the egg may have just been laid and the male has not yet stored feces.

Delivering More of the Sample to the Female.—Because 0.04 ml of fluid remains in the tip of a tuberculin syringe (Russman 1980) after the plunger is pushed:

1. Extend all samples, no matter how small, to 0.10–0.15 ml by using semen extender. This ensures that at least 60–70% of the sample will be delivered, even if 0.04 ml remains in the syringe.
2. Position air behind the sample before inseminating, so that little of the sample remains after the plunger is pushed. To do this, pull the plunger back halfway with the syringe tip uncapped, place a cap on the syringe tip, and quickly flick the syringe forward so that the tip ends up down. Practice this with water in the syringe so that you can do it without making substantial semen go too far into the tip and adhere to the cap.
3. It is not necessary to save a semen sample before inseminating, because enough sample remains in the syringe for microscopic examination of quality even after the most successful insemination.

LITERATURE CITED

ARCHIBALD, G. W. 1974. Methods for breeding and rearing cranes in captivity. Int. Zoo Yearbook 14:147-155.

______, G. W., and D. L. VIESS. 1979. Captive propagation at the International Crane Foundation 1973–78. Pages 51-73 *in* J.C. Lewis, ed. Proc. 1978 crane workshop. Colorado State Univ. Printing Serv., Fort Collins.

BAKST, M. R., and D. M. BIRD. 1987. Localization of oviductal sperm-storage tubules in the American kestrel (*Falco sparverius*). Auk 104:321-324.

BIRKHEAD, T. R., and F. M. HUNTER. 1990. Numbers of sperm-storage tubules in the zebra finch (*Poephila guttata*) and Bengalese finch (*Lonchura striata*). Auk 107:193-197.

GEE, G. F. 1983. Crane reproductive physiology and conservation. Zoo Biol. 2:199-213.

______, and C. M. MIRANDE. 1996. Special techniques, Part A: crane artificial insemination. Pages 205-217 *in* D. H. Ellis, G. F. Gee, and C. M. Mirande, eds. Cranes: their biology, husbandry, and conservation. Natl. Biol. Serv., Washington, D.C., and Int. Crane Found., Baraboo, Wis.

______, and T. J. SEXTON. 1980. Artificial insemination and semen preservation of nondomestic birds. Annu. Proc. Am.

Assoc. Zoo Vet. 1980:53-58.

______, and S.A. TEMPLE. 1978. Artificial insemination for breeding non-domestic birds. Symp. Zool. Soc. London 43:51-72.

HEREFORD, S. G. 1987. Semen volumes and peak production periods for cranes at the International Crane Foundation. Pages 525-534 *in* G. W. Archibald and R. F. Pasquier, eds. Proc. 1983 int. crane workshop. Int. Crane Found., Baraboo, Wis.

LAKE, P. E. 1971. The male in reproduction. Pages 1411-1447 *in* D. J. Ball and B. M. Freeman, eds. Physiology and biochemistry of the domestic fowl. Vol. 3. Academic Press, New York, N.Y.

PUTNAM, M. S. 1983. An ultrasonographic study of oviducal egg formation in cranes: implication for artificial insemination programs. Unpubl. Rep., Int. Crane Found., Baraboo, Wis. 17pp.

RUSSMAN, S. R. 1980. Techniques in the artificial insemination of cranes. Unpubl. Rep., Int. Crane Found., Baraboo, Wis. 19pp.

______. 1987. Timing inseminations to maximize fertility in cranes. Pages 398-405 *in* J. C. Lewis, ed. Proc. 1985 crane workshop. Platte River Whooping Crane Maintenance Trust, Grand Island, Nebr.

SMYTH, J. R., JR. 1968. Poultry. Pages 258-300 *in* E. J. Perry, ed. The artificial insemination of farm animals. Rutgers Univ. Press, New Brunswick, N.J.

VAN DRIMMELEN, G. C. 1946. "Spermnests" in the oviduct of the domestic hen. J. S. African Vet. Med. Assoc. 17(1):42-52.

THE EFFECT OF DISTURBANCE ON THE REPRODUCTION AND MANAGEMENT OF CAPTIVE CRANES[1]

CLAIRE M. MIRANDE, International Crane Foundation, P.O. Box 447, Baraboo, WI 53913, USA
JAMES W. CARPENTER,[2] Patuxent Wildlife Research Center, 11510 American Holly Drive, Laurel, MD 20708, USA
ANN M. BURKE,[3] International Crane Foundation, P.O. Box 447, Baraboo, WI 53913, USA

Abstract: A retrospective analysis of egg laying histories and observations of crane pairs at the International Crane Foundation and the Patuxent Wildlife Research Center indicates that disturbance associated with captivity has adverse impacts on captive breeding. Females on public display laid significantly fewer eggs than those off display ($P < 0.05$). Crane pairs moved to other pens (other than adjacent pens) or between facilities laid slightly fewer eggs than those birds which were not moved, although the results were not statistically significant ($P = 0.188$). Pen design and construction also appeared to have adverse impacts on breeding. In addition, human activities, intraspecific interactions, and rearing methods influence a bird's response to disturbance, and, therefore, these factors must also be considered in an effective crane management program. Modifying procedures to minimize disturbance, timing necessary disturbances after the breeding and molt seasons, and carefully monitoring birds for signs of stress can result in increased reproductive potential for captive, endangered cranes.

PROC. NORTH AM. CRANE WORKSHOP 7:56-61

Key words: behavior, disturbance, reproduction, cranes, *Grus*.

From the inception of propagation programs at the International Crane Foundation (ICF) and the Patuxent Wildlife Research Center (Patuxent Center), captive breeding was accepted as a technique that could prevent extinction and contribute to the recovery of certain species by providing a controlled protective captive environment, fostering coordinated research, and providing stock for bolstering or reestablishing wild populations (Archibald 1974, Derrickson and Carpenter 1981). To develop propagation and reintroduction techniques for endangered cranes, ICF maintains 110 individuals of all 15 species of cranes with emphasis on the endangered species, including Siberian (*Grus leucogeranus*), red-crowned (*G. japonensis*), white-naped (*G. vipio*), hooded (*G. monachus*), wattled (*Bugeranus carunculatus*), and black-necked (*G. nigricollis*) cranes. The Patuxent Center maintains 275 individuals including whooping (*G. americana*), Mississippi sandhill (*G. canadensis pulla*), greater sandhill (*G. c. tabida*), and Florida sandhill (*G. c. pratensis*) cranes.

Maintaining birds in captivity for breeding purposes requires considerable knowledge of their reproductive biology and behavior. For example, it is known that the stimulus that initiates the breeding cycle in birds is a complex combination of innate physiological rhythms and environmental stimuli (Gee 1983). Most of these external stimuli, e.g., photoperiod, light intensity, temperature, rainfall, humidity, nutrition, presence of a mate, territory, and nesting material, have been identified and their influence varies among species. Less is known, however, about factors such as stress (i.e., disturbance) which may have an adverse influence on the breeding season.

Stress is the effect of physical, physiological, or emotional factors (stressors) that induce an alteration in an animal's homeostasis or adaptive state (Anonymous 1987). Stress responses often involve changes in the neuroendocrinologic function, autonomic nervous system, and mental state of the animal, as well as its behavior. If stressors are short-term, an animal's response usually will not result in long-term harmful effects. Prolonged or excessive distress may result in harmful responses, such as abnormal feeding and social behavior and inefficient reproduction.

However, the ability to correlate a stressor with a specific effect can be very difficult. Failure to establish an acceptable clinical measure of stress results from the (1) inability to determine which biological value is the best measure of stress, (2) lack of nonspecific stress response that characterizes all types of stressors, and (3) failure to establish a correlation between stress measures and a meaningful impact on the animal's well-being (Moberg 1987). In addition, finding a measure of that stressor is further complicated by inter-animal variability in stress response. The biological responses depend on a variety of factors, such as prior experience, genetics, age, or physiological state (Moberg 1987). The role that stress plays in preventing apparently healthy birds from breeding in captivity is poorly understood (Cooper 1978), although stress has been shown to affect reproductive success in birds by altering hormone

---

[1]Originally accepted for publication in but omitted from Proc. 1988 North Am. crane workshop.

[2]Present address: Department of Clinical Sciences, College of Veterinary Medicine, Kansas State University, Manhattan, KS 66506, USA.

[3]Present address: College of Natural Resources, University of Wisconsin-Stevens Point, Stevens Point, WI 54481, USA.

levels (Gee 1983, Stoskopf 1983).

The purpose of this paper is to retrospectively evaluate the effects of disturbance associated with public display, pen transfers, pen design, and construction on egg production in captive cranes. Based in part on these findings or observations, management recommendations are presented to reduce the impact of disturbance on crane breeding. Recommendations are also made for managing human activities, neighboring cranes, and rearing methods to increase crane reproduction in captivity.

We thank G. Pasdirtz (University of Wisconsin, Madison), R. Rogers (University of Wisconsin, Stevens Point), and C. Bunck (Patuxent Center) for their advice on statistical analyses. We also appreciate the efforts of G. Archibald, R. Gabel, G. Gee, D. Hoffman, S. Swengel, and M. Wellington in reviewing this manuscript.

METHODS

Public Display and Pen Transfers

Cranes at ICF are housed in 1 of 2 facilities. One facility is on public display, whereas the second facility is off display in an isolated area in which visitors are generally not permitted. Pen transfers include the movement of pairs within the same facility or between facilities; it does not include the movement of birds to adjacent pens for disease management.

To evaluate the effects of disturbance on reproduction, we examined the laying histories of cranes at ICF in relation to the effects of public display and pen transfers. Nine of 21 females were selected for study of their individual transfer history and egg laying history because they (1) had a history of pen transfers, (2) produced eggs for at least 2 seasons between 1974 and 1987, and (3) were managed for optimum reproduction. Data were analyzed starting with the female's first year of egg production and ending with her 1987 laying season, death, or shipment from ICF. Total number of eggs laid was selected as the most suitable measure of productivity.

Egg production during years in which pairs were on display was compared with production during years in which the same cranes were off display. Because pen transfers could confound the effects of public display, the data were analyzed separately for years in which the birds were moved or were not moved. A move year was one in which a pen transfer occurred between the date on which the last egg was laid in the previous season and the first egg of the following season. A non-move year was designated when there were no pen transfers during the selected time period. Years in which cranes were moved to adjacent pens were not included in the data set.

To evaluate the effects of pen transfers, we compared egg production during years in which pairs were moved with production during years in which cranes were not moved. The data were separated for years in which the birds were on display or were off display to correct for confounding effects of the cranes' display status. All analyses used the Wilcoxon signed rank test, a nonparametric test, with a 1-sided alternative (Hollander and Wolfe 1973). Significance levels of $P < 0.05$ were considered sufficient to reject statistical hypotheses.

Pen Design and Construction

By examining records at both ICF and the Patuxent Center, we identified a number of cases where disturbance associated with pen design or facility construction appeared to have an impact on crane breeding. Two examples of these disturbances were examined.

RESULTS

Public Display and Pen Transfers

Annual egg production at ICF in relation to public display and pen transfers is presented in Table 1. Six of the 9 females in the sample that were maintained both on and off display consistently produced more eggs while off display, independent of pen transfer effects. Females which were moved to other pens produced more eggs while off display than while on display ($P = 0.016$; see columns 1 and 3, Table 1). Females which were not moved also produced more eggs while off display than while on display ($P = 0.031$; see columns 2 and 4, Table 1).

Analysis of the data for birds off public display indicates that females not moved produced slightly more eggs than females which were moved, although the results were not statistically significant ($P = 0.188$; see columns 1 and 2, Table 1). The data were insufficient to compare egg production by cranes on public display during years in which they were moved and not moved (see columns 3 and 4, Table 1).

Pen design and Construction

Disturbance Associated with Pen Design.—At ICF, 4 pairs of cranes with prior laying histories were placed in a new exhibit area. The birds appeared healthy and displayed normal behaviors but did not lay eggs. The exhibit was equipped with a variety of features to promote breeding. such as visual barriers between pairs, photoperiod control, and sprinkler systems. By closely observing the birds and examining the pen design, human disturbance appeared to

Table 1. Effect of public display and pen transfers[a] on mean egg production (production years) of female cranes at the International Crane Foundation, 1974–87.

	Off display		On display	
Female	Bird moved	Bird not moved	Bird moved	Bird not moved
A	3.5 (2)	5.8 (5)		
B	15.0 (1)	12.7 (3)	0.0 (1)	0.0 (1)
C	2.0 (1)	10.0 (2)		
D	9.5 (2)	9.5 (2)	4.0 (1)	6.5 (2)
E	10.0 (1)	13.7 (7)	9.0 (1)	
F	2.0 (1)	2.0 (4)	0.3 (3)	0.0 (1)
G	2.0 (1)	2.5 (4)	0.0 (1)	0.0 (1)
H	5.0 (2)	3.0 (1)	0.0 (2)	0.0 (1)
I	0.0 (1)	1.2 (5)		
$\bar{x}$	5.6(12)	6.9(33)	1.5 (9)	2.2 (6)

[a] Includes transfer of birds within or between facilities; does not include production years in which birds were moved between adjacent pens.

cause the lack of reproduction. The individual enclosures were designed for public viewing from 1 end and access for bird care and maintenance from the other. Although the aviculturists only entered the pens briefly in the morning, the birds were effectively disturbed from both ends of their pen, causing disruption of territorial security. Consequently, the interior shelter was sealed off and the birds were fed and watered from the public entrance. This consolidated all disturbance at 1 location and provided the birds with a secure nesting site. As a result, 3 females which had been on display for several years and never laid began laying. Another female just reaching sexual maturity laid her first eggs under this management.

Disturbance Associated with Construction.—In October 1979, the Patuxent Center initiated the construction of a new perimeter fence around the crane breeding facility that required 4 to 5 workers and the use of heavy equipment. Because of inclement weather, the construction was intermittent and not completed until late winter. Construction activities occurred within 25 m of the whooping crane breeding pens. As an apparent response to the construction-related disturbance, a marked decrease in production occurred. Four whooping crane pairs produced 21 eggs in 1979, whereas only 2 whooping crane pairs produced 6 eggs in 1980.

In 1985, 24 chain-link breeding pens were constructed at the Patuxent Center to replace a pen complex built as a temporary structure in the early 1970's. Construction and associated disturbance at the new whooping crane breeding complex, timed to interfere as little as possible with either the 1985 or 1986 breeding season, was initiated at the end of the 1985 breeding season and was not completed until late October. In early July 1985, prior to the initiation of the construction, whooping crane pairs were moved approximately 50 m to a complex housing sandhill crane pairs. The initial phase of construction consisted of a team of 7 to 8 workers supported by heavy equipment. The cranes were moved to their new pen complex in early November 1985. However, minor disturbances, such as installation of visual barriers, filling in holes, and installation of gate guards, continued until early January. Although 3 pairs of whooping cranes bred in 1985, producing a total of 13 eggs, none of the pairs was productive in 1986. This decrease in productivity can probably be attributed to a combination of disturbance and the movement of the cranes to smaller pens. In 1987, the 3 previously productive pairs of whooping cranes and 2 new pairs came into production.

DISCUSSION

It is known that wild sandhill cranes will abandon their territories and fail to breed if they are repeatedly disturbed during the breeding season (Derrickson and Carpenter 1981). Similarly, frequent disturbance can significantly alter the breeding behavior of captive pairs and can ultimately result in reproductive failure (Derrickson and Carpenter 1981). Unless captive birds have adapted to consistently high levels of disturbance, it is important to prevent and reduce disturbance during the breeding season to enhance breeding success.

Although disturbance is not the only stress that may interfere with reproduction in cranes, disturbance does appear to have been instrumental in reducing the productivity of cranes at ICF and the Patuxent Center. Unfortunately, it is difficult to quantify the impact because other sources of stress and factors such as weather, fright, subclinical disease, lack of territorial integrity, and changes in animal care personnel also influence reproduction. The impact of more subtle disturbances on a bird's behavior and productivity is even more difficult to evaluate. However, based on an analysis of bird moves versus breeding results at ICF, and observations at both ICF and the Patuxent Center on the adverse effects that disturbance appears to have on egg production, it does appear that certain types or levels of disturbance can affect crane breeding. Therefore, the following management procedures are recommended to reduce disturbance to cranes.

Display

The evaluation of the data indicates that captive cranes

produce fewer eggs while on display. Therefore, it is desirable to maintain valuable breeding birds off display. Unfortunately, this option does not exist for many zoos involved in breeding programs. These zoos, therefore, should consider constructing off-exhibit enclosures for breeding. Cranes could then be rotated between areas based on breeding priorities. Subadults, nonbreeders, or well-represented pairs make up a significant portion of a captive population and are generally suitable for display.

If it is necessary to maintain breeding cranes on display, facilities and management practices should be designed to promote breeding. The presence of other species in the exhibit should be carefully considered. Species should be selected that are not aggressive to cranes, that solicit minimal aggression from cranes, and that are unlikely to disturb nesting sites.

Pen Transfers

Analysis of the egg production data suggests that moving birds to other pens, other than adjacent pens, or facilities has a slight negative effect on egg production. Although the data in this comparison are limited and not statistically significant, differences of even 1 to 2 eggs per year can be important in the preservation of genetic diversity in small captive populations. Therefore, efforts should be made to place pairs in suitable breeding pens and to minimize transfers as long as reproduction is desired. If a pair does not breed in a particular pen, the pair should be closely observed to determine if environmental or behavioral disturbances are affecting dominance or territorial security. The status of the pair bond should be evaluated by observations for proximity, synchronized maintenance activities, guard and unison calls, synchronized display, male defense of the female and territory, and aggression between pair members (Mirande and Archibald 1990). Attempts should be made to minimize disturbance. If efforts are unsuccessful, transfer to a different enclosure should be considered.

Although it is better to minimize disturbance by moving cranes within a complex, pen transfers to new areas will periodically be necessary. For example, old facilities may need to be replaced or birds may need to be moved because of pair incompatibility or to establish new pair bonds after the death of a mate.

Facility Design

Optimally, crane pairs should be housed in units, each containing 2 pens (each minimally 15 × 18 m) that are generally arranged in a linear series (Archibald 1974, Archibald and Veiss 1979, Carpenter 1986). Isolation for breeding pairs and disease management can be ensured or facilitated by allowing only 1 compartment to be occupied at a given time so that each pair is flanked by empty pens. In effect, this provides a psychological territory that is 3 times as large as the occupied pen. Ideally, buffer zones (approximately 9 m) should be placed between rows of pens.

Visual barriers should be provided along fences to define territorial borders and to provide cover for easily stressed birds (Mirande and Archibald 1990). The provision of small buildings (4.3 × 3-4 m) is desirable because birds can be locked in familiar enclosures during severe weather or in case of illness or injury. If the cranes are fed and watered from the shelters, the birds can be serviced with minimal disturbance.

Facilities should be designed to provide a secure, undisturbed section in each pen for nesting. Entry into these areas should be avoided. If birds must be handled, they should be herded and caught in other areas of the pen.

Construction

All unavoidable or planned disturbances, such as major construction projects, should take into account the ability of the crane pair to tolerate stress, their proximity to disturbance, and their physiological stage in the annual breeding cycle. Necessary disturbances should be timed to occur in the fall after parent-reared chicks are several months old and molt is completed. In addition,, these activities should be completed as far as possible in advance of the next breeding season.

Other Recommendations

In addition to the sources of disturbance previously discussed in this paper, other captive management activities should be evaluated to minimize disturbance effects on birds.

Human Activities.—Disturbance can be reduced by acclimating pairs to necessary activities before breeding commences to minimize potential stress effects. This can be accomplished largely by restricting access to breeding areas to essential personnel only and by completing all routine husbandry and propagation activities, i.e., nest checking, egg collecting, feeding, and artificial insemination, on a regular daily schedule (Derrickson and Carpenter 1981). Regular contact, however, by personnel who care for the birds (i.e., aviculturists) can be advantageous because contact is less stressful when it is necessary.

It is important to maintain permanent, experienced staff in captive breeding programs because cranes easily recognize individuals and adjust more easily to the presence of familiar people. In addition, aviculturists who are familiar with crane behavior and the likely reactions of individual birds to various stimuli or stressors are the most qualified to work with, and be a minimal disturbance to, the birds. The ability of aviculturists to identify sources of stress can influence management decisions designed to eliminate or reduce these sources.

Although procedures have been developed to minimize stress in birds, individuals vary in their response to human activities, and procedures may need to be modified for these birds. For example, artificial insemination procedures are generally initiated 2–4 weeks prior to expected egg laying to condition the birds to handling procedures. However, observations at the Patuxent Center on a compatible pair of whooping cranes during the 1981 breeding season revealed that normal pair activities were disrupted for several hours following routine capture and handling for artificial insemination (Derrickson and Carpenter 1987). Because both birds were extremely stressed following handling, all attempts at artificial insemination were halted. Although the pair resumed normal behavior and progressed to nest building, the female did not lay. The following year the pair was not captured or handled for artificial insemination early in the breeding season, and the problem was avoided. Stress-related behavior was not observed, and the birds came into breeding condition rapidly. Semen collection from the male and artificial insemination of the female were resumed after the laying of their first egg without apparent ill-effect, and the female subsequently produced an additional 4 fertile eggs (Derrickson and Carpenter 1987).

A similar pattern was observed in a pair of common cranes (*G. grus*) at ICF. Attempts at artificial insemination early in the breeding season resulted in extreme stress and the disruption of normal behaviors. After attempts were ceased, the birds gradually reestablished their dominance and laid. Artificial insemination was resumed after the laying of the first egg without any apparent adverse effects.

Intraspecific Interactions.—Crane pairs are highly territorial and are bred most successfully in isolated pens. Cranes in adjacent pens can inhibit breeding because pair members, especially males, often spend much of their time along bordering fences trying to drive off the "intruding" birds. This has several negative effects, including (1) involvement of the male in aggressive display to the degree that normal pair interactions are inhibited, (2) compromise of territorial security, and (3) injuries resulting in physiological stress. Therefore, it is important to provide buffer zones or visual barriers between aggressive birds. It is also important to observe pairs closely when new neighbors are introduced. Occasionally males are stimulated to such high levels of aggression that they displace their aggression and kill their mates (Mirande and Archibald 1990). Pairs should be watched very closely for an increase in submissive behavior by the female and members may need to be separated temporarily to avoid injury to the female.

Rearing Method.—The response of cranes to disturbance will vary greatly with their rearing history. Wild-caught cranes are the most sensitive to disturbance. Gradually conditioning these birds to the presence of people and encouragement of the expression of dominant behavior toward people are important to promotion of breeding and reducing the risk of injury. Aviculturists can minimize stress by assuming submissive posture around the birds, by avoiding direct eye contact, and by making soft sounds to let the birds know when they are approaching and where they are working. Daily offerings of small amounts of choice food items such as corn can also help to tame birds and reduce disturbance effects. Parent-reared birds are generally very wary and exhibit behavior similar to wild-caught cranes. There is a greater risk of these birds injuring themselves as a result of disturbance than with hand-reared birds. Conditioning these birds to the presence of humans is also very important. Hand-reared birds are generally very tolerant of people and, if not severely imprinted on humans, breed very well in captivity. These individuals are the least affected by handling and breed more readily on display than cranes reared by other methods.

Rearing method should be carefully selected based on the management goals for each individual crane. Cranes reared for captive management should either be hand-reared by methods to reduce imprinting or parent-reared but exposed to people to reduce inhibition of breeding from human-related stress later in life (Mirande and Archibald 1990).

CONCLUSIONS

A retrospective analysis of egg laying histories and observations of crane pairs at ICF and the Patuxent Center indicates that disturbance associated with captivity has an adverse impact on captive breeding. Females on public display laid significantly fewer eggs than those off display ($P < 0.05$), independent of crane moves. Crane pairs moved to other pens (other than adjacent pens) or between facilities laid slightly fewer eggs than those birds that were not moved, although the results were not statistically significant. Therefore, to promote reproduction and maximize the success of captive propagation programs, aviculturists should maintain valuable crane pairs off display whenever possible. Pairs on display should be managed to minimize disturbance. Until more data are obtained, pen transfers should be minimized or timed to reduce possible impact on breeding goals for genetic management, although some pen transfers may be beneficial if birds are moved to less disturbed pens.

Facilities should be designed to promote dominance and territorial security, provide a secure nesting area, and minimize aggression toward neighbors. Disturbance levels in the captive environment should be reduced as much as possible in advance of the breeding season (Cooper 1978, Carpenter and Gabel 1984, Derrickson and Carpenter 1987). Necessary disturbances, such as pen construction, should be timed shortly after the molt season or when parent-reared chicks are several months old to minimize disturbance effects. In addition, human activities, intraspecific interactions, and rearing methods influence a bird's response to disturbance, and, therefore, these factors must also be considered in an effective crane management program.

Captive pairs should be monitored throughout the breeding season for changes in behavior such as frequent pacing, excessive preening, decreased frequency of precopulatory behavior or unison calling, and egg breaking, which indicate stress. When such behavioral changes are noted, propagation procedures should be immediately re-evaluated and modified (Derrickson and Carpenter 1987).

LITERATURE CITED

ANONYMOUS. 1987. Panel report on the colloquium on recognition and alleviation of animal pain and distress. J. Am. Vet. Med. Assoc. 19:1186-1191.

ARCHIBALD. G. W. 1974. Methods for breeding and rearing cranes in captivity. Pages 147-155 *in* N. Duplaix-Hall, ed. International zoo yearbook. Zool. Soc. London, London, U.K.

______, and D. L. VEISS. 1979. Captive propagation at the International Crane Foundation 1973-78. Pages 51-73 *in* J. C. Lewis, ed. Proc. 1978 crane workshop. Colorado State Univ. Printing Serv., Fort Collins.

CARPENTER, J. W. 1986. Cranes (Order Gruiformes). Pages 315-326 *in* M. E. Fowler, ed. Zoo and wild animal medicine. Second ed. W. B. Saunders Co., Philadelphia, Pa.

______, and R. R. GABEL. 1984. Establishment and management of a breeding program for captive endangered birds. Proc. Annu. Conf. Assoc. Avian Vet. 1984:237-265.

COOPER, J. E. 1978. Veterinary aspects of captive birds of prey. Steadfast Press, Soul, Gloucestershire, U.K. 265pp.

DERRICKSON, S. R., and J. W. CARPENTER. 1981. Ethology and management of captive sandhill cranes. Annu. Proc. Am. Assoc. Zoo Vet. 1980:76-81.

______, and ______. 1987. Behavioral management of captive cranes-factors influencing propagation and reintroduction. Pages 493-511 *in* G. W. Archibald and R. F. Pasquier, eds. Proc. 1983 international crane workshop. Int. Crane Found., Baraboo, Wis.

GEE, G. F. 1983. Crane reproductive physiology and conservation. Zoo Biol. 2:199-213.

HOLLANDER, M., and D. A. WOLFE. 1973. Nonparametric statistical methods. John Wiley and Sons, New York, N.Y. 503pp.

MIRANDE, C. M., and G. ARCHIBALD. 1990. Sexual maturity and pair formation in captive cranes at the International Crane Foundation. Proc. Annu. Conf. Am. Assoc. Zool. Parks and Aquariums 1990:216-225.

MOBERG, G. P. 1987. Problems in defining stress and distress in animals. J. Am. Vet. Med. Assoc. 191:1207-1211.

STOSKOPF, M. K. 1983. The physiological effects of psychological stress. Zoo Biol. 2: 179-190.

NESTING SUCCESS AND PRODUCTION OF GREATER SANDHILL CRANES DURING EXPERIMENTAL PREDATOR CONTROL AT MALHEUR NATIONAL WILDLIFE REFUGE, OREGON, 1982-83

CARROLL D. LITTLEFIELD,[1] U.S. Fish and Wildlife Service, Malheur National Wildlife Refuge, HC 72 Box 245, Princeton, OR 97721, USA

JOHN E. CORNELY,[2] U.S. Fish and Wildlife Service, Malheur National Wildlife Refuge, HC 72 Box 245, Princeton, OR 97721, USA

Abstract: Greater sandhill cranes (*Grus canadensis tabida*) nest commonly on Malheur National Wildlife Refuge, Oregon, but depredations by common ravens (*Corvus corax*) and coyotes (*Canis latrans*) contributed substantially to low recruitment during the 1970's. An experimental predator removal study was initiated in 1982 to determine if nesting success and recruitment would increase if predator populations were reduced. We monitored 120 crane nests located in 2 treatment areas (ravens and coyotes removed, and only ravens removed) and 1 control area (40 nests per area). Nesting success between the 2 predator control areas differed significantly during the 2-year study, but not between either predator control area and the area where no predators were removed. Two years of predator removal did not sufficiently reduce the number of predators in the study area to increase nest success to a level that would maintain a stable nesting population.

PROC. NORTH AM. CRANE WORKSHOP 7:62-66

Key words: *Grus canadensis tabida*, Malheur NWR, Oregon, greater sandhill crane, predator control, nesting success, common raven, *Corvus corax*, coyote, *Canis latrans*.

The Central Valley Population (CVP) of greater sandhill cranes is 1 of 5 populations of this subspecies. Because of its small size and low productivity, the CVP has been classified as a sensitive population, receiving high priority for management by the U.S. Fish and Wildlife Service in the Pacific Flyway. Most cranes of the CVP nest in southcentral and eastern Oregon and in northeastern California. Malheur National Wildlife Refuge (NWR), Harney Co., Oregon, is the most important nesting area for the CVP (Littlefield and Thompson 1979; Littlefield 1982, 1989; Stern et al. 1987).

Depredations by common ravens and coyotes on sandhill crane eggs and unfledged young at Malheur NWR were believed to have contributed substantially to the low number of chicks fledged (Littlefield and Thompson 1987). Of 674 crane nests monitored during 1966-81, 48.5% were destroyed by predators. Annual mortality of crane chicks for 11 years (1970-81) averaged 84.8% (C. D. Littlefield, unpubl. data).

Reduction of annual cattle grazing on Malheur NWR after 1973 improved nesting habitat and resulted in increased nest success (Littlefield and Paullin 1990), but the percentage of fledged chicks in the autumn population has averaged only about 6% since then (1974-81). The number of nesting crane pairs peaked at 236 in 1975 and declined to 214 in 1982 and 1983.

This paper reports on an experimental predator control study conducted at Malheur NWR in 1982 and 1983 to test the effectiveness of (1) common raven control alone for improving sandhill crane nest success, and (2) simultaneous coyote and raven control on crane nest success and recruitment. This paper summarizes results of the 2-year study.

We thank S. Melvin and A. Sargeant for critical reviews of the manuscript. In addition, many U.S. Fish and Wildlife Service personnel participated in this study. Important field assistance was provided by J. Anderson, M. Barnes, B. Ehlers, E. Kelley, C. Miller, D. Paullin, D. Severson, and S. Thompson. Special thanks go to J. Mazzoni, Malheur NWR Manager in 1982-83, for permission to conduct the study on the refuge, and D. Paullin, who provided materials and logistic support. M. Stern of The Nature Conservancy and F. Ramsey of Oregon State University provided assistance with statistical analysis. The efforts of ADC personnel D. Frye and E. Dunlap for coyote removal are appreciated. R. Schneider assisted with figure production.

STUDY AREA

Malheur NWR includes more than 72,000 ha of interspersed meadows, marsh, and shrub-covered uplands, the latter dominated by sagebrush (*Artemisia* spp.). Most sandhill crane nesting occurs in emergent vegetation in depressions in the meadows (Littlefield and Ryder 1968) of the narrow Blitzen Valley. The valley is surrounded by rimrocks that provide nesting sites for common ravens.

Common ravens are abundant throughout the Malheur NWR and adjacent areas (Littlefield and Thompson 1987).

---

[1]Present address: HCR 4 Box 212, Muleshoe, TX 79347, USA.

[2]Present address: U.S. Fish and Wildlife Service, Refuges and Wildlife, P.O. Box 25486, DFC, Denver, CO 80225, USA.

More than 1,300 ravens have been counted at winter roost sites in the area. Coyote densities are unknown, but observations by refuge staff and others indicate that coyotes are generally abundant. Coyote predation on sandhill crane eggs was less than that by ravens or raccoons (*Procyon lotor*), but coyotes, unlike ravens, kill crane chicks (Littlefield and Lindstedt 1992).

Ravens and coyotes are common on the refuge during the period when cranes are nesting. Each year a pronounced post fledging movement of ravens onto the refuge occurs in late May and early June, when nesting of many other avian species peaks (Stiehl 1978, Littlefield and Thompson 1987). Coyotes are commonly encountered during the crane nesting period (April-June), but numbers increase in late July after meadows dry and hay mowing activities begin (C. D. Littlefield, unpubl. data).

The 8,153-ha Raven Removal Study Area (RRSA) was located in the southern Blitzen Valley (Fig. 1). Common ravens were killed in this area. A Raven and Coyote Removal Study Area (RCRSA) located in the Double-O area on the western portion of the refuge was 7,976 ha. A Control Study Area (CSA), 7,494 ha in size, was established near the center of the Blitzen Valley where no experimental predator management occurred. The CSA was about 8 km from the RRSA, and the RCRSA was about 26 km northwest of the CSA (Fig. 1). Study sites were selected on the basis of predator populations and the number of nesting crane pairs. The RRSA had the greatest number of nesting crane pairs on the refuge, whereas the RCRSA had the highest predator densities and lowest crane nest success from 1978 to 1982 (C. D. Littlefield, unpubl. data).

METHODS

Common ravens were killed with DRC-1339 (3 chloro-p-toluidine hydrochloride), a selective toxicant that is not lethal to mammals or raptorial birds at dosages lethal to ravens (Decino et al. 1966, Larson and Dietrich 1970). In 1982, common raven control was conducted from 12 February to 10 June in the RRSA and 25 February to 14 June in the RCRSA. A total of 7,335, 1-cm^3 meat baits treated with 7.5 mg/kg of DRC-1339 (Larsen and Dietrich 1970) was placed on carcasses of mule deer (*Odocoileus hemionus*), black-tailed jack rabbit (*Lepus californicus*), pronghorn (*Antilocapra americana*), and carp (*Cyprinus carpio*) between 12 February and 16 April 1982. Depending on carcass size, 5 to 50 meat baits were used daily per carcass. From 16 April to 14 June, 1,828 chicken eggs, injected with 1 ml of 10% solution of DRC-1339 in water, were placed throughout the RRSA and the RCRSA. Usually, 4 injected eggs were placed in each artificial nest. All known common raven nests

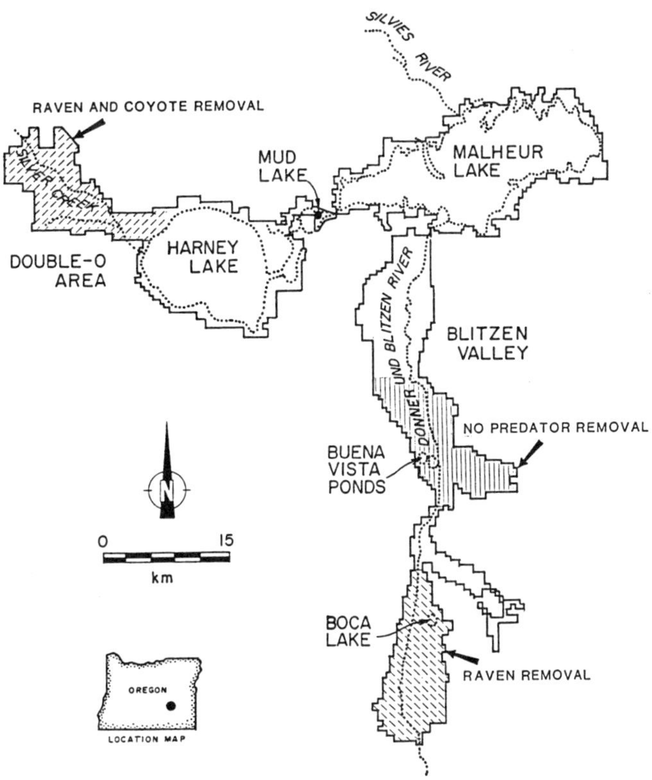

Fig. 1. Location of 3 study areas for evaluation of predator removal on sandhill crane nest success and production, Malheur National Wildlife Refuge, Oregon, 1982–83.

were examined annually by the refuge staff. In 1982 and 1983, 42 raven nests were surveyed, and activity and fates were determined. Most of the nests were on or adjacent to the 3 study areas.

Coyotes were removed from the RCRSA between 2 December 1981 and August 1982 by aerial gunning from a helicopter and by trapping. Removal in 1982 was most intense in the spring and only minor control took place after May. In the RCRSA, Animal Damage Control (ADC) personnel and private trappers removed 214 coyotes during the 2-year period. Of 147 coyotes killed in 1982, only 8 were taken during summer. However, coyotes were seen regularly on all study areas during summer. In 1983, 67 coyotes and 1 red fox (*Vulpes vulpes*) were removed. Red foxes were previously unrecorded from Malheur NWR. Nearly all coyotes were killed before summer. One or 2 coyotes were seen daily in August, but their numbers were fewer than in August 1982 (C. D. Littlefield, unpubl. data).

From 25 February through 21 May 1983, 3,685 chicken eggs treated with DRC-1339 were placed in the RRSA and the RCRSA. Dosages were the same as in 1982. Coyotes were removed from the RCRSA by trapping and aerial gunning from November 1982 through 31 July 1983. Only 1 helicopter flight was made in 1983 compared to 3 in 1982.

Searches for greater sandhill crane nests were conducted from 13 April through 27 May 1982 and 15 April through 20 May 1983. Searches were continued until 60 active crane nests (20 per study area) were found. Most nests were located during ground searches, but some were spotted from fixed-winged aircraft and helicopter. Egg incubation stage was estimated by the flotation method described by Westerskov (1950). Nests were reexamined after the estimated hatching date and nest success determined (success = at least 1 hatched egg). Predators were identified by examining the nest and egg remains (Sooter 1946, Einarson 1956).

To assess crane productivity, ratios based on counts of adult and young sandhill cranes (age class based on plumage) were attempted from ground surveys conducted from 21 to 25 September 1982 and from 22 to 28 September 1983. Annual production was calculated as:

$$\frac{\text{No. of young}}{\text{No. of young} + \text{No. of breeding adults}} \times 100\%$$

Young were counted on refuge grain fields where they concentrated during feeding in September, before most migrant cranes had arrived. Grain fields were present in or near all 3 study areas, so production per study area was usually possible to determine, particularly in the RRSA and the RCRSA.

RESULTS AND DISCUSSION

Common Raven and Coyote Removal

Although the number of common ravens killed during the study was not determined, populations were greatly reduced, especially on predator removal areas. At Malheur NWR traditional raven nests are generally occupied year after year. By May 1983 none of 20 known traditional raven nests near the predator removal areas was active. Near the CSA, 4 of 8 traditional raven nesting sites were used in 1982 and 3 were active in 1983, indicating that some ravens from that area were killed by treated baits set out in 1 of the predator removal areas.

Nest Success and Predation-1982

A total of 60 (20 per study area) greater sandhill crane nests was examined in 1982 (Table 1). Thirteen (65%) of 20 nests in the RRSA were successful. Coyotes destroyed 2 nests, ravens destroyed 1, and the others were lost to undetermined predators. Of the 20 nests examined in the RCRSA, 12 (60%) were successful. Destruction of 8 nests was attributed to common ravens (3), raccoons (2), coyote (1), and undetermined predators (2).

Table 1. Success of greater sandhill crane nests after predator removal, Malheur National Wildlife Refuge, Oregon, 1982-83 (*n* = 20 nests examined in each study area per year).

Study area	1982 No.	1982 %	1983 No.	1983 %
Raven Removal (RRSA)	13	65	9	45
Raven and Coyote Removal (RCRSA)	12	60	16	80
Control (CSA)	14	70	13	65

The highest nesting success in 1982 was in the CSA (70%). Destruction of 3 nests was attributed to ravens, and another 3 nests were destroyed by undetermined predators. Water management in the untreated area was different than in the other 2 study areas in 1982. By mid-April water was abundant in most of the crane nesting habitat of that area and remained relatively deep throughout the spring and early summer. Coyote activity in that area likely was reduced because of the deep water.

Average nest success in 1982 ($n = 60$) was 65%, ranging from 60% to 70%, but was not significantly different among the 3 study areas ($\chi^2 = 0.44$, $P > 0.05$). Common ravens destroyed 7 clutches, coyotes 3, raccoons 2, and unknown predators 9. Nesting success, however, was higher in all study areas when compared to the 16-year (1966-81) average (C. D. Littlefield, unpubl. data): 43.9% for RRSA ($n = 460$ nests), 50.0% for RCRSA ($n = 34$), and 45.9% for CSA ($n = 137$).

Unfortunately, irrigation was delayed on a portion of the crane nesting habitat until mid-May in the RRSA. Delayed irrigation facilitated hunting of marsh and meadow habitat by coyotes while crane nesting was in progress. In addition to the 2 nests known to have been destroyed by coyotes, we believe the 4 nests attributed to unknown predators likely were destroyed by coyotes.

Nest Success and Predation-1983

Sixty crane nests (20 per study area) also were examined in 1983 (Table 1). Success was only 45% in the RRSA for the 20 nests examined, but 3 nests were lost to flooding and 1 was abandoned. Ample water was available after mid-April, but breached dikes prevented water stabilization in some areas; the Blitzen River overflowed in late May, flooding 3 monitored nests. Fortunately most clutches already had hatched when the flooding occurred. No nest losses were attributed to ravens, but raccoons destroyed 3, coyote 1, and undetermined predators 3.

In the RCRSA, 80% of the 20 nests examined were successful. One clutch was lost to a common raven, 1 to a coyote, 1 to a raccoon, and 1 was abandoned.

Sandhill cranes achieved 65% nest success (13/20) in the CSA. Most nest sites had water by mid-April, and cranes initiated nesting shortly after their territories received water. Most nesting was completed before high water from local streams inundated much of the area. Common ravens destroyed 2 nests, coyotes 2, raccoon 1, undetermined predator 1, and 1 was abandoned.

Nest success was significantly different among the 3 areas in 1983 ($\chi^2 = 9.13$, $P \leq 0.05$). The RCRSA had significantly higher nest success than the RRSA ($\chi^2 = 7.30$, $P \leq 0.01$). Success was not significantly different between the RRSA and the CSA ($\chi^2 = 0.92$, $P < 0.05$) or between the RCRSA and the CSA ($\chi^2 = 2.30$, $P < 0.05$). In all 3 areas combined, raccoons destroyed 5 nests, coyotes 4, ravens 3, and undetermined predators 4. Three nests were flooded, 2 were abandoned, and 1 contained infertile or addled eggs.

Nesting Summary, 1982-83

Nest success in the RRSA averaged 55%. In this area only 1 nest was destroyed by ravens in 2 years. Undetermined predators destroyed 7 nests, raccoons 3, and coyotes 3.

In the RCRSA nest success averaged 70%. Four clutches were lost to ravens, 3 to raccoons, 2 to coyotes, and 2 to unknown predators. Nest success in the CSA averaged 68%. Common ravens destroyed 5, coyotes 3, unknown predators 3, raccoons 1, and 1 was abandoned. For 120 nests monitored during the 2-year study, overall nest success averaged 64%.

Differences in nest success among the 3 study areas were not statistically significant during the 2-year study ($\chi^2 = 2.25$, df=5, $P > 0.05$). However, where both common ravens and coyotes were removed, nest success during the second year was 80%. This was the highest success rate since the 1940's, when predators were routinely killed throughout Malheur NWR.

Productivity

Data on productivity were inconclusive because crane pairs with chicks from outside the study area moved onto the grain field before brood counts were conducted in both 1982 and 1983. Therefore, the actual number produced in the study area was not determined. However, of the 25 chicks produced on or adjacent to the refuge in 1982, only 2 were fledged from the 93 crane pairs in the RRSA and 1 from the 27 pairs in the RCRSA. In 1983, 39 young were fledged, including 4 from the RRSA and 5 from the RCRSA. Of the young cranes hatched on Malheur NWR in 1982 and 1983, 90.1% and 84.8%, respectively, died before fledging, primarily from predation.

The length of this study was to be 3 years, but funding was not provided for the final year. Malheur NWR encompasses a large area where predators are abundant. Under such circumstances predator removal may have to take place for more than 2 years before there is a significant increase in sandhill crane nest success or recruitment.

LITERATURE CITED

DECINO, T. J., D. J. CUNNINGHAM, and E. W. SHAFER. 1966. Toxicity of DRC-1339 to starlings. J. Wildl. Manage. 30:249-253.

EINARSON, A. S. 1956. Determination of some predator species by field signs. Oregon State Univ. Monogr. Stud. Zool. 10. 34pp.

LARSON, K. H., and J. H. DIETRICH. 1970. Reduction of a raven population on lambing grounds with DRC-1339. J. Wildl. Manage. 34:200-204.

LITTLEFIELD, C. D. 1982. The status and distribution of greater sandhill cranes in California, 1981. California Dep. Fish Game, Wildl. Branch, Admin. Dep. 82-1:1-27.

______. 1989. Status of greater sandhill crane breeding populations in California, 1988. Unpubl. Rep., Wildl. Manage. Div., California Dep. Fish Game, Sacramento. 40pp.

______, and S. M. LINDSTEDT. 1992. Survival of juvenile greater sandhill cranes at Malheur National Wildlife Refuge, Oregon. Pages 21-31 *in* D. A. Wood, ed. Proc. 1988 North Am. crane workshop. Florida Game Fresh Water Fish Comm. Nongame Wildl. Program Tech. Rep. 12.

______, and D. G. PAULLIN. 1990. Effects of land management on nesting success of sandhill cranes in Oregon. Wildl. Soc. Bull. 18:63-65.

______, and R. A. RYDER. 1968. Breeding biology of the greater sandhill crane on Malheur National Wildlife Refuge, Oregon. Trans. North Am. Wildl. Nat. Resour. Conf. 33:444-454.

______, and S. P. THOMPSON. 1979. Distribution and status of the Central Valley Population of greater sandhill cranes. Pages 113-120 *in* J. C. Lewis, ed. Proc. 1978 crane workshop. Colorado State Univ. Printing Serv., Fort Collins.

______, and ______. 1987. Greater sandhill cranes and common ravens on Malheur National Wildlife Refuge, Oregon: their history and relationship. Pages 156-165 *in* J.C. Lewis, ed. Proc. 1985 crane workshop. Platte River Whooping Crane Maintenance Trust, Grand Island, Nebr.

SOOTER, C. A. 1946. Habits of coyotes in destroying nests and eggs of waterfowl. J. Wildl. Manage. 10:33-38.

STERN, M. A., C. D. LITTLEFIELD, and G. PAMPUSH. 1987. The status and distribution of greater sandhill cranes in Oregon,

1986. Unpubl. Rep., Nat. Conserv., Portland, Oreg. 28pp.

STIEHL, R. E. 1978. Aspects of the ecology of the common raven in Harney Basin, Oregon. Ph.D. Thesis, Portland State Univ., Portland, Oreg. 178pp.

WESTERSKOV, K. 1950. Methods for determining the age of game bird eggs. J. Wildl. Manage. 14:56-67.

USE OF DREDGED MATERIAL TO CONSTRUCT WINTER WHOOPING CRANE HABITAT

DARRELL E. EVANS, U.S. Army Engineer Waterways Experiment Station, Natural Resources Division, 3909 Halls Ferry Road, Vicksburg, MS 39180, USA
THOMAS V. STEHN, U.S. Fish and Wildlife Service, Aransas National Wildlife Refuge, P.O. Box 100, Austwell, TX 77950, USA

Abstract: Aransas National Wildlife Refuge (ANWR) and nearby coastal marshes serve as the winter home for the only natural flock of whooping cranes (*Grus americana*). In recent years shoreline erosion and the subsequent loss of wintering habitat have been observed on the refuge adjacent to the Gulf Intracoastal Waterway (GIWW). In 1988 the U.S. Army Corps of Engineers (USACE) District, Galveston, Texas, and U.S. Fish and Wildlife Service (USFWS) entered into an informal agreement to attempt to slow shoreline/habitat loss on the refuge. Efforts to curtail habitat loss have included armoring the most erosive reaches with temporary concrete erosion control structures and using articulated concrete mats to armor severely eroded reaches. Most recent efforts have been directed at determining if dredged material removed from the GIWW during routine channel maintenance could be used to construct winter crane habitat. One experimental site was constructed in 1991 by Mitchell Energy Corporation and 2 in 1993 by USACE. Current plans call for the long-term monitoring of the sites to determine the relative success of the habitat creation effort. A comprehensive biomonitoring program is being developed by researchers at the USACE Waterways Experiment Station (WES) to track the long-term development and to characterize habitat conditions and wildlife use of the experimental sites.

PROC. NORTH AM. CRANE WORKSHOP 7:67-71

Key words: Aransas National Wildlife Refuge, dredged material, ecological monitoring, *Grus americana*, habitat creation, whooping cranes.

The establishment of ANWR in 1937 helped protect and conserve the wintering habitat of the whooping crane. In the winter of 1938, the whooping crane population was estimated at 29 individuals and only 2 small flocks survived in the wild. During the past 57 years the population has increased slowly to total 158 individuals in the 1995 winter (T. V. Stehn, unpubl. data). The protection, conservation, and management of the wintering grounds at Aransas is and will continue to be paramount to survival of the species in the wild.

The GIWW is a federal navigation canal 38 m wide by 4.7 m deep for waterborne commerce located between Texas and Florida; it was constructed through Aransas in 1944. Construction of the GIWW, past dredged material disposal practices, and shoreline erosion created a net loss of 465 ha of whooping crane habitat at ANWR (Ramirez et al. 1988). Net losses inside the critical habitat from 1930 to 1986 have amounted to 841 ha, equaling an 11% loss in a study corridor 1,830 m wide centered on the GIWW (Sherrod and Medina 1992).

Critical habitat on the wintering grounds adjacent to the GIWW is being lost to erosion at a rate of 1.0-1.6 ha annually (Stehn 1987, USACE 1988) with total losses at ANWR exceeding 93 ha (Stehn 1987). Rate of shoreline erosion is 0.7-1.2 m per year along some of the more erosive reaches (Stehn 1987, USACE 1988). Since 1940, the GIWW shoreline has receded approximately 28 m on either side of the channel (Stehn 1987).

Loss of critical habitat in these areas represents a potentially threatening situation to the survival of the species on the refuge because of the proximity of winter crane territories to the GIWW. In winter 1994-95, 19 of 46 (41%) of the total adult crane territories adjoined the GIWW (Stehn 1995). Exact cause and rate of shoreline/habitat loss are presently unclear, but loss is believed to be caused by natural wave (fetch) and current action, wakes thrown up by marine traffic using the GIWW, and, to some extent, past dredged material disposal practices (Stehn 1987, Zhang et al. 1993). This paper describes efforts taken to (1) stop habitat loss on the wintering grounds caused by shoreline erosion and maintenance dredging disposal practices, (2) create additional marshlands for whooping cranes by using dredged material, and (3) formulate long-term plans for the ecological monitoring of the created marshes.

STUDY AREA

The GIWW passes through 68 km of whooping crane critical habitat in areas regularly used by wintering cranes. This includes an 11-km reach by the Welder Flats Coastal Preserve next to Shoalwater Bay and a 19-km reach through the ANWR between Aransas and San Antonio Bays (Fig. 1). Dredged material was used to create habitat in Mesquite and San Antonio Bays adjacent to ANWR.

RESULTS AND DISCUSSION

Shoreline Erosion

The USFWS and USACE-Galveston conducted several activities from 1989 to 1992 aimed at stopping or slowing erosion of critical whooping crane habitat. The most significant of these was the initiation of an on-going study under

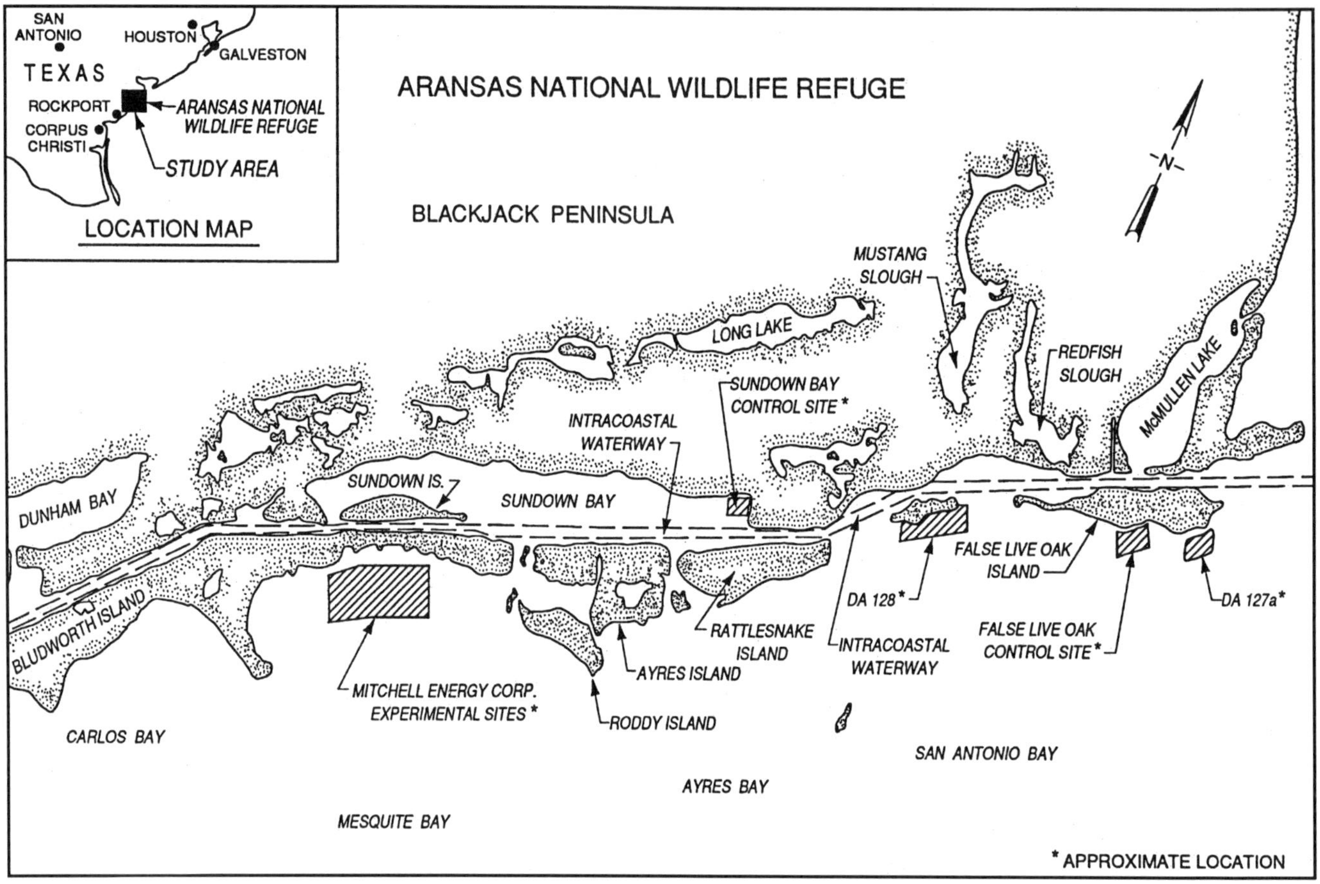

Fig. 1. Location of study sites for evaluation of use of dredged material to construct winter habitat for whooping cranes along the Gulf Intracoastal Waterway, Aransas National Wildlife Refuge, Texas.

Section 216 of the 1970 Flood Control Act, which is investigating the feasibility of relocating the GIWW to a different route. Temporary erosion control measures (e.g., concrete bag retaining walls, posting signs within the confines of the refuge urging boat operators to operate at speeds which would produce little or no wakes) were initiated in 1989 as a stopgap measure on several of the more erosive reaches (i.e., those that would be irreversibly modified before the completion of the Section 216 study). Approximately 2,652 m of erosive shoreline were protected from 1989 to 1992 by use of more than 57,000 bags of cement in a huge volunteer effort involving government, private corporations, and hundreds of individuals.

In May 1993 an agreement was reached between the USFWS and USACE-Galveston, which called for the armoring of some to the more erosive reaches with technology regularly used by USACE on inland waterways. By 1995 approximately 5,486 m of shoreline had been protected with articulated concrete mats laid on geotextile fabric. Mats were anchored/linked with polyester cables which allow the mats to move and conform to existing land features. Current plans call for the armoring of an additional 610 m annually until the Section 216 study is completed and implemented.

Creation of Habitat from Dredged Material

More permanent solutions to the problem of habitat loss included investigating the use of dredged material removed from the navigable channel during routine, scheduled maintenance to (1) construct additional winter habitat and (2) construct erosion control structures for protecting existing habitat. In 1989 a cooperative study between USACE-Galveston, USFWS-ANWR, and WES was begun to survey the area to determine the feasibility of using dredged material removed from the GIWW during routine channel maintenance to protect some of the more erosive reaches. The goals of the initial study were to (1) characterize vegetation and wildlife use on some of the more erosive reaches and (2) conduct the preliminary engineering studies to determine the feasibility of using dredged material in a beneficial manner to construct additional winter habitat. A 50-year dredged material disposal plan is being developed cooperatively between the USFWS and USACE-Galveston. This plan will call for use of dredged material to create an additional 653 ha of winter habitat.

The first large-scale attempt to create winter whooping crane habitat with dredged material was undertaken in 1991 by the Mitchell Energy Corporation (MEC). A 5.3-ha site was constructed on the bayside of Bludworth Island (Fig. 1) by use of material removed as part of a dredging operation (channel establishment) in the Mesquite Bay area. The site was planted with species native to the Texas Coastal Bend. In 1993, MEC constructed a 3.6-ha site which adjoined the original construction effort. Both sites were protected with articulated concrete mats connected with polyester cables. One-year monitoring on the first site indicated ground coverage of approximately 99% and 89% for low and high marsh species, respectively (C. Belaire, Belaire Consulting, Inc., Rockport, Tex., unpubl. data). Smooth cordgrass (*Spartina alterniflora*) was the dominant species on the low marsh areas of the project and saltmeadow cordgrass (*S. patens*), saltgrass (*Distichlis spicata*), seaside paspalum (*Paspalum vaginatum*), and saltwort (*Batis maritima*) dominated the drier, high marsh areas. Preliminary observations suggest that the MEC wetland is maturing well, and whooping cranes were observed in the created habitat 4 times from 1991 to 1994 (2 cranes in January 1991, 5 cranes in March 1993, 2 cranes in November 1994, and 2 cranes in December 1994) (T. V. Stehn, unpubl. data).

A fisheries study of the created Mitchell habitat compared with natural marshes, seagrass beds in shallow bays, and unvegetated shallow bay bottom showed the new marsh to benefit fisheries (Rozas et al. 1994). Salt marsh and seagrass habitats supported significantly greater densities of most species. Therefore, replacing open bay bottom with marsh and seagrass habitats should have a positive effect on most species that were dominant in the study area. Even though some open bay habitat will be lost by creating new marsh, the area replaced by marsh is small relative to the total area of open bay habitat in the vicinity, and species that use shallow unvegetated bottom will likely find suitable habitat near constructed marshes (Rozas et al. 1994). If marshes that are functionally equivalent to natural marshes can be constructed, the increased benefit of enlarging the habitat area for fishery and forage species that use marsh systems should outweigh the loss of open bay habitat (Rozas et al. 1994).

In summer 1993, the USACE-Galveston began construction of 2 additional sites by using material obtained from ongoing dredging operations in the area. The 2 sites were located east of False Live Oak Point (Disposal Area [DA] 127a) and east of an unnamed dredged material island across from the opening to Mustang Slough (DA 128) (Fig. 1). The False Live Oak 9.3-ha site was contained within an earthen levee. Three openings were constructed through the levees to facilitate water exchange on the site and a riprap breakwater (stone) was constructed around the bay side to protect the developing marsh from fetch. The second 8.5-ha site, DA 128, was constructed at an existing dredged material disposal site and armored with large diameter, geotextile grout tubes filled with dredged material. Experimental sites were planted in 1993 by consultants under contract to the USACE-Galveston with native salt marsh species common to the intertidal marshes of the Texas Coastal Bend. Both sites were allowed to consolidate and dewater for 2 years while plans for the long-term ecological monitoring of the sites were developed by personnel from USFWS-ANWR, USACE-Galveston, and WES.

Long-term Monitoring

Biologists from the USFWS-ANWR, the USACE-Galveston, and the WES have developed a long-term program aimed at monitoring 3 biological aspects of the experimental sites. Monitoring data will be used to evaluate the success or failure of the habitat creation effort and to provide information for scientists and engineers charged with future habitat creation efforts. Long-term ecological monitoring of the project area will involve sampling efforts to assess (1) the development of vegetation on the experimental sites, (2) the use of the experimental sites by avian species, and (3) macrobenthic and invertebrate abundance and composition on experimental sites.

Control sites include open bay habitat in San Antonio and Sundown Bays and saltmarsh on the north end of Sundown Bay (Fig. 1). These natural communities will be compared with the created marshes and open bay habitats that existed prior to marsh creation.

Vegetation

Development of vegetation on the experimental sites will be monitored annually during the study by use of remotely sensed data and conventional field sampling techniques. Habitat types will be manually delineated on aerial photos and digitized into a Geographic Information System (GIS) for spatial analyses. Percent coverage of the various habitat elements/components (e.g., open water, planted vegetation, bare ground, tidal flats, and tidal channels) will be determined for each year of the study and used to track the development of the vegetation on the experimental sites.

Avian Use

Avian use of the experimental sites will be monitored bimonthly on both control and experimental sites. Composition and abundance of avian species will be determined for comparisons of avian diversity. Two fixed-width transects have been established on the control and experimental sites to identify (1) and tally all avian species occurring on each site, (2) how the sites are being used (e.g., feeding, loafing, or resting), and (3) what habitat types (e.g., tidal flats, open water, or low marsh) within the sites are being used. Richness and evenness indices will be calculated and used to compare avian diversity on the sites and evaluate whether the experimental sites are mimicking the control sites.

Macrobenthos/Invertebrates

Quarterly macrobenthic/invertebrate sampling will be aimed at (1) investigating taxonomic composition, taxa richness, and total abundance of macrobenthic/invertebrate species on control and experimental sites and (2) determining if the types and numbers of macrobenthic/invertebrate species on and around experimental sites are similar to those found on control sites. Taxonomic composition, taxa richness, and total abundance (animals per m^2) will be estimated for each of the sites (experimental and control) and used to determine if significant differences exist.

The 2 main concerns that will be addressed in our field sampling are (1) are macrobenthic and infauna abundance and composition on an experimental site comparable to macrobenthic and infauna abundance and composition on a control/existing site, and (2) how did the loss of open water, shallow bay habitat affect macrobenthic and infauna abundance and composition (how many and what kinds of species did we impact by building the experimental site?). Primary efforts will be directed towards the macrobenthic component (e.g., fish, crabs, and shrimp) because of its importance to wintering cranes as a potential food resource.

Permanent macrobenthic trap stations were established during summer 1995 and permanently located by use of a Global Positioning System. Macrobenthic/invertebrate data will be collected quarterly and will involve the use of 1.5- × 1.5-m portable drop nets, a 7.5-cm PVC cylindrical push corer/sampler, and commercial crab traps. We will run 3 drop net traps per site per day and trap for 4 consecutive days each quarter. This design will provide us with a minimum of 12 samples per site per quarter (36 total samples per quarter) and 144 samples per year.

We have established 27 permanent infauna sampling points on the 3 study sites. Infauna samples will be collected with the push corer, processed through a 0.5-mm sieve, and preserved in the field. Identification and analysis of macrobenthic and infaunal samples will be done by invertebrate biologists in the Coastal Ecology Group of the Environmental Laboratory at WES.

Crab Abundance

Crab abundance and distribution on the 3 sites will also be monitored quarterly with commercially available crab traps. Four permanent trap sites were established in August 1995 on each of the 3 study sites to determine the species, size, and numbers of crabs that will be available to wintering cranes. Crab traps were baited with raw chicken each morning and checked once daily for 4 consecutive days. Crabs were identified to species, sexed, measured, and released back on the site at the end of each day.

Home Range

Another important aspect of the study will involve use of observation data collected by ANWR biologists to calculate winter home range and territory size for cranes using the refuge. Observation data will be digitized into a GIS and used to calculate home range/territory size and shape according to several accepted methods. Vegetation sampling (i.e., species dominance, species frequency, vegetation height, density, and coverage) within home ranges/territories and core areas will be conducted in the third and fourth years of the study to contribute to an understanding of the structural characteristics of winter habitat. Intensive field sampling within home ranges/territories will provide insight into the structural characteristics of winter habitat and provide a model/template to guide future habitat creation efforts.

Preliminary Field Sampling

Preliminary analysis of drop net data collected in summer 1995 suggests that a diverse assemblage of species (vertebrate and invertebrate) is already present on the study sites. Dominant species on the 3 sites included white shrimp (*Penaeus setiferous*), grass shrimp (*Palaeomenetes pugio*), and blue crabs (*Callinectes sapidus*). Analyses of drop net data indicated that DA 128 had the largest number of species (7) but the DA 127a site had the highest number of individuals (86). The Sundown Bay control site had the fewest species (5) but the greatest number of white shrimp. DA 127a had the greatest number of crabs (23).

Analyses of infauna data indicated that the Sundown Bay control site had the greatest number species (9) and the greatest number of individuals (20). Species identified in infauna samples from the 3 sites included clam worms

(*Nereis succinea*), an amphipod (*Ampelisca abdita*), polychaete (*Laeonereis culveri*), lunar dove shell (*Mitrella lunata*), mud shrimp (*Callianassa* spp.), polychaete worms (Capitellidae), unidentifiable polychaetes, blue crabs, shrimp (*Acetocina [Retusa] caniculata*), dwarf surf clam (*Mulinia lateralis*), and tube-building worm (*Diopatra cuprea*).

Sixty-nine blue crabs were captured on the 3 study sites during the sampling period. Crab abundance was greatest on the DA 128 site (31) and least on the Sundown Bay control site (13). Twenty-five blue crabs were trapped on DA 127a during the sampling period.

CONCLUSIONS

Development of the MEC sites appears promising and the sites appear to be progressing as expected (T. V. Stehn, unpubl. data). Information obtained from data collected on both the MEC and USACE sites in subsequent years will be used to better understand seasonal and annual developmental variation among the sites. Baseline data collected in the first year of this study will be used to make comparisons among sites and years to gain insight into the development of the experimental sites and ultimately to evaluate the success/failure of the habitat creation effort. Data collected as part of this study will also be used to determine the feasibility of conducting similar efforts in the future. We hope that USFWS and USACE researchers can establish some fundamental design criteria that can be used to guide future habitat creation efforts. A successful habitat creation effort at Aransas would provide much needed winter habitat for the wild whooping crane population and would benefit both the USFWS by identifying a mechanism for protecting erosive shoreline and the USACE-Galveston by confirming a cost-effective mechanism for the safe, beneficial use of dredged material.

LITERATURE CITED

RAMIREZ, P., JR., T. V. STEHN, G. A. JACKSON, and T. C. MAUER. 1988. Physical and chemical impacts on whooping crane wintering habitat-the role of the Gulf Intracoastal Waterway. Unpubl. Rep., U.S. Fish Wildl. Serv., Austwell, Tex. 20pp.

ROZAS, L., T. MINELLO, R. BURDITT, S. WHALEY, J. BOYD, M. PATTILLO, and T. BAUMER. 1994. Comparison of estuarine habitat utilization by nekton at Aransas National Wildlife Refuge. Unpubl. Rep., Galveston Dist., U.S. Army Corps Eng., Galveston, Tex. 36pp.

SHERROD, C., and R. MEDINA. 1992. Whooping crane habitat alteration analysis at Aransas National Wildlife Refuge, Texas. Pages 87-100 *in* D. A. Wood, ed. Proc. 1988 North Am. crane workshop. Florida Game Fresh Water Fish Comm. Nongame Wildl. Program Tech. Rep. 12.

STEHN, T. V. 1987. Impacts of the Gulf Intracoastal Waterway on the endangered whooping crane. Unpubl. Rep., U.S. Fish Wildl. Serv., Aransas Natl. Wildl. Refuge, Austwell, Tex. 13pp.

______. 1995. Whooping cranes during the 1994-95 winter. Unpubl. Rep., U.S. Fish Wildl. Serv., Aransas Natl. Wildl. Refuge, Austwell, Tex. 54pp.

U.S. ARMY CORPS OF ENGINEERS, GALVESTON DISTRICT. 1988. Study of erosion along the Gulf Intracoastal Waterway in the Aransas National Wildlife Refuge. Unpubl. Rep., Galveston Dist., U.S. Army Corps Eng., Galveston, Tex. 45pp.

ZHANG, J., F. TING, D. HERSHBERGER, H. YU, and C. A. SPELL. 1993. Bank erosion of the GIWW at the Aransas National Wildlife Refuge. Texas A&M Res. Found. Rep. 332, College Station. 138pp.

HOME RANGE AND HABITAT USE BY ISOLATION-REARED SANDHILL CRANES

WENRUI DUAN,[1] Ohio Cooperative Fish and Wildlife Research Unit, The Ohio State University, 1735 Neil Ave., Columbus, OH 43210, USA
THEODORE A. BOOKHOUT,[2] Ohio Cooperative Fish and Wildlife Research Unit, The Ohio State University, 1735 Neil Ave., Columbus, OH 43210, USA
RICHARD P. URBANEK, U.S. Fish and Wildlife Service, Seney National Wildlife Refuge, Seney, MI 49883, USA

Abstract: We isolation-reared, placed radio transmitters on, and released 38 greater sandhill cranes (*Grus canadensis tabida*) on Seney National Wildlife Refuge (Seney) in 1988–90 to develop procedures for initiating new populations of cranes. Here we report on habitat selection by breeding birds with functional radio transmitters. Home ranges established by 6 cranes in 1992, when they were 2 to 4 years old, averaged 199±50.8 (SE) ha (harmonic mean method, 75% utilization) and were of 2 types: feeding grounds separated from nesting habitat and feeding grounds adjacent to nesting habitat. Home ranges consisted of 36% emergent palustrine wetlands, 28% forested upland, 11% open upland, and 11% forested palustrine; the remainder was scrub-shrub, upland shrub, bog, or open water. Four of 5 cranes monitored in 1993 nested–all in emergent palustrine wetlands, also the nesting habitat of wild sandhill cranes at Seney. Open field and mudflat were the major feeding habitats, as they were for wild cranes. Five of the 6 cranes did not use habitats in proportion to availability in the home range (χ^2, $P < 0.05$). Three cranes significantly selected emergent palustrine wetlands and no crane avoided this habitat type; 2 cranes selected open upland, and no crane avoided this type; and 4 cranes avoided forested palustrine, and no crane preferentially selected this habitat type. The nesting areas of the 4 cranes in 1993 were the same areas they used in 1992, when they were paired but were not known to have nested. Feeding grounds changed according to their availability in the 2 years.

PROC. NORTH AM. CRANE WORKSHOP 7:72-78

Key words: *Grus canadensis*, isolation-rearing, release, home range, habitat use, sandhill crane, reintroduction technique.

As part of an effort to effect recovery of the endangered whooping crane (*Grus americana*), and other endangered cranes, the Ohio Cooperative Fish and Wildlife Research Unit isolation-(costume-)reared 38 sandhill cranes and released them into a migratory population of sandhill cranes at Seney in 1988–90. Hereafter, unless otherwise indicated, all reference to sandhill cranes means the isolation-reared cranes. Among the cranes, 16 (9 males, 7 females) were released in 1988, 13 (4 males, 9 females) in 1989, and 9 (5 males, 4 females) in 1990. These birds exhibited high survival (84% 1 year after release) and return (74%) to the natal area after migration to winter sites (Urbanek and Bookhout 1992*a*, 1994). Before isolation-rearing can be declared successful as a reintroduction technique, the cranes must exhibit normal sexual behavior, establish a breeding territory, mate successfully with wild cranes, and produce and fledge chicks. In this paper we describe home ranges and habitat use for 6 male cranes, 4 of which we monitored in both 1992 and 1993.

We are grateful to the Hiawatha National Forest (USDA Forest Service) and the United Board of Christian Higher Education in Asia for financial support. We thank M. Tansy, Manager, Seney National Wildlife Refuge, and other refuge staff for cooperation and assistance during this study. T. Reuther, J. Reuther, E. Bunker, K. Boyd, and D. Decaire assisted in the field. This paper is a contribution of the Ohio Cooperative Fish and Wildlife Research Unit, cooperatively supported by the National Biological Service, Ohio Division of Wildlife, The Ohio State University, and the Wildlife Management Institute.

METHODS AND STUDY AREA

Seney consists of about 38,000 ha in Schoolcraft County in the east-central Upper Peninsula of Michigan. Vegetation cover is about 54% marsh, grassland, and shrub; 30% forested upland; 9% forested wetland; and 7% open water (U.S. Department of the Interior 1978). About 65% is wetland. There are 26 major pools, 21 of which have controllable water levels.

Dominant marsh plants are cattail (*Typha latifolia*) and sedges (*Carex* spp.). Dominant shrubs are willow (*Salix* spp.), leatherleaf (*Chamaedaphne calyculata*), and alder (*Alnus rugosa*). Dominant tree species are jack pine (*Pinus banksiana*) and red pine (*P. resinosa*) on forested sand ridges in wetland and sugar maple (*Acer saccharum*) and red pine in forested upland.

All cranes were radio-tagged and color-marked (for details see Urbanek and Bookhout 1992*a*). Cranes on the refuge with functional radios (n = 5 in 1992, 5 in 1993, 4

[1]Present address: Department of Zoology, The Ohio State University, 1735 Neil Ave., Columbus, OH 43210, USA.
[2]Retired.

birds common to both years) were monitored daily or weekly from a vehicle (truck with Yagi antenna and a Telonics TS-1/TR-2 scanner/receiver) in summer. When a radio signal was detected, the triangulation method (Kenward 1987:151-156) was used to ascertain the location of the cranes. If the crane could not be observed, the triangulation data were calculated by program Locate II (Nams 1990) to obtain the Universal Transverse Mercator (UTM) coordinates of the crane's location and 95% error ellipses. Each crane location was triangulated from 2 or 3 different detection sites in open areas and from 3 to 6 sites in forested areas. All locations with a calculated error of >1 ha were discarded. All cranes visually detected were observed with a 15-60× spotting scope to ascertain identification according to color bands.

Crane locations were marked on 1:24,000 U.S.G.S. topographic maps with UTM coordinates, or recorded on 1:10,000-1:12,000 aerial photos taken in August 1992. The locations of any crane for which more than 50 samples were located, the minimum sample size recommended by Ackerman et al. (1990), were calculated with the program Home Range (Ackerman et al. 1990). The activity center of the roosting or nesting territory and the activity center of the feeding grounds also were calculated with the Home Range program. For the home range study, an important goal was to obtain unbiased location samples. Directly observing the crane and recording habitat type and UTM coordinates provided the best accuracy.

Home range shape and size were determined by 2 different methods: convex polygon and harmonic mean. The harmonic mean method (Dixon and Chapman 1980) was judged by McMillen (1988) to be the better one to represent crane home range, because the method does not assume any a priori distribution and it also can define a home range of any shape; therefore, we selected this method for analysis. The convex polygon (Mohr 1947) is one of the most common methods used to represent animal home ranges, so for comparison purposes this method also was used (for details of harmonic mean and minimum convex polygon methods see White and Garrott 1990:145-180; Samuel and Garton 1985, 1987). The 75% utilization harmonic mean home range was used to identify home range outline, because this pattern best represented what we thought was actually occurring in the field.

Habitat of cranes was classed as open water, upland, and palustrine. All 6 crane pairs selected their roosting or nesting habitats in palustrine wetland. The 3 palustrine classes used in the analysis were emergent palustrine (E-P), scrub-shrub (S-S), and forested palustrine (F-P) (after Cowardin et al. 1979). Upland areas were classified as open upland (O-U), forested upland (F-U), and upland shrub (U-S).

Information on habitats used by cranes was collected based on even distribution sampling and aerial photos. The sample points were selected every 100 m and based on the UTM coordinates base line on 1:10,000-1:12,000 aerial photos. We took vegetation samples after cranes had left the use area. The number of samples was approximately equal to the number of hectares of the use area except in deep marshes, open water, and private property, where sampling effort was less intensive. The plot size was 20 × 20 m (400 m^2) in upland, 5 × 5 m (25 m^2) in scrub-shrub wetland and forested palustrine, and 2 × 2 m (4 m^2) in other wetland areas. The dominant tree species and shrubs, tree diameter, and coverage of tree canopy were recorded in upland and forested palustrine habitat. In scrub-shrub habitat, dominant shrub species, height of shrubs, and depth of water were recorded. In emergent palustrine wetland, dominant plant species were documented.

A use area cover map was made for each crane according to the habitat samples and 1:10,000-1:12,000 aerial photos. The preference of habitat used by the cranes was calculated independently by χ^2 analysis (Neu et al. 1974). A χ^2 test was used to test for the goodness-of-fit of utilized habitat to availability within a crane home range. Two specific hypotheses were (1) cranes used habitat in proportion to availability, all habitat types considered simultaneously, and (2) cranes used habitat in proportion to availability of habitat type, each habitat type considered separately. To test hypothesis 1, we compared the total χ^2 value with χ^2 critical values ($P <$ 0.05). To test hypothesis 2, we compared the confidence interval on the observed sample locations for each habitat type with availability of this habitat within a crane home range.

RESULTS

During the 2 study years (1992 and 1993), 17 cranes were observed. We computed home ranges for 6 cranes for which more than 50 summer locations were identified. Average home range sizes computed by the harmonic mean method were 384 ± 109 (SE) ha for 95% utilization, 199 ± 51 ha for 75% utilization, and 131 ± 36 ha for the core area (the maximum area where the observed utilization distribution exceeds a uniform utilization distribution [Ackerman 1990]). Average home range sizes depicted by the convex polygon method were 254 ± 65 ha for the 100% convex polygon and 198 ± 47 ha for the 95% convex polygon (Table 1).

Home range patterns were of 2 types: roosting and nesting habitat separated from the feeding ground, and feeding ground adjacent to the roosting or nesting area. The largest percentage area (about 36%) of the 6 home ranges was emergent palustrine wetland (Fig.1). The home range of

Table 1. Home range sizes (ha) of isolation-reared cranes, Seney National Wildlife Refuge, Michigan, 1992-93.

Crane	Harmonic mean		Convex polygon	
	95%[a]	75%	100%	95%
88-05	652	329	467	391
88-13	432	232	266	186
89-02[b]	728	349	409	249
89-06	270	146	191	185
89-14	90	54	62	50
90-09	132	84	128	125
$\bar{x}$	384	199	254	198
SE	109	51	65	48

[a] Percentage in harmonic mean is the percentage of utilization volume; percentage in convex polygon is actual percentage of sample points included in the polygon.

[b] 1992 data only; this bird died on Kanapaha Prairie, Florida, in the winter of 1992-93.

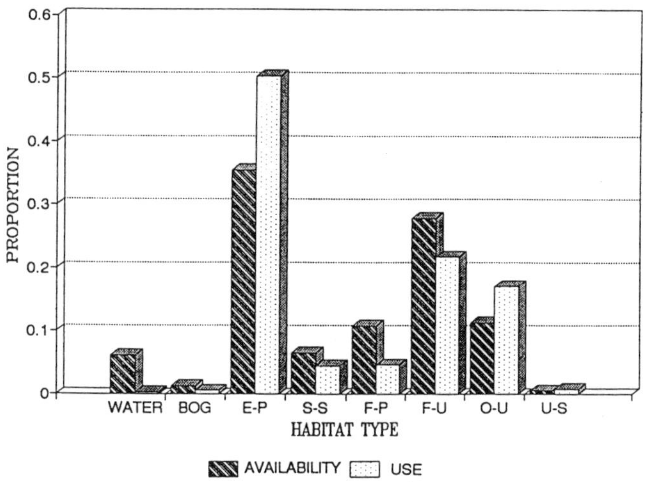

Fig. 1. Average habitat availability and use for 6 isolation-reared sandhill cranes, Seney National Wildlife Refuge, Michigan, 1992-93. E-P = emergent palustrine, S-S = scrub-shrub, F-P = forested palustrine, F-U = forested upland, O-U = open upland, U-S - upland shrub.

crane 89-06 contained 63% emergent palustrine wetland, but the home range of crane 90-09 contained only 10%. The emergent palustrine wetland was dominated by cattail or sedge, and most area of the emergent palustrine wetland was dotted by willow or alder. Four of the 5 cranes monitored in 1993 nested, all in emergent palustrine wetland. Forested upland was the second largest percentage area of crane home ranges. The home ranges contained 28% forested upland on average.

Feeding centers of 3 of the 6 cranes (88-05, 89-14, 90-09) were in open hayfield or at the border area between open field and forested upland. Two crane feeding grounds (88-05, 90-09) were separated from nesting habitat; those feeding grounds contained only open upland (hayfield) and forested upland. Cranes 88-13 and 89-06 selected feeding grounds at mudflats created by drawn-down pools. The home range of crane 89-02 contained no open fields; he spent about 90% of his time in emergent palustrine (40%) and forested upland (48%).

A sample of 777 locations for which sample intervals were ≥1 hour was obtained in the summers of 1992 and 1993. Habitat use was variable (Table 2), but 50% of the total samples of the 6 cranes were located in emergent palustrine wetland dominated by cattail or sedge and often dotted by willow or alder. Upland contained 40% of the total samples, which included forested upland (22%), open upland (17%), and upland shrub (1%). Open water and bog were rarely used (1%). Cranes 88-05, 88-13, and 89-06 spent more than one-half (58-77%) of their time in emergent palustrine wetland, and cranes 89-02 and 89-14 spent about 40% of their time in that habitat type. Crane 90-09, however, spent most (72%) of his time in upland rather than in emergent wetland (14%).

The 4 cranes that nested in 1993 used emergent palustrine wetland as their nesting territories, and all 6 monitored cranes roosted in emergent palustrine wetland or shallow water. About 22% of samples were in forested upland, all near the border between forested upland and open upland. On average about 17% of the total samples were in open uplands (hayfield).

A χ^2 test of use (observed) to availability (expected), considering all habitats simultaneously, showed (Table 2) that 5 of the 6 cranes did not use habitats in proportion to availability (Hypothesis 1 was rejected), but crane 89-02 did use habitat in proportion to its availability (Hypothesis 1 was not rejected). Results of habitat preference and avoidance analysis (Table 3) showed that 3 cranes (88-05, 88-13, 89-06) significantly preferred emergent palustrine wetland. The other 3 cranes (89-02, 89-14, 90-09) did not select emergent palustrine, but they used this type commonly. Cranes with open water (88-05, 89-06, 89-14) or bogs in their home ranges (88-02 and 90-09) avoided those types.

No crane preferred forested palustrine within its home range, and 4 cranes avoided forested palustrine wetland. These sites were dominated by red pine and black spruce (*Picea mariana*). Basal areas ranged from 1.4 to 4.1 m²/ha. Three cranes avoided scrub-shrub habitat, and no crane preferred this habitat. The scrub-shrub habitat in the 6 crane home ranges was dominated by alder (mean height = 260 ± 15.7 [SE] cm) and willow (mean height = 165 ± 5.5 [SE] cm). Two cranes (89-06, 89-14) showed preference for

Table 2. Calculation of the χ^2 statistic on habitat[a] availability and habitat use (%) for isolation-reared sandhill cranes, Seney National Wildlife Refuge, Michigan, 1992–93.

Crane		Water	Bog	E-P	S-S	F-P	F-U	O-U	U-S	Total
88-05	O	0.5	1.1	69.3	1.1	0.5	11.1	11.6	4.8	1.0
	E	**16.5**	**2.8**	**41.3**	**5.9**	**18.7**	**10.0**	**1.4**	**3.5**	**1.0**
	χ^2	15.5	1.0	19.0	3.9	17.7	0.1	74.3	0.5	132.1
88-13	O	0.5	2.2	58.6	0.0	10.2	27.4	1.1		1.0
	E	**0.5**	**4.2**	**31.2**		**18.4**	**45.4**	**0.2**		**1.0**
	χ^2		1.0	24.1		3.7	6.0	4.1		38.7
89-02	O			40.4	9.0		48.3	2.2		1.0
	E			**29.4**	**10.7**		**57.8**	**2.4**		**1.0**
	χ^2			4.1	0.3		1.6			6.0
89-06	O			77.2	4.3	2.2	16.3			1.0
	E	**10.4**	**0.8**	**62.9**	**14.0**	**7.2**	**4.7**			**1.0**
	χ^2	10.4	0.8	3.3	6.7	3.5	28.6			53.3
89-14	O	0.8		43.0	12.4	0.8	8.3	34.7		1.0
	E	**9.7**		**37.9**	**8.0**	**6.1**	**11.7**	**26.7**		**1.0**
	χ^2	8.2		0.7	2.4	4.6	1.0	2.4		19.3
90-09	O			14.0		14.0	19.0	53.0		1.0
	E			**10.1**	**0.5**	**14.8**	**37.6**	**37.1**		**1.0**
	χ^2			1.5	0.5		9.2	6.8		18.1

[a] E-P = emergent palustrine, S-S = scrub-shrub, F-P = forested palustrine, F-U = forested upland, O-U = open upland, U-S = upland shrub, O = observed (use), E = expected (availability).

forested upland, but 2 cranes (88-13, 90-09) showed an avoidance for forested upland. Two cranes (88-05, 90-09) preferred open upland, and no crane avoided open upland. Results of this analysis indicated that Hypothesis II also was to be rejected.

A comparison of home ranges between 1992 and 1993 was made for 3 of the nesting cranes (the fourth nesting crane had a non-functional transmitter in 1992 that was replaced in autumn 1992). There was little difference in location of the home range of crane 88-05 (Fig. 2A) in the 2 years, although size was larger in 1993. In the summer of 1992, M Pool was filled with water so that the home range of 88-13 (Fig. 2B) included only the edge of M Pool. This pool was drawn down in spring 1993, creating a large mudflat. Crane 88-13 fed on a portion of that mudflat. Thus the size of home range of 88-13 was larger and shifted more southwesterly in 1993 than in 1992. Crane 89-06 selected the G Pool mudflat, created by drawdown of that pool, as his feeding ground in the summer of 1992 (Fig. 2C). But in 1993 G Pool was filled with water, so the home range of 89-06 was smaller in 1993 than in 1992.

DISCUSSION

Home range is defined as the area occupied by the individual in its normal activities of food gathering, mating, and caring for young (Burt 1943). The term "home range" is used to describe a normal movement of an animal so that exploratory movements outside that area should not be included. In this study the "summer home range" was

Table 3. Selection[a] of habitat types within home ranges of 6 isolation-reared sandhill cranes, Seney National Wildlife Refuge, Michigan, 1992–93.

	Crane					
Type	88-05	88-13	89-02	89-06	89-14	90-09
Open water	–	0	n	–	–	n
Bog	–	–	n	–	n	n
Emergent palustrine	+	+	0	+	0	0
Scrub-shrub	–	n	0	–	0	–
Forested palustrine	–	–	n	–	–	0
Forested upland	0	–	0	+	+	–
Open upland	+	0	0	n	0	+
Upland shrub	0	n	n	n	n	0

[a] + = selected, – = avoided, 0 = no evidence for selection or avoidance, n = not present in home range.

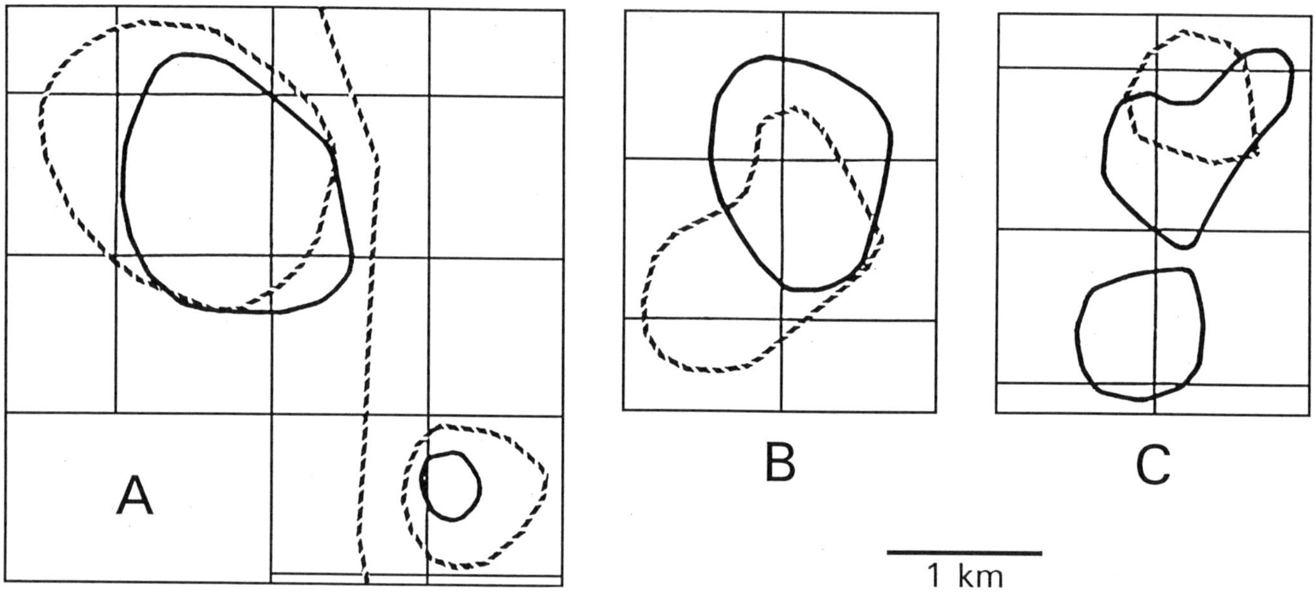

Fig. 2. Comparison of the 75% harmonic mean home ranges of cranes (A) 88-05, (B) 88-13, and (C) 89-06 at Seney National Wildlife Refuge, Michigan, in 1992 (solid lines) and 1993 (dashed lines).

defined as the area occupied by a crane (or a pair of cranes) in its normal activities from its arrival at the home range area to the time that it started congregating with other cranes prior to migration. Most Great Lakes sandhill cranes spend the winter in Florida and southern Georgia, so congregation and migration must be excluded from home range calculations. For the isolation-reared cranes, it is important to know the size of home range, the habitat they use for nesting and foraging, and how their time was spent in each habitat, i.e., the habitat time budget. This information can be used to formulate a habitat management plan and also may be useful to selection of whooping crane release sites.

McMillen (1988) chose the 95% harmonic mean utilization distribution to estimate home ranges of crane chicks. We observed that the area used by adult cranes was much larger than the area used by crane chicks, and the use area pattern of adult cranes was different from that of chicks because some feeding grounds of adult cranes were isolated from the nesting area by highways or towns. Three crane home ranges (cranes 88-05, 89-02, 90-09) calculated by the 95% harmonic mean utilization included highway M-77 or the town of Germfask, which cranes never used. For this reason we chose 75% harmonic mean home range instead of 95% harmonic mean home range for habitat analysis. The former better represented what we thought was actually occurring in the field.

Although average size of the 6 home ranges was about 200 ha (Table 1), extremes were 50 ha to about 300 ha (75% harmonic mean). Home ranges of wild cranes also vary greatly in size. Littlefield and Ryder (1968) reported that the territory size of western greater sandhill cranes varied with density of birds in a particular area. The smallest crane territory they recorded was 1.2 ha, and average size was 25.1 ha. In Okefenokee Swamp in Georgia the average annual home range (convex polygon) of non-migratory, adult Florida sandhill cranes was 93 ± 26 (SE) ha (Bennett 1989). However, the average home range (convex polygon) of Florida sandhill cranes in Florida was about 1,370 ha for territorial adults and 550 ha for resident adult crane pairs (Nesbitt 1990). Many factors affect size of home range, such as food availability, habitat type, and crane density in the area. At Seney the smallest home range (crane 89-14) included the nesting marsh adjacent to the feeding ground, so the crane pair could walk to its feeding ground. On the other hand, 2 pairs of wild cranes also used this same marsh as their territory, which limited the activity area of crane 89-14. Crane 89-02 established a territory on the east side of highway M-77, off Seney but abutting the refuge. In this home range predominant habitats were forested upland and emergent palustrine dotted by willow or alder. Lack of open hayfield for feeding in the home range and lower crane density near the crane territory may have been the main reasons for the larger size of this home range. The home range of 88-05 had an isolated feeding ground. The separate feeding ground was about 2.4 km from the nesting marsh, so this home range was the second largest among home ranges of the 6 cranes.

On average the 6 crane home ranges contained 36%

emergent palustrine wetland (range 10–63% of total area of home ranges). In different places sandhill cranes use different habitats. For example, in southern Michigan breeding territories of sandhill cranes are mostly located in sedge marsh, but in the Upper Peninsula of Michigan they are located mostly in sphagnum moss (*Sphagnum* spp.)-leatherleaf habitat except at Seney (Walkinshaw 1933, 1978). Although sandhill cranes use habitat variably, almost all of them select breeding territories in wetlands (Walkinshaw 1981, Johnsgard 1983). At Seney the most extensive study of wild cranes was in 1985–87 (McMillen 1988, Urbanek and Bookhout 1992*b*), and most of them nested in emergent palustrine wetland that contained cattail or sedge. They selected nest sites in or near emergent (non-woody) wetlands and avoided forested uplands (Baker et al. 1995). Comparison of the confidence interval on the proportion of habitat used to that available (home range components) indicates that the cranes we studied, like other cranes at Seney, selected their breeding territory in emergent palustrine wetland.

Home ranges of the 6 cranes contained 40% upland. Taylor (1976) reported upland openings within several kilometers of the nest site were important feeding grounds for wild sandhill cranes in the Upper Peninsula of Michigan and that openings were used increasingly as summer progressed. Our observations support his. Feeding grounds of cranes 89-14 and 90-09 were open hayfield (2.5–30 ha) surrounded by forested upland, and the feeding ground of crane 89-02 contained an open field (about 8 ha) dominated by bracken fern (*Pteridium aquilinum*) and surrounded by forested upland. Although feeding grounds of cranes 88-13 and 89-06 did not contain a hayfield, they did contain open mudflats created by drawn-down pools. By comparing the confidence interval on the proportion of observed habitat use to the proportion of habitat available (home range components), we conclude the cranes preferred to use open upland.

According to vegetation sampling data and vegetation cover maps, dominant tree species of the forested upland in home ranges of cranes were red pine and sugar maple. Cranes 89-06 and 89-14 significantly preferred forested upland, but cranes 88-13 and 90-09 significantly avoided forested upland. One possible reason for the difference is that the home range of crane 89-06 was located in a tourist area, and the home range of crane 89-14 was located on Seney near a main entrance to the refuge. Human activities and vehicles disturbed the 2 cranes more often, so they stayed at the edge of forested upland instead of in the open area. The home range of crane 88-13 was located near M Pool where human activities were rare, and the feeding ground of crane 90-09 was located at a private hayfield and forested upland with no human activities. Thus cranes 88-13 and 90-09 spent more time in open upland rather than in forested upland. The distribution of locations indicated that the cranes preferred to use the border between forested upland and open upland, especially where hayfield was adjacent to hardwood forested upland. Cranes rarely visited deep forested upland without open ground.

We conclude that the cranes we studied were selective in their choice of habitat types and that their primary preference for emergent palustrine matched that of wild sandhill cranes at Seney. Further, location of breeding territory was fixed as long as the selected site continued to be available; feeding grounds changed according to availability and were particularly influenced by pool drawdowns.

LITERATURE CITED

ACKERMAN B. B., F. A. LEBAN, M. D. SAMUEL, and E. O. GARTON. 1990. User's manual for program home range. Univ. Idaho For. Wildl. Range Exp. Stn. Contrib. 259. 80pp.

BAKER, B. W., B. S. CADE, W. L. MANGUS, and J. L. MCMILLEN. 1995. Spatial analysis of sandhill crane nesting habitat. J. Wildl. Manage. 59:752-758.

BENNETT, A. J. 1989. Movements and home ranges of Florida sandhill cranes. J. Wildl. Manage. 53:830-836.

BURT, W. H. 1943. Territoriality and home range concepts as applied to mammals. J. Mammal. 24:346-352.

COWARDIN, L. M., V. CARTER, F. C. GOLET, and E. T. LAROE. 1979. Classification of wetlands and deepwater habitats of the United States. U.S. Fish. Wildl. Serv. FWS/OBS-79/31. 131pp.

DIXON, K. R., and J. A. CHAPMAN. 1980. Harmonic mean measure of animal activity areas. Ecology 61:1040-1044.

JOHNSGARD, P. A. 1983. Cranes of the world. Indiana Univ. Press, Bloomington. 257pp.

KENWARD, R. 1987. Wildlife radio tagging. Academic Press, Inc., London, U.K. 222pp.

LITTLEFIELD, C. D., and R. A. RYDER. 1968. Breeding biology of the greater sandhill crane on Malheur National Wildlife Refuge, Oregon. Trans. North Am. Wildl. Nat. Resour. Conf. 33:444-454.

MCMILLEN, J. L. 1988. Productivity and movements of the greater sandhill crane population at Seney National Wildlife Refuge: potential for an introduction of whooping cranes. Ph.D. Thesis, Ohio State Univ., Columbus. 240pp.

MOHR, C. O. 1947. Table of equivalent populations of North American small mammals. Am. Midl. Nat. 37:223-249.

NAMS, V. O. 1990. Locate II user's guide. Pacer, Truro, N.S. 84pp.

NESBITT, S. A. 1990. Home range and habitat use of Florida sandhill cranes. J. Wildl. Manage. 54:92-96.

NEU, C. W., C. R. BYERS, J. M. PEEK, and V. BOY. 1974. A technique for analysis of utilization-availability data. J. Wildl. Manage. 38:541-545.

SAMUEL, M. D., and E. O. GARTON. 1985. Home range: a weighted normal estimate and tests of underlying assumptions. J. Wildl. Manage. 49:513-519.

______, and ______. 1987. Incorporating activity time in harmonic home range analysis. J. Wildl. Manage. 51:254-258.

TAYLOR, W. E. 1976. Sandhill crane habitat management on the Hiawatha National Forest. Pages 44-50 *in* J. C. Lewis ed. Proc. [1975] int. crane workshop. Oklahoma State Univ. Publ. Printing, Stillwater.

U.S. DEPARTMENT OF THE INTERIOR. 1978. Technical report for Seney National Wildlife Refuge R-1963. Commonwealth Associates, Inc., Jackson, Mich. 296pp.

URBANEK, R. P., and T. A. BOOKHOUT. 1992*a*. Development of an isolation-rearing/gentle release procedure for reintroducing migratory cranes. Proc. North Am. Crane Workshop 6:120-130.

______, and ______. 1992*b*. Nesting of greater sandhill cranes on Seney National Wildlife Refuge. Pages 161-167 *in* D. A. Wood, ed. Proc. 1988 North Am. crane workshop. Florida Game Fresh Water Fish Comm. Nongame Wildl. Program Tech. Rep. 12.

______, and ______. 1994. Performance of captive-reared cranes released into a migration route in eastern North America. Pages 121-129 *in* H. Higuchi and J. Minton, eds. The future of cranes and wetlands. Wild Bird Soc. Japan, Tokyo.

WALKINSHAW, L. H. 1933. The sandhill crane in a Michigan marsh. Wilson Bull. 45:99-106.

______. Sandhill crane studies in Michigan's Upper Peninsula. Jack-pine Warbler 56:107-121.

______. 1981. Sandhill crane. Pages 151-162 *in* J. C. Lewis and H. Masatomi, eds. Crane research around the world. Int. Crane Found., Baraboo, Wis.

WHITE, G. C., and R. A. GARROTT. 1990. Analysis of wildlife radio-tracking data. Academic Press Inc., San Diego, Calif. 383pp.

AN EXPERIMENTAL RELEASE OF WHOOPING CRANES IN FLORIDA - THE FIRST THREE YEARS

STEPHEN A. NESBITT, Florida Game and Fresh Water Fish Commission, 4005 South Main Street, Gainesville, FL 32601, USA
MARTIN J. FOLK, Florida Game and Fresh Water Fish Commission, 4825 Citrus Drive, St. Cloud, FL 34772, USA
MARILYN G. SPALDING, Department of Pathobiology, College of Veterinary Medicine, University of Florida, Gainesville, FL 32611, USA
JAMES A. SCHMIDT, Florida Game and Fresh Water Fish Commission, 3250 Rustic Drive, Kissimmee, FL 34744, USA
STEPHEN T. SCHWIKERT, Florida Game and Fresh Water Fish Commission, 4005 South Main Street, Gainesville, FL 32601, USA
JANE M. NICOLICH, Endangered Species Research Program, Patuxent Wildlife Research Center, U.S. Geological Survey-Biological Resources Division, Laurel, MD 20708, USA
MARIANNE WELLINGTON, International Crane Foundation, E-11376 Shady Lane Road, Baraboo, WI 53913, USA
JAMES C. LEWIS, U.S. Fish and Wildlife Service, P.O. Box 1306, Albuquerque, NM 87103, USA
TOM H. LOGAN, Florida Game and Fresh Water Fish Commission, 620 South Meridian Street, Tallahassee, FL 32399-1600, USA

Abstract: Fifty-two juvenile whooping cranes (*Grus americana*) were soft-released in Florida from February 1993 to April 1995. The birds were released in groups ranging in size from 5 to 14 individuals. The first-year survival rate was 0.42 for all years combined. First-year rates for each year were 0.36 for 1993, 0.32 for 1994, and 0.47 for 1995. Most mortality (62%) occurred during the first 3 months following release. The survival rate of 0.81 the second year after release was similar to that of Florida sandhill cranes (*G. canadensis pratensis*) (0.87). Predation by bobcats (*Lynx rufus*) was the only known source of mortality. We attempted to modify the roosting behavior and habitat use of released birds. Before and immediately after release, efforts were made to condition birds to roost in water and avoid rank, overgrown habitat. Most dispersal distances were similar to those of local populations of sandhill cranes, but there were 2 episodes that exceeded the normal range of local cranes. Four pairs have formed among birds that survived for greater than 1 year. Two pairs defended territories in spring 1995, and 1 completed a nest platform within the defended territory.

PROC. NORTH AM. CRANE WORKSHOP 7:79-85

Key words: captive-rearing, cranes, Florida, *Grus americana*, soft-release, whooping crane.

The minimal criteria for downlisting the whooping crane from endangered to threatened status are (1) 2 self-sustaining wild populations, additional to the Wood Buffalo-Aransas population (AWBP), that (2) must be reproducing at an acceptable rate for 10 years (U.S. Fish and Wildlife Service 1994). Feasibility studies of establishing a population of whooping cranes in Florida in order to downlist the species began in 1980 (Nesbitt and Carpenter 1993), and the first experimental release of whooping cranes occurred in 1993. Florida offers a unique opportunity to establish a nonmigratory population of whooping cranes similar to the population that occurred in Louisiana until the late 1940's (Lowery 1974, Gomez 1992). The Florida peninsula provides extensive areas of suitable crane habitat that supports a stable population of 4,000 to 6,000 Florida sandhill cranes (Nesbitt 1996).

The goal of the Florida release is to establish a population of ≥25 breeding pairs by the year 2020 (U.S. Fish and Wildlife Service 1994). Techniques for raising birds and introducing them into the wild were developed with sandhill cranes and have been used successfully to release Mississippi sandhill cranes (*G. c. pulla*) (Ellis et al. 1992).

Previous studies of sandhill cranes in Florida (Nesbitt and Carpenter 1993) and Mississippi (Ellis et al. 1992) indicated that initial mortality might be high, 40 to 60% during the first year after release, and that predation would be a major factor. Because there had been mortality at Patuxent Wildlife Research Center (Patuxent) due to eastern equine encephalitis (EEE) (Carpenter and Dein 1987), we were also concerned about the birds' reaction to this or other diseases in the wild.

It was presumed, when the idea for establishing whooping cranes in Florida was initially discussed in the late 1970's, that cross-fostering of whooping crane eggs into sandhill crane nests would be the preferred method of introduction. This technique successfully produced wild, migratory cranes in Idaho (Drewien and Bizeau 1978). However, by 1984 there was no evidence of pair bonds developing among the Idaho birds, and doubts about the behavioral consequences of foster rearing by non-conspecifics began to arise (Drewien et al. 1989, Mahan and Simmers 1992, U.S. Fish and Wildlife Service 1994). Because of these behavioral concerns, we decided that soft-release (Ellis et al. 1992) of captive-reared birds would be the primary method of introduction. Whooping cranes for release originated from 2 captive flocks. One flock was located at the Patuxent Wildlife Research Center (formerly U.S. Fish and Wildlife Service), National Biological Service, Laurel, Maryland. The other, at the International Crane Foundation (ICF), Baraboo, Wisconsin, is maintained under direction of the U.S. Fish and Wildlife Service.

This establishment effort is cooperatively funded by the

Florida Game and Fresh Water Fish Commission and Regions 2 and 4 of the U.S. Fish and Wildlife Service. Administration of the project is covered by a memorandum of understanding between the 2 agencies. We wish to thank A. Adams, R. Overstreet, and R. Gerali for maintaining their property in a condition that the cranes have found so attractive and for allowing us access to their property. Transportation of some of the whooping cranes from Maryland and Wisconsin to Florida was provided by Windway Capitol Corporation. Their generosity has been of great benefit to this project. B. Wagner has assisted with soft-release and monitoring the cranes from 1993 to 1995. M. Nagendran assisted with releases in 1995 and provided several helpful suggestions.

STUDY AREA

Previous studies identified the Kissimmee Prairie area of Florida as the area with the best potential to support a new nonmigratory population of whooping cranes (Bishop 1988). The Kissimmee Prairie consists of some 500,000 ha of freshwater marsh and open grasslands in Osceola and Polk Counties (Fig. 1) associated with the floodplain of the Kissimmee River. Marshes were dominated by pickerelweed (*Pontederia cordata*) and maidencane (*Panicum hemitomon*). Most grasslands were in improved pasture used for livestock grazing or sod production; some were allowed to revert to native prairie. Most areas preferred by the local crane population were managed for grazing. Releases were centered at the Three Lakes Wildlife Management Area (TLWMA) in south-central Osceola County (Fig. 1).

METHODS

Whooping cranes were hatched the spring prior to release at the captive-rearing facilities and were reared especially for release with established methods (Wellington et al. 1996). They were flown to Florida in groups of 5 to 14 birds on commercial carriers or on private aircraft from November to April. Shipments were scheduled so the birds would be introduced into the acclimation/release pen after dark to reduce stress and aid in acclimation to the new environment. Birds were examined for transportation injuries and banded with numbered aluminum USFWS bands and a unique series of colored plastic leg bands before being introduced to the pen. They were also fitted with leg-band-mounted radio transmitters (Melvin et al. 1983). Each radio transmitter (Advanced Telemetry Systems, Isanti, Minn.) had a mortality indicator switch, battery life of 24 months, and weighed approximately 65 g. One wing was brailed (Ellis and Dein 1991) to preclude flying during the 2- to 4-week acclimation phase. Brails were changed to the alternate wing after 2 weeks unless release was to occur before 4 weeks of acclimation. Capturing the bird for brail change or removal was accomplished after dark with lights and long-handled nets. Birds were returned to the pen after brail removal and were then free to leave.

The primary release enclosure was built in 1992 on the Sunset Ranch unit of the Three Lakes Wildlife Management Area and was similar to those used for the release of Mississippi sandhill cranes (Ellis et al. 1992). It was constructed of galvanized chainlink fence, 30 × 120 m and 3 m high. Temporary (portable) satellite release pens were used during the third year. These pens were 15 × 18 m, and 2.5 m high, and made of welded wire supported by PVC posts; each could be erected on site in ≤40 man-hours. A metal detector was used in and around the perimeter of the pens to remove any metal construction waste.

A quarantine protocol, developed by representatives from the National Wildlife Health Center, Madison, Wisconsin, Patuxent, ICF, and University of Florida School of Veterinary Medicine, required that the birds be quarantined for 60 days before shipment to Florida. They were also quarantined after they arrived in Florida for 2–4 weeks during their acclimation period.

We cleared and mowed 10–20 ha around the release pen to reduce stalking opportunity for bobcats and reduce post-release predation. We also attempted to capture and relocate bobcats from the release vicinity.

Each bird was checked daily for the first year post-release. Monitoring was reduced to ≥twice per week after the first year. Birds that dispersed (movement ≥10 km) from the release area were monitored from fixed-wing aircraft. Necropsies of dead birds were conducted by wildlife pathologists of the University of Florida School of Veterinary Medicine or at the National Wildlife Health Center. First-year post-released survival was based on the percent of birds that survived for the 12 months following the date of brail removal for each release group. We used the staggered entry design, a modified Kaplan-Meier procedure (Pollock et al. 1989), to estimate survival after the first year to compare results with estimates for the Florida crane population studied earlier (Nesbitt 1992). This is a simple nonparametric method for deriving a running survival rate that can accommodate individuals being added to or lost from the study population. Survival results are preliminary at this time, and an in-depth analysis or comparisons of rearing and release techniques would be premature.

The whooping cranes released in Florida were designated as "experimental nonessential" (Lewis and Finger 1993) and thus not subject to several aspects of the Endangered Species Act. This designation increased management flexibility.

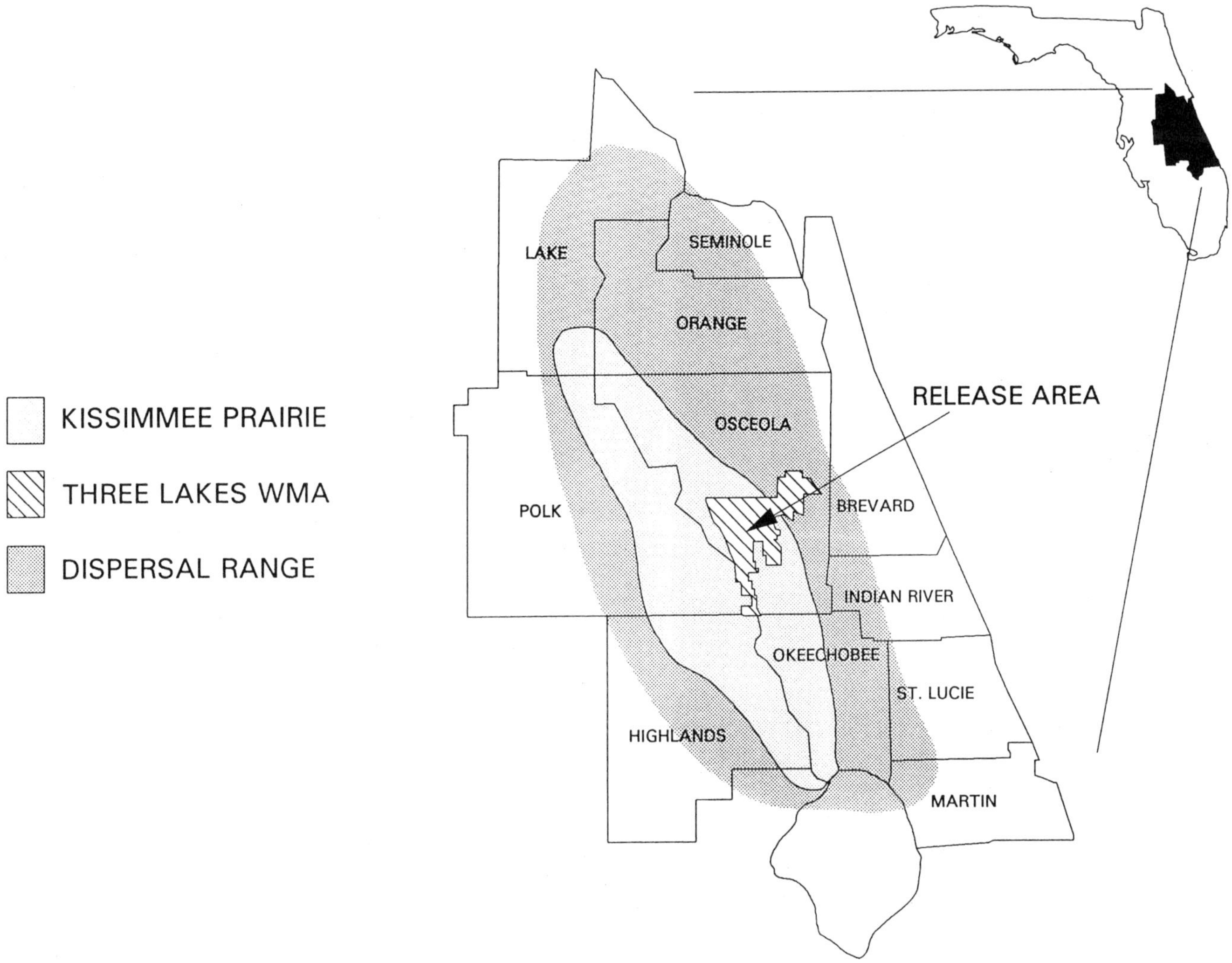

Fig. 1. Kissimmee Prairie, Three Lakes Wildlife Management Area, and release site for whooping cranes introduced in Florida, 1993–95.

Another advantage of this designation was that the owners of the private property that the cranes were likely to frequent would not be constrained in their land use practices due to the presence of the whooping cranes.

RESULTS

Releases and Habitat Management

Birds from the initial releases in 1993 and 1994 tended to roost on dry ground and to use heavily vegetated areas. Birds that did not roost in water or were prone to use overgrown areas were more likely to be killed by bobcats. Efforts were made to modify the release area habitat after these first releases by mechanically removing brush and mowing an additional area of pasture (25–30 ha) around the release pen. Changes were made at the rearing facilities to better prepare the birds for release. Standing water was provided in most rearing pens to encourage the birds to roost in water, and cranes were housed in pens with an open, unobstructed view of the surroundings. Some birds were exposed to food items and feeding sites similar to those they might find in Florida during "marsh walks" with their costumed caretakers. Birds were allowed minimal flight experience in a covered flight pen prior to release. We hoped that limiting flight experience at rearing sites would encourage site loyalty to the Florida release site.

We began considering alternative release locations, mainly on private property, in the fall of 1994. A temporary satellite release pen was constructed on 1 of these locations, but the exact sites for the pens were not selected until just before the birds were to be introduced. This allowed us to

Table 1. Characteristics of 52 whooping cranes released in Florida, 1993–95, and their status on 15 April 1996.

Date of release	Radio band colors[a]	Rearing location[b]	Sex	Hatch year	Rearing method[c]	Cause and date of death	
10 Feb 93	[GO]	I	M	92	CT	Unknown	Sep 94
	YG	I	M		CT		
	[YW]	I	M		CT	Bobcat	2 Mar 93
	BB	I	M		CT		
	[OB]	I	F		PR	Bobcat	23 Feb 93
	[OW]	I	F		CT	Bobcat	10 May 93
	[GB]	I	F		CT	Bobcat	22 Feb 93
	[SW]	I	F		CT	Bobcat	16 Feb 93
	[RG]	P	M		CT	Bobcat	16 May 95
	[BS]	P	M		CT	Bobcat	4 Jul 93
	[WY]	P	F		CT	Bobcat	22 Feb 93
	[SY]	P	F		PR	Bobcat	3 Jun 93
	[BO]	P	F		CT	Bobcat	20 Jun 93
	SO	P	F		CT		
14 Dec 93	[Bk]	I	M	93	PR	Bobcat	10 Feb 94
	[B]	I	F		PR	Bobcat	25 Dec 93
	[G]	P	M		PR	Bobcat	16 Dec 93
	[W]	P	M		PR	Bobcat	30 Sep 94
	Y	P	F		PR		
15 Feb 94	RG	P	M	93	CT		
	[R]	P	M		CT	Bobcat	28 Mar 94
	[RR]	P	M		CT	Bobcat	17 Feb 94
	[RBk]	P	M		CT	Bobcat	16 Mar 94
	RW	P	F		CT		
	RB	P	F		CT		
20 Apr 94	[YY]	P	M	93	CT	Bobcat	2 Nov 94
	[WYY]	P	M		CT	Bobcat	6 May 94
	[WY]	P	M		CT	Bobcat	6 May 94
	R	P	F		CT		
	BkY	P	F		CT		
	[RY]	P	F		CT	Bobcat	2 Nov 94
	[B]	P	F		CT	Bobcat	6 May 94
	[GY]	P	F		CT	Bobcat	2 Nov 94
23/25 Dec 94	[BkBk]	I	F	94	CT	Bobcat	23 Sep 95
	[RR]	I	M		CT	Bobcat	4 Jan 95
	GYY	I	M		CT		
	[OB]	I	M		PR	Bobcat	21 Feb 95
	OY	I	M		CT		
	BkWW	I	M		CT		
	[BkW]	I	F		CT	Bobcat	4 Jan 95
	GRR	I	F		CT		
12/20 Mar 95	BY	P	M	94	CT		
	[BB]	P	M		CT	Bobcat	22 Oct 95
	BBk	P	M		CT		
	BR	P	M		CT		
	BRR	P	M		CT		
13 Apr 95	[WW]	P	M	94	CT	Bobcat	4 Oct 95
	WY	P	F		CT		
	WB	P	F		CT		
	[WBk]	P	M		CT	Bobcat	10 Jul 95
	[WR]	P	M		CT	Bobcat	4 Oct 95
	WRR	P	F		CT		

[a] G= green, O = orange, Y = yellow, W = white, B = blue, S = silver, R = Red, Bk = black, [Mortalities].
[b] I = ICF, P = Patuxent.
[c] CT = costume-reared, PR = parent-reared.

Table 2. Comparison of survival (%) by sex, rearing method, and release method of whooping cranes released in Florida, 1993-95.

Number of cranes	Sex		Rearing method		Release method	
	Male	Female	Costume	Parent	Soft-release	Satellite
Released	29	23	44	8	36	16
Surviving 3-12 months	62.1	69.6	70.5	37.5	52.8	81.3
Surviving >12 months	37.9	43.5	45.5	25.0	27.8	62.5

respond to any recent changes in water levels, habitat conditions, or local crane use, and select the site that provided optimal crane habitat with lower predator density. Birds were first held in the large acclimation pen; then after brails were removed, they were moved to the satellite release pens at the selected site. They were introduced to these pens after dark and could leave as soon as they were able to fly over the fence. If they did not leave the pen after 3 days, 1 side of the pen was removed.

Fifty-two whooping cranes were soft-released between 10 February 1993 and 13 April 1995 (Table 1). Forty-one cranes were released in 5 cohorts (14, 5, 6, 8, and 8 birds) from the large acclimation pen on TLWMA and 11 were released in 3 cohorts (2, 3, and 6 birds) from a satellite pen. The time of day following brail removal when the first bird from each of the first 3 release groups left the release pen ranged from 0732 to 1600 hours on the following day. Among the group released in April 1994, the first bird did not leave the pen until 1200 hours 3 days after brail removal. Only 1 of the birds released from the satellite pen departed prior to the pen being opened, 3 days after they were moved to the pen.

Mortality and Survival

Thirty-five of the released cranes survived for ≥3 months after release. The first-year survival of 52 cranes released was 38.2%. Survival rate by release year was 35.7% (1993), 31.6% (1994), and 47.4% (1995). There were no major differences in first-year survival when results of releases were compared for sex or rearing method (Table 2). There was improved survival for birds released from satellite pens. First-year survival was 30.6% for birds released by the standard soft-release method and 50.0% for releases by the satellite method. All mortality during the first year post-release was the result of bobcat predation. Disturbance from private airboat use near the release site may have contributed to at least 1 mortality event. Most (62%) of the first-year mortality occurred within 3 months of release (Fig. 2). One bird that was found caught in a barbed-wire fence in December 1994 was rehabilitated and returned to the wild.

The survival rate during the second year after release improved to 0.81, similar to that for a normal Florida sandhill crane population in north-central Florida (0.87, Nesbitt 1992).

Use of metal detectors prior to arrival of the first cranes did not avoid ingestion of metal pen construction scraps. Six birds from the initial group of 14 ingested metal scraps and 2 exhibited lethargy and weight loss (Spalding et al. 1997). These birds were captured and the metal surgically removed. They recovered from the surgery in a few days and were released, but both birds were eventually killed by predators. Subsequent release groups were checked before and after arrival in Florida for ingested metal. Birds that were discovered to have ingested metal at the rearing facilities were not shipped for release unless the metal was removed or judged not to be a threat.

Dispersal and Movements

The mean time between leaving the release pen to initial dispersal from the release area for the birds released in 1993 and 1994 was 89.75 days (range 43-163 days). Five of 18 birds that survived more than 89 days never dispersed, but

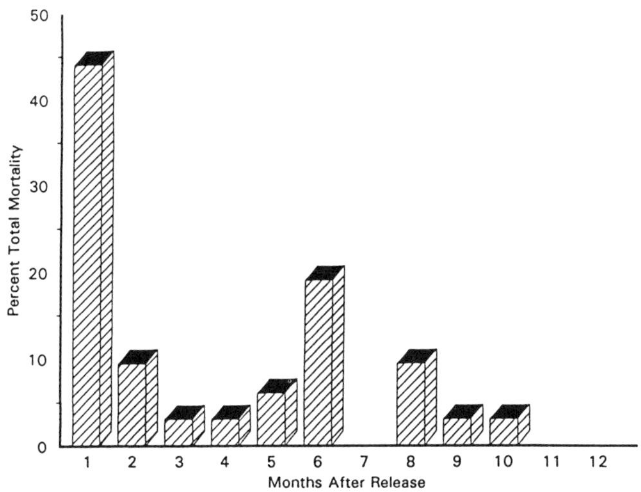

Fig. 2. Distribution of mortality by month after release for 32 of 52 whooping cranes released in Florida, 1993-95.

only 1 of these survived >1 year.

Three from the first release group left the release area on 16 May 1993, initially flying south-southeast and then north, covering a large part of central Florida during a period of 2 weeks (Fig. 1). Subsequent dispersing birds have traveled no farther from the release area than this first group and have, in some instances, followed the same general route. In all cases, dispersing birds have selected habitat that supported Florida sandhill cranes. We never saw whooping cranes flying with sandhill cranes, but they could have followed cranes we did not see or simply selected habitat that was used by sandhill cranes. There have been 4 dispersal events during the first 3 release years. Birds left the release area in groups of 2–5. Only 1 group failed to return to the release area within 5 months. The exception was the pair color-marked RG/RB, which left the release area 30 March 1994, traveled north to Lake County, and joined 3 older birds, color-marked YG, SO, OG, that had been there for several weeks. Although the 3 older birds returned to the release area, the pair (RG/RB) has remained (as of November 1995) in Lake County.

Pairing Behavior

Four pairs developed within groups surviving from the 1993 and 1994 releases. Formation of the first pair bond was initiated in December 1993 between YG (♂) and SO (♀), 18 and 20 months old, respectively. A second pair formed in March 1994 and consisted of RG (♂) and RB (♀), 11 and 9 months old, respectively. The third became apparent in October 1994 between BB (♂), 28 months old, and Y (♀), 17 months old. A fourth pair formed between the 2 birds remaining from the April 1994 release (BkY [♀] and R [♀]) perhaps a result of these being the only survivors of a cohort that was unusually cohesive. This pair consisted of 2 females; 1 bird (BkY) exhibited male-like behavior when the pair was displaying dominance and when unison calling. These pairs have remained together (as of November 1995), separated from other whooping cranes in the area. The first pair attempted copulation several times during January and February 1994, constructed a nest platform in a marsh 300 m from the release pen in January 1995, and defended this marsh from local sandhill crane pairs. The third pair (BB and Y) also appeared to be defending a territory during spring 1995.

DISCUSSION

The size of release groups may have had an influence on the post-release behavior of these cranes. Aside from the advantage to us of handling a smaller number of birds for brail changes and release, smaller groups (6–8 birds) seemed to persist as cohesive units longer following release than the initial larger group of 14. Smaller groups remained identifiable in larger flocks where 2 or more release groups were together. Release groups comprised of birds of more than 1 rearing technique were not as cohesive as a group of birds all raised similarly. Therefore, it seems best to assess experiments in rearing strategies with groups of birds all reared under the same conditions. Releasing groups reared under mixed strategies might obscure advantages or disadvantages of techniques unless effects of rearing method were controlled by experimental design.

We noticed little difference in site attachment between birds released from the larger, permanent pens and those released from the satellite pens. It appeared, however, that the satellite release method promoted longer association with the release area, perhaps because the satellite pen was in better habitat. One consideration for pen site selection was use by Florida sandhill cranes and previously-released whooping cranes. Improved habitat with less woody vegetation, more natural foods, more Florida sandhill crane use, and fewer predators may explain why birds released with the satellite method had a higher first-year survival rate. Increased effort to improve conditioning for release at the rearing site also may have contributed to improved survival of the 1995 release group. Our efforts to reduce predation at the TLWMA soft-release site by improving the release habitat (mowing, burning, and predator removal) and to modify the behavior of birds after release (i.e., to roost in water, select appropriate habitat) were not as successful.

Sandhill cranes released without an adequate acclimation period ("abrupt releases") resulted in premature dispersal and lower post-release survival in other trials (Drewien et al. 1982, Ellis et al. 1992). Birds released by the satellite technique did not disperse sooner or farther than those released by the original soft-release method. No birds dispersed until ≥40 days after release by either method. The satellite soft-release method initiated in 1995 has resulted in reduced early mortality (first 3 months) without premature dispersal. Moving birds to a satellite site near the acclimation pen appears to produce a survival-dispersal result similar to the more traditional soft-release method developed in Mississippi (Ellis et al. 1992).

We expected that the whooping cranes released in Florida would remain nonmigratory, as did greater sandhill cranes (*G. c. tabida*) experimentally released in 1986–87 (Nesbitt and Carpenter 1993). We also expected that they would interact with the local Florida sandhill cranes and perhaps adopt their daily and seasonal movements. They have remained nonmigratory, but several weeks elapsed before we noticed any influence on their movements by local sandhill cranes.

Studies of sandhill cranes have shown that birds socialized for extended periods in the same rearing group were unlikely to pair with members of their release group (Nesbitt and Carpenter 1993). One of 4 pairs of released whooping cranes that have formed consisted of birds from different release years; the others were between members of the same release year. However, the 2 pairs that have shown the greatest inclination to breed consist of members of different rearing groups released together (YG/SO) or birds released in different years (BB/Y). Continuing to release several small groups annually rather than 1 large group may improve pair formation opportunities between birds released in the same year.

We plan to release ≥20 birds annually during the next 10 years and hope to develop a self-sustaining non-migratory population of whooping cranes before 2020. It is our hope that with continued refinement of rearing and release techniques we can increase first-year survival to greater than 60%. The threat of disease, particularly EEE, is still a concern, and we will continue to monitor the health of the experimentally released whooping cranes, as well as that of local Florida sandhill cranes.

LITERATURE CITED

BISHOP, M. A. 1988. Factors affecting productivity and habitat use of Florida sandhill cranes (*Grus canadensis pratensis*): an evaluation of three areas in central Florida for a nonmigratory population of whooping cranes (*Grus americana*). Ph.D. Thesis, Univ. Florida, Gainesville. 190pp.

CARPENTER, J. W., and F. J. DEIN. 1987. An outbreak of eastern equine encephalitis virus in captive whooping cranes. Pages 123-127 *in* J. C. Lewis, ed. Proc. 1985 crane workshop. Platte River Whooping Crane Maintenance Trust, Grand Island, Nebr.

DREWIEN, R. C., and E. G. BIZEAU. 1978. Cross-fostering whooping cranes to sandhill crane foster parents. Pages 201-222 *in* S. A. Temple, ed. Endangered birds: management techniques for preserving threatened species, Univ. Wisconsin Press, Madison.

_____, W. M. BROWN, and E. G. BIZEAU. 1989. Whooping crane cross-fostering experiment. Unpubl. Rep., Hornocker Wildl. Res. Inst., Univ. Idaho, Moscow. 10pp.

_____, S. R. DERRICKSON, and E. G. BIZEAU. 1982. Experimental release of captive parent-reared greater sandhill cranes at Grays Lake Refuge, Idaho. Pages 99-111 *in* J. C. Lewis, ed. Proc. 1981 crane workshop. Natl. Audubon Soc., Tavernier, Fla.

ELLIS, D. H., and F. J. DEIN. 1991. Flight restraint techniques for captive cranes. Pages 447-451 *in* J. Harris, ed. Proc. 1987 int. crane workshop. Int. Crane Found., Baraboo, Wis.

_____, G. H. OLSEN, G. F. GEE, J. M. NICOLICH, K. E. O'MALLEY, M. NAGENDRAN, S. G. HEREFORD, P. RANGE, W. T. HARPER, R. P. INGRAM, and D. G. SMITH. 1992. Techniques for rearing and releasing nonmigratory cranes: lessons from the Mississippi sandhill crane program. Proc. North Am. Crane Workshop 6:135-141.

GOMEZ, G. M. 1992. Whooping cranes in southwest Louisiana: history and human attitudes. Proc. North Am. Crane Workshop 6:19-23.

LEWIS, J. C., and L. FINGER. 1993. Endangered and threatened wildlife and plants, establishment of an experimental nonessential population of whooping cranes in Florida. Fed. Register 58:5647-5658.

LOWERY, G. H., JR. 1974. Louisiana birds. Louisiana State Univ. Press, Baton Rouge. 651pp.

MAHAN, T. A., and B. S. SIMMERS. 1992. Social preference of four cross-foster reared sandhill cranes. Proc. North Am. Crane Workshop 6:114-119.

MELVIN, S. M., R. C. DREWIEN, S. A. TEMPLE, and E. G. BIZEAU. 1983. Leg-band attachment of radio transmitters for large birds. Wildl. Soc. Bull. 11:282-285.

NESBITT, S. A. 1992. First reproductive success and individual productivity in sandhill cranes. J. Wildl. Manage. 56:573-577.

_____. 1996. Florida sandhill crane. Pages 219-229 *in* J. A. Rodgers, Jr., H. W. Kale, II, and H. T. Smith, eds. Rare and endangered biota of Florida. Vol. 5: birds. Univ. Florida Press, Gainesville.

_____, and J. W. CARPENTER. 1993. Survival and movements of greater sandhill cranes experimentally released in Florida. J. Wildl. Manage. 57:673-679.

POLLACK, K. H., S. R. WINTERSTEIN, C. M. BUNCK, and P. D. CURTIS. 1989. Survival analysis in telemetry studies: the staggered entry design. J. Wildl. Manage. 53:7-15.

SPALDING, M. G., S. A. NESBITT, M. J. FOLK, L. R. MCDOWELL, and M. S. SEPÚLVEDA. 1997. Metal consumption by whooping cranes and possible zinc toxicosis. Proc. North Am. Crane Workshop 7:237-242.

U. S. FISH AND WILDLIFE SERVICE. 1994. Whooping crane recovery plan. U.S. Fish Wild. Serv., Albuquerque, N.M. 92pp.

WELLINGTON, M., A. BURKE, J. M. NICOLICH, and K. O'MALLEY. 1996. Chick rearing. Pages 77-104 *in* D. H. Ellis, G. F. Gee, and C. M. Mirande, eds. Cranes: their biology, husbandry, and conservation. Natl. Biol. Serv., Washington, D.C., and Int. Crane Found., Baraboo, Wis.

USE OF CROSS-FOSTERED WHOOPING CRANES AS GUIDE BIRDS

RODERICK C. DREWIEN, Hornocker Wildlife Research Institute, University of Idaho, P.O. Box 3246, Moscow, ID 83843, USA
WENDY L. MUNROE,[1] International Crane Foundation, E-11376 Shady Lane Road, P.O. Box 447, Baraboo, WI 53913, USA and Hornocker Wildlife Research Institute, University of Idaho, P.O. Box 3246, Moscow, ID 83843, USA
KENT R. CLEGG, Hornocker Wildlife Research Institute, University of Idaho, P.O. Box 3246, Moscow, ID 83843, USA
WENDY M. BROWN,[2] Hornocker Wildlife Research Institute, University of Idaho, P.O. Box 3246, Moscow, ID 83843, USA

Abstract: We tested the use of wild, cross-fostered, adult whooping cranes (*Grus americana*) as guide birds to adopt and lead young whooping cranes on a predetermined migration route in the Rocky Mountains. We captured 3 wild adults (1 male and 1 female in 1993, 1 male in 1994) during winter at the Bosque del Apache National Wildlife Refuge (Bosque NWR), New Mexico, and moved them to captive facilities 80 km from Grays Lake National Wildlife Refuge (Grays Lake), Idaho, where they had originally been raised by sandhill crane (*G. canadensis*) foster parents. Adults were held for 6.75-7 months without complications. Five isolation-reared whooping crane chicks (chicks), 14-21 days old, were placed in enclosures next to penned adults, and were released with adults when 24-47 days old. Adults and chicks commingled in the pen until chicks fledged in late August. Their activities were recorded throughout the captive period. Although adults and chicks appeared to develop bonds in captivity, adults did not remain with chicks after release and migrated without them. Chicks did not permanently associate with any cranes after release in Idaho or, in 1 case, after transport to New Mexico. Injuries and deaths of 3- to 5-month-old chicks during vehicle transportation limited evaluation of the guide bird technique.

PROC. NORTH AM. CRANE WORKSHOP 7:86-95

Key words: adoption, isolation-reared, Grays Lake, *Grus americana*, guide bird, Idaho, New Mexico, whooping crane.

The Grays Lake cross-fostering experiment (1975-88) produced whooping cranes capable of migrating and surviving in the wild in the Rocky Mountain region of the United States (Drewien and Bizeau 1978; Drewien and Kuyt 1979; U.S. Fish and Wildlife Service 1986, 1994). However, because none paired and nested, the technique failed to create a self-sustaining migratory population. Cross-fostering may have interfered with normal sexual imprinting of whoopers on conspecifics (Drewien et al. 1989, Mahan and Simmers 1992, U.S. Fish and Wildlife Service 1994).

A potential use for surviving cross-fostered adult whooping cranes (adults) was as guide birds to adopt and introduce young whoopers to the wild. Unpaired adults had exhibited parental behavior during summers at Grays Lake and in a pen that we maintained for experimental purposes near Grays Lake during 1990-91 (R. C. Drewien and W. M. Brown, unpubl. data; U.S. Fish and Wildlife Service 1994). Solitary territorial males had occasionally built nests and helped neighboring sandhill crane pairs raise sandhill chicks, including protecting, feeding, and brooding young. In 1986, a male built several nests and intermittently incubated an empty nest and sandhill eggs that we placed in that nest. In 1988, 2 males assisted a sandhill crane foster-parent pair in raising a whooping crane chick. Both males and females associated with, fed, and temporarily reared sandhill crane chicks in the pen during 1990-91, suggesting that adults might adopt and rear whooping crane chicks (R. C. Drewien and W. M. Brown, unpubl. data). In addition, we (R. C. Drewien and W. M. Brown, unpubl. data) have observed adult sandhill cranes adopting chicks in the wild.

In 1992, the U.S. Fish and Wildlife Service (USFWS) and the Canadian Wildlife Service (CWS) recommended initiating a guide bird experiment at Grays Lake (U.S. Fish and Wildlife Service 1994). The objective was to have chicks imprint on conspecifics, learn survival skills from adults, and follow a predetermined migration route. Success of the experiment was dependent upon adults adopting flightless young, rearing them to fledging, and leading them in migration from Grays Lake south to the New Mexico winter area. In January 1993, 4 males (9-17 yr old) and 4 females (7-13 yr old) remained in the Rocky Mountain flock. Three adults that wintered at Bosque NWR were selected to serve as guide birds to introduce isolation-reared whooping crane chicks (Horwich 1989) into the wild. In this paper we report results of the guide bird experiment conducted in 1993 and 1994.

We appreciate financial support provided by the USFWS, World Wildlife Fund Canada, and the Whooping Crane Conservation Association. The project benefitted from administrative assistance of J. Lewis, USFWS, and R. Edwards, CWS. The CWS provided whooping crane eggs for the experiment. Others deserving special thanks include G. Archibald, S. Bowersox, A. Burke, J. Langenberg, C. Mirande, and M. Wellington, International Crane Foundation (ICF); D. Ellis, G. Gee, J. Nicolich, and G. Olsen, Patuxent

---

[1]Present address: 919 13th Ave., Seattle, WA 98122, USA.

[2]Present address: U.S. Fish and Wildlife Service, P.O. Box 1306, Albuquerque, NM 87103, USA.

Wildlife Research Center; M. Albaugh, S. Derrickson, and S. Leathery, National Zoological Park; M. Fisher, T. Nuffer, C. Peck, G. Sibbett, R. Stoor, and volunteer L. Brinkerhoff, Grays Lake NWR; and P. Norton, T. Tadano, J. Taylor and staff, and volunteer D. Ensminger, Bosque del Apache NWR. We thank T. Kohler, Windway Capital Corporation, Sheboygan, Wisconsin, for providing his aircraft to transfer whooping crane eggs and chicks from Wisconsin to Idaho. Veterinarians J. Langenberg, ICF; G. Olsen, Patuxent; and T. Moe and staff, Eastside Veterinary Clinic, Idaho Falls, Idaho, provided crane health care and advice. Pilot J. Winship, USFWS, flew cranes from Idaho to New Mexico. We appreciate reviews of the manuscript by J. Lewis, S. Nesbitt, R. Shea, R. Urbanek, and an anonymous reviewer.

METHODS

Capturing Guide Birds and Shipping Eggs and Chicks to Idaho

Adults were captured in January 1993 (2) and February 1994 (1) in New Mexico by night-lighting (Drewien and Clegg 1992). They were shipped from New Mexico by commercial air to Salt Lake City, Utah, and transported by vehicle to pen facilities near Grace, Idaho. Whooping crane eggs or chicks, originating from wild nests in Canada, were shipped by air from ICF to Idaho in June 1993 and 1994.

Hatching Chicks at the International Crane Foundation and in Idaho

The CWS provided 4 whooping crane eggs annually in May to ICF from nests in Wood Buffalo National Park, Canada. In 1993, 4 chicks were isolation-reared with a puppet/costume technique (Horwich 1989, Horwich et al. 1992, Archibald and Archibald 1992, Urbanek and Bookhout 1992) at ICF until they were self-feeding at 16–21 days, given health checks and tested for diseases, and then shipped to Idaho.

In 1994, 4 eggs were shipped by air to Idaho and hatched at Grays Lake. Chicks were reared in a similar but more simplified fashion. They were maintained in individual pens in a refuge building. Cloth models of adult whoopers in a brooding position were placed in pens. Chicks were exercised daily by a costumed caretaker in a marsh 50 m from the facilities.

Captive Facilities in Idaho

A 15.2- × 24.4-m holding pen, built in 1989 on the Clegg Ranch, near Grace, Idaho, 80 km southwest of Grays Lake, was renovated in 1992 and used for the experiment. The pen was 3 m high in the center with 2.1-m sides and was enclosed with nylon mesh netting (5 × 5 cm); the lower 0.67 m was covered with 5- × 5-cm welded wire set in 0.2 m of cement footing. A 3- × 6.1-m building was erected at 1 end of the pen for shelter during inclement weather. The building was subdivided into rooms with separate openings to the pen, and bedding consisted of chipped wood. A year-round spring-fed creek which did not freeze and had been enlarged to form a shallow pool traversed the width of the pen and provided drinking water, night roosts, and bathing sites (Fig. 1). The pen was enclosed by a New Zealand design 1.2-m-high anti-predator electric fence, although predators were uncommon in the area. The facility was >200 m from most human activities and buildings.

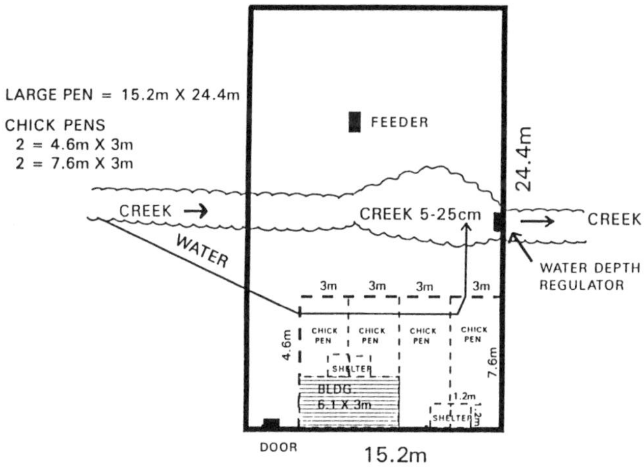

Fig. 1. Guide bird pen for associating captured, wild, cross-fostered whooping cranes with captive-reared young, Grace, Idaho, 1993–94.

Four small enclosures were constructed inside the pen (Fig. 1) to initially rear individual chicks and separate them from adults. This also prevented intraspecific aggression among chicks but allowed them to be near adults for socialization. Water and food were separated to maximize exercise opportunities. Small sheds (Fig. 1) with heat lamps were placed in each enclosure for use at night and during inclement weather. Enclosures were separated from the larger pen by 1.3- × 2.5-cm welded wire and nylon mesh netting. Chicks were fed commercial pellets (chick starter and maintenance diets) formulated for cranes and containing medication to control coccidiosis; adults ate corn, barley, wheat and natural foods found in the pen. Occasionally, we placed rainbow trout (*Oncorhynchus mykiss*) fingerlings in the pond for supplemental food. Before releasing chicks with adults, we placed a flightless sandhill crane chick (≤35 days old) with adults annually to assess their aggressiveness. Whooping crane chicks were placed with adults only when lack of aggression was confirmed.

Diurnal activities of cranes were monitored by a remote control video camera. The camera (Panasonic WV-BL204) and 10× zoom lens were placed in an outdoor, all-weather, domed housing unit and suspended from a 2.4-m-high tripod outside the pen; the video monitor was in a trailer 200 m from the pen. Diurnal activities were observed on the video monitor and recorded at 15-second intervals to develop time budgets. Most samples were recorded during 2-hour periods (range 0.5–4.0 hr) from dawn to dusk. Major activity categories recorded were drinking, feeding, vigilance (alert), resting, locomotion, comfort movements (body maintenance), agonistic, and vocalization. We annually divided time budget data collection into 3 periods: Period 1-late June/early July when chicks were in individual enclosures adjacent to but separated from adults, Period 2-early to mid-July when adults and young were first put together, and Period 3-late July-late August when adults and older chicks were together.

Release at Grays Lake

Cranes were transported in plywood boxes by vehicle to release sites at Grays Lake when chicks fledged (<90 days). Release sites were at fall staging areas of sandhill and whooping cranes where refuge grain fields, meadows, and wetland roost sites were in close proximity and protected from human disturbance. Families were placed in a subdivided holding pen (6.8 × 9.1 m) that separated adults and young for ≤1 day. This allowed for a gentle release and provided birds an opportunity to adjust to a new environment. The pen had drop doors on pulleys operated with a rope from a blind. The pen site was mowed to reduce standing vegetation to enhance predator detection. Supplemental food (barley and commercial pellets) was placed near the pen.

Predator control (trapping-12 traps for 2 weeks annually), primarily for coyotes (*Canis latrans*), was conducted near release sites before or during the release period by Animal Damage Control agents (ADC), U.S. Department of Agriculture. In 1993, 2 coyotes and 10 red fox (*Vulpes vulpes*) were removed; 2 red fox were trapped in 1994.

To identify individuals and monitor movements, we radiomarked chicks (Melvin et al. 1983) or marked them with colored plastic leg bands before release; adults had previously been marked with colored leg bands (Drewien and Bizeau 1978). If family bonds were not established by fall migration, plans called for recapturing chicks by night-lighting or bait trapping and transporting them by USFWS aircraft to the Bosque NWR winter site.

RESULTS

Capturing and Maintaining Adult Whooping Cranes In Captivity

In January 1993, we captured 2 adults (11-yr-old male, 10-yr-old female) by night-lighting at Bosque NWR. Adults were held in captivity for 7 months (27/28 Jan–28 Aug) without problems or injuries. In 1994, we made 7 capture attempts on 4 individuals but caught only 1 male (18 yr old). This male was confined for 6.75 months (11 Feb–31 Aug) without problems except that it was successfully treated (ivermectin–0.12 cc) for gapeworm (*Syngamus* sp.) after exhibiting signs of respiratory distress in late July.

Transporting Eggs and Chicks to Idaho; Hatching, Rearing, and Socializing Chicks

On 21 June 1993, 4 chicks, 16–21 days old, were transferred from ICF to Idaho without incident and placed in individual enclosures next to 2 adults. Although chicks were raised with conspecifics and caretakers wearing whooping crane costumes, upon arriving in Idaho they exhibited no aversion to humans at distances of ≥1 m. However, after chicks associated with adults for 18 days, they would not approach a costume caretaker playing whooping crane brood calls. Instead, they moved away from the caretaker and remained with adults. After associating with adults >2 weeks, chicks also responded in a similar fashion to humans approaching the pen.

On 3 June 1994, 4 eggs were transported from ICF to Grays Lake and they hatched 4–6 June. By 7 June, the 3 older chicks were being walked daily by a costumed caretaker to a nearby wetland for exercise and socialization.

Three chicks died during 9–14 June when 3, 8, and 10 days old. The 3-day-old chick was accidentally stepped on by the caretaker when it approached unseen through tall grass from behind and underneath the caretaker's long costume. The costume was shortened after the accident to increase visibility around the feet and legs. Cause of death for the 8-day-old chick was undetermined; that of the 10-day-old chick was possibly acute aspiration/pulmonary congestion (N. Thomas, National Wildlife Health Center [NWHC], Madison, Wisconsin, pers. commun.). The surviving chick was moved to an enclosure next to an adult when 14 days old and capable of self-feeding.

When chicks were introduced to the pen, adults exhibited no aggression towards them. In both years, we placed single sandhill crane chicks with adults for 21 and 20 hours, respectively. In 1993 the female immediately approached the chick, but it did not follow and remained alone until dark.

The following morning, it was roosting with the adults. No aggression was noted and the sandhill was removed. In 1994 the chick was following within 5 hours, and the male captured insects and fed it. They roosted together in the pond, and the sandhill was removed the following morning.

In 1993, whooping crane chicks were socialized with each other to assess aggressiveness before they were placed with adults. Chicks 2 and 3 were 21 days old when placed together. Two days later, 3 attacked 2 and we separated them. When 27 days old, they again were put together without problems and were released with 2 adults; no additional hostilities occurred. Chicks 1 and 4 were placed together when 31 and 26 days old, but aggressive behavior was more frequent than with the other 2 chicks. The next day, 1 attacked 4 and we separated them. When 35 and 30 days old, they were released with the 4 whoopers. The following day we removed 4, the youngest chick, because it was being harassed by other chicks. When 35 days old, again it was placed with the other 5 whoopers, but it was harassed by several chicks and we removed it. It was successfully introduced when 47 days old to the other 5 whoopers. In 1993, chicks were successfully integrated with each other and released with adults at 27, 27, 35, and 47 days, respectively. In 1994, the single chick was placed with the adult at 24 days.

Activities of Adults and Chicks in Pen

We recorded 422 hours (181 adult, 241 chick) of diurnal time budget data. During Period 1, we recorded 77:05 hours (31:55 adult, 45:12 chick) of activities (Table 1). During both years, adults and chicks spent similar time feeding, although in 1994 the 2 birds spent only about half the time feeding compared to the 6 cranes in 1993. Chicks spent more time drinking, but adults spent more time in alert behavior and preening and body maintenance (comfort movements). Time spent resting was more variable among individuals and suggested no obvious trend.

Feeding differed between adults and chicks. Adults directed 86.9% of their foraging effort to natural sites and 13.1% to feeders. Chicks foraged 52.7% in natural sites and 47.3% at feeders. Time spent in locomotion (walking, running) varied among individuals and between adults and chicks, and accounted for 13.0-30.7% of diurnal activities. Pacing edges of pens accounted for a mean of 41.3% (range 12.3-61.9, n = 3) for all locomotion in adults and 30.2% (SD = 28.0, range 1.4-74.2, n = 5) in chicks. Agonistic behavior recorded for chicks in 1993 included attempts to peck each other through plastic mesh or interacting when 2 or more were placed together for socialization.

During period 2 we recorded 103:20 hours (46:25 adult, 56:55 chick) of time budget data (Table 2). Adults spent more time in alert behavior ($\bar{x}$ = 42.2% vs. 27.4%) and locomotion ($\bar{x}$ = 25.9% vs. $\bar{x}$ = 13.0%), including pacing the pen, but less time feeding ($\bar{x}$ = 4.5% vs. $\bar{x}$ = 15.4%) and resting ($\bar{x}$ = 9.1% vs. $\bar{x}$ = 24.9%) than chicks. The few agonistic interactions in 1993 were minor confrontations between chicks.

In July 1994, the adult captured a northern pocket gopher (*Thomomys talpoides*) and presented it to the 44-day-old chick, which made several unsuccessful attempts to swallow it; the adult finally ate the gopher. During both years, trout were placed in the pond and the cranes eagerly captured and consumed them. In 1993, adults were not observed feeding chicks. In 1994, the male frequently presented or pointed out insects and earthworms to the chick, and 16% of the chick's foraging activities consisted of consuming items presented by the adult (Table 2).

Table 1. Diurnal activities (%) of 3 wild adult and 5 prefledged (14-22 to 24-29 days old) whooping cranes. Adults (Ad) were in a large pen but separated from chicks (yg) that were confined in separate enclosures inside the pen, Grace, Idaho, 23-29 June 1993[a] (2 Ad, 4 yg) and 18-29 June 1994 (1 Ad, 1 yg).

	Adults			Chicks			
	1993		1994	1993			1994
Activity	Male	Female	Male	$\bar{x}$	min.	max.	
Feeding	8.6	10.0	4.6	8.6	6.3	10.8	3.5
Upland[b]	45.6	56.6	50.0	52.7	39.2	66.3	65.0
Wetland[b]	40.9	38.3	29.4				
Feeder[b,c]	13.5	5.1	20.6	47.3	33.7	60.8	35.0
Drinking	0.2	0.1	0.6	2.4	1.5	3.3	3.4
Alert	38.1	30.1	32.6	24.5	20.1	36.4	21.3
Resting	12.8	23.5	19.7	25.9	15.2	36.7	14.2
Comfort movements	21.8	19.1	13.5	11.9	10.1	13.5	7.4
Locomotion	18.5	17.1	28.9	22.8	14.6	30.7	13.0
Pacing pen[d]	49.7	12.3	61.9	37.4	1.4	74.2	37.7
Agonistic behavior				0.4	0.0	0.9	
Not obs[e]				3.5	0.6	8.6	37.0
Other		tr	0.1				0.2
Total %	100.0	99.9	100.0	100.0			100.0
No. hours	8:15	9:05	14:35	28:12			17:00
No. samples	1,980	2,420	3,500	6,766			4,080

[a] Data from 2 chicks were included through 5 July 1993 because they were in separate enclosures to this date.
[b] % time feeding in different areas; no wetlands in chick enclosures.
[c] Feeders contained mixed grains for adults and commercial pellets for chicks.
[d] % locomotion pacing edges of pen.
[e] Inside shelters and not visible.

Table 2. Diurnal activities (%) of 3 wild adult and 5 prefledged (24–29 to 44–51 days old) whooping cranes in a pen together, 29 June–21 July 1993 and 29 June–18 July 1994, Grace, Idaho.

	Adults			Chicks			
	1993		1994	1993			1994
Activity	Male	Female	Male	$\bar{x}$	min.	max.	
Feeding	5.1	4.9	3.5	16.6	11.0	20.3	11.7
Upland[a]	37.3	19.6	31.2	29.1	25.2	34.1	26.7
Wetland[a]	34.7	58.1	13.7	43.7	20.0	55.2	17.7
Feeder[a,b]	28.0	22.3	42.9	27.2	19.6	45.9	39.6
Chick[c]			12.2				16.0
Drinking	0.1	0.1	0.3	1.2	0.9	1.4	1.5
Alert	40.0	36.4	50.1	25.0	24.0	27.5	33.3
Resting	13.4	12.0	1.8	28.0	16.7	37.3	19.3
Comfort movements	23.5	21.5	9.4	14.1	12.9	15.0	14.2
Locomotion	17.7	25.0	34.9	12.3	11.3	13.4	15.9
Pacing pen[d]	27.0	26.0	67.8	2.8	0.0	5.3	16.0
Agonistic behavior				0.1	0.1	0.2	
Not obs[e]				2.7	0.2	9.7	4.0
Other	0.2	0.1		tr			0.1
Total %	100.0	100.0	100.0	100.0			100.0
No. hours	9:35	12:30	24:20	31:35			25:20
No. samples	2,300	3,000	5,840	7,580			6,080

[a] % time feeding in different areas.
[b] Feeders contained mixed grains for adults and commercial pellets for chicks.
[c] Adult feeding chick.
[d] % locomotion pacing edges of pen.
[e] Inside shelters and not visible.

Table 3. Diurnal activities (%) of 3 wild adult and 5 prefledged (45–52 to 84–89 days old) whooping cranes in a pen, 22 July –27 August 1993 and 19 July–31 August 1994, Grace, Idaho.

	Adults			Chicks			
	1993		1994	1993			1994
Activity	Male	Female	Male	$\bar{x}$	min.	max.	
Feeding	5.5	6.4	3.9	18.6	12.2	25.9	11.6
Upland[a]	38.0	23.3	21.0	36.2	25.9	44.6	28.0
Wetland[a]	36.6	54.9	41.9	35.7	24.3	50.6	36.7
Feeder[a,b]	25.4	21.8	29.1	28.1	23.5	38.0	34.5
Chick[c]			8.0				0.8
Drinking	0.1	0.1	0.1	0.6	0.6	0.7	0.6
Alert	36.5	34.5	44.0	29.6	25.9	33.8	40.9
Resting	12.1	9.7	2.8	15.6	6.4	20.8	7.6
Comfort movements	24.6	23.0	13.7	17.9	13.6	21.3	18.1
Locomotion	21.0	26.1	35.4	15.9	13.0	18.0	21.0
Pacing pen[d]	28.9	11.8	74.7	15.8	5.3	24.1	42.6
Agonistic behavior				0.2	0.1	0.2	
Not obs[e]	0.1	0.1		1.6			0.1
Other	0.1	0.1	0.1	tr			0.1
Total %	100.0	100.0	100.0	100.0			100.0
No. hours	16:20	12:35	73:59	62:20			76:30
No. samples	3,920	3,020	17,755	14,942			18,360

[a] % time feeding in different areas.
[b] Feeders contained mixed grains for adults and commercial pellets for chicks.
[c] Adult feeding chick.
[d] % locomotion pacing edges of pen.
[e] Inside shelters and not visible.

During period 3 we recorded 241:39 hours (102:54 adult, 138:45 chick) of time budget data (Table 3). Adults continued to exhibit more alert behavior ($\bar{x}$ = 38.3% vs. $\bar{x}$ = 33.7%) and locomotion ($\bar{x}$ = 27.5% vs. $\bar{x}$ = 17.0%) than chicks, but spent less time feeding ($\bar{x}$ = 5.3% vs. $\bar{x}$ = 17.2%), drinking ($\bar{x}$ = 0.1% vs. $\bar{x}$ = 0.6%), and resting ($\bar{x}$ = 8.2% vs. $\bar{x}$ = 13.7%). Comfort movements were similar between adults and chicks ($\bar{x}$ = 20.4% vs. $\bar{x}$ = 18.0%). Chicks started eating grain in August but preferred feeding in natural areas as did adults (Table 3). After early August 1994, the adult discontinued feeding the chick when it became more efficient at feeding and started consuming grain.

Comparing activities during the 3 time periods (Table 4) revealed differences and trends between activities of adults and chicks. In feeding, adults were generally similar over time, whereas chicks increased food intake with age. Chicks consistently drank more than adults (Tables 1–3) but were less alert, although vigilance increased with age (Table 4). Adults generally spent less time resting but more time in locomotion; adults paced the pen more than chicks (Tables 1–3). Adults generally were consistent in body maintenance (comfort movements), but chicks increased these activities with age.

There were noticeable differences in activities between years. The single adult in 1994 spent less time feeding ($\bar{x}$ = 4.0% vs. $\bar{x}$ = 6.8%), resting ($\bar{x}$ = 8.1% vs. $\bar{x}$ = 13.9%), and in comfort movements ($\bar{x}$ = 12.2% vs. $\bar{x}$ = 22.3%), and more time alert ($\bar{x}$ = 42.3% vs. $\bar{x}$ = 35.9%) and in locomotion ($\bar{x}$ = 29.7% vs. $\bar{x}$ = 20.9%) than 2 adults in 1993. The single chick in 1994 spent less time feeding ($\bar{x}$ = 8.9% vs. $\bar{x}$ = 14.6%) and resting ($\bar{x}$ = 13.7% vs. $\bar{x}$ = 23.2%), more time alert ($\bar{x}$ = 31.8% vs. $\bar{x}$ = 26.4%), and similar time in comfort movements ($\bar{x}$ = 13.2% vs. $\bar{x}$ =14.6%) and locomotion ($\bar{x}$ = 16.6% vs. $\bar{x}$ = 17.0%) as chicks in 1993.

All prefledged chicks showed inconsistent behavior for

roosting in water. In 1993, adults roosted in the pond, but chicks often roosted in adjacent uplands. In 1994, the single chick always roosted in uplands and the adult finally abandoned its pond roost site in early July and joined the chick nightly in uplands. We often flushed chicks from upland roosts at night and they normally ran to the pond. At fledging, chicks were still inconsistent at roosting in water.

Chicks first flew short distances (< 10 m) when 72-79 days old in 1993 and 87 days in 1994. In 1994, the single chick may have developed more slowly because it was infected with gapeworms; the infection was successfully treated with ivermectin (0.07 cc) when the chick was 59 days old. One 81-day-old chick in 1993 broke a wing in the pen after it started flying. The wing was successfully treated by a veterinarian but release to the wild was delayed.

In 1993, the male and female were in captivity for 7 months. They appeared compatible, and no aggression occurred between them. From late April through mid-May they danced and unison called frequently; dancing and unison calling continued until their release in late August. During the summer, considerable synchrony in their activities was observed, suggesting that a pair bond was developing. During both years, adults and chicks were compatible; chicks exhibited much following behavior and often mimicked activities of adults. During 1994, the adult spent considerable time finding natural food items and presenting them to the chick. In both years, adults and chicks danced together, suggesting that parental bonds had developed.

Table 4. Comparison of diurnal activities (mean %) of 3 wild adult and 5 prefledged whooping cranes while confined in a pen during 3 periods[a], summers 1993-94, Grace, Idaho.

Activity	Period 1 Adult	Period 1 Chick	Period 2 Adult	Period 2 Chick	Period 3 Adult	Period 3 Chick
1993						
Feeding	9.3	8.6	5.0	16.6	6.0	18.6
Alert	34.1	24.5	38.2	25.0	35.5	29.6
Resting	18.2	25.9	12.7	28.0	10.9	15.6
Comfort movements	20.5	11.9	22.5	14.1	23.8	17.9
Locomotion	17.8	22.8	21.4	12.3	23.6	15.9
Other	0.1	6.3	0.2	4.0	0.2	2.4
1994						
Feeding	4.6	3.5	3.5	11.7	3.9	11.6
Alert	32.6	21.3	50.1	33.3	44.0	40.9
Resting	19.7	14.2	1.8	19.3	2.8	7.6
Comfort movements	13.5	7.4	9.4	14.2	13.7	18.1
Locomotion	28.9	13.0	34.9	15.9	35.4	21.0
Other	0.7	40.6[b]	0.3	5.6	0.2	0.8

[a] Period 1 (late June)-chicks were confined to separate enclosures inside a larger pen with adults, Period 2 (29 June to 18-21 July)-chicks and adults were together, Period 3 (18-21 July to 31 August)-adults and older chicks to fledging were together.

[b] 1994 chick spent 37.0% of time inside shelter and was not visible.

Release at Grays Lake

1993.—On 28 August, all whoopers except the chick with the wing injury were transported to Grays Lake. During the 1.3-hour trip, the 87- and 88-day-old chicks became agitated and struggled, although adults remained relatively calm. One chick died shortly after arrival at the release site and a second limped. Chicks exhibited signs of stress, including higher pulse rates, gaping, and trembling; adults appeared normal.

Adults and chicks were separated in the pen. The following morning, 2 chicks were released first and they remained next to the pen. The door was opened 1 hour later for the adults but it took 1.7 hours for the male to find the opening and he flew 0.5 km. The female left 16 minutes later and flew 0.4 km; the chicks remained at the pen. For several nights the female returned to roost near the chicks, but the male did not join them.

The male exhibited little interest in the chicks and confined most of his activities within 0.5-1.3 km of the release site for nearly 2 months before migrating. He fed daily in a refuge grainfield 0.5 km from the release site. The female showed more interest in the chicks and the male than the male exhibited toward any of them. She divided much of her time between the male and the chicks but favored the male.

Chicks remained within 0.4 km of the release site for nearly 1 month. They fed on pellets and barley at the feeder and in adjacent meadows.

After release, 1 chick limped, and we captured it by night-lighting. The chick had a leg fracture for which it was treated; it was then placed in the pen with the chick recuperating from the wing injury.

A single chick remained at the release site until the 2 rehabilitated chicks (104-107 days old) were released in mid-September. A sandhill pair (former foster parents) frequented the release area and showed interest in the chick, but the chick did not reciprocate.

After release, chicks would not walk into taller vegetation nor roost in the marsh beyond the mowed area. We flattened emergent vegetation, making trails to open water ponds. Chicks then walked into the marsh and roosted in more secure sites.

The 3 chicks remained together as a cohesive unit. By late September, they started feeding in a refuge grainfield

Table 5. Number of cranes using the area frequented by 5 whooping cranes (2 adults, 3 young) released for guide bird experiment, September–October 1993, Grays Lake, Idaho. The decline in numbers reflects fall migration from the area.

Date	No. of sandhill cranes	No. of whooping cranes	
		Adults	Juveniles
September			
10	907	3	1
17	889	3	1
23	228	3	1
25	146	3	1
27	121	3	3[a]
30	164	3	3
October			
1	148	3	3
5	129	3	3
10	93	3	3
14	142	3	3
20	19	2	3
21	8	2	3
24	9	2	3
26	6	2[b]	3
27	1	0	3
30	1	0	3[c]

[a] Two juveniles recuperating from injuries in pen were released at Grays Lake.
[b] Migrated.
[c] The 3 juvenile whooping cranes did not migrate and were captured on 4 November.

with other cranes and waterfowl 0.5 km from the release site. Although chicks joined the adults almost daily in late September and October, they did not remain together as a family.

More than 900 sandhill cranes and 3 adult whoopers gathered for fall migration near the release site (Table 5). Most had migrated by mid-October; the 2 adults departed on 26 October, but the 3 chicks did not follow. On 4 November, we captured the chicks and transported them to the pen at the Clegg Ranch. During the trip the chicks became extremely agitated and struggled. One died of capture myopathy, and a second chick exhibited symptoms of shock and died 5 days later. Veterinarians suggested using tranquilizers before future moves.

1994.—On 1 September, we moved the male and 89-day-old chick to Grays Lake. Both were vaccinated for avian cholera (0.5 ml, subcutaneously) (Price 1985; J. Price, NWHC, pers. commun.), and the chick was given a tranquilizer (2.25 ml, valium); the chick slept most of the trip. The adult was placed in a pen and the chick was released outside. When released, the chick limped slightly, but remained next to the pen, resting and preening. After 3 hours, we released the adult; he walked 30 m from the pen and began feeding, including capturing and consuming an American coot (*Fulica americana*). The chick followed but was limping and finally lay down. The adult approached it 7 times during the next 3 hours but it did not respond; we captured the chick and took it to a veterinarian.

The chick walked with a limp the following morning and the diagnosis was a possible bruised ligament. We returned it to the pen at the Clegg Ranch for rehabilitation. The cause of the leg injury was unknown.

On 9 September, the chick was released on a refuge grainfield used by the adult and >300 sandhill cranes. The chick approached and followed the adult but was ignored. Sandhill cranes harassed the chick. The chick attempted to follow the adult when it flew from the grainfield, but it was not a strong flyer, and landed alone in a nearby meadow. At sunset, the chick was recaptured and returned to the pen.

On 19 September, we released the chick a third time with a sandhill chick that had recovered from a broken leg and was in the pen with the whooper. The 2 chicks were released at sunrise on a refuge grainfield used by the adult the previous day. Although >90 sandhill cranes arrived to feed, the adult flew to another grainfield 3 km away. After eating in the grainfield, the 2 chicks flew to a nearby meadow and foraged, but did not socialize with other cranes. They roosted that night in a nearby marsh. The next morning, the 2 chicks fed together, but the sandhill chick joined a sandhill flock in the afternoon. The whooper chick remained alone and roosted that night within 150 m of a sandhill crane roost.

On 21 September the adult flew to the grainfield where the chick was released 2 days previously. The chick observed the adult land 0.6 km away and started walking toward it. A coyote, hidden in the grass, attacked the chick. From our observation point, 1 km distant, we sounded the vehicle horn and finally fired a rifle shot and the coyote left. We retrieved the chick, which had a broken and dislocated leg, multiple lacerations, and bite marks, and took it to a veterinarian. It was flown by the USFWS to New Mexico for intensive care. Because of severe and debilitating injuries, including inability to walk, the chick was euthanized on 5 October.

Status of Adults after Release At Grays Lake

In 1993, the 2 adults migrated from Grays Lake on 26 October but did not remain together. On 8–9 November, the male was observed alone near Vernal, Utah (K. Day, Utah Division of Wildlife Research [DWR], pers. commun.). The male wintered with 5 sandhill cranes near Hyline Lake, Fruita, Colorado, from 27 November 1993 to 7 March 1994 (B. Osmundson, USFWS, Grand Junction, Colorado, pers. commun.). This male had wintered in New Mexico during the previous 11 years. Hyline Lake, in western Colorado, is on the traditional spring-fall migration route of Rocky Mountain cranes (Drewien and Bizeau 1974, 1978, 1981; Peterson and Drewien 1997). The female arrived at Bosque NWR, New Mexico, on 29 November 1993 and departed on 2 March 1994.

In 1994, the male migrated from Grays Lake on 26 October, but no confirmed sightings were obtained after that date. We received a report of a whooping crane on 13 November near Bear River City, Utah (T. Aldrich, Utah DWR, pers. commun.), which was possibly this bird. The male had wintered for the previous 18 years at the Bosque NWR.

Juvenile Whooping Crane at the New Mexico Winter Area

On 10 November 1993, the surviving juvenile was given a tranquilizer (1 cc Diazepam, subcutaneously), flown by the USFWS to New Mexico, and placed in a temporary pen at Bosque NWR. The effects of the tranquilizer subsided in about 6 hours and the crane was released. A feeder with grain and pellets was placed near the pen.

The juvenile spent 2 days near the pen. Due to the abundance of coyotes, we captured and moved it to another wetland, where it remained for nearly 1 month. After 1 month, the juvenile left the area, moved >1 km, and joined flocks of sandhill cranes, feeding in farm units during the day and roosting with them at night.

The juvenile was attracted to white birds, approaching snow geese (*Chen caerulescens*) and 4 adult whooping cranes at the refuge. The juvenile temporarily stayed with various whoopers for several days or longer but no associations were permanent. During December and January, the juvenile intermittently associated with the female guide bird, but on 8 February 1994, the juvenile died. A necropsy at the NWHC revealed that the chick died from avian cholera; overall body condition was judged to be excellent (N. Thomas, pers. commun.).

DISCUSSION

Our experiments in 1993–94 demonstrated that we could successfully maintain wild-caught adult whooping cranes in captivity up to 7 months, socialize isolation-reared chicks to each other to reduce aggression, and introduce ≥24-day-old chicks to adults without major complications. Chicks exhibited following behavior from 5 to 48 hours after release. As chicks accompanied adults around the pen, they learned alert and feeding behaviors, and to respond adversely to human presence. In 1994, the adult captured invertebrates and fed them to the chick. In 1993, invertebrates were rare in the pen and adults did not feed chicks. Weather data from eastern Idaho showed that the 1993 summer was the coldest on record during this century. Freezing temperatures occurred during June and on 24 July, eliminating most insects. The lack of invertebrates in 1993 probably explained why adults did not feed chicks and why whoopers spent more time foraging than in 1994 when insects and other natural foods were more abundant.

The remote video camera system was indispensable for monitoring cranes, recording activities, assessing health and status, and minimizing human intrusion. We followed their activities for days without entering the pen. With the zoom lens we could determine needs of chicks in their enclosures and later when they were with adults. We could also monitor aggressive interactions and check food availability in feeders. Our equipment operated well under all diurnal weather conditions but did not function at night. When a VCR was used, activities could also be recorded. Activity data collected in this study provide comparisons with activities of whooping cranes maintained in other captive flocks and under different management regimes.

Pacing behavior in the pen provided a measure of boredom and nervousness, and varied among individuals. Among adults, the female consistently paced less than the 2 males. Chicks generally did not pace as much as adults, and most of their pacing resulted from following adults. However, all chicks increased pacing behavior as they approached fledging age.

A major factor influencing survival in the wild is roosting at night in water to avoid terrestrial predators. Apparently, lack of early (<30 days) experience adversely impacted the ability of chicks to seek water roosts. In contrast, sandhill chicks that we have reared sought water roosts without special training. Captive-rearing procedures should be modified to encourage roosting in water at an early age for whooping cranes scheduled for wild release. Inconsistent use of appropriate roost sites in Florida has contributed to high losses of released captive-reared whoopers to predators (Nesbitt et al. 1997).

Injuries and death of chicks during vehicle transportation limited the evaluation of this experiment. Although no problems occurred while adults were moved, 4 of 5 chicks sustained injuries or died during or following >1 hour vehicle transportation, and all chicks exhibited symptoms of stress. Although Drewien and Clegg (1992) had previously captured and transported 19 adult whooping cranes and numerous sandhill cranes (unpubl. data) without incident, moving 3- to 5-month-old whooping cranes presented unforeseen problems. Attempting to raise chicks in captivity while encouraging wild behaviors, including aversion to human presence, had drawbacks; chicks proved difficult to handle and transport. This unexpected problem would not have occurred had the rearing facility been located at or near release sites. However, no suitable location existed at Grays Lake to maintain overwintering adults with open water, due to severe winters with heavy snowfall and extremely cold temperatures. Future experiments of this nature should be conducted at or adjacent to preselected release sites to minimize problems associated with transportation; sedation of young birds should be considered if transportation is required.

Our observations of adults and prefledged chicks in captivity during both years indicated that parental bonds had developed and we anticipated that they would persist after release. Behaviors of the adult male and female in 1993 also suggested that a pair bond was developing, as there was considerable synchrony in activities and unison calling. However, activities after releases indicated that firm bonds did not exist between the 2 adults in 1993 or between any adults and chicks.

In 1993, 3 juveniles released at Grays Lake survived for 2 months in the wild. They did not maintain parental bonds with 2 adults and they showed no inclination to migrate with any of 3 adult whooping cranes or more than 900 sandhill cranes using the release area. We attribute their survival mainly to removal of 2 coyotes that frequented the release area.

The single surviving chick in 1993 was moved to the Bosque NWR winter area where it survived for 3 months before succumbing to avian cholera. During the first month, it was mainly sedentary and alone. Later, it moved up to 4.5 km from the release site, foraging in refuge wetlands and agricultural fields, and roosting with other cranes at night. Although it temporarily joined whooping cranes and sandhill cranes, it did not associate with any individuals permanently. Losses from avian cholera at Bosque NWR during the winter 1993–94 were the highest recorded, involving more than 1,500 birds, including 103 sandhill cranes (Taylor 1994). The juvenile died 2–4 weeks before spring migration.

Only 1 of 3 adults maintained in captivity for 6.75–7 months arrived at its traditional winter area in New Mexico. One male vanished after migrating from Grays Lake, and another wintered along the spring-fall migration route in western Colorado, and then disappeared. These findings suggest that confinement may have hindered their physical ability to migrate long distances and at high altitudes along the Rocky Mountain corridor (Drewien and Bizeau 1981, Peterson and Drewien 1997).

Isolation-reared juvenile sandhill cranes released with wild sandhill cranes have experienced high post-release survival in both migratory and nonmigratory situations (Horwich 1989, Archibald and Archibald 1992, Ellis et al. 1992, Urbanek and Bookhout 1992). However, juvenile whooping cranes released in this study and in Florida (Nesbitt et al. 1997) have experienced lower survival rates. The U.S. Whooping Crane Recovery Plan (U.S. Fish and Wildlife Service 1994:44) recommended testing the release of captive-reared whooping cranes into nesting or staging areas of wild sandhill cranes with the expectation that whooping cranes will learn survival techniques and migration patterns from sandhill cranes. Results from this study and in Florida (Nesbitt et al. 1997) showed that juvenile whooping cranes did not readily join sandhill cranes flocks. Observations at Grays Lake and at Bosque NWR showed that sandhill cranes had little or no interest in associating with juvenile whooping cranes. Instead, they often harassed and occasionally chased juveniles from sandhill crane flocks. Similar agonistic behavior by sandhill cranes, other than foster parents, toward juvenile whooping cranes was noted in the cross-fostering experiment and resulted in some foster-parent families remaining along edges of flocks or isolated during fall and early winter (Drewien 1975, 1976, Drewien and Bizeau 1978). By midwinter, however, some juveniles were able to defend against sandhill crane harassment (Drewien 1976, Drewien and Bizeau 1978). Because of these observations, we suspect that releasing juvenile whooping cranes on summer or fall staging areas of sandhill cranes to learn survival skills and migration patterns will meet with minimal success.

LITERATURE CITED

ARCHIBALD, K., and G. ARCHIBALD. 1992. Releasing puppet-reared sandhill cranes into the wild: a progress report. Pages 251-254 *in* D. A. Wood, ed. Proc. 1988 North Am. crane workshop. Florida Game Fresh Water Fish Comm. Nongame Wildl. Program Tech. Rep. 12.

DREWIEN, R. C. 1975. Whooping crane transplant experiment. Unpubl. Prog. Rep. 4, Idaho Coop. Wildl. Res. Unit, Moscow. 17pp.

______, 1976. Whooping crane transplant experiment. Unpubl. Prog. Rep. 8, Idaho Coop. Wildl. Res. Unit, Moscow. 32pp.

______, and E. G. BIZEAU. 1974. Status and distribution of greater sandhill cranes in the Rocky Mountains. J. Wildl. Manage. 38:720-742.

______, and______. 1978. Cross-fostering whooping cranes to sandhill crane foster parents. Pages 201-222 *in* S. A. Temple, ed. Endangered birds: management techniques for preserving threatened species. Univ. Wisconsin Press, Madison.

______, and ______. 1981. Use of radiotelemetry to study movements of juvenile whooping cranes. Pages 130-134 *in* J. C. Lewis and H. Masatomi, eds. Crane research around the world. Int. Crane Found., Baraboo, Wis.

______, W. M. BROWN, and E. G. BIZEAU. 1989. Whooping

crane cross-fostering experiment. Unpubl. Rep., Hornocker Wildl. Res. Inst., Univ. Idaho, Moscow. 10pp.

______, and K. R. CLEGG. 1992. Capturing whooping cranes and sandhill cranes by night-lighting. Proc. North Am. Crane Workshop 6:43-49.

______, and E. KUYT. 1979. Teamwork helps the whooping crane. Natl. Geogr. 155:680-693.

ELLIS, D. H., G. H. OLSEN, G. F. GEE, J. M. NICOLICH, K. E. O'MALLEY, M. NAGENDRAN, S. G. HERFORD, P. RANGE, W. T. HARPER, R. P. INGRAM, and D. G. SMITH. 1992. Techniques for rearing and releasing nonmigratory cranes: lessons from the Mississippi sandhill crane program. Proc. North Am. Crane Workshop 6:135-141.

HORWICH, R. H. 1989. Use of surrogate parental models and age periods in a successful release of hand-reared sandhill cranes. Zoo Biol. 8:379-390.

______, J. WOOD, and R. ANDERSON. 1992. Release of sandhill crane chicks hand-reared with artificial stimuli. Pages 255-261 *in* D. A. Wood, ed. Proc. 1988 North American crane workshop. Florida Game Fresh Water Fish Comm. Nongame Wildl. Program Tech. Rep. 12.

MAHAN, T. A., and B. S. SIMMERS. 1992. Social preference of four cross-foster reared sandhill cranes. Proc. North Am. Crane Workshop 6:114-119.

MELVIN, S. M., R. C. DREWIEN, S. A. TEMPLE, and E. G. BIZEAU. 1983. Leg-band attachment of radio transmitters for large birds. Wildl. Soc. Bull. 11:282-285.

NESBITT, S. A., M. J. FOLK, M. G. SPALDING, J. A. SCHMIDT, S. T. SCHWIKERT, J. M. NICOLICH, M. WELLINGTON, J. C. LEWIS, and T. H. LOGAN. 1997. An experimental release of whooping cranes in Florida–the first three years. Proc. North Am. Crane Workshop 7:79-85.

PETERSON, M. K., and R. C. DREWIEN. 1997. Harts Basin, Colorado: a spring migration stopover area for Rocky Mountain sandhill cranes and whooping cranes. Proc. North Am. Crane Workshop 7:153-156.

PRICE, J. I. 1985. Immunizing Canada geese against avian cholera. Wildl. Soc. Bull. 13:508-515.

URBANEK, R. P., and T. A. BOOKHOUT. 1992. Development of an isolation-rearing/gentle release procedure for reintroducing migratory cranes. Proc. North Am. Crane Workshop 6:120-130.

U.S. FISH AND WILDLIFE SERVICE. 1986. Whooping crane recovery plan. U.S. Fish Wildl. Serv., Albuquerque, N.M. 283pp.

______. 1994. Whooping crane recovery plan. U.S. Fish Wildl. Serv., Albuquerque, N.M. 92pp.

TAYLOR, J. P. 1994. 1993-94 winter season disease report, Bosque del Apache National Wildlife Refuge, New Mexico. Unpubl. Rep., U.S. Fish Wildl. Serv., Socorro, N.M. 5pp.

A REINTRODUCTION TECHNIQUE FOR MIGRATORY BIRDS: LEADING CANADA GEESE AND ISOLATION-REARED SANDHILL CRANES WITH ULTRALIGHT AIRCRAFT

WILLIAM A. LISHMAN, Operation Migration, 2731 Durham Regional Road 19, Blackstock, ON L0B 1B0, Canada
TIGHE L. TEETS,[1] Operation Migration, 2731 Durham Regional Road 19, Blackstock, ON L0B 1B0, Canada
JOSEPH W. DUFF, Operation Migration, 2731 Durham Regional Road 19, Blackstock, ON L0B 1B0, Canada
WILLIAM J. L. SLADEN, Environmental Studies, Airlie Center, 6809 Airlie Road, Warrenton, VA 20187, USA
GAVIN G. SHIRE, Environmental Studies, Airlie Center, 6809 Airlie Road, Warrenton, VA 20187, USA
KIRK M. GOOLSBY, Environmental Studies, Airlie Center, 6809 Airlie Road, Warrenton, VA 20187, USA
WAYNE A. BEZNER KERR,[2] Operation Migration, 2731 Durham Regional Road 19, Blackstock, ON L0B 1B0, Canada
RICHARD P. URBANEK, U.S. Fish and Wildlife Service, Seney National Wildlife Refuge, Seney, MI 49883, USA

Abstract: No successful method for establishing self-sustaining populations of whooping cranes *(Grus americana)*, particularly in a migration situation, has been proven. This research initiated development of a reintroduction technique using ultralight aircraft to lead cranes from a natal area along a desired route to a predetermined wintering site. Canada geese *(Branta canadensis)* were used in initial migration efforts. Ultralight aircraft and pilots successfully led 86 juvenile geese on 3 southbound migrations from Ontario to winter sites 640–1,312 km from the natal area. Of 16 1993-hatched geese that survived their first winter in Virginia and 35 1994-hatched geese that were successfully led to South Carolina, 46 (90%) returned unassisted to their natal area in Ontario on their first spring migrations. Only 15 (50%) of 30 1995-hatched geese trucked to New York to begin aircraft-led migration returned to the Ontario rearing area the following spring. Of 16 geese trucked the entire route to Virginia but allowed to fly freely at predetermined stops, none returned to Ontario. In 1995, isolation-(costume-)reared sandhill cranes *(G. canadensis)* were trained to follow the aircraft in flights within 50 km of the Ontario rearing area. Planned future research will involve leading sandhill cranes, and then whooping cranes, on an actual migration.

PROC. NORTH AM. CRANE WORKSHOP 7:96-104

Key words: reintroduction, Canada goose, sandhill crane, whooping crane, migration, ultralight aircraft, isolation-rearing, costume-rearing, *Grus americana*, *Grus canadensis*, *Branta canadensis*.

The experiments of Lorenz (1978) with graylag geese (*Anser anser*) showed that in the first days after hatching, goslings become attached to large moving objects and that these young birds maintain this attachment to the surrogate parents through the juvenile period. Had Lorenz been able to fly, he could have carried his research further and led the birds in flight. In the 1970's Bartlett and Bartlett (1973) led imprinted snow geese (*A. caerulescens*) southbound during migration with a truck, and W. Carrick (Toronto, Ont., unpubl. data) trained Canada geese imprinted on a model aircraft to follow a boat. From 1986 to 1990, the senior author, in association with Carrick, trained Canada geese to fly with a motorcycle and a specially designed aircraft (Lishman 1989, 1991). In July 1988 these efforts resulted in the first flight of a flock of birds in formation behind an aircraft. This event supported the concept that techniques using aircraft-led birds could be developed to establish migration routes for specific threatened or endangered species.

Meanwhile, isolation-rearing techniques, using puppets and costumed humans as surrogate parents, were being developed to produce cranes suitable for release into the wild (Horwich 1989; Horwich et al. 1992; Urbanek and Bookhout 1992, 1994). The goal of this research was to develop a reintroduction technique for migratory populations of the endangered whooping crane. Unlike geese and swans (*Cygnus* spp.), successful reintroduction of cranes is dependent on birds that will avoid humans after release. Although isolation-reared sandhill cranes were successfully induced to migrate in the wild, their ultimate wintering areas were determined mainly by random association with wild sandhill flocks. In reintroduction of whooping cranes, particularly on a proposed migration route where sandhill cranes are not present, a specific, predetermined wintering area is desirable.

If the techniques of (1) isolation-(costume-)rearing to produce birds capable of surviving in the wild and (2) use of ultralight aircraft to teach birds an appropriate migration route and wintering area can be combined, an effective method for establishing migratory populations of the whooping crane may be created. This paper describes 3 major aircraft-led migrations with Canada geese and 1 flying experiment with isolation-reared sandhill cranes during 1993–95.

We thank W. Carrick, G. Lishman, J. Dickens, K.

---

[1]Present address: Seney National Wildlife Refuge, Seney, MI 49883, USA.

[2]Present address: 249 Royal St., Waterloo, ON N2J 2J1, Canada.

Richards, M. Simpkinson, and J. Monohan for assistance with geese and crane rearing; R. Van Heuvelen for aircraft and pen construction and ground crew assistance; D. Woodhouse for piloting and aircraft maintenance; and G. Archibald and J. Langenberg, International Crane Foundation, for advice and encouragement. We are grateful to owners of the stopover sites used during migration, to R. Joyner for facilitating use of Tom Yawkey Wildlife Center as a wintering site, to Airlie Foundation, Yawkey Foundation, and to corporate sponsors Honda Canada, Ivoprop, and Bell Mobility. T. Bookhout, J. Clem, J. Kaplan, and G. Olsen provided useful comments on the manuscript.

STUDY AREA

Geese in 1993 and cranes in 1995 were reared in facilities maintained by the senior author at Purple Hill near Blackstock, Ontario. In 1994, geese were reared on a sod farm near Nestleton, about 11 km from Purple Hill. In 1995, 3 flocks of geese were reared: 2 at Nestleton and a third at Airlie Center, near Warrenton, Virginia. Geese were led along a predetermined migration route to Airlie in each autumn, 1993–95, and onward to the Tom Yawkey Wildlife Center (South Carolina Department of Natural Resources) near Georgetown, South Carolina, in 1994–95 (Fig. 1). Cranes were not led on a migration but were flown locally within 50 km of the rearing site in Ontario.

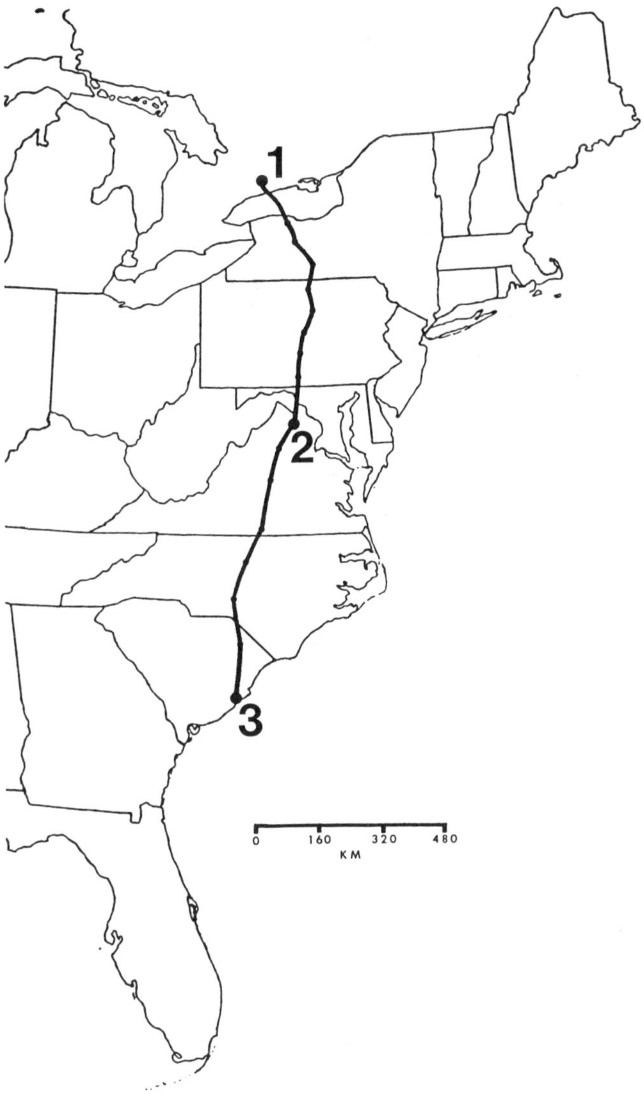

Fig. 1. Route on which captive-reared, juvenile Canada geese were led on autumn migration by ultralight aircraft, 1993–95: (1) Rearing area near Blackstock, Ont., (2) Airlie Center, Va., (3) Tom Yawkey Wildlife Center, S.C.

METHODS

Canada Geese

Canada geese were hatched in incubators from locally collected eggs. A taped recording of the particular aircraft that would be used after the birds fledged was played during the last week of incubation. Goslings were imprinted on caretakers who also carried and played the recording during walks (Lishman 1989). Normal human appearance of caretakers was not concealed, and the geese were not conditioned to fear humans.

When geese were 2 weeks of age, a wearable model of the aircraft, referred to as a "goose toddler," was used by the pilot/leader in conjunction with the tape-recorded sound of the engine, and the goslings were encouraged to follow. At 4–5 weeks the actual aircraft was introduced to the birds, and the engine was started and left to idle for short periods. Once geese could adequately avoid the tires, they were induced to follow the taxiing aircraft on a regular schedule at least twice daily. Other caretakers also ran with the birds alongside the aircraft.

During 1993–94, goslings were raised in groups of 6–12 birds of similar age. Colored bands were used to readily identify groups. Groups were penned and exercised separately but in close proximity. After fledging, the birds were integrated into larger groups and then combined into a single flock just prior to migration. Geese were tarsal-banded in Ontario and then neck-banded when they reached Airlie Center, Virginia. Neckbands were of gray tubular plastic design with black 4-digit codes (Sladen and Kistchinski 1977, Sladen and Limpert 1988).

In 1993, 2 identical aircraft were used to train the birds. These were Cosmos Echo trikes with Ghost 16-m^2 wings (10-m wingspan). This aircraft consisted of a hang glider type

Fig. 2. The attraction of chicks to the costumed parent was used in initial training of sandhill cranes to follow the ultralight aircraft (photo by R. P. Urbanek).

wing with a tricycle cockpit and gear assembly suspended below. Both aircraft were powered by Konig 2-stroke, 4-cylinder, 28-horsepower radial engines and 4-blade propellers. In 1995, 4 identical Cosmos Phase II aircraft with Zoom 19-m^2 wings were used due to the requirements of filming for the motion picture *Fly Away Home* by Columbia Pictures. These were powered by Rotax 503, 2-stroke, 50-horsepower engines with reduction gears and 6-blade props. All of the trike aircraft used a pusher propeller fitted with an aluminum ring or "goose guard" around the perimeter to minimize danger to the birds. The geese were also habituated to, but not imprinted on, 2 other aircraft for the purpose of filming. These were a modified Easy Riser powered by a Konig 3-cylinder, 2-stroke engine with a 2-blade propeller and a Max Air Drifter powered by a Rotax 503 engine and a 3-blade propeller. The aircraft used were capable of sustained flight between 45 and 72 km/h. No ground vehicles were used in the imprinting process.

Sandhill Cranes

Sandhill cranes were isolation-reared (Horwich 1989, Urbanek and Bookhout 1992) from eggs collected by helicopter along the north shore of the North Channel of Lake Huron. Eggs were incubated in a separate building, and then hatchlings were transferred to the rearing facility. The facility consisted of 9 adjacent 0.8- × 1.2-m indoor compartments, each connected by a sliding door to a 0.8- × 2.4-m outdoor run. Each compartment could be accessed by a door opening into a common workroom. Outdoor runs were positioned along a 275- × 15-m grass runway on which chicks could be trained with the Cosmos Echo ultralight aircraft. The immediate area also contained grassy fields and a small pond where chicks were given additional exercise.

In isolation-rearing of cranes, caretakers were dressed in costumes that concealed the human form, particularly the face and hands (Horwich 1989, Urbanek and Bookhout 1992). The costume included a puppet head that resembled an adult crane and that was used to interact with the chicks. Chicks were walked approximately 2 hours each day to reinforce following response and reduce the possibility of developmental leg deformities that could result from inadequate exercise (Wellington et al. 1996). During these exercise periods, taped sound recordings of the ultralight

Table 1. Summary of flights in ultralight aircraft-led migration of Canada geese, 1993–95. False starts and local flights are not included.

Group	Year	Dates	Route segment[a]	Distance (km)	No. of flight days	No. of flights	Flight duration (min) Range	Flight duration (min) Mean	Total flight time (hr)
UG1	1993	19–25 Oct	Ont.–Va.	640	4	6	60–180	122	12.2
UG2	1994	11–16 Oct	Ont.–Va.	640	5	9	48–125	84	12.6
		1–10 Dec	Va.–S.C.	672	7	12	27–100	66	13.2
UG3b	1995	30 Oct–10 Nov	N.Y.[b]–Va.	640	6	9	51–120	88	14.7
UG3a,b		16–20 Nov	Va.–S.C.	672	4	8[c]	70–160	117	15.6

[a] Refer to Fig. 1.
[b] Trucked from rearing site in Ontario around Lake Ontario to Gaines, New York, on 28 October.
[c] On 16 November, 2 groups of geese were flown separately for 70 and 105 min., respectively, during the same flight. Mean for these 2 groups was used in calculation of mean flight duration and total flight time.

engine were continuously played to accustom chicks to the sound of the aircraft. Then, at 2–3 weeks of age, the smaller "goose toddler" model craft was introduced to help the chicks overcome fear of the overhead wing. This model could be easily carried during exercise periods and recordings of the engine were played from it to help the chicks associate movement with sound of the aircraft. This process continued until the birds could be grouped together and introduced to the actual aircraft at about 5 weeks of age.

The aircraft was introduced from a distance. The costumed parent sat next to it so that chicks would acclimate to the larger aircraft and wing. Eventually the engine was started for short periods, and later, with the engine running, a parent sat in the cockpit to propel the craft. This continued during the pre-fledging period. Once the chicks were comfortable and acclimated to the aircraft, the aircraft was taxied on the runway. During taxiing, 1 costumed parent operated the aircraft while another walked or ran directly under mid-wing. This technique accustomed the chicks to close proximity and association with the taxiing aircraft by use of the more desirable walking parent to evoke the following response (Fig. 2). Because the young chicks could be controlled more easily by the walking parent, the same method was used later as the birds reached fledging age and the aircraft left the runway. Flights were generally attempted in early morning or evening during optimal conditions for flying aircraft. The pilot was costumed during all flights.

RESULTS

1993 Goose Cohort

In 1993, 23 geese in 2 groups were taken southbound on a predetermined, 640-km route from southern Ontario to winter at Airlie Center, Virginia. In the first experiment, 18 geese (UG1's) were led by 2 ultralight aircraft 19–25 October with 4 stops en route; in the second, 5 geese (TG1's) were trucked in a closed vehicle on 10 December (Tables 1 and 2). At Airlie, the 2 flocks were penned separately alongside different lakes in good habitat frequented by resident and migratory Canada geese. Each flock was released daily under supervision but stayed in the vicinity of its pen. On 1 January, the close of migratory goose hunting season, both flocks were no longer penned. The UG1's were flown with the aircraft about once per month. Both flocks remained in the general vicinity of their release sites and were given supplementary feed of corn.

Two UG1's died during the winter. The remaining 16 UG1's were last observed at Airlie on 1 April 1994. In mid-April, at least 13 (81%) of the UG1's returned to or near the natal area in Ontario; 12 were observed at Purple Hill and another 80 km away. Detailed accounts were provided by Sladen and Lishman (1994), Sladen et al. (1994), and Sladen and Lishman (1995). The geese soon dispersed to presumably better habitat, and none were subsequently seen in their natal area, although reports of 2 being shot were received. The 5 trucked geese (TG1's) remained at their winter area at Airlie, and all survived to molt, when 1 was killed by a fox. The remainder disappeared, presumably shot, on the first day of the resident goose hunting season on 6 September 1994.

1994 Goose Cohort

On 11 October 1994, a flock of 38 geese (UG2's) was led south by 2 ultralight aircraft to Airlie along the same migration route used by the UG1's. Three dropped out en route, 1 of which was recaptured and added to the flock at Airlie. After 6 weeks at Airlie, 36 UG2's were led by the same aircraft on a 672-km predetermined route to the Tom Yawkey Wildlife Center, near Georgetown, South Carolina.

Table 2. Number of captive-reared Canada geese led by ultralight aircraft or transported by truck on autumn migration to winter in Virginia or South Carolina and number returning to the rearing area in Ontario, 1993–95.

	1993		1994	1995	
Migration progression	Aircraft (UG1)	Truck[a] (TG1)	Aircraft (UG2)	Aircraft (UG3b)	Truck[b] (TG2)
Departed from Ontario	18	5	38	30[d]	16
Arrived at Airlie	18	5	35[c]	29	16
Departed from Airlie			36	29	
Arrived at Tom Yawkey			35	29	
Survived winter	16	5	≥33		
Returned to Ontario	13	0	33	15	0

[a] Enclosed truck; birds were not allowed to fly during transport.
[b] Birds trucked from Ontario to Virginia were periodically released and allowed to fly freely at predetermined stops en route.
[c] An additional goose dropped out en route, was retrieved, and then trucked to Airlie for a total of 36 birds at that site.
[d] Trucked first leg of migration around Lake Ontario.

One goose dropped out on the first leg and returned to Airlie. This bird returned to its natal area alone the next spring. The remaining 35 birds arrived at the South Carolina destination on 10 December 1994. Total distance of the UG2 migration was 1,312 km (Fig. 1). Six days after release in South Carolina, the UG2's disappeared and were not observed again until they started appearing at their natal area in Ontario the following spring. In all, 33 of 35 (95%) geese that arrived at Tom Yawkey Wildlife Center returned to their natal area on 19–20 April. The UG2's left the sod farm on 12 June, prior to molting. Three returned when barely able to fly before the molt was completed, and by 9 September, 19 geese had returned to Nestleton. These birds came and went from the Nestleton location during the autumn of 1995. On 27 September, 3 of the UG2's were shot 2 km east of Nestleton, and on 16 November the remaining birds departed from Nestleton. A detailed account was provided by Sladen and Lishman (1995).

1995 Goose Cohort

Three separate flocks of geese were reared in 1995, partially to provide imprinted geese for the fictional feature film *Fly Away Home*. Two groups were raised at the sod farm in Ontario, and the third group was reared at Airlie, Virginia. One group of 30 was imprinted to fly with aircraft (UG3b's) and was led to Airlie in late October after having been trucked around Lake Ontario (Tables 1 and 2). Transport by truck was necessary to avoid crossing the lake by aircraft during persistent poor weather. Two geese dropped out in New York. One was retrieved to rejoin the flock. Thus 29 geese arrived at Airlie on 10 November.

The other group of 16 birds raised in Ontario (TG2's) was not taught to fly with aircraft but was likewise imprinted on humans and was trucked at the same time, stopping at several locations en route to overnight, like the UG3b, in pens open to the sky, but, unlike the UG3b, allowed to fly freely at each stop. When at Airlie they were penned at night for 2 weeks and then released to fend for themselves, being provided with only a minimum of supplementary feed in cold weather. As of September 1996, all but 3 which disappeared in March and 1 that died remained at Airlie.

The third group of 31 geese (UG3a's) was hatched and trained to fly with an identical aircraft at Airlie. These geese joined with UG3b's for a total flock of 60 birds, which, on 16 November, was led by 4 ultralight aircraft from Airlie to the Tom Yawkey Wildlife Center along the same route used by the UG2's the previous year. A total of 59 geese arrived at their winter destination on 20 November. The migration and daily study of these birds and the TG2's were described by Sladen (1996).

Only 15 of the UG3b's returned to the Ontario natal area the following spring. None of the 16 TG2's has been sighted in Ontario (Table 2).

1995 Sandhill Crane Cohort

Rearing and Initial Flights at the Rearing Facility.—Eight chicks hatched during 20–31 May, 2 of which died when less than 1 week old. The remaining chicks were trained with the aircraft. All responded well to the aircraft during daily sessions. However, 2 of these chicks developed severe leg deformities and were subsequently euthanized at 4–5 weeks of age. Another chick, learning to fly at age 10 weeks but not yet fledged, fractured its leg while following the taxiing aircraft and had to be removed from the experi-

Table 3. Flight summary for isolation-reared juvenile sandhill cranes trained to follow an ultralight aircraft, Blackstock, Ontario, 27 July–14 September (*n* = 3) and 17–28 September (*n* = 2), 1995.

Period or date	Number of flights	Flight departure runway[a]	Flight duration (min)	Distance (km)	Distance from start (km)	Maximum altitude (m)
27 Jul–14 Sep	18	PH	0–18	0– 3	0– 2	0– 60
17 Sep	3	FCF	31–34	5– 8	1– 3	90–350
18 Sep	2	FCF	16–65	15–27	8–13	140–350
21–24 Sep	3	PH	24–70	13–35	1–11	150–390
25 Sep	1	PH	84	64	48	560
25 Sep	1	Omemee	86	56	40	450
27–28 Sep	3	PH	30–80	32	10–13	490
28 Sep	1	SF	100	15	6	1,300

[a] PH = Purple Hill, FCF = Frew's cornfield, SF = Nestleton sod farm.

ment. Because the accident was not witnessed, whether the fracture was due to collision with the aircraft or merely ambulatory (Olsen 1994) was not determined. The remaining chicks fledged and successfully followed the aircraft in flight, but were reluctant to leave the vicinity of the rearing facility and would only remain airborne a short time (Table 3). On 16 September, as caretakers prepared to move the birds to a temporary pen erected at a different runway, 1 of the birds was startled when costumed parents carrying cardboard transport boxes approached the facility. Birds appeared excited because conditions were windy and they had not been released for exercise that day. The crane jumped in its run, struck some part of the interior pen structure, and broke its neck.

Flight Performance After Birds Were Moved to an Unfamiliar Runway.—On 17 September the 2 remaining cranes were moved to an unfamiliar runway in Frew's cornfield, 12 km from the runway at the Purple Hill rearing facility. Much improved performance was immediately achieved at the new runway. Flight times increased from 15–18 minutes to 30 and 65 minutes within 4 days. Altitudes increased from less than 60 m prior to 17 September to more than 300 m. Regular flights were made at 300 m and later to more than 1,200 m for increasing periods of time (Table 3).

On the initial flights from Frew's runway, the cranes, particularly the dominant No. 5, broke from the aircraft in the short-distance flights and returned to the runway area from several kilometers away, demonstrating the ability to home on a previous, though relatively unfamiliar, point of flight origin. Cranes were also led to land and then take off again within a short period of time from new runways; for example, on 18 September the birds were led airborne for 16 minutes from Frew's runway, landed, and then were led airborne again without difficulty.

Examples of Imprinting Response to the Aircraft.—On 21 September the pilot made an unscheduled landing at the Nestleton sod farm because of high winds. While he waited for the ground crew to arrive with transport boxes for trucking the cranes back to Purple Hill, an unfamiliar caretaker stayed with the birds as the aircraft was put in the hangar, out of sight. The birds then took flight and flew above the hangar, while repeatedly calling, and would not land on the runway. Only when the aircraft was removed from the hangar and moved into view more than 100 m away did the birds land near the caretakers and plane. The aircraft could not be put away until the ground crew arrived and the birds were boxed for transport to the rearing area.

During an otherwise routine morning flight on 22 September, both birds left the Purple Hill runway, but crane No. 5, apparently disturbed by high winds, was reluctant to follow the aircraft away from the rearing facility. As winds increased, No. 5 landed in front of the facility, where he began to call loudly. Crane No. 2 responded by flying farther behind the plane. After approximately 15 minutes of attempting to entice No. 5 back into the air, the pilot flew low over the rearing facility, at which time No. 2 landed beside No. 5. After confirming that both birds were on the ground, the pilot returned to the Nestleton sod farm, remaining airborne approximately 20 minutes at 400–500 m while the birds called loudly and flew in tight circles 100 m over the aircraft hangar. The pilot returned to an area of the runway free from uncostumed humans, and No. 2 responded by immediately landing and approaching the pilot/aircraft. Apparently No. 2 had left No. 5 to pursue the aircraft soon after the pilot departed from the rearing area at Purple Hill. Because the pilot rapidly climbed to a high altitude and was not watching for the birds, he did not know that No. 2 had followed until the bird was observed shortly after landing at the sod farm.

Cross-country Flight.—On 25 September, a round-trip flight to Omemee gliderport, 64 km from Purple Hill, was made to evaluate long-distance flight performance and thermal flying behavior. The initial 48-km leg was initiated at 0930 hours during a period of relatively intense low level thermal convection. As the pilot attempted to leave Purple Hill, both birds left their customary flying position astern and beside the aircraft. They were found circling in a narrow but strong thermal at approximately 250 m. When the aircraft joined them and began circling aggressively in the thermal core, the birds resumed their normal positions off the wingtip. A slow ($\bar{x}$ = 0.3 m/sec) climb was continued to 600 m, where a level of smooth air remained above the convective layer. The remainder of the flight proceeded in normal flying formation in smooth air. Upon approach at Omemee, aircraft and birds began descent from 600 m and all glided downward for approximately 15 minutes. Upon landing, the birds and pilot rested for approximately 1 hour with the costumed principal caretaker. Total flight time was 84 minutes at a mean ground speed of 45 km/h (Table 3).

During initiation of the return flight, the birds demonstrated a desire to follow the principal caretaker, who remained on the ground, but once the plane was in the air the birds were eager to follow and left this new runway without reluctance. On this flight, to permit examination of thermal flying behavior of the cranes, the pilot reduced engine RPM to a level below that required to maintain level flight. After an initial climb in a thermal to approximately 500 m, an average sink rate of 0.5 m/sec was established, analagous to gliding flight. To avoid having to add power (analagous to flapping flight), the pilot used thermals for climb to regain altitude lost during forward progress. Rather than navigate in straight lines back to the natal area, a course was selected based on estimations of where thermals were most likely to be encountered. The cranes quickly showed their superiority in thermal flying, easily outclimbing the aircraft by circling tightly in the thermal core. Although the cranes were able to climb more quickly than the aircraft, they rarely climbed more than 50 m above it, and they left thermals to continue following the aircraft on course. As the flight proceeded, thermal strength increased such that the pilot was unwilling to climb above 500 m because of control concerns. Vertical gusts as strong as ± 5 m/sec were outside the control range of the aircraft but seemed to attract the birds. Because of strong headwinds, a precautionary landing was made at the Nestleton sod farm after a flight of 56 km and 86 minutes at a mean ground speed of 39 km/h. The birds had readily followed the aircraft to land at new runways, but their behavior after landing at these runways differed from that observed at Purple Hill. They appeared hungry immediately after flight and did not wander more than 10 m from the caretaker or plane while on the ground.

Terminus.—The sandhill crane project ended on 1 October when the birds were returned to captivity in accordance with Canadian Wildlife Service permit requirements.

DISCUSSION

Spring Migration and Return of the Geese

Of 16 1993-hatched geese that survived their first winter in Virginia and 35 1994-hatched geese that were successfully led to the Yawkey Center in South Carolina, 46 (90%) returned unassisted from their first spring migration to the natal area in Ontario. This high return rate bodes well for use of ultralight aircraft to teach birds a migration route in reintroductions. The geese returned despite flying in a migration route occupied by many migratory and resident Canada geese. They found their way back despite diversions from other geese, much as had isolation-reared sandhill cranes previously released into wild sandhill flocks at Seney National Wildlife Refuge (NWR) (Urbanek and Bookhout 1992, 1994).

The lower return rate (50%) of geese that were trucked around Lake Ontario to begin aircraft-led migration from New York and the failure of any of the geese trucked the entire route from Ontario to Virginia to return are consistent with results of sandhill crane studies at Seney NWR. In a release of 16 isolation-reared cranes in 1988 (Urbanek 1990), 8 began migration correctly from Seney by following wild cranes. The other 8 birds formed a self-guiding flock, did not depart with the wild cranes, and were subsequently transported in boxes for release on staging areas in Wisconsin, 370 km away. Of the 8 cranes that began migration correctly, all returned to Upper Michigan the following spring. Of the 8 cranes released in Wisconsin, only 3 (38%) were observed at Seney the following spring, although transmitter failures in this group reduced probability of finding all of these birds. These studies indicate the important relationship between captive-reared birds beginning migration from the natal area on their own power and subsequent successful migration.

The UG1's and UG2's did not have transmitters attached for following their migration north, so we have no data on the route they used. We only know that 90% successfully returned to their natal area.

Sandhill Cranes

Like geese, sandhill cranes were readily trained to follow the aircraft. We believe that long-distance flight behind the aircraft could have been accomplished sooner, but aircraft and pilot availability was limited by conflicts with the concurrent goose work. Cranes were initially reluctant to leave their rearing area but did so without hesitation after some long-distance flying experience. Moving the birds to a different runway was useful to overcome this initial reluctance. As the study continued, the cranes grew physically stronger and more eager to fly with the plane. They flew in an energy-efficient formation slightly behind the plane, just off the wing tips, as they would in a flock of wild cranes. Birds followed at all distances and altitudes tested. Flights were limited by capabilities of the aircraft rather than of the birds. Like wild cranes, and unlike geese, the cranes found thermal lift, and when the plane remained in the thermal, the birds also stayed and gained additional altitude. Cranes could be led to land and then take off again without difficulty from new runways. The innate ability of cranes to remain aware of their geographic location was demonstrated by cranes returning unassisted to runways used only 1 or few times.

The small number of cranes available for flights resulted from mortality during rearing. Most, perhaps all, of the crane deaths were avoidable. Captive-rearing of cranes is labor-intensive and requires special facilities, but the procedures have been thoroughly developed (Horwich 1989, Urbanek and Bookhout 1992, Nagendran et al. 1996, Wellington et al. 1996) and need only be followed to ensure high survival. However, this project represented the first effort at rearing cranes at this site, where previously only geese had been raised. Delay in completion of the facility, inadequate design safeguards, and insufficient manpower were all problems which can be corrected in future rearing attempts. Only 1 death was related to the aircraft, and this occurred because of inadequate assistance. In that incident a single caretaker attempted to release the chicks, which were just learning to fly, and then taxi the aircraft with the birds following and not in view. The apparent collision of the bird with the plane was not witnessed by the pilot and probably could have been avoided by use of another costumed parent to maneuver the birds, which was the usual protocol.

Differences Between Flight of Geese and Cranes with Aircraft

Differences in flight of geese and cranes were related primarily to wing loading, i.e., the total weight of the bird divided by the area of the wing. Wing loading in geese is relatively high, which requires the bird to work harder to maintain flight. Both species took advantage of the vortices created by the aircraft wing. However, cranes, because of their larger wing area, could more easily sail behind the wing. They were able to ride this air current in a greater range of conditions and at a greater distance from the wing. Cranes therefore displayed greater flight endurance shortly after fledging than did geese.

Cranes also demonstrated more efficient climbing and descent ability than geese. Flying with well rested geese under the best conditions yielded the greatest rate of climb recorded, 30 m/min. Under the same conditions, climb rates of 100–130 m/min were achieved with sandhills. Descent rates were also greater with cranes. On several occasions, descent from high altitude by aircraft with geese proved to be a slow process. Geese commonly wiffled, or flipped while dropping; however, altitude lost was minimal. For geese, wiffling was effective at positioning but was not a method the birds would use to descend from more than 1,000 m. Sandhills, because of their light wing loading, descended rapidly. We estimated that cranes descended at a rate of 160–180 m/min, although precise data were not recorded. Final approach of cranes could be nearly vertical, unlike the more horizontal, aircraft-style, running landing of geese. In general, the ultralight aircraft that was used could more easily fly with cranes than with geese.

Development of the Technique for Reintroducing Cranes: Current Progress and Future Needs

In 1995 use of ultralight aircraft to lead sandhill cranes on a southbound migration route was tested in the Rocky Mountains when Clegg et al. (1997) successfully trained 15 sandhill cranes to follow an aircraft and led 7 birds from Idaho to New Mexico. It has now been conclusively demonstrated that geese and cranes can be successfully taught to follow an ultralight aircraft for long distances. A successful autumn migration was also achieved when Ellis et al. (1997) trained 10 sandhill cranes to follow a truck and then led 9 of them successfully from northern to southern Arizona. As of September 1996, no studies have confirmed that cranes or geese will return of their own volition to a previously taught wintering area. Work with cranes has not yet progressed sufficiently to address this objective, and none of the experimental geese led by aircraft to predetermined wintering areas has been reported in those areas during subsequent winters. The small number of geese, the resident goose hunt in Virginia, and our inability to determine the winter location of the UG2's after they left the Yawkey Center have confounded assessment of return to wintering area. Successful development of this technique will require return of the birds to the wintering area that they were taught on their first migration. In future work it is imperative that birds are monitored, e.g., by radiotelemetry, on subsequent migrations so that their wintering sites are documented and, if not the desired site, can be studied to determine why the birds are wintering there.

Inducing cranes to winter on a predetermined site will

depend on building and reinforcing their affinity for that site during their first winter. Because of the benefits of control by and interaction with the costumed parent, these required techniques can best be developed with isolation-reared birds. Procedures could be largely site dependent and will most effectively be developed at an actual whooping crane reintroduction site.

The use of ultralight aircraft as a reintroduction tool, with possible support from ground vehicles, should next be tested with whooping cranes isolation-reared at a northern reintroduction site and then led to a predetermined wintering area. In accordance with recovery plans for the whooping crane (Edwards et al. 1994, U.S. Fish and Wildlife Service 1994), this trial could represent the initial step in establishing a population of whooping cranes that breed in a prairie province of Canada and winter in the southeastern United States.

LITERATURE CITED

BARTLETT, D., and J. BARTLETT. 1973. Beyond the north wind with the snow goose. Natl. Geogr. 144:822-847.

CLEGG, K. R., J. C. LEWIS, and D. H. ELLIS. 1997. Use of ultralight aircraft for introducing migratory crane populations. Proc. North Am. Crane Workshop 7:105-113.

EDWARDS, R., S. BRECHTEL, R. BROMLEY, D. HJERTAAS, B. JOHNS. E. KUYT, J. LEWIS, N. MANNERS, R. STARDOM, and G. TARRY. 1994. National recovery plan for the whooping crane. Rep. 6. Recovery Nat. Endangered Wildl. Comm., Ottawa, Ont. 39pp.

ELLIS, D. H., B. CLAUSS, T. WATANABE, R. C. MYKUT, and M. KINLOCH. 1997. Results of an experiment to lead cranes on migration behind motorized ground vehicles. Proc. North Am. Crane Workshop 7:114-122.

HORWICH, R. H. 1989. Use of surrogate parental models and age periods in a successful release of hand-reared sandhill cranes. Zoo Biol. 8:379-390.

______, J. WOOD, and R. ANDERSON. 1992. Release of sandhill crane chicks hand-reared with artificial stimuli. Pages 255-262 *in* D. A. Wood, ed. Proc. 1988 crane workshop. Florida Game Fresh Water Fish Comm. Nongame Wildl. Prog. Tech. Rep. 12.

LISHMAN, W. A. 1989. C'mon geese. Cooper-Lishman Prod., Blackstock, Ont. Video. 30min.

______. 1991. Conditioning migratory birds to fly with ultralight aircraft. Unpubl. Rep., Blackstock, Ont. 20pp.

LORENZ, K. Z. 1978. The year of the greylag goose. Harcourt, Brace, and Jovanovch, New York, N.Y. 199pp.

NAGENDRAN, M., R. P. URBANEK, and D. H. ELLIS. 1996. Reintroduction techniques. Pages 231-240 *in* D. H. Ellis, G. F. Gee, and C. M. Mirande, eds. Cranes: their biology, husbandry, and conservation. Nat. Biol. Serv., Washington, D.C., and Int. Crane Found., Baraboo, Wis.

OLSEN, G. H. 1994. Crane orthopedics: pediatrics and adults. Semin. Avian and Exotic Pet Med. 3:73-80.

SLADEN, W. J. L. 1996. Ultralight/Canada goose migration experiment. Unpubl. Prog. Rep. 4, Airlie Cent., Warrenton, Va. 2pp.

______, and A. A. KISTCHINSKI. 1977. Some results from circumpolar marking programs on northern swans and geese. Pages 498-507 *in* T. J. Peterle, ed. Proc. XIII Int. Congr. Game Biol., Wildl. Manage. Inst. and The Wildl. Soc., Washington, D.C.

______, and R. LIMPERT. 1988. Methods of marking penguins and waterfowl in captivity. Proc. Annu. Conf. Am. Assoc. Zool. Parks and Aquariums 1988:672-674.

______, and W. A. LISHMAN. 1994. Progress report on ultralight/Canada goose migration experiment. Trumpeter Swan Soc. Newsl. 23:22-23.

______, and ______. 1995. Ultrageese #2 (UG2). Unpubl. Prog. Rep. 4, Airlie Cent., Warrenton, Va. 2pp.

______, ______, and G. G. SHIRE. 1994. Ultralight/Canada goose migration experiment. Unpubl. Prog. Rep. 3, Airlie Cent., Warrenton, Va. 10pp.

U.S. FISH AND WILDLIFE SERVICE. 1994. Whooping crane recovery plan. U.S. Fish Wildl. Serv., Albuquerque, N.M. 92pp.

URBANEK, R. P. 1990. Behavior and survival of captive-reared juvenile sandhill cranes introduced by gentle release into a migratory flock of sandhill cranes. Unpubl. Rep., Ohio Coop. Fish Wildl. Res. Unit, Columbus. 98pp.

______, and T. A. BOOKHOUT. 1992. Development of an isolation-rearing/gentle release procedure for reintroducing migratory cranes. Proc. North Am. Crane Workshop 6:120-130.

______, and ______. 1994. Performance of captive-reared cranes released into a migration route in eastern North America. Pages 121-129 *in* H. Higuchi and J. Minton, eds. The future of cranes and wetlands. Wild Bird Soc. Japan, Tokyo.

WELLINGTON, M., A. BURKE, J. M. NICOLICH, and K. O'MALLEY. 1996. Chick rearing. Pages 77-95 *in* D. H. Ellis, G. F. Gee, and C. M. Mirande, eds. Cranes: their biology, husbandry, and conservation. Nat. Biol. Serv., Washington, D.C., and Int. Crane Found., Baraboo, Wis.

USE OF ULTRALIGHT AIRCRAFT FOR INTRODUCING MIGRATORY CRANE POPULATIONS

KENT R. CLEGG, 550 Bench Lago Road, Grace, ID 83241, USA
JAMES C. LEWIS, U.S. Fish and Wildlife Service, 500 Gold Ave. SW, Albuquerque, NM 87103, USA
DAVID H. ELLIS, Patuxent Wildlife Research Center, U.S. Geological Survey-Biological Resources Division, HCR 01, Box 4420, Oracle, AZ 85623, USA

Abstract: Greater sandhill cranes (*Grus canadensis tabida*) were used as the research surrogate for whooping cranes (*Grus americana*) to determine if captive-reared cranes could be led by an ultralight aircraft (UL) along a migration route and if, after release on a wintering area, they would integrate with wild cranes and migrate north in spring to their natal area without assistance. In 1995, KRC raised 15 cranes to fledging and trained them to respond to his vocal imitation of a sandhill crane brood call. Chicks learned to follow him as he walked, drove an all-terrain vehicle (ATV), or piloted a UL. The caretakers were not in crane costumes. Cranes were tame but allowed to roam at will without accompanying humans part of the day and were penned at night. Daily excursions provided exposure to habitats, foods, and predators the birds would encounter after release into the wild. In mid-October 1995, 11 radio-tagged cranes were led in migration from Grace, Idaho, to Bosque del Apache National Wildlife Refuge (BdANWR), central New Mexico, and released near wild wintering sandhill cranes. The 1,204-km migration took 11 days, including 1 day when the aircraft were grounded due to a winter storm. Hazards encountered enroute included mountainous terrain, turbulent air, and attacks by golden eagles (*Aquila chrysaetos*). On the wintering grounds hazards included crane hunters and coyotes (*Canis latrans*). Within 2 days after release at the BdANWR wintering site, the research cranes were associating with and imitating the behavior of wild cranes. The 4 surviving birds migrated north in spring 1996, and in May 1996, 2 were within 53 km of their Idaho natal area.

PROC. NORTH AM. CRANE WORKSHOP 7:105-113

Key words: introduction, sandhill crane, whooping crane, migration, ultralight aircraft, *Grus americana*, *Grus canadensis*, Idaho, New Mexico.

The endangered whooping crane numbers approximately 200 in the wild and 100 in captivity. The only self-sustaining wild population nests in the Northwest Territories, Canada, and winters on the Gulf Coast of Texas. This population is at risk because its Texas winter habitat borders the Gulf Intracoastal Waterway, one of the most heavily used waterways in the world. Much of the associated boat tonnage consists of petroleum and chemical products. An accidental spill could pollute much of the wintering area and jeopardize this population.

To promote survival in the wild, the recovery plans of Canada and the United States recommend establishing 2 additional wild populations (Edwards et al. 1994, U.S. Fish and Wildlife Service 1994). An experimental introduction is underway in Florida to start a nonmigratory population (Nesbitt et al. 1997). The Canadian Wildlife Service and U.S. Fish and Wildlife Service hope to introduce a migratory population in the Canadian prairie provinces at the end of this decade.

Juvenile cranes learn a migration route and wintering location from their parents. The source of whooping cranes for the Canadian reintroduction will be captive-reared birds conditioned for release into the wild. But there is no tested technique for introducing cranes in an area where migration must be learned and there are no parents or other adults to teach the route to captive-reared birds. Further, it would be advantageous in the introduction to lead whooping cranes along a chosen route, to preselected stopover areas, and to a predetermined wintering area. Such control would increase the potential protection, simplify monitoring of birds, and make it more acceptable to citizens concerned about introductions of endangered species. A technique is needed to satisfy these objectives.

Beginning in 1983, KRC led tame sandhill cranes behind ATV's and automobiles. In 1994, 1 such flight covered 61 km behind a pickup truck. From 1990 through 1992, DHE and other personnel at Patuxent Wildlife Research Center trained 4-6 sandhill crane chicks each year to follow a truck or automobile so their flight performance could be observed while they were wearing backpack harnesses carrying satellite transmitters. In 1993, Lishman reared Canada geese (*Branta canadensis*) in Ontario, trained them to follow a UL, and in fall led 18 for 600 km to Virginia where they spent the winter (Lishman 1995, Lishman et al. 1997). The following spring, 13 returned unassisted to Ontario. In 1994 he led Canada geese to South Carolina and most of them returned to Ontario in spring 1995. In 1994, KRC reared 6 sandhill cranes and taught them to follow a UL in local flights within Idaho.

Two reintroduction circumstances have been discussed by the U.S. Whooping Crane Recovery Team: (1) sandhill crane populations would be resident in the nesting and wintering areas or would be established there before the whooping cranes were introduced, and (2) the whooping cranes would be introduced in breeding and wintering habitat where no other cranes existed. The migration route used by the Rocky Mountain crane populations was selected for our research because a 20-year data base on sandhill crane and whooping crane habitats and behaviors exists which provides a comparative baseline. Also, requirements of the Endangered Species Act and National Environmental Policy Act have been fulfilled for research with whooping cranes in this area. Most of the suitable habitat in the Rocky Mountains is already occupied by sandhill cranes, so this first phase of our research tests those circumstances in which whooping cranes would be released in areas occupied by sandhill cranes.

Our primary objectives were to determine if (1) young captive-reared sandhill cranes could be trained to follow a UL along a 1,200-km migration route in fall to a specific winter area, (2) cranes reared in a semi-wild manner would develop behavior typical of wild cranes after release at a wintering site, and (3) these birds would then return unassisted to their summering area in spring.

We thank the following for financial support of the research: U.S. Fish and Wildlife Service; Windway Capital Corporation of Sheboygan, Wisconsin, that provided funds through the International Crane Foundation; World Wildlife Fund-Canada; Canadian Wildlife Service; and National Biological Service. Some telemetry equipment was provided by R. P. Urbanek and the Ohio Cooperative Fish and Wildlife Research Unit, Ohio State University, Columbus, Ohio. Special thanks are due M. G., P. M., S. G., and F. M. Clegg for their assistance in many aspects of rearing and migration of the birds. We also appreciate the substantial assistance of E. Spalding, who piloted the Challenger UL during migration. T. Watanabe assisted in the ground crew during the migration. M. V. Connolly, BdANWR, monitored the birds while they wintered at the refuge. R. J. Garcia, Alamosa/Monte Vista National Wildlife Refuge, monitored the cranes while they were staging in southern Colorado in March-April.

STUDY AREA

Sandhill cranes were reared on the Clegg Ranch near Grace, southeastern Idaho. The ranch includes small ponds, pastures, alfalfa, and barley fields where the birds foraged during the day. The 1,200-km migration route extended southward from Grace through eastern Utah to Moab, across the corner of southwestern Colorado to Cortez, south to Gallup, New Mexico, southeasterly to Los Lunas, New Mexico, and south along the Rio Grande to BdANWR, south of San Antonio, New Mexico. The 140,790-ha BdANWR includes approximately 9,880 ha of managed wetlands and several thousand hectares of corn and alfalfa. Annual peak winter populations of 15,000-20,000 sandhill cranes occur on the refuge. The route was selected to avoid some high-elevation, rugged terrain traversed by most cranes in the Rocky Mountains because their route was more hazardous to UL aircraft. Also, only small numbers of cranes migrate along the route selected, consequently there would be less likelihood of the research cranes leaving the UL to join wild cranes.

METHODS

Rearing and Training

Searches for wild sandhill crane nests occurred in late April and May at Grays Lake National Wildlife Refuge, 55 km from the Clegg ranch. Eggs were floated at the nest site to estimate the hatching date. Eggs were removed from nests about 4 days before hatching and grouped by hatching date to minimize aggression among chicks. Eggs were kept in an incubator until pipped, then moved to a container with natural nesting materials where an infrared heat lamp provided warmth. Nest material was replaced every 4 days to minimize fungal growth. Water in a pan beneath the nesting material helped maintain the high humidity appropriate for hatching. When the chick began pipping the egg, a tape recording of KRC's imitation of the crane brood call was played. The call, lasting about 5 seconds, occurred at 30-sec intervals on the tape which was played for up to 30 minutes. The tape was played periodically over the approximately 30 hours required for the bird to hatch. The tape was not used after the chicks hatched because there was regular contact with KRC and his calling.

After hatching, 6-8 chicks of similar age (a several day age spread) were placed in each group pen (2.12 m long, 0.912 m wide, and 0.6 m high) at night until they were 5-10 days of age. Then the 12 oldest individuals were placed on a small island in a pond inside a 15.2-m × 30.4-m pen. The youngest chicks were gradually added to this large group. Food was placed in bowls at several sites on the island and a transparent windbreak and full-sized crane decoy were placed there. A heat lamp suspended over the island provided a source of warmth. Chicks were viewed periodically through the evening, via video camera, to monitor their well-being. The island was progressively flooded over an interval of 50 days so the juvenile birds were roosting in water by the time they were fledging age.

A Dragonfly UL (high wing, push propeller, single seat, open cockpit, min. air speed 32 km/h, max. air speed 104 km/h, capable of 5 hours aloft with an extra fuel tank, and designed to tow gliders) was modified specifically for this project. A 4 × 4 Polaris Explorer ATV was used to lead prefledged chicks.

The juvenile cranes were led daily across an open field to a ditch bank and allowed to forage, without a caretaker present, for natural foods in the water and uplands. The birds fed in grain and alfalfa fields on seeds, insects, and earthworms (*Lumbricus* sp.). They were observed from a distance with binoculars at 15-30-min intervals. The decoy was also used as a daytime attractant in fields when the caretaker left the birds. The birds were penned at night. Potential predators in the area included golden eagles, coyotes, raccoons (*Procyon lotor*), and red-tailed hawks (*Buteo jamaicensis*).

Training the birds to follow KRC on foot, on the ATV, and in the UL capitalized on the following behavior of chicks. He called the birds to follow by imitating the sandhill crane brood call, and the birds walked behind him. At age 20 days the chicks were exposed to the sounds and appearance of the ATV and UL. The UL was flown over the pen at 2-3-day intervals and left idling nearby for short periods. The plane was also parked near the pen so the birds grew accustomed to its appearance, the wind from the propeller, and engine noise. The birds were trained to run after the ATV or the taxiing UL. After they fledged the cranes were encouraged to follow the flying UL.

Migration

Birds were radiotagged with solar-powered transmitters; some also had a Ni/Cad battery for operation when solar energy was not available. Bright yellow leg bands (7.5 cm high) with black numbers were affixed shortly before the migration flight. The cranes were led south by the Dragonfly UL on 16 October 1995. A second faster Challenger UL (single seat, open cockpit, high wing, push propeller, max. air speed 161 km/h) accompanied them to help monitor wind conditions ahead of the Dragonfly UL, select suitable landing sites, and to protect the flock from attacking golden eagles. Landing locations were not predetermined but were selected when the birds became tired, night approached, fuel supplies were low, weather became unfavorable, or the cranes scattered when golden eagles attacked.

The pilots usually had radio contact with each other and with a member of the ground crew. Four personnel with 3 vehicles comprised the ground crew. One pickup truck towed an aircraft trailer used to transport a portable crane pen to New Mexico and to return to Idaho with a UL. Another pickup truck and a lightweight 4WD truck were used as chase vehicles. The cranes were placed in a net-covered, portable pen at night. Six panels (each 4.56 m × 2.1 m) tied together provided a pen 4.56 m × 9.1 m × 2.1 m. At least 2 persons camped near the cranes at night.

Birds were penned at midday and overnight to protect them from predators. In some situations wetlands were available, so the birds could feed and roost in water. On other occasions surface water was not available near landing sites and the birds were penned in uplands. They were offered food and water 2 to 3 times daily. Food ingestion was limited in early morning on flight days.

One day after arrival at BdANWR, New Mexico, the aircraft, pen, and foods were removed. The research cranes were allowed to integrate with the wild cranes on their own initiative and to learn appropriate behavior for survival at the winter site. KRC remained at the refuge for a week to monitor the birds' integration with the wild sandhill cranes. The research cranes were monitored daily through the winter months by BdANWR personnel.

RESULTS

Rearing and Training

Twenty wild sandhill crane eggs were acquired and 20 hatched. There was a 10-day age difference between the youngest and oldest chicks, and 2 of the youngest were killed by the others. Three chicks died of unknown causes and 15 fledged. During initial flights only 1 or 2 birds flew with the UL, but within 5 days all 15 were following the aircraft. Gradually the flight distance was increased. The pilot observed the flying cranes to determine when they were tired (i.e., panting or gaping) and should be allowed to rest. The cranes adapted well to the Dragonfly UL but during initial flights their attempts to follow were uncoordinated. With experience, some cranes learned to minimize energy expenditure by riding air vortices off the UL wings. Several hazards existed for the young cranes as they learned to fly. Powerlines adjacent to the ranch property proved fatal for 1 young bird shortly after it learned to fly. One crane was killed when it flew into the UL propeller. Thereafter, a metal guard was uniquely engineered and installed to protect the cranes and propeller. Two golden eagles attacked the cranes during a flight training session near the ranch, and 1 crane received a wing injury that prevented it from participating in the migration. Another crane lost a foot when it became entangled in a wing strut while flying. The injured birds healed and could still fly, but they were excluded from the migration because of concern that they would not be able to keep up with the other birds.

Migration

Each fledged, captive-reared sandhill crane learned to follow the UL. By migration time the flock was flying about 40 km daily. Migration began on 16 October with 11 cranes leaving the Clegg Ranch and ended on 26 October when 8 arrived at BdANWR in central New Mexico (Figs. 1 and 2). The 1,204-km migration required 11 days including 23 hours of flight time (Table 1). During most of the migration, the birds flew in formation off the UL wingtips to take advantage of vortices off the wings. The UL's were grounded 1 day due to snow and strong winds. Six cranes flew with the UL the entire migration. Crane No. 50 disappeared 16 October just north of Logan, Utah, after accompanying the aircraft about 76 km and returned to the Clegg Ranch 3 days later.

Daily distances flown varied from 43 to 217 km (Fig. 1) depending on wind conditions, temperature, terrain, fuel supplies, golden eagle attacks, stamina of the cranes, and diminishing daylight. Mean flight altitude was about 300 m. The highest pass traversed was about 2,590 m above sea level. The UL crossed it at about 150 m altitude. Average flight speed was 52 km/h with occasional bursts of up to 70 km/h. The typical morning flight began about 0900 hours, or later if ice was on the planes, and continued for 1-2 hours. If winds moderated sufficiently, another flight was made in late evening and continued as long as light was adequate. Only once were the birds reluctant to fly. On that occasion their reluctance may have been due to an attack by eagles that morning or excessive feeding in the afternoon.

Some cranes had difficulty keeping up with the aircraft, especially at higher altitudes in rugged terrain. No. 43 had difficulty flying after the first day and at the end of the migration was diagnosed as having coccidiosis. No. 42 stayed with the group the first day and the last 3.5 days but had difficulty keeping up with the flock through some of the most rugged terrain. Whenever 1 bird tired and landed, others endeavored to join it. It was necessary to transport non-flying birds in a trailer to keep the main flock flying. No. 42 was transported 400 km, and No. 43 was transported most of the migration.

Other than the difficult flying conditions, the most persistent problem was attacks by golden eagles. On 17 October the cranes crossed a pass at 1,829 m in the Bear River Mountain Range. A pair of golden eagles attacked Nos. 41 and 49, lagging about 90 m behind and 45-60 m below the Dragonfly UL. The pilot of the Challenger UL saw 1 eagle strike No. 41 and carry it about 460 m to the ground where its carcass was later recovered. The second crane was last seen being pursued by the second eagle.

Attacks by 1-3 eagles occurred 5 times during the migration. Single eagles were observed on 2 other occasions when the cranes were not attacked. When an attack was detected, both aircraft intercepted and chased the eagles. The eagles typically attacked from above. Sometimes the first indication of an impending attack was when the cranes moved close around or beneath the UL. The progress of migration was disrupted for several hours after some of these attacks. As the pilots gained experience they learned to anticipate eagle attacks, to interrupt the attack by intercepting the eagles, and by shooting shell crackers to deter the attacks. As they gained experience, the pilots became more adept at using the UL's to protect the cranes, and the cranes became more adept at avoiding the eagles.

The cranes responded to approaching unfamiliar aircraft as they would to an approaching eagle. They flew under the wings, alongside the Dragonfly UL, or they scattered. The Challenger UL was not flown with the birds until the morning of migration, and it took several days before the cranes seemed to recognize and lose their fear of it.

Fig. 1. Migration route of the ultralight aircraft and sandhill cranes, October 1995. Numbers indicate overnight stops: (1) Hyrum, Utah; (2) Deer Creek Reservoir below Heber City, Utah; (3) Daniel's Pass, Utah; (4) 8 km north of Helper, Utah; (5) Green River, Utah; (6) and (7) Lisbon Valley, Utah; (8) south of Shiprock, New Mexico; (9) Twin Lakes, New Mexico; and (10) west of Los Lunas, New Mexico. Daily flight segments are described in Table 1.

Fig. 2. A flock of greater sandhill cranes follows the ultralight aircraft past a volcanic escarpment in northern New Mexico (photo by D. H. Ellis).

No. 43, the crane transported by trailer most of the way, was taken to the Rio Grande Zoo in Albuquerque, New Mexico, for treatment. The diagnosis was coccidiosis, an illness common among sandhill cranes of the Rocky Mountains (Parker and Duszynski 1986), and one that can cause severe respiratory effects. No. 43 was retained at the zoo for treatment.

Adaptation to the Wild and the Wintering Site

The migration was completed at about 1300 hours on 26 October. Many wild sandhill cranes were in fields adjacent to the field where the UL's and research cranes landed. After the media and the welcoming personnel left the area, the 7 research birds walked towards wild birds. Wild pairs, unaccompanied by young, approached the research cranes and called to them. Research vehicles and aircraft were removed from the vicinity. At dusk the wild birds began flying to their water roosting sites. The research birds flew with some departing wild birds, but when they gained enough altitude to view the Dragonfly UL at a distance they separated from the wild birds and landed by the aircraft. The birds were then penned for the night.

On the morning of 27 October, the 7 research birds were released from the pen to join wild birds feeding in the vicinity. The pen and aircraft were removed from the area. That evening research birds failed to fly with wild birds when they departed for their water roost. We attempted to frighten them so they would join the wild birds. They flew, but in the fading light we were unable to determine where they roosted because their transmitters did not operate without sunlight.

By 0900 hours on 28 October, most research birds were back with wild cranes feeding in alfalfa or on the edge of standing cornfields. Between 1000 and 1200 hours we

Table 1. Migration history of research sandhill cranes led by ultralight aircraft from Grace, Idaho, to Bosque del Apache National Wildlife Refuge, New Mexico, October 1995 (see Fig. 1).

October	Segment location	Distance (km)	Crane nos. and performance[a]				
			40,44–48	41,49[b]	42	43[c]	50[d]
16	Grace, Id., to Richmond, Ut.	76	+	+	+	+	+
	Richmond to Logan, Ut.	11	+	+	+	+	(10)
	Logan to Hyrum, Ut.	16	+	+	+	+	
17	Hyrum to Eden Pass, Ut.	14	+	(12,K,O)	+	T	
	Eden Pass to Park City, Ut.	89	+		T	T	
	Park City to Deer Creek, Ut.	31	+		T	T	
18	Deer Creek to Daniel Pass, Ut.	24	+		T	T	
19	Daniel Pass to Helper Jct., Ut.	75	+		+	T	
20	Helper Jct., to Green River, Ut.	116	+		T	T	
21	Green River to La Sal Jct., Ut.	109	+		+	T	
	La Sal Jct., to Lisbon Valley, Ut.	16	+		T	T	
22	Lisbon Valley	0					
23	Lisbon Valley to Cortez, Colo.	108	+		T	T	
	Cortez to Shiprock, N.M.	69	+		+	T	
24	Shiprock to Tohatchi, N.M.	101	+		+	T	
	Tohatchi to Twin Lakes, N.M.	16	+		+	T	
25	Twin Lakes to Mesita, N.M.	177	+		+	T	
	Mesita to Los Lunas, N.M.	40	+		+	T	
26	Los Lunas to Bosque, N.M.	116	+		+	T	
Total distance flown		1,204	1,204	115	803	88	86

[a] Flew with aircraft (+), transported in truck (T), km flown in portion of segment (No.), missing (O), killed by eagle (K).

[b] No. 41 was ridden to the ground by a golden eagle and the carcass found 18 October. No. 49 was simultaneously pursued by a second eagle and is presumed dead; no transmission signal was received during a ground search.

[c] No. 43 was unable to keep up with the flock and was later diagnosed as having coccidiosis.

[d] No. 50 disappeared on the second flight, was not detected in ground searches, but returned to Grace, Idaho, on 18 October.

recovered carcasses of Nos. 46 and 48 in the corn. It appeared they had been killed by coyotes; a common coyote hunting tactic is to use dense weed cover and the corn to conceal its approach to prey on snow geese or cranes entering the standing corn to feed. We do not know if the cranes were killed early that morning or during the previous night.

On the evening of 28 October, the surviving birds accompanied wild birds to water roosts. On 29 October, only 2 of the surviving birds could be found on the refuge. By searching north of the refuge, we found 1 of the 3 missing birds (No. 42) hiding in dense vegetation. It was captured and examination indicated it had a minor flesh wound in 1 wing caused by a shotgun pellet. This bird was returned to the refuge and released. On 31 October, project personnel were notified that the 2 missing birds, Nos. 40 and 47, had been killed by a crane hunter on 29 October and delivered to a hunter checking station. The cranes were flying with a flock of wild cranes when killed. After the 3 research cranes were shot, information sheets (fliers) were prepared and given to all hunters before the 2 remaining 2-day hunts. These fliers described the research, the bright yellow leg markers on the research birds, and requested that hunters not shoot these cranes.

No. 50, the crane which left the group on the first day of migration and returned to the Clegg Ranch, was transported to BdANWR on 13 November. It was released in the vicinity of No. 42, which it immediately joined. No. 50 was last located on the evening of 15 November and could not be found during a subsequent search of the refuge and the Rio Grande Valley 30 km north of the refuge. We do not know if No. 50 died or left the area.

The 4 surviving research cranes spent the winter associating with wild cranes, feeding in uplands during the day and roosting in wetlands at night. Their movements were limited to the refuge most of the winter. The distance traveled between the roost and feeding area was typically less than 3 km. The research cranes associated with the same groupings of wild birds all winter. They responded to the approach of humans by taking flight, just like the wild cranes. The transmitter of No. 44 failed in late November, but No. 44

continued to accompany No. 45 and wild birds and was regularly observed. No. 42 remained with wild birds all winter and seldom associated with the other research birds. No. 43, treated for coccidiosis, was released on 21 November and associated with wild birds through the remainder of the winter. All research birds were captured in February 1996 by night-lighting (Drewien and Clegg 1992), and new radio transmitters were affixed.

Spring Migration

The 4 research cranes migrated from New Mexico to the San Luis Valley of southern Colorado in mid-March where they staged with other cranes of the Rocky Mountain population. The cranes did not retrace their southward migration route but followed the migration route typically used by cranes wintering in New Mexico. In late April they migrated north, and Nos. 44 and 45 were found 53 km from the Clegg Ranch in mid-May. The location of the other 2 birds is unknown.

DISCUSSION

Fifteen cranes were reared together to fledging age, thereby reducing labor and pen requirements. Several factors were considered important to minimize aggression between chicks. Food was available at all times, and they were taught to self-feed as soon as possible, so they were never very hungry. Aggressive behavior seemed more evident when chicks were hungry. Access to live prey may also have reduced aggression. Exercise kept them occupied, and pursued chicks were able to flee (rather than be cornered against an artificial structure), which reduced the potential for injuries incurred by other group members. When young chicks became aggressive (3–10 days of age), their beaks were lightly trimmed (i.e., sufficiently to bring a drop of blood but not to injure the growth plates). The beaks were then tender and birds were reluctant to peck each other.

The frequency of attacks by golden eagles was not anticipated. Only 1 attack on a flying crane, presumed to be by a golden eagle, had previously been documented in North America (Drewien and Bizeau 1981). Eagle attacks were also a frequent problem in the trucking-migration research reported in this proceedings (Ellis et al. 1997). Knowledge gained about eagle attack strategy in the first few days of migration allowed the pilots to anticipate later attacks and avoid further losses. For the future, we believe improved use of the second aircraft and continued use of shell crackers will diminish the number of attacks and the loss of cranes, and reduce the flight delays caused by attacking eagles.

Cranes "surfing" the airflow over the UL wing or its wing tip vortices received some advantage, but generally cranes had to maintain a steady flapping flight to keep up with the UL. Thermals (rising currents of heated air) in the mountains are usually too violent for a UL to safely utilize to gain altitude. Consequently, research cranes following the UL could not take advantage of the natural lift provided by thermals, which wild cranes use to conserve energy. Wild cranes migrate using a combination of flapping flight, spiralling, and gliding.

Daily flight times and distances of research cranes (Table 1) were less than that of migrating wild cranes, which flew from 48 to 740 km in daily flights that lasted from 1 to more than 10 hours (Melvin and Temple 1982). Sandhill cranes migrating between Colorado and New Mexico flew 225–345 km in 1 day (Stahlecker 1992). A juvenile whooping crane and its sandhill crane foster parents migrated 262–276 km/day (Drewien and Bizeau 1981). Whooping cranes have made nonstop flights of 9–10 hours covering 700–800 km (Kuyt 1992).

Some individuals lacked stamina for the long flapping flight migration and flying conditions in mountainous terrain. These birds tired sooner and lagged behind the main flock. They seemed to be the preferred target for attacking golden eagles. The lack of stamina exhibited by No. 43 resulted from an active case of coccidiosis, which can have severe respiratory effects (Carpenter et al. 1984).

Water-roosting behavior is essential to survival of cranes after they are released in the wild. Released sandhill cranes reared at a field facility showed no reluctance to roost in water (Urbanek and Bookhout 1992). However, in the absence of water-roosting opportunities in captivity, whooping cranes did not quickly learn water-roosting after release in the wild and were vulnerable to predation (Nesbitt et al. 1997). This problem was corrected by prerelease exposure to water-roosting in captivity (Drewien et al 1997, Nesbitt et al 1997). Since our techniques are ultimately designed for use with whooping cranes, KRC reared the research cranes in a manner that would promote appropriate water-roosting behavior. The research sandhill cranes roosted in water after they joined wild cranes. Nos. 43 and 50 were released at wetland sites in New Mexico and roosted in water the evening of release.

Group rearing the cranes in semi-wild circumstances on the Clegg Ranch helped prepare them for life independent of a caretaker when they arrived at BdANWR. We believe that survival and integration to the wild can be improved at the wintering site by releasing birds in the wetlands, rather than uplands, where they can roost overnight and, on their own initiative, join the wild flocks in daily feeding flights to uplands. In one respect, quick integration of research cranes with wild cranes was a disadvantage because it led to the shooting of 3 project cranes that flew with wild birds and left the refuge to feed on private croplands. The hunters reported that the cranes were flying in a flock of wild cranes when they were shot.

In 1995, 3 crane hunts (each involving 60 hunters for 2 days) were scheduled on private lands in the Rio Grande Valley of New Mexico. The first hunt was 28–29 October, and the others were scheduled for early December and early January. The departure of the research cranes, with wild cranes, from the abundant food (alfalfa, insects, corn, wetland plants and animals) and relative safety of the refuge was not anticipated. Unfortunately, our research birds arrived on the refuge 2 days before the first crane hunt on private lands and 4 days before personnel began knocking down corn on the refuge. The knocked-down corn is more accessible to cranes and diminishes the likelihood that cranes will leave the refuge to feed on adjacent private croplands. Further, the danger of coyote predation is less than when the corn is standing. In the future, corn can be knocked down before the research birds arrive to encourage them to remain on the refuge. The chance of shooting losses can also be further diminished by distributing fliers to hunters as they pass through a checking station operated by the New Mexico Department of Fish and Game before the hunt. The fliers will request hunter cooperation in protecting the research cranes. The hunter who shot the cranes in 1995 said he was able to see the bright yellow leg markers while the cranes were flying but was not aware of their association with the UL research project.

We also had problems with our radio telemetry equipment. Signals from the solar-cell transmitters were not as strong as standard battery-powered transmitters used on other crane projects. Our solar-powered units operated marginally or not at all when it was cloudy, at dusk, when the birds sat on the ground, or when the instrumented leg was folded into belly feathers as the bird stood on the other leg. The unsatisfactory transmitters prevented recovery of No. 49, believed to have been killed by an eagle. Our inability to find where the cranes landed on the evening of 27 October may have contributed to the deaths of cranes No. 46 and 48. The failure to find No. 50 in November was also probably due to poor radio performance. Subsequently we captured the 4 surviving cranes by night-lighting (Drewien and Clegg 1992) and replaced the solar-powered transmitters with battery-powered ones.

The 2 known surviving research cranes returned to within 53 km of the Clegg Ranch. Wild yearling sandhill cranes typically return to their natal area later in spring than the adults, and some disperse to sites other than their natal area; the magnitude and pattern of such dispersal in the Rocky Mountains were not determined (Drewien 1973). Captive-reared sandhill cranes, released as juveniles in fall in the Upper Peninsula of Michigan, migrated to Florida, then spent the following summer within 155 km of the natal area (Urbanek and Bookhout 1992). Excluding 3 birds that summered at extreme distances, the average male summered 3.5 km and the average female 26.6 km from their original rearing/release site. About one-third of all marked whooping cranes returned to their natal area in subsequent years. Ninety-five percent of the whooping cranes in the Rocky Mountain introduced population could be found within a 200-km radius of the natal area, and those remaining outside the normal summer range were yearlings or 2-year-olds. A higher percentage of the breeding-age whooping cranes, especially males, returned to the specific natal site (Drewien et al. 1989). Thus, the return of the UL cranes to within 53 km of their natal area was within documented parameters.

CONCLUSIONS

Basic techniques of training, migration, and introduction to the wild were suitable and show promise for improved use in future reintroductions. Although some disappointing losses of project birds were experienced, such losses can be diminished in the future as techniques are refined. Additional testing of the UL technique with sandhill cranes is proposed for 1996 to provide a more complete evaluation of its merits and suitability for reintroducing cranes.

LITERATURE CITED

CARPENTER, J. W., M. N. NOVILLA, R. FAYER, and G. C. IVERSON. 1984. Disseminated visceral coccidiosis in sandhill cranes. J. Am. Vet. Med. Assoc. 185:1342-1347.

DREWIEN, R. C. 1973. Ecology of Rocky Mountain greater sandhill cranes. Ph.D. Thesis, Univ. Idaho, Moscow. 168pp.

_____, and E. G. BIZEAU. 1981. Use of radiotelemetry to study movements of juvenile whooping cranes. Pages 130-134 *in* J. C. Lewis and H. Masatomi, eds. Crane research around the world. Int. Crane Found., Baraboo, Wis.

_____, and K. R. CLEGG. 1992. Capturing whooping cranes and sandhill cranes by night-lighting. Proc. North Am. Crane Workshop 6:43-49.

_____, W. M. BROWN, and E. G. BIZEAU. 1989. Whooping crane cross-fostering experiment. Unpubl. Rep., Hornocker Wildl. Res. Inst., Univ. Idaho, Moscow. 10pp.

_____, W. L. MUNROE, K. R. CLEGG, and W. M. BROWN. 1997. Use of cross-fostered whooping cranes as guide birds. Proc. North Am. Crane Workshop 7:86-95.

EDWARDS, R., S. BRECHTEL, R. BROMLEY, D. HJERTAAS, B. JOHNS, E. KUYT, J. LEWIS, N. MANNERS, R. STARDOM, and G. TARRY. 1994. National recovery plan for the whooping crane. Rep. 6. Recovery of Nat. Endangered Wildl. Comm., Ottawa, Ont. 39pp.

ELLIS, D. H., B. CLAUSS, T. WATANABE, R. C. MYKAT, M.

KINLOCH, and C. H. ELLIS. 1997. Results of an experiment to lead cranes on migration behind motorized ground vehicles. Proc. North Am. Crane Workshop 7:114-122.

KUYT, E. 1992. Aerial radio-tracking of whooping cranes migrating between Wood Buffalo National Park and Aransas National Wildlife Refuge, 1981-84. Can. Wildl. Serv. Occas. Pap. 74. 53pp.

LISHMAN, W. 1995. Father goose. Little, Brown and Company (Canada) Limited, Toronto, Ont. 229pp.

______, T. L. TEETS, J. W. DUFF, W. J. L. SLADEN, G. G. SHIRE, K. M. GOOLSBY, W. A. BEZNER KERR, and R. P. URBANEK. 1997. A reintroduction technique for migratory birds: leading Canada geese and isolation-reared sandhill cranes with ultralight aircraft. Proc. North Am. Crane Workshop 7:96-104.

MELVIN, S. M., and S. A. TEMPLE. 1982. Migration ecology of sandhill cranes: a review. Pages 73-87 *in* J. C. Lewis, ed. Proc. 1981 crane workshop. Natl. Audubon Soc., Tavernier, Fla.

NESBITT, S. A., M. J. FOLK, M. G. SPALDING, J. A. SCHMIDT, S. T. SCHWIKERT, J. M. NICOLICH, M. WELLINGTON, J. C. LEWIS, and T. H. LOGAN. 1997. An experimental release of whooping cranes in Florida-the first three years. Proc. North Am. Crane Workshop 7:79-85.

PARKER, B. B, and D. W. DUSZYNSKI. 1986. Coccidiosis of sandhill cranes (*Grus canadensis*) wintering in New Mexico. J. Wildl. Dis. 22:25-35.

STAHLECKER, D. W. 1992. Crane migration in northern New Mexico. Pages 1-12 *in* D. A. Wood, ed. Proc. 1988 North Am. crane workshop. Florida Game Fresh Water Fish Comm. Nongame Wildl. Program Tech. Rep. 12.

URBANEK, R. P., and T. A. BOOKHOUT. 1992. Development of an isolation-rearing/gentle release procedure for reintroducing migratory cranes. Proc. North Am. Crane Workshop 6:120-130.

U.S. FISH AND WILDLIFE SERVICE. 1994. Whooping crane recovery plan. U.S. Fish Wildl. Serv., Albuquerque, N.M. 92pp.

RESULTS OF AN EXPERIMENT TO LEAD CRANES ON MIGRATION BEHIND MOTORIZED GROUND VEHICLES

DAVID H. ELLIS, Patuxent Wildlife Research Center, U.S. Geological Survey–Biological Resources Division, Laurel, MD 20708-4019, USA
BRIAN CLAUSS, Patuxent Wildlife Research Center, U.S. Geological Survey–Biological Resources Division, Laurel, MD 20708-4019, USA
TSUYOSHI WATANABE, Patuxent Wildlife Research Center, U.S. Geological Survey–Biological Resources Division, Laurel, MD 20708-4019, USA
R. CURT MYKUT, Patuxent Wildlife Research Center, U.S. Geological Survey–Biological Resources Division, Laurel, MD 20708-4019, USA
MATTHEW KINLOCH, Patuxent Wildlife Research Center, U.S. Geological Survey–Biological Resources Division, Laurel, MD 20708-4019, USA
CATHERINE H. ELLIS, Patuxent Wildlife Research Center, U.S. Geological Survey–Biological Resources Division, Laurel, MD 20708-4019, USA

Abstract: Ten greater sandhill cranes (*Grus canadensis tabida*), trained to enter and ride in a specially-equipped truck, were transported at ca 80 days of age from their rearing site at Patuxent Wildlife Research Center (Patuxent), Maryland, to a reintroduction site located within the species' former breeding range in northern Arizona. After 5 additional weeks of training, these juvenile cranes were led south ca 600 km to a wintering area on the Arizona/Mexico border. Nine of the 10 survived the trek, 495 km of which were flown, although only a few cranes flew every stage of the route. Their longest flight was 77 km. Major problems during the migration were powerline collisions (3, 1 fatal), eagle attacks (none fatal), and overheating (when air temperatures exceeded ca 25 C). All cranes that entered training quickly learned to follow the truck, and their tenacity when following under unfavorable conditions (e.g., poor light, extreme dust, or heat) showed that cranes could consistently be led over long distances. We cannot predict if the cranes will retrace their route unassisted when adults, but 2 cranes returned 130 km to the starting point of the migration after the flock was scattered by an eagle during our migration south. Three other cranes were recovered 55 km from the attack site and on course toward the starting point.

PROC. NORTH AM. CRANE WORKSHOP 7:114-122

Key words: reintroduction techniques, migration, sandhill crane, whooping crane, *Grus* spp.

Because of the need to develop techniques for establishing additional migratory flocks of whooping cranes (*Grus americana*), 2 experiments were initiated in 1995 to lead sandhill cranes, as research surrogates, on migration. An ultralight experiment is described in a companion paper (Clegg et al. 1997). The first year of a 2-year experiment to lead cranes with a truck (Fig. 1) is described below.

Training birds to follow humans is far from new. For more than 5,000 years, falconers have trained raptors to return to their trainers (Burnham 1990). Sometimes this resulted in birds following pedestrian or equestrian trainers for several kilometers. By the mid-20th century, scientists understood the following response for some avian species and had trained neonatal waterfowl to follow humans (Lorenz 1952; see Immelman [1972] for a review of imprinting).

During the 1980's, 2 remarkable experiments were conducted in Washington state. In the first, Kenneth Franklin trained 2 prairie falcons (*Falco mexicanus*) to circle up and join an ultralight aircraft trailing a food-garnished lure (Franklin and Franklin 1985). Franklin later trained peregrine falcons (*F. peregrinus*) to follow him in free fall as he and the falcon sky-dived from a small fixed-wing aircraft (K. Franklin, Friday Harbor, Wash., pers. commun.). Less spectacular, but more pertinent to our crane experiments, in 1979 and 1980 John McNeely (Sharon, Conn., pers. commun.) trained a red-tailed hawk (*Buteo jamaicensis*) to fly with his hang glider. The bird also landed on the glider and received food from McNeely as he flew. A movie, *John McNeely and the Hawk*, resulted from the experiment. Also pertinent are the waterfowl training experiments of William H. Carrick (Toronto, Ont., pers. commun.), who trained geese and swans (1970–present) to follow boats and other craft. This pioneering work led to the spectacular successes of Bill Lishman and Joe Duff in leading Canada geese (*Branta canadensis*) on 3 migrations (1993–95) from eastern Canada to Virginia or to South Carolina (Lishman et al. 1997).

Probably all aviculturists who have reared cranes have noted that chicks are prone to follow their human caretakers. Beginning in 1973, George Archibald (International Crane Foundation, Baraboo, Wis., pers. commun.) trained sandhill cranes to fly and return to him. He added Eurasian cranes (*Grus grus*) in 1975 until he had a group of up to 20 cranes that would launch, climb to 100 m or more, fly for 5 minutes or more, then return. Probably the first to lead a crane chick with a motorized vehicle were Des and Jen Bartlett in 1971 (Bartlett and Bartlett 1975:69). They reared a sandhill crane chick with a flock of snow geese (*Anser caerulescens*) and trained all to follow an automobile. Because their primary goal was to photograph the birds in flight, they made no attempt to actually lead the birds along a migration route following the vehicle. They did, however, release the crane

Fig. 1. A group of 10 greater sandhill cranes following the cranemobile on an early morning flight past the Vulture Mountains, 16 October 1996 (photo by D. H. Ellis).

and the geese periodically at refuges from southern Canada to Texas.

From 1990 through 1992, we at Patuxent, each year, trained 4–6 sandhill crane chicks to follow a motorized vehicle so we could observe their flight performance while they wore various backpack harnesses carrying satellite transmitters. Due to space constraints, these flights never exceeded 1 km. Beginning in 1983, Kent Clegg (Grace, Id., pers. commun.) led tame sandhill cranes behind all-terrain vehicles and automobiles. In 1994, 1 such flight extended to 61 km behind a pickup truck.

Occasionally sandhill cranes have been transported by truck to migration stopover sites after they failed to migrate on their own (R. H. Horwich, pers. commun.; Nagendran 1992; Urbanek and Bookhout 1992). In some of these attempts, the trucked cranes proceeded on their migration without further assistance. In none of these attempts did the birds actually follow the truck, and in all instances the trucked cranes had wild cranes to follow after their release.

We are greatly indebted to Colonel L. Triphahn for allowing our training on the Navajo Army Depot near Belmont, Arizona. Sergeant D. Hack coordinated all our activities on the depot and sponsored our stay. W. Shifflett, Manager of the Buenos Aires National Wildlife Refuge (BANWR), hosted our wintering cranes, and P. Landin, G. Jontz, S. Gall, and W. Kuvlesky assisted with bird care at the refuge. We greatly appreciate T. Kohler for transporting 6 of our cranes to Arizona, gratis. The Arizona Game and Fish Department (AGFD) greatly assisted us during the migration. They passed along locations of wayward (i.e., eagle-scattered) birds and provided an aircraft to search for birds by radiotelemetry. J. G. Goodwin, Jr., coordinated all of these efforts. P. Smith (AGFD) advised us in setting up the project. Important participants in the field include F. B. P. Trahan, who volunteered for the 1995 migration, and S. Thomas, who volunteered to help rear and train cranes in the 1994 pilot project. C. Van Cleave, Arizona's most noted raptor rehabilitator, donned another hat (i.e., gruid flock

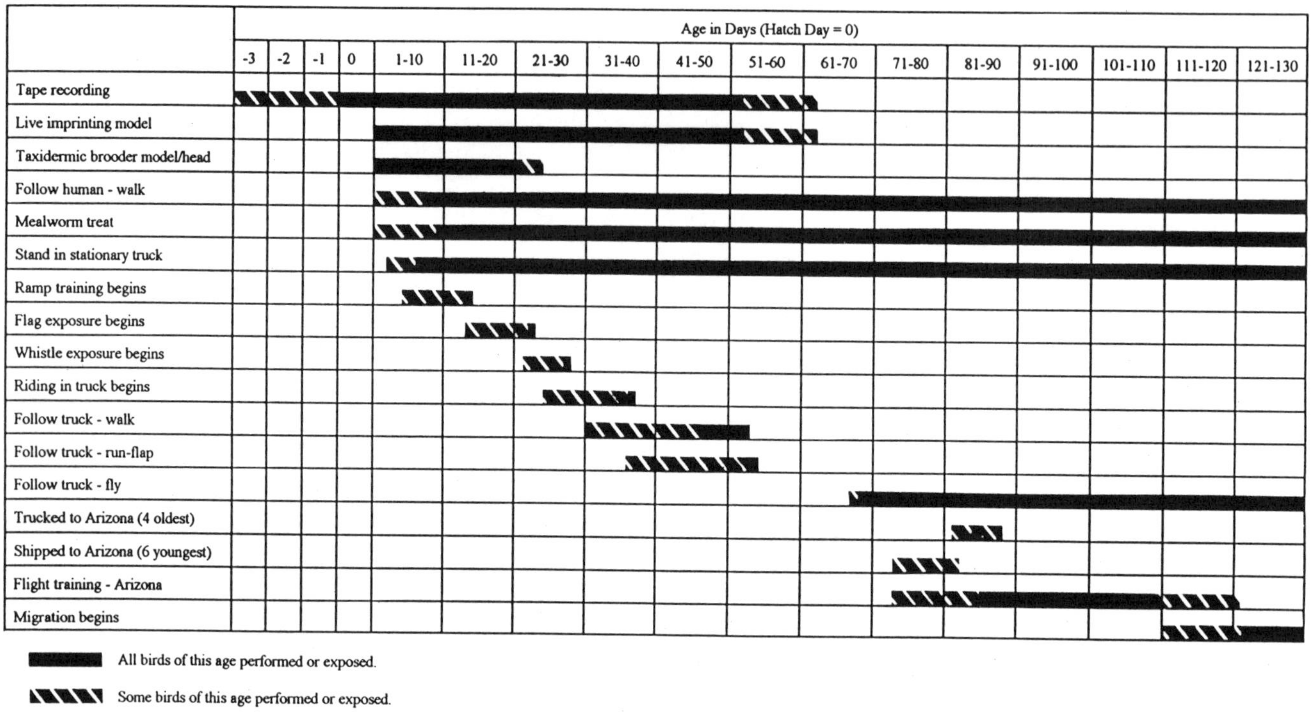

Fig. 2. Training timeline for greater sandhill cranes, trucking experiment, Arizona, 1995.

reassembler) during our odyssey. Most fundamental of all, we thank G. F. Gee, T. Fontaine, K. O'Malley, J. Nicolich, and all of the Patuxent rearing staff for making the project fly.

METHODS

In our 1994 pilot study, we trained 5 sandhill cranes to walk a ramp into a pickup truck. To lure the chicks up the ramp, we scattered mealworms (larvae of *Tenebrio* sp.) on the ground, up the ramp (2.4 m long with a 72-cm rise), and into the truck bed. The trainer on the ramp and in the truck bed also lured the chicks by imitating an adult brood call. Ramp training began at about 10 days of age. All chicks quickly learned to walk up and down the ramp and showed very little aversion to doing so or to riding in the truck. The only difficulty was preventing flightless chicks from walking off the edge of the ramp. The propensity of the chicks to do so necessitated stationing a caretaker on each side of the ramp. None of these birds was trained to fly behind the truck; however, from about 65 days of age, they did ride in the truck.

Also in 1994, we searched for a suitable migration corridor for a long-distance trucking study. We needed a site with good crane habitat, where sandhill cranes no longer bred, and, if possible, removed from sandhill crane migration areas. We reasoned that, if our sandhill crane migration study was to approximate conditions in the future whooping crane migration project, it was important that our sandhill crane chicks not have other sandhill cranes available to lead them on migration, either north or south. We considered and examined potential routes starting at northern termini in Maine, New York, and Ontario and leading south to the Atlantic Coast marshes, but finally settled on a route leading south from the mountains of northern Arizona to the BANWR on the Mexican border. The northern terminus was ideal because sandhill cranes had formerly bred there (Phillips et al. 1978:30), but they were now absent as breeders and unlikely even as migrants in this area. The winter terminus (BANWR) was selected because sandhill cranes never winter there and seldom migrate through. Also, facilities were available to overwinter the birds in a pen, safe from predation.

The Rearing Process

Birds used in the 1995 experiment were subjected to stimuli according to the schedule in Fig. 2. They were exposed to live adult imprinting models and were otherwise reared as described for hand-reared Mississippi sandhill cranes (*G. c. pulla*) intended for release (Ellis et al. 1992), except we substituted a blue and white jacket and red baseball cap for the amorphous crane costume developed by Horwich et al. (1992). The senior author concluded that it was unreasonable to subject the crane rearing crew to the discomfort of wearing the full costume while rearing and training the chicks in Maryland and for 7 weeks in Arizona before we demonstrated that the southward migration was possible. Another departure from the Mississippi sandhill crane rearing regime was that the tape-recorded calls normally played for the chicks during hatching and rearing now included human imitations of a sandhill crane brood call, engine noises of our cranemobile, and the flapping sounds made by the flags that were later attached to the cranemobile during travel.

Training

Figure 2 presents most aspects of the training process. In 1994, our transport chamber for the birds consisted of the bed of the pickup truck with the floor lined with rubber matting, with 45-cm-tall sidewalls, and with white painted wood sidewall panels 41 cm tall. The box was covered by a nylon net (11-cm mesh length) to prevent birds from escaping while in transit.

During transport, cranes were thrown off balance by sudden changes in momentum, but drivers quickly learned to accelerate and decelerate gradually and to avoid severe bumps and rapid turns so cranes could be transported moderate distances with little discomfort. The transport maxima during the pilot year were ca 45 minutes and ca 100 km/h.

In 1995, our transport vehicle was a modified U.S. Army ambulance. The entry ramp was 3.1 m long, with a 1.3-m rise. When working with small chicks, we used a 43-cm-tall, vinyl-coated welded-wire fence attached to both sides of the ramp to prevent falling. After the chicks were about 30 days old, the ramp fence was removed, but trainers continued to guide the chicks as they traversed the ramp. During the migration, we manually lifted cranes into the truck bed, so the ramp was seldom used.

All aspects of training, except flying behind the truck, were initiated prior to transporting the birds west. Because leg and wing injuries regularly occur in groups of recently fledged chicks at Patuxent (Wellington et al. 1996), we decided to wait to transport the birds west until the youngest was 74 days old. Sandhill crane long bones are fully grown and growth plates are closed at about 70–84 days of age (J. Langenberg, International Crane Foundation, Baraboo, Wis., pers. commun.).

From 24 to 27 August, 4 cranes were trucked from Maryland to Arizona unfettered in the rear of the cranemobile. All were weak and wobbly toward the end of the trip. One bird (No. 86) sat most of the last 2 days. From his appearance upon arrival, we expected him to die. Within 2 days of arrival, all other birds were well-coordinated, active, and had begun flight training. No. 86 only gradually commenced flight training, and it was not until 9 days after arrival that he participated fully with the other cranes. Further, all 4 trucked birds subsequently showed a broad band of imperfectly formed barbs (fault bars or stress bars) on their back and scapular feathers, apparently coinciding with the time of transit and evidencing the physiological stress (Grubb 1989) associated with the journey west.

On 28 August 1995, the remaining 6 study birds were flown west in a small jet aircraft and spent only 8 hours in special crane shipping boxes (Swengel and Carpenter 1996). None of these birds exhibited physical damage after shipment other than temporarily mussed bustles. All were alert and ready to begin training upon arrival.

Table 1. Details of training sandhill cranes to fly behind a truck, Navajo Army Depot, Belmont, Arizona.

Total number of flights	53 on 23 days during 34-day period
Dates	30 Aug to 1 Oct 1995
Total flight distance	225 km
Total flight time	5 hr, 1 min
Average flock size	8 birds
Average altitude	19 m
Highest altitude	60 m
Lowest altitude	1 m
Average speed	49 km/h
Highest speed	72 km/h
Lowest speed	10 km/h

Flight Training

Training was conducted at the Navajo Army Depot, 20 km west of Flagstaff, Arizona. Because we hoped that our cranes would eventually breed where sandhill cranes had bred a century earlier (Phillips et al. 1978), on 13 September we trucked them to Mormon Lake and allowed them to wander in the marshes. We also encouraged them to capture grasshoppers (Locustidae) and aquatic animals in the marshes at the Navajo Army Depot.

Flight training is summarized in Table 1. At the onset, flights were short (1–4 km), but by the end of September the birds were making 15-km flights followed by a ca 10-minute rest, then a return flight of up to 15 km. The birds were trained to fly past fences and powerlines and through developed areas, prairies, and forests. During training, we locked the rear doors of the cranemobile open and 1 or more trainers sat on the elevated truck bed calling and waving to encourage the birds. Further visual and audio stimulation was provided by 2 flags affixed above roof level.

The Migration

Prior to the migration, all 6 crew members traversed all or nearly all of the route once, and the 3 primary drivers had covered the route 2 or more times. Each vehicle carried a set of maps (scale 8:1,000,000 with powerlines marked in bold) and a 2-way radio. Coordination of widely separated vehicles and with cooperators was done by each of us calling C. Ellis at our Oracle, Arizona, base. Each crane was fitted with a VHF radio transmitter (solar powered) prior to migration. Radios were mounted on plastic leg bands with average total package weight of about 60 g.

Three vehicles participated in the migration. With few exceptions, the cranemobile was the middle vehicle during travel. Occupants of the lead vehicle, using their 2-way radio, advised of traffic conditions, announced hazards, and cleared intersections (so the cranemobile could drive through with speed unabated). The navigator in the third car announced, into his 2-way radio, data (e.g., concerning ground speed and altitude of cranes) for the recorder. This vehicle also dropped back to search for delinquent cranes.

RESULTS

The migration began midday on 2 October and ended on 14 October at 0915 hours (Table 2). Our southward passage included many unique and bizarre interactions between trainers, cranes, railroad trains, police, a jogger, eagles, bystanders, and oncoming vehicles. Brevity requires that we focus only on those experiences that best provide lessons for future attempts. The route is shown in Fig. 3.

Day 1: 2 October.—We tried to walk the flock through 4 km of forest to join a dirt road and begin the first flight. The cranes, already familiar with the area, flew back to camp, and we, left with only 1 crane, chose to regather the birds and truck them to a spot ca 7 km from camp. The birds then followed the cranemobile for 3 km until we stopped because of approaching darkness. As we began to set up camp, we inadvertently flushed the birds before they were penned and were able to call only some of them back to camp, so we spent the night with only 6 cranes, bemoaning our decision to use solar transmitters that disallowed nighttime searching.

Day 2: 3 October.—At dawn, all of the lost cranes flew back to camp. By 0800 hours we had gathered our search teams and began the flight into the forest. The first forest flight proceeded well, but the second terminated when the birds scattered and all but 1 were lost. Our telemetry equipment led us back to the open prairie where the migration began and there we camped with only 1 bird.

Table 2. Details of southward migration of sandhill cranes across Arizona behind a truck, October 1995.

Migration dates	2 Oct to 14 Oct 1995
Total number of migration days	13
Total number of flight days	8 (7 days of >5 km flights)
Total number of flights	29
Total migration length	589 km[a]
Total flight distance	495 km
Total flight time	11 hr, 58 min
Average altitude	30 m
Highest altitude	ca 600 m
Lowest altitude	1 m
Average speed	46 km/h
Fastest speed	78 km/h
Slowest speed	24 km/h
Longest flight	77 km
Total distance birds were trucked	124 km

[a] Garland Prairie to Buenos Aires NWR.

Day 3: 4 October.—The day was spent retrieving cranes (all but 1, No. 116, returned) and replacing a tire. In the evening, we drove our 9 cranes through the forest to camp beyond the tall ponderosa pines (*Pinus ponderosa*). In retrospect, our observations of golden eagles (*Aquila chrysaetos*) at 3 locations along the birds' path and subsequent observations of eagles attacking cranes led us to suspect that the scattering on 3 October was due to an eagle attack.

Day 4: 5 October.—Our cranes completed several excellent flights. All 9 birds were anxious to stay near the cranemobile. At 78 km/h, a 13-km flight descending from the mountains into Chino Valley was our fastest flight of the entire migration. To facilitate the transfer of the cranes across a canyon with an unsafe bridge, 2 trainers remained with the flock while the rest of the crew drove 2 vehicles on the long circuitous route to the other side. The 2 trainers then ran toward the canyon (with the cranes running in pursuit), hopped over the rim, and hid behind boulders. The cranes lifted off and circled down to join the team across the canyon ca 500 m away.

Our only fatality occurred about 110 km into the migration when 1 of the 3 males (No. 89) died upon impact with a powerline. About 10 km after the collision, an adult golden eagle swooped through the flock, scattering the cranes. No birds were injured, all regrouped by landing at the cranemobile, and the eagle was dispelled by firing 5 rifle slugs (22 caliber) near it.

During the final flight of the day, the birds soared up to about 300 m, followed us 23 km, then were lured down to the evening camp site in Skull Valley by 4 trainers calling

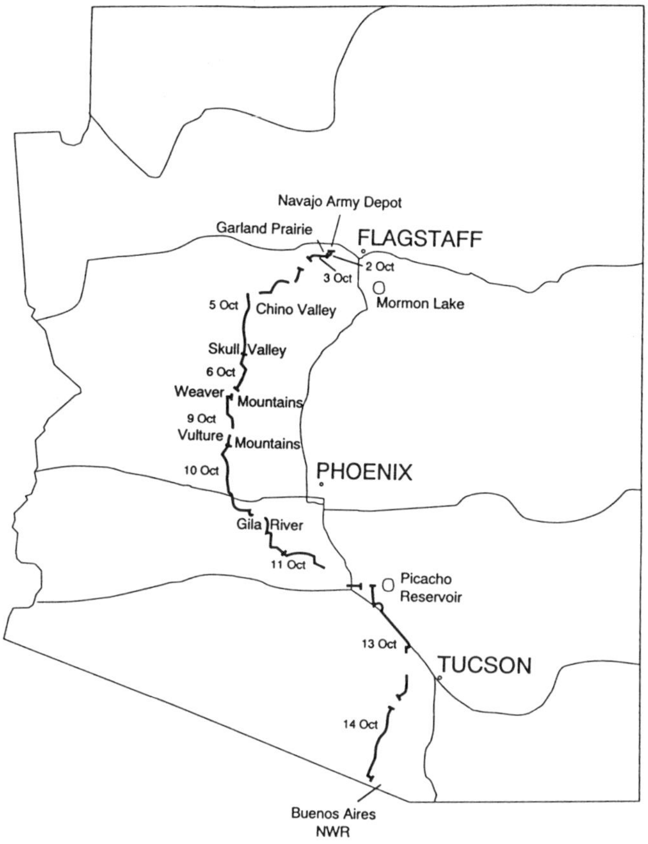

Fig. 3. Flown segments of truck-led sandhill crane migration, Arizona, October 1995.

and flapping.

Day 5: 6 October.—The day began with two excellent flights as the 8 cranes climbed over the Weaver Mountains. However, the second flight was terminated by another eagle attack scattering the birds. Only 1 crane stayed with the cranemobile. One more was recovered in the evening.

Day 6: 7 October.—Our telemetry search for the lost cranes continued; some signals were received. The crane lost on Day 2 (No. 116) was retrieved, and 3 more cranes were reported ca 55 km from the previous day's attack site.

Day 7: 8 October.—We recovered the 3 cranes (Nos. 94, 109, and 114) reported on 7 October. Later, we found 2 more cranes (Nos. 117 and 123) by radiotelemetry. Remarkably they had returned to Garland Prairie, near where we began the migration and ca 130 km from the eagle attack site.

Day 8: 9 October.—At noon, the migration resumed, but we were now in the Sonoran Desert where midday temperatures exceeded 30 C. The birds only managed 37 km in 5 short flights. More golden eagles were expected in the Vulture Mountains, so we passed below the beetling peaks after sunset with a shell cracker-loaded shotgun out the window of the lead car and a rifle at the ready.

Day 9: 10 October.—Our first flight (and our longest flight of the migration) began before sunrise and continued 77 km to the Gila River. One bird (No. 117), whose feathers had been damaged during retrieval operations the previous day, was lost at the onset of the flight (it reappeared by 28 October near where it was lost). South from the Gila River, the migration stalled due to heat. At sunset, we flew the birds again, and kept them moving until it was too dark to count them.

Day 10: 11 October.—Migration began at dawn with a 40-km flight. Two more flights fizzled during the heat of the day. We decided to truck the birds the last 13 km to Picacho Reservoir and rest them until 13 October.

Day 11: 12 October.—The final crane (No. 112), lost during the second eagle attack, was retrieved and joined the flock at Picacho Reservoir.

Day 12: 13 October.—The migration began again with 2 strong morning flights but terminated when all 8 cranes overheated. An evening flight resulted in 1 crane (No. 96) being lost near dusk. It was recovered 16 October.

Day 13: 14 October.—By 0915 hours, we completed the migration after 2 flights (totaling 60 km). The cranes, however, were overheated for the last 30 km. Only 6 birds actually arrived together at the southern terminus. The seventh (No. 86) dropped out, overheated, only 5 km from the BANWR headquarters but was immediately recovered.

DISCUSSION AND CONCLUSIONS

The migration is summarized in Fig. 4. The primary hazard during the migration was powerlines. We counted a minimum of 120 sets of powerlines along the route. Many near collisions, and 3 collisions, were observed: 1 fatal, 2 without injury.

We also encountered a minor difficulty in following winding roads through tall dense forest where it was nigh impossible to keep the birds in view. The result was that we slowed down when the birds were out of sight, only to have the cranes reappear ahead of the vehicles. These difficulties were compounded because the forested area occurred near the onset of the migration when the team was inexperienced and before the cranes were tenacious in following.

Another difficulty was eagle attacks. Our birds seemed able, from the onset, to avoid capture, but the second attack required 3 days to reassemble our flock. With experience, we probably could have avoided this problem, and, with more reliable and more powerful radio transmitters, we probably could have gathered our birds before they scattered so widely. Future routes should be chosen to minimize the potential for eagle attacks.

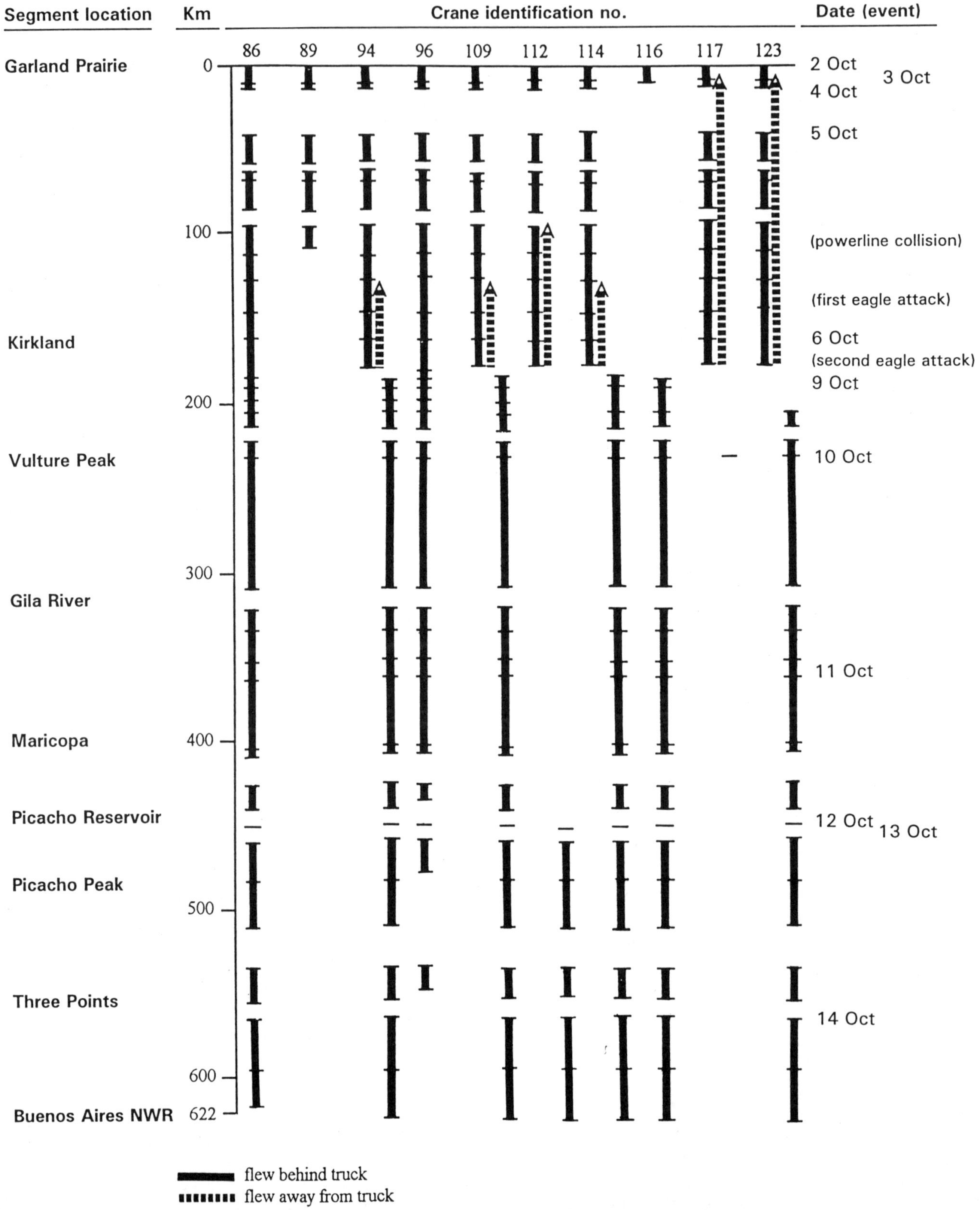

Fig. 4. History of greater sandhill crane truck-led migration: Garland Prairie to Buenos Aires NWR, Arizona, October 1995.

Our greatest difficulty was the unseasonably warm temperatures encountered as we traversed the Sonoran Desert, the final two-thirds of the route. We timed our departure from the high-elevation (ca 2,300 m) summering area by an appropriate environmental cue, the first hard frost. But, we should have led the birds to a mid-elevation staging area and maintained them there until November or at least until midday temperatures dropped below 20 C in the Sonoran Desert.

Other lessons included: (1) avoid flying fatigued birds through powerlines, (2) use a minimum crew of 7 people during migration (each of 3 drivers needs a navigator and a seventh person rides in the rear of the cranemobile calling to the birds), (3) transport cranes to the release site by air, and (4) costume-rear cranes unless they are to be integrated into a wild flock immediately after the migration.

A final lesson is that we should probably have given our birds "abandonment training" prior to departure. During flight training at the Navajo Army Depot, crane No. 109 was lost for 5 hours. After it was recovered, it changed from being one of our most adventuresome cranes to being the most tenacious (i.e., most attached to the trainers). Had we stranded each crane, solo, for a few hours in a pen guarded from a blind, and then arranged for it to "find" its keepers, we believe that all birds would have been more attentive at the onset of the migration. As the migration progressed, the cranes became more and more attentive (an observation consistent with the experiences of Urbanek and Bookhout [1992] and Lishman et al. [1996]), but the frustrations we experienced on Days 1 and 2 probably could have been avoided with such training.

Although our migration extended from 2 to 14 October (13 calendar days), we actually traversed our 622-km route (495 km of which were flown) in 8 days (days in which more than 3 km were flown). Only 3 days involved flights totaling more than 2 hours. Without the handicap of traveling in desert heat, we believe the same route could be traversed in 5 flight days.

In closing, we found that leading cranes long distances was feasible at least in unforested areas. We experienced 100% survival during training and good survival on migration (90%, although our birds had to be recovered following eagle attacks). With the lessons learned in 1995, we believe it is possible to have even greater success with fewer interruptions in future experiments.

The success of this project can be judged only in part by our ability to train cranes to follow a truck. Ultimately, we need to know if trained cranes can return unassisted to breed in the area where they fledged. Ironically, one of the greatest difficulties of our expedition (i.e., the second eagle attack) produced some premature, but favorable, results. Two of the birds were relocated on Garland Prairie where the migration started and ca 130 km from the attack site. Three others were found ca 55 km from the attack site and within 7° of direct course for Garland Prairie. A sixth bird was recovered ca 95 km from the attack site and 25° east of a direct vector to Garland Prairie. Although it is dangerous to make too much of these observations, the fact remains that 2 of 8 cranes scattered by the eagle returned to Garland Prairie and 3 others were nearly on course to that location when recovered.

LITERATURE CITED

BARTLETT, D., and J. BARTLETT. 1975. The flight of the snow geese. Stein and Day, New York, N.Y. 189pp.

BURNHAM, W. A. 1990. Raptors and people. Pages 170-188 *in* I. Newton, ed. Birds of prey. Weldon Owen, Inc., San Francisco, Calif.

CLEGG, K. R., J. C. LEWIS, and D. H. ELLIS. 1997. Use of ultralight aircraft for introducing migratory crane populations. Proc. North Am. Crane Workshop 7:105-113.

ELLIS, D. H., G. H. OLSEN, G. F. GEE, J. M. NICOLICH, K. E. O'MALLEY, M. NAGENDRAN, S. G. HEREFORD, P. RANGE, W. T. HARPER, R. P. INGRAM, and D. G. SMITH. 1992. Techniques for rearing and releasing nonmigratory cranes: lessons from the Mississippi sandhill crane program. Proc. North Am. Crane Workshop 6:135-141.

FRANKLIN, K., and S. FRANKLIN. 1985. Falcon training techniques using an ultralight aircraft. North Am. Falconer's Assoc. J. 24:21-29.

GRUBB, T. C., JR. 1989. Ptilochronology: feather growth bars as indicators of nutritional status. Auk 106:314-320.

HORWICH, R. H., J. WOOD, and R. ANDERSON. 1992. Release of sandhill crane chicks hand-reared with artificial stimuli. Pages 255-261 *in* D. A. Wood, ed. Proc. 1988 North Am. crane workshop. Florida Game Fresh Water Fish Comm. Nongame Wildl. Program Tech. Rep. 12.

IMMELMAN, K. 1972. Sexual and other long-term aspects of imprinting in birds and other species. Pages 147-174 *in* D. S. Lehrman, R. A. Hinde, and E. Shaw, eds. Advances in the study of behavior. Vol. 4. Academic Press, New York, N.Y.

LISHMAN, W. A., T. L. TEETS, J. W. DUFF, W. J. L. SLADEN, G. G. SHIRE, K. M. GOOLSBY, W. A. BEZNER KERR, and R. P. URBANEK. 1997. A reintroduction technique for migratory birds: leading Canada geese and isolation-reared sandhill cranes with ultralight aircraft. Proc. North Am. Crane Workshop 7:96-104.

LORENZ, K. Z. 1952. King Solomon's ring: new light on animal ways. Thomas Y. Crowell Co., New York, N.Y. 225pp.

NAGENDRAN, M. 1992. Winter release of isolation-reared greater sandhill cranes in south Texas. Proc. North Am. Crane Workshop 6:131-134.

PHILLIPS, A., J. MARSHALL, and G. MONSON. 1978. The birds of Arizona. Univ. Arizona Press, Tucson. 220pp.

SWENGEL, S. R., and J. W. CARPENTER. 1996. General husbandry. Pages 31-43 *in* D. H. Ellis, G. F. Gee, and C. M. Mirande, eds. Cranes: their biology, husbandry, and conservation. Nat. Biol. Serv., Washington, D.C., and Int. Crane Found., Baraboo, Wis.

URBANEK, R. P., and T. A. BOOKHOUT. 1992. Development of an isolation-rearing/gentle release procedure for reintroducing migratory cranes. Proc. North Am. Crane Workshop 6:120-130.

WELLINGTON, M., A. BURKE, J. M. NICOLICH, and K. O'MALLEY. 1996. Chick rearing. Pages 77-95 *in* D. H. Ellis, G. F. Gee, and C. M. Mirande, eds. Cranes: their biology, husbandry, and conservation. Nat. Biol. Serv., Washington, D.C., and Int. Crane Found., Baraboo, Wis.

HABITAT USE BY MIGRANT WHOOPING CRANES IN SASKATCHEWAN

BRIAN W. JOHNS, Environment Canada, Canadian Wildlife Service, 115 Perimeter Road, Saskatoon, SK S7N 0X4, Canada
ERIC J. WOODSWORTH, Environment Canada, Canadian Wildlife Service, 115 Perimeter Road, Saskatoon, SK S7N 0X4, Canada
ED A. DRIVER, Environment Canada, Canadian Wildlife Service, 115 Perimeter Road, Saskatoon, SK S7N 0X4, Canada

Abstract: We investigated habitat use by migrating whooping cranes (*Grus americana*) in Saskatchewan between 1986 and 1990. At foraging and roosting sites and a sample of randomly selected sites we measured various habitat parameters. Palustrine and lacustrine wetlands were the most frequently used wetlands. Temporary and seasonal wetlands were the primary roost habitat during spring migration, and semi-permanent and permanent wetlands were frequently used during the fall. Cranes were attracted to areas of higher than average wetland density. Land use within 2 km of roost sites was intensively modified by man. Cereal crops were the most used feeding areas. Mean distances to potential disturbance were lower in spring than in fall. Because the wetlands used were primarily in private ownership, a cooperative approach to wetland preservation is needed between wildlife managers and landowners.

PROC. NORTH AM. CRANE WORKSHOP 7:123-131

Key words: *Grus americana*, habitat use, migration, Saskatchewan, whooping crane.

Though the current whooping crane breeding area lies within Wood Buffalo National Park (WBNP), Canada, the historical nesting area encompassed the prairies of southern Canada and the north-central United States (Allen 1952). Saskatchewan was once the center of the former nesting area in Canada. The last recorded nesting site outside the WBNP area was at Luck Lake, Saskatchewan, in 1927 (Hjertaas 1994).

Whooping cranes do not currently breed in Saskatchewan, but individuals of the Wood Buffalo/Aransas population spend several days to weeks in the province each year (Johns 1992). Each spring and fall, this remnant population migrates through Saskatchewan between the breeding area in WBNP, Northwest Territories, and wintering grounds on the Aransas National Wildlife Refuge, Texas (Allen 1956).

From a radio tracking study of whooping crane migration (Howe 1989), Saskatchewan ranked third in reported crane spring stopovers (12.8%), behind Kansas and Nebraska. However, the Saskatchewan stopovers comprised 43.7% of all crane use-days during the spring migration. In the fall migration period, whooping cranes made more extended stopovers in central Saskatchewan than anywhere else along their migration route (Johnson and Temple 1980, Howe 1989, Armbruster 1990). Forty percent (40.4%) of fall stopovers occurred in Saskatchewan, accounting for 68.4% of crane use-days during fall migration.

Though the value of south-central Saskatchewan as a fall staging area for whooping cranes was described by Howe (1989), Johns (1992), and Kuyt (1992), the lack of detailed information on how this staging area is used by migrating whooping cranes prompted the Canadian Wildlife Service to conduct this investigation. This report documents habitat use by whooping cranes on the Saskatchewan staging area for the period 1986 to 1990.

We thank all the observers who reported whooping crane sightings, and the landowners who allowed us access to their land. Thanks also go to T. Stehn and D. Nieman for reviews of earlier drafts of the paper.

STUDY AREA

Known and potential whooping crane habitat in Saskatchewan exists in a 300-km-wide band which crosses the province diagonally for 1,000 km from just south of Lake Athabasca in the northwest to Estevan in the southeast. We examined traditional and non-traditional sites from the Mixedwood Forest and Aspen Parkland interface (Bird 1961, Harris et al. 1983, Kabzems et al. 1986) to the Canada-U.S.A. border. This encompasses the agricultural portion of the province and all of the historical breeding area in Saskatchewan. If staging or stopover sites (Armbruster 1990) exist on the northern igneous (granitic) shield of Saskatchewan (Richards and Fung 1969), these sites remain to be reported.

The landscape of agricultural Saskatchewan is comprised of a series of low, gently rolling hills, deposits of terminal moraine from the last glaciation, interspersed with river valleys and flat glacial lacustrine plains between the southern edge of the Mixedwood Forest (54° 30'N) and the international border (49° 00'N). Cultivated fields, small groves of aspen (*Populus tremuloides*), and hundreds of thousands of wetlands dot the farmland landscapes. Eighty percent of this land is cultivated; the dominant crop, wheat, represents 65% of the cultivated acreage while barley and canola are the next 2 most common crops. The area is crisscrossed with paved and secondary roads in a 2- by 1-mile grid. Historically, 2 of the 4 quarters of each section were settled, a maximum of 140,000 farmsteads by 1945 (Driver 1991). Subsequently, the number of farmsteads dropped to the current 59,000 active farms (Saskatchewan Agriculture and Food 1991). Many of the 80,000 abandoned farmsteads now are incorporated into cultivated fields.

Table 1. Characteristics of whooping crane roost sites (%) in Saskatchewan, 1986–90.

Variable	Spring	Fall	Random	Variable	Spring	Fall	Random
Wetland system (*n*)	(55)	(46)	(28)	Site security	Stable	Stable	Stable
Riverine	0	6	4	Turbidity (*n*)	(48)	(47)	(26)
Lacustrine	16	61	36	Clear	88	66	85
Palustrine	84	33	61	Turbid	6	21	4
Wetland type (*n*)	(48)	(29)	(26)	Saline	6	13	12
I	21	0	8	Substrate (*n*)	(44)	(42)	(28)
III	52	14	31	Soft mud	52	83	50
IV	19	31	19	Hard mud	32	10	32
V	8	55	42	Detritus	11	0	11
Site ownership (*n*)	(55)	(46)	(27)	Sand	2	0	7
Private	96	85	96	Ice	2	0	0
Provincial	4	13	4	Other	0	2	0
Federal	0	2	0	Dominant emergents (*n*)	(47)	(35)	(23)
Site description (*n*)	(47)	(48)	(28)	*Scholochloa*	40	6	26
Semi-permanent marsh	62	48	43	*Scirpus*	19	43	26
Flooded cropland	21	2	7	*Carex*	19	29	22
Lake	9	40	21	*Typha*	2	11	13
River	0	4	0	Other	19	11	13
Reservoir	6	6	4	Vegetation density (*n*)	(53)	(47)	(28)
Dugout	0	0	11	None	17	32	21
Flooded pasture	0	0	4	Scattered	19	13	14
Wet meadow	2	0	4	Clumped	11	17	7
Creek	0	0	7	Choked	28	0	18
Other species (*n*)	(56)	(48)		Ringed	25	38	39
None	22	20		Wetland class (*n*)	(56)	(46)	(26)
Sandhill cranes	15	33		Emergent wetland	59	22	35
Geese	24	33		Unconsolidated bottom	25	59	46
Ducks	24	5		Aquatic bed	14	4	8
Swans	5	10		Unconsolidated shore	2	9	0
Shorebirds	10	0		Rocky shore	0	2	0
Adjacent habitat (*n*)	(54)	(44)	(24)	Streambed	0	4	4
Stubble	41	32	13	Forested wetland	0	0	8
Fallow	33	16	46	Wetland modifier (*n*)	(56)	(47)	(28)
Cultivated stubble	13	0	8	Permanently flooded	5	32	50
Swathed crop	0	5	0	Semi-permanently flooded	16	53	18
Flooded cropland	2	0	0	Seasonal	57	9	21
Pasture	9	32	17	Saturated	0	0	4
Flooded pasture	0	2	0	Temporarily flooded	7	0	4
Hay meadow	2	5	0	Intermittently flooded	11	0	0
Wet meadow	0	9	4	Artificially flooded	4	4	4
Other	0	0	13	Intermittently exposed	0	2	0

METHODS

To determine locations and to assess the characteristics of the habitat whooping cranes use in prairie Saskatchewan, we obtained reported sightings of cranes from the Whooping Crane Hot Line (Johns 1992) as well as from other observers. All locations were plotted on 1:250,000 and 1:50,000 Geological Survey of Canada (GSC) topographic map sheets and 1:20,000 aerial photo-mosaics which were produced by Central Survey and Mapping, the Province of Saskatchewan.

Habitat composition and land use characteristics for spring and fall migrating cranes were evaluated for all observed roost and feeding locations between fall 1986 and spring 1990. Thirty-seven habitat attributes (Appendix A) were recorded for these sites; several are similar to those described in Howe (1987) and Armbruster (1990). Major measurements at upland foraging locations included 17 categories for site description and adjacent habitat, topographic slope, visibility in 4 cardinal directions, security of the site, associated bird species, and the extent of similar habitat. Visibility was measured at 1.4 m above ground, the average crane head height (Armbruster 1990). Water depth, water quality, substrate type, shoreline slope, dissolved solid conductance, dominant emergent vegetation, and wetland classification were noted for observed roosting and potential roost sites selected at random (Appendix A). Wetland classification followed Stewart and Kantrud (1971) and Cowardin et al. (1979). The area of wetlands and upland habitats (fields, aspen groves, farmsteads) within 2 km of roost and potential roost sites was measured electronically from 1:20,000 photo-mosaics.

Roost site data were divided into traditional and nontraditional categories, the former representing townships with multiple sightings or adjacent townships each with at least 1 sighting, the latter being single sightings in isolated townships. Roost site data were also divided into sites used by family groups and non-family groups. All variables were evaluated to determine whether their distributions were normal before proceeding with comparative statistical analyses. Distance to buildings, roads, and powerlines; shoreline slope; and pH, all normally distributed, were analysed with canonical discriminant analysis (SAS Institute Inc. 1990). Logistic regressions were performed on non-normally distributed quantitative data of sets of variables including topographic slope, wetland area, distance to feeding site, distance to trees, specific conductance, and visibility. Measures of central tendency were made for most mensural values; these values provide the initial description of roost and feeding sites.

RESULTS AND DISCUSSION

Roost Site Characteristics

Two wetland systems, palustrine and lacustrine, were frequented by whooping cranes migrating through Saskatchewan. Palustrine wetlands were most frequently used during spring (84% of the sites observed, n = 55) (Table 1). Use of palustrine sites in fall was only 33% (n = 46). During spring, 16% of whooping crane observations were from lacustrine wetlands compared with 61% use in the fall. The remaining 6% of fall roosting sites were riverine. Distribution of potential wetland systems available to cranes differed substantially from the observed results. In the random sample, 61% of the wetlands were classified as palustrine and 36% belonged to the lacustrine system.

The wetlands within the 2 systems generally belong to 1 of 4 wetland classes (Stewart and Kantrud 1971). Class I and III represent the more ephemeral wetland type and are referred to as temporary and seasonal wetlands, respectively. These classes were most common during the period April through July and had a 73% visitation rate (n = 48) from cranes migrating northward. Temporary and seasonal wetlands undergo considerable or complete water drawdown primarily from evapo-transpiration processes. With the disappearance of more than 95% of Class I and III wetlands, fall migrating cranes shift to favourable Class IV (semi-permanent) and Class V (permanent) wetlands (Tables 1 and 2). During the fall period, these wetlands represented 86% of the observed roosting wetlands. The remaining fall roosting habitat was composed of Class III wetlands. The degree to which Class IV and V wetlands are used in successive migrations is extremely small. We found few instances of repeated use of the same or nearby wetlands. Thirteen localities were used by a number of different cranes during the study period but never more than once by the same crane. Four additional sites were used by the same birds on a subsequent migration. Spring "staging" areas are much less predictable because of the vagaries of fall precipitation and spring run-off that help create temporary and seasonal wetlands.

Table 2. Frequency of wetland types used by migrating whooping cranes in Saskatchewan, 1986-90.

	Family groups		Non-family groups		
Type	Spring	Fall	Spring	Fall	Random
Class I	3	0	7	0	2
Class III	7	1	18	3	8
Class IV	0	5	8	5	5
Class V	2	7	2	9	11

The majority of wetlands are relatively fresh (Stewart and Kantrud 1971) with salinities less than 5,000 micromhos (Table 3, Fig. 1) and slightly alkaline, pH averaging 7.7 in late spring and 7.9 in fall (Table 3). On average, wetlands used for roosting were large. Spring sites averaged 36 ha (±136.9 SD), whereas fall roosts were 14 times as large, averaging 508 ha (Table 3). Crane use of wetlands depends primarily on wetland size and permanency. Fewer than 5% of the approximately 1-2 million wetlands in the spring landscape are greater than 1 ha (U.S. Department of Interior and Environment Canada 1995). Large wetlands exist primarily on glacial lake basins, on glacial lake deltas, and at the edges of terminal moraine deposits.

Family groups were attracted to areas of relatively high wetland density. Within a 2-km radius of a roosting site, approximately 16% of the area was wetlands compared to 9% for randomly selected sites (Fig. 2). However, this trend was not significant. A similar trend was noted for cross-

Table 3. Characteristics of whooping crane roost sites ($\bar{x}$) in Saskatchewan, 1986-90.

Variable	Spring			Fall			Random		
	n	$\bar{x}$	SD	*n*	$\bar{x}$	SD	*n*	$\bar{x}$	SD
Specific conductance (micromhos/cm^3)	30	4,080	13,294	30	6,178	12,880	22	1,825	1,852
pH	35	7.7	0.48	32	7.9	0.56	22	8.0	0.55
Wetland area (ha)	39	36.0	136.9	38	508	1,806	28	974	4,687
Water depth (cm)	30	15.8	7.2	32	12.9	7.5			
Shoreline slope (°)	46	2.2	1.02	32	2.1	1.18	28	5.7	8.23
Visibility[a]	220	1.9	1.05	188	2.5	1.38	112	1.98	1.30
Distance to nearest trees (m)	43	237	267	39	303	379	24	150	223
Distance to upland feeding site (m)	23	419	782.1	37	992	1,248.6	23	177	239.7
Distance to nearest buildings (m)	52	714	468	45	1,037	488	24	578	491
Distance to nearest roads (m)	43	616	631	41	654	488	26	474	327
Distance to nearest powerlines (m)	29	687	437	22	845	508	10	319	287
Topographic slope (°)[b]	224	4.5	4.8	192	3.3	4.7	112	4.0	4.1

[a] 1 = <100m, 2 = 0.1-0.25 km, 3 = 0.25-0.5 km.
[b] Average within 500 m.

fostered whooping cranes in southern Colorado (Shenk and Ringelman 1992).

Armbruster (1990) considered the optimum water depth at roosting sites to be less than or equal to 30 cm. In Saskatchewan, water depth at spring roost sites averaged 15.8 cm in depth (range 3-30), whereas depths at fall sites were on average 2.9 cm shallower (Table 3). These results are similar to those reported by Ward and Anderson (1987) and Howe (1989).

Regardless of season, sites selected by cranes had very gently sloping littoral zones ($\bar{x}$ = 2.15°) (Table 3) which were significantly different from the 5° average of the shoreline slopes (P < 0.001) in the random sample of the Saskatchewan migration corridor. These shallower basins used by migrating cranes may afford greater protection from predators. Shallow basins of moderate area may allow cranes to roost at greater distances from shore. This may provide a distance discouragement to predators while providing a visibility (Table 3) advantage to the crane. A basin with a more abrupt shoreline reduces the width of safety zone.

Ninety-six percent of spring roosts and 85% of fall roosts were on private land (Table 1). Because most of these are

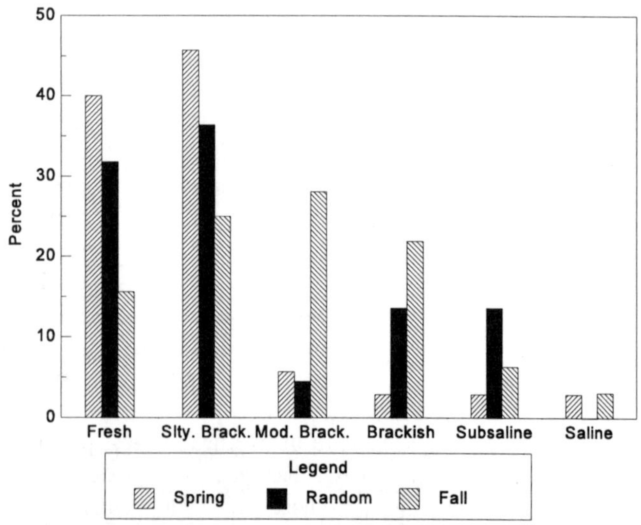

Fig. 1. Water quality of whooping crane roost wetlands in Saskatchewan, 1986-90 (Stewart and Kantrud 1971).

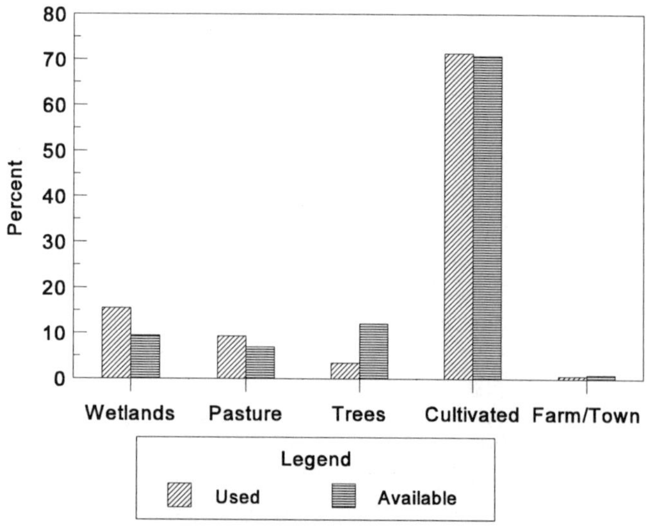

Fig. 2. Land use within 2 km of whooping crane family roost wetlands in Saskatchewan, 1986-90.

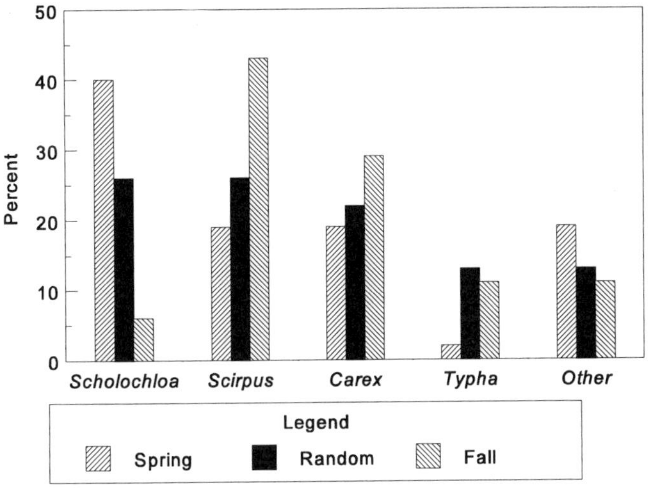

Fig. 3. Dominant emergent vegetation at whooping crane roosting sites in Saskatchewan, 1986-90.

located within cultivated fields with vegetation <0.3 m in height, they provide an increased safety factor to cranes. This surveillance factor also appeared to influence how close the roost was to trees. Roost sites were at greater distances from trees ($\bar{x}$ = 237 m in spring and 303 m in fall) than were random sites ($\bar{x}$ = 150 m) (Table 3).

On most large lacustrine wetlands, small, widely dispersed stands of *Typha* spp. and *Scirpus* spp. occur. This emergent vegetation does not appear to impede cranes' surveillance of their surroundings. In spring, growth has yet to begin and many of the overwintering stems are bent over and compressed by snow and/or broken by strong winds. By fall this emergent vegetation also is reduced in height from its summer stature. Most of these stems measure 0.6 to 1.2 m; the flower spike of *Typha* is 1.2-1.6 m in height, compared to an average height of 1.4 m for cranes. Class I and III habitat is dominated by grassy species of *Scholochloa* spp. and *Carex* spp. (Fig. 3).

In the southern half of Saskatchewan, the potential for feeding sites within 1 km of a roost site was greater (P = 0.0001) during spring migration (94.9%) than fall (72.9%). This difference was reflected in the distance the cranes fed from their roosts, averaging 419 m in spring and 992 m in fall (Table 3). Within the entire migratory corridor, Howe (1989) found foraging site distance from roost ranged from 100 to 8,000 m, 56.2% of these within 1 km of a roost.

There were no significant differences between roost sites used and random sites in the number ($\bar{x}$ = 3.0 and 3.6, respectively) and percent land area (0.6% and 0.9%) of farmyards and towns within 2 km of the roost site (Fig. 2). Roosts were, however, generally more isolated from human disturbance (buildings) ($\bar{x}$ = 714 m in spring and 1,037 m in fall) than randomly selected wetlands ($\bar{x}$ = 578); whooping cranes seemed less tolerant of human disturbance in fall than in spring, but differences were not significant (Table 3). Differences in tolerance to human disturbance between spring and fall may be related to the amount of disturbance the cranes experience in the intervening months before migration. The cranes' northern breeding grounds are isolated from human intrusion, unlike their southern wintering area. In a study of cross-fostered whooping cranes in Colorado, mean distance from roosts to buildings during fall migration was 738 m (Shenk and Ringelman 1992), similar to spring distances in Saskatchewan.

Whether cranes are aware of overhead wires, posing a threat to them while flying, has yet to be determined. However, the roost sites they chose were significantly farther (P < 0.0001) (Table 8) from overhead wires (687 m in spring, 845 m in fall) than were random sites (319 m) (Table

Table 4. Characteristics of upland feeding sites (%) used by whooping cranes in Saskatchewan, 1986-90.

Variable	Spring	Fall
Site description (n)	(99)	(98)
Stubble	82	77
Cultivated stubble	18	8
Pasture	0	6
Swathed grain	0	5
Fallow	0	1
Hay meadow	0	1
With other species (n)	(105)	(103)
None	80	58
Sandhill cranes	14	35
Geese	3	3
Ducks	3	1
Cattle	0	3
Crop type (n)	(98)	(87)
Wheat	62	57
Barley	28	36
Durham	7	3
Oats	2	3
Adjacent habitat (n)	(98)	(96)
Stubble	53	45
Fallow	37	21
Cultivated stubble	5	0
Swathed crop	0	4
Pasture	3	19
Flooded pasture	0	1
Hay meadow	0	7
Wet meadow	0	1
Other	2	1
Site security	Stable	Stable
Site ownership		
Private	100	100

Table 5. Characteristics of upland feeding sites ($\bar{x}$) used by whooping cranes in Saskatchewan, 1986-90.

Variable	Spring			Fall		
	n	$\bar{x}$	SD	n	$\bar{x}$	SD
Visibility[a]	408	2.4	1.1	404	2.9	1.38
Distance to nearest trees (m)	58	277	166	81	258	172
Distance to nearest buildings (m)	96	682	421	96	767	441
Distance to nearest roads (m)	56	399	3,223	81	528	611
Distance to nearest powerlines (m)	45	535	336	49	895	900
Topographic slope (°)[b]	408	2.1	3.6	404	1.8	3.1

[a] 1 = <100m, 2 = 0.1-0.25 km, 3 = 0.25-0.5 km.
[b] Average within 500 m.

3). In a sample of 35 roost sites in Colorado, Shenk and Ringelman (1992) found mean distances to be only 471 m from powerlines during fall migration.

Feeding Site Characteristics

Mean values and variation for the studied habitat characteristics of observed spring and fall feeding sites are given in Tables 4 and 5. This study and Howe (1989) found that the principal feeding sites for migrating whooping cranes were croplands adjacent to roost sites. The primary feeding sites in Saskatchewan were nearly all upland sites. Harvested cereal crop fields comprised 100% of all spring feeding sites and 85% of fall foraging sites (Table 4). Wheat stubble fields in Saskatchewan are the primary spring foraging sites for sandhill cranes (Iverson et al. 1982). There was no significant difference between spring and fall use of crop types (Table 6). Wheat fields (primarily red spring wheat) made up 60.5% of all feeding sites in croplands, and barley accounted for 31.5% (Fig. 4). During the study period, wheat crops averaged 3.6 million ha per year or 59.4% of the seeded cropland in the migration corridor through Saskatchewan and barley constituted 12.4% (Saskatchewan Agriculture and Food 1991) (Fig. 4).

As demonstrated with roost sites, visibility at feeding sites was also greater in fall than in the spring (Table 5). Upland feeding sites provided a greater degree of visibility ($\bar{x}$ = 2.4 in spring, 2.9 in fall) than did roost sites ($\bar{x}$ = 1.9 in spring, 2.5 in fall). Roost sites are in basins that tend to

Table 6. Comparison between spring and fall feeding site use by whooping cranes in Saskatchewan, 1986-90.

Variable	Significance	Test
Overall	$P = 0.48$	
Parameter estimates		
Average slope	$P = 0.47$	Logistic
Visibility	$P = 0.30$	regression[a]
Distance to trees	$P = 0.26$	
Other species	$P = 0.003$	Categorical
Crop type	$P = 0.36$	analysis[b]
Overall multivariate	$P = 0.26$	
Univariate[c]		
Distance to buildings	$P = 0.07$	Canonical
Distance to roads	$P = 0.45$	discriminant
Distance to powerlines	$P = 0.10$	analysis

[a] Significance of maximum likelihood estimates of parameters in the full model, from PROC LOGISTIC; overall significance is assessed by the score statistic.
[b] CATMOD procedure; method used is logistic regression for categorical variables. Significance of parameter estimates are shown for the full maximum likelihood model on these variables.
[c] Univariate results produced by CANDISC procedure, using same data as the multivariate test, i.e., only observations with no missing data for these variables.

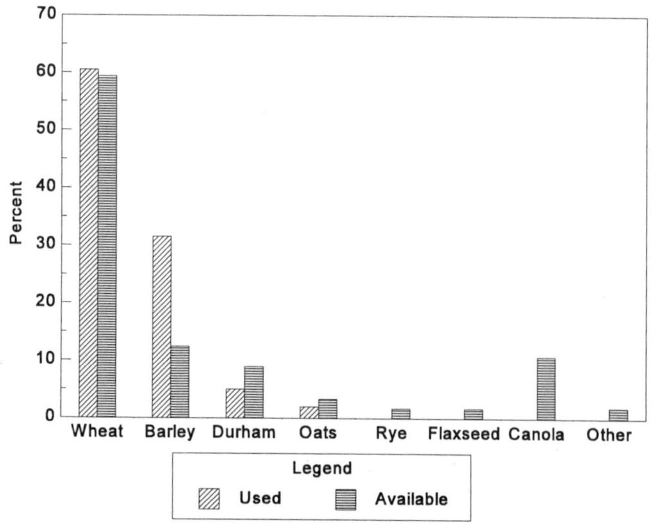

Fig. 4. Crop types used by whooping cranes as feeding sites in Saskatchewan, 1986-90 (Saskatchewan Agriculture and Food 1991).

Table 7. Comparison between spring and fall roosting site use by whooping cranes in Saskatchewan, 1986–90.

Variable	Significance	Test
Overall	$P = 0.54$	
Average slope	$P = 0.64$	Logistic
Visibility	$P = 0.61$	regression
Area	$P = 0.45$	
Distance to feeding site	$P = 0.58$	
Specific conductivity	$P = 0.41$	
Distance to trees	$P = 0.35$	
Disturbed vs. undisturbed	$P < 0.0001$	Categorical
Other species	$P < 0.0001$	analysis[a]
Overall multivariate	$P = 0.78$	
Building distance	$P = 0.12$	Canonical
Road distance	$P = 0.96$	discriminant
Powerline distance	$P = 0.94$	analysis
Shore slope	$P = 0.89$	
pH	$P = 0.89$	

[a] No multivariate categorical analysis was possible. Results shown are for analysis of single variables in PROC CATMOD.

provide reduced horizontal visibility, whereas feeding locations are primarily in harvested cropland virtually devoid of any tall vegetation. The exceptions were isolated groves of aspen and willow (*Salix* spp.). There were no seasonal differences in proximity to trees ($\bar{x}$ = 277 m in spring and 258 m in fall) (Tables 5 and 6), distances being similar to those of roosts (Table 3).

There were slight differences between seasons in distance to nearest buildings ($P = 0.07$) and roads ($P = 0.45$), but none of these differences was significant (Tables 5 and 6). During fall migration, however, the mean distance from family groups to nearest buildings was 904 m, whereas non-family groups appeared more tolerant and averaged only 615 m to nearest buildings. There was no significant difference between family and non-family groups in distance to nearest road (421 and 426 m, respectively). The distance to the nearest powerline averaged 535 m in spring and 895 m during fall migration ($P = 0.10$).

There was a significant difference ($P = 0.003$) between seasons in whether whooping cranes fed alone or with other species (Table 6). When whooping cranes fed with other species, they most frequently were observed with sandhill cranes. Fourteen percent of spring feeding sites were in association with sandhills, and 35% of all fall feeding locations also contained sandhill cranes (Table 4).

Differences Between Roost and Random Sites

Roost site characteristics did not differ between seasons (Table 7), so the data were combined and then compared with a sample of random sites. Overall significant differences ($P = 0.0001$) between roost and random sites in distance to buildings, roads, and powerlines were found through canonical discriminant analysis (Table 8). Shore slope was shallower and pH was lower at used sites. Univariate tests showed that shore slope and distance to powerlines and buildings were significantly different between roost and random sites (Table 8). Joint comparisons of topographic slope, visibility, wetland area, distance to feeding fields, specific conductance, and distance to trees were not significant, either as a group or individually within the logistic regression analysis. Univariate tests yielded significant differences between roost and random sites for distance to feeding fields and trees and for shoreline slope; these results do not contradict the logistic regression result, but were due to larger non-missing sample sizes.

Table 8. Comparison between sites used by migrating whooping cranes and random sites in Saskatchewan, 1986–90.

Variable	Significance	Test
Overall	$P = 0.28$	
Average slope	$P = 0.13$	Logistic
Visibility	$P = 0.13$	regression
Area	$P = 0.61$	
Distance to feeding site	$P = 0.46$	
Specific conductivity	$P = 0.81$	
Distance to trees	$P = 0.39$	
Overall multivariate	$P < 0.0001$	
Distance to powerlines	$P < 0.0001$	Canonical
Shore slope	$P = 0.0006$	discriminant
pH	$P = 0.29$	analysis
Distance to roads	$P = 0.17$	
Distance to buildings	$P = 0.003$	

MANAGEMENT IMPLICATIONS

Choice of roost site appears to be influenced by several natural and manmade landscape features. Whooping cranes use wetlands that are at a greater distance from disturbance (buildings) and threats (powerlines). These wetlands generally have shallower slopes and lower pH values. Roost sites in close proximity (<1 km) to feeding fields appear to be preferred. This preference to feed close to the roost is beneficial in terms of reduced energy costs and accumulation of potential energy reserves for migration. In addition, the farther the cranes have to fly, if flight is at low altitudes, the greater the chance of encountering an electrical transmission line.

Since most of the roost sites occur on private land, a

cooperative approach between wildlife managers and farmers must be taken to preserve wetlands for cranes. Incentives to preserve wetlands on private lands through programs such as the North American Waterfowl Management Plan will benefit not only whooping cranes but all species associated with those wetlands.

LITERATURE CITED

ALLEN, R. P. 1952. The whooping crane. Natl. Audubon Soc. Res. Rep. 3. 246pp.

______. 1956. A report on the whooping crane's northern breeding grounds. Natl. Audubon Soc. Suppl. Res. Rep. 3. 60pp.

ARMBRUSTER, M. J. 1990. Characterization of habitat used by whooping cranes during migration. U.S. Fish. Wildl. Serv. Biol. Rep. 90(4). 16pp.

BIRD, R. D. 1961. Ecology of the aspen parkland of western Canada in relation to land use. Agric. Can. Publ. 1066. Ottawa, Ont. 155pp.

COWARDIN, L. M., V. CARTER, F. C. GOLET, and E. T. LAROE. 1979. Classification of wetlands and deepwater habitats of the United States. U.S. Fish Wildl. Serv. FWS/OBS-79/31. 131pp.

DRIVER, E. A. 1991. Migratory bird-agriculture interactions: research needs. Unpubl. Rep., Can. Wildl. Serv., Saskatoon, Sask. 73pp.

HARRIS, W. C., A. KABZEMS, A. L. KOSOWANS, G. A. PADBURY, and J. S. ROWE. 1983. Ecological regions of Saskatchewan. Sask. Parks Renewable Resour. Tech. Bull. 10. 57pp.

HJERTAAS, D. G. 1994. Summer and breeding records of the whooping crane in Saskatchewan. Blue Jay 52:99-115.

HOWE, M. A. 1987. Habitat use by migrating whooping cranes in the Aransas-Wood Buffalo corridor. Pages 303-311 *in* J. C. Lewis, ed. Proc. 1985 crane workshop. Platte River Whooping Crane Maintenance Trust, Grand Island, Nebr.

______. 1989. Migration of radio-marked whooping cranes from the Aransas-Wood Buffalo population: patterns of habitat use, behavior, and survival. U.S. Fish Wildl. Serv. Fish Wildl. Tech. Rep. 21. 33pp.

IVERSON, G. C., T. C. TACHA, and P. A. VOHS, JR. 1982. Food contents of sandhill cranes during winter and spring. Pages 95-98 *in* J.C. Lewis, ed. Proc. 1981 crane workshop. Natl. Audubon Soc., Tavernier, Fla.

JOHNS, B. W. 1992. Preliminary identification of whooping crane staging areas in Prairie Canada. Pages 61-66 *in* D. A. Wood, ed. Proc. 1988 North Am. crane workshop. Florida Game Fresh Water Fish Comm. Nongame Wildl. Program Tech. Rep. 12.

JOHNSON, K. A., and S. A. TEMPLE. 1980. The migration ecology of the whooping crane. Unpubl. Rep., U.S. Fish Wildl. Serv., Washington, D.C. 87pp.+33pp. app.

KABZEMS, A., A. L. KOSOWANS, and W. C. HARRIS. 1986. Mixedwood section in an ecological perspective. Second ed. Sask. Parks Renewable Resour. Tech. Bull. 8. 122pp.

KUYT, E. 1992. Aerial radio-tracking of whooping cranes migrating between Wood Buffalo National Park and Aransas National Wildlife Refuge, 1981-84. Can. Wildl. Serv. Occ. Pap. 74. 53pp.

RICHARDS, J. H., and K. I. FUNG. 1969. Atlas of Saskatchewan. Modern Press, Saskatoon, Sask. 236pp.

SASKATCHEWAN AGRICULTURE AND FOOD. 1991. Agricultural statistics 1990. Unpubl. Rep., Sask. Agric. Food, Econ. Branch., Regina, Sask. 159pp.

SAS INSTITUTE INC. 1990. SAS/STAT user's guide: statistics. Version 6 ed. SAS Inst. Inc., Cary, N.C. 890pp.

SHENK, T. M., and J. K. RINGELMAN. 1992. Habitat use by cross-fostered whooping cranes in Colorado. J. Wildl. Manage. 56:769-776.

STEWART, R. E., and H. A. KANTRUD. 1971. Classification of natural ponds and lakes in the glaciated prairie region. U.S. Fish. Wildl. Serv. Resour. Publ. 92. 57pp.

U.S. DEPARTMENT OF THE INTERIOR, FISH AND WILDLIFE SERVICE, and ENVIRONMENT CANADA, CANADIAN WILDLIFE SERVICE. 1995. Waterfowl breeding population survey, southern Saskatchewan, 1995. Unpubl. Rep., U.S. Fish Wildl. Serv., Washington, D.C. 15pp.

WARD, J. P., and S. H. ANDERSON. 1987. Roost site use versus preference by two migrating whooping cranes. Pages 283-288 *in* J.C. Lewis, ed. Proc. 1985 crane workshop. Platte River Whooping Crane Maintenance Trust, Grand Island, Nebr.

Appendix A. Variables measured at used and potential whooping crane migration stopover sites in Saskatchewan, 1986–90.

Site use by season	Feeding sites Spring	Feeding sites Fall	Roost sites Spring	Roost sites Fall	Random sites
Site description	X	X	X	X	X
Adjacent habitat	X	X	X	X	X
Extent of similar habitat	X	X	X	X	X
Crop type	X	X			
Food source	X	X			
Foods eaten	X	X			
Slope to the north	X	X	X	X	X
Slope to the south	X	X	X	X	X
Slope to the east	X	X	X	X	X
Slope to the west	X	X	X	X	X
Visibility to the north	X	X	X	X	X
Visibility to the south	X	X	X	X	X
Visibility to the east	X	X	X	X	X
Visibility to the west	X	X	X	X	X
Site security	X	X	X	X	X
Site ownership	X	X	X	X	X
Distance to buildings	X	X	X	X	X
Distance to trees	X	X	X	X	X
Distance to powerlines	X	X	X	X	X
Distance to roads	X	X	X	X	X
Road type	X	X	X	X	X
Presence of other species	X	X	X	X	
Water depth			X	X	X
Turbidity			X	X	X
Substrate			X	X	X
Dominant emergents			X	X	X
Vegetation density			X	X	X
Wetland system[a]			X	X	X
Wetland subsystem[a]			X	X	X
Wetland class[a]			X	X	X
Wetland modifier[a]			X	X	X
Wetland type[b]			X	X	X
Shoreline slope			X	X	X
Wetland area			X	X	X
Distance to upland feeding			X	X	X
Specific conductance			X	X	X
pH			X	X	X

[a] Cowardin et al. (1979).
[b] Stewart and Kantrud (1971).

AVAILABILITY OF STOPOVER HABITAT FOR MIGRANT WHOOPING CRANES IN NEBRASKA

DALE W. STAHLECKER, Eagle Ecological Services, 30 Fonda Road, Santa Fe, NM 87505, USA

Abstract: Four stratified random samples of 512 National Wetland Inventory (NWI) maps within the central Nebraska portion of the Wood Buffalo-Aransas whooping crane (*Grus americana*) migration corridor were used to evaluate the availability of wetland roost sites. Wetlands were eliminated as potential roosts if visibility was obscured by vegetation or slope, if certain human activities occurred within 100–800 m, or if water <30 cm deep was not available. Seasonal emergent wetlands, available as roosts primarily in spring, dominated all samples, particularly in the north. Sixty-five percent of all wetlands >0.04 ha passed map review and 52% passed when ground-truthed. NWI map review was a good predictor of both suitability (63% correct) and unsuitability (73% correct). More than one-half of all open and emergent wetlands >1 ha passed both map and field review. Four of 6 major east-flowing rivers provide additional roosting habitat of varying quality. Stopover habitat is available throughout the migration corridor in Nebraska, though quality is best in the northern sandhill region and along major rivers. NWI map review, with adequate ground-truthing and observer experience, can provide good estimates of roost availability in open prairie where woody vegetation is limited.

PROC. NORTH AM. CRANE WORKSHOP 7:132-140

Key words: *Grus americana*, habitat maps, migration, Nebraska, whooping crane.

The whooping crane was one of the most perilously endangered of species; all individuals alive today are descended from 16 cranes that wintered at Aransas National Wildlife Refuge, Texas, in 1941 (U.S. Fish and Wildlife Service 1986). Research and management from the 1930's through the mid-1970's were concentrated on wintering and breeding (Wood Buffalo National Park, Canada) areas. Documentation of the migration route (Howe 1989, Kuyt 1992) and stopover roosting habitat requirements (Lingle et al. 1984, Howe 1987, Ward and Anderson 1987) came in the 1980's and is ongoing. As information has accrued on roost sites of migratory cranes, suitability criteria and indices for non-traditional stopover roosts (sites generally used only 1 night [Melvin and Temple 1982]) were developed (Armbruster 1990). Concern about the availability of suitable wetland roosts within the migration corridor (Lingle 1987) prompted efforts to inventory potential roost sites.

Nebraska has long been known as an important stopover for migrating whooping cranes (Swenk 1933, Johnsgard and Redfield 1977). Research and management efforts there have concentrated on the Platte River, where conflicts have been high (Krapu et al. 1982, Lingle 1982, Faanes 1992). Suitable riverine roosting habitat for whooping cranes in the Big Bend region of the Platte has declined an average of 55% since 1938 (Currier et al. 1985) to less than 5,000 ha. Such detailed information on habitat availability elsewhere in Nebraska is not currently available.

NWI, the U.S. Fish and Wildlife Service's (USFWS) catalog of the nation's wetlands, provides baseline data on the physical characteristics and extent of all United States wetlands (Montanari and Townsend 1977, Cowardin et al. 1979). Though NWI has not yet been completed throughout the Wood Buffalo-Aransas whooping crane migration corridor, it does provide a potentially large data base for analysis of the availability of stopover habitat to this crane population. Stahlecker (1992) found that NWI maps, when used in conjunction with aerial photo maps and suitability criteria (Armbruster 1990), were poor predictors (33% correct) of suitable whooping crane roosts in Oklahoma, but good predictors (97% correct) of unsuitability of potential crane roosts. Experience gained in Oklahoma was used in the second phase of this project to estimate roost availability in Nebraska.

This study was funded by USFWS; I thank J. C. Lewis for support and guidance throughout the project, C. R. Elliott for NWI maps and data summaries, D. H. Johnson for advice and assistance on statistical tests, and J. G. Sidle for a flight over the Platte River. G. R. Lingle gave insight into the unique access problems of the Nebraska sandhills. T. G. Smith produced Figure 1. D. H. Johnson, G. R. Lingle, J. C. Lewis, R. A. Lock, and J. G. Sidle reviewed a draft of this manuscript.

STUDY AREA AND METHODS

Central Nebraska is largely privately owned and highly agricultural. Cattle ranching dominates the northcentral third (sandhills) of the state, where sandy soils on rolling hills are largely unsuitable for tillage but very productive of grass and hay. The high water table in the sandhills produces many natural lakes and streams, and few man-made wetlands have been constructed. Though pastures and cattle are also present in the heavier-soiled southcentral two-thirds of Nebraska, farming of grains, particularly corn, is the dominant land use and man-made stock ponds are common.

Table 1. Wetland classifications and modifiers (Cowardin et al. 1979) commonly encountered on Nebraska National Wetlands Inventory maps and their stopover roost categories.

Roost type	Map code	Wetland classification: System (subsystem), class	Modifiers: Water regime[a]	Modifiers: Special[b]
Open	L2AB	Lacustrine(2 = Littoral), Aquatic Bed	A = temporary	d = partially drained
	PAB	Palustrine, Aquatic Bed	C = seasonal	
	PAB/EM	Palustrine, Aquatic Bed/EMergent	F = semipermanent	f = farmed
	PUS	Palustrine, Unconsolidated Shore,	G = intermittently exposed	h = impounded
	PUB	Palustrine, Unconsolidated Bottom		x = excavated
Emergent	PEM	Palustrine, EMergent	H = permanent	
Riverine	R2UB	Riverine (2 = Lower Perennial), Unconsolidated Bottom		
	R2US	Riverine (2 = Lower Perennial), Unconsolidated Shore		
No good[c]	R4SB	Riverine (4 = Intermittent), Stream Bed		
	PFO	Palustrine, FOrested		
	PSS	Palustrine, Shrub/Scrub		

[a] Modifiers indicate length of time wetland has surface water.
[b] Modifiers indicate mostly human activities.
[c] Intermittent streams are seldom suitable roosts; wooded wetlands most likely have horizontal visibility problems.

I used NWI maps, U.S. Geological Survey (USGS) topographic maps, whooping crane roost suitability criteria (Armbruster 1990), and ground truthing to estimate the density of potential whooping crane roosts in central Nebraska. There are 11 common NWI wetland classes (Cowardin et al. 1979) in the study area. Eight have potential as whooping crane roost sites and can be grouped into 3 general wetland types: open water, emergent, and riverine (Table 1).

Two USGS map scales are pertinent to this study: 1:24,000 and 1:100,000. NWI maps are 1:24,000 topographic maps (157.5 km^2) of categorized wetlands. There are 32 (4 by 8) NWI maps within each smaller scale 1:100,000 map and 16 (15 full maps plus 2 half maps) 1:100,000 maps within the whooping crane migration corridor in central Nebraska (Fig. 1). Sandhills dominate 7 of the northern and westernmost 1:100,000 maps.

I conducted 4 stratified random samples of the 512 (16 × 32) NWI maps available as detailed below. In Sample I all wetlands were evaluated and in Sample II only wetlands ≥1 ha were examined. Both samples were first evaluated from data on NWI maps (subscript ME = Map Exam); later a subsample of each was ground-truthed (subscript GT = Ground Truth). In addition, the USFWS has totaled all wetlands by type on 8 of the 17 1:100,000 maps used in this study. I summarized those totals by potential roost type (open, emergent, or riverine) to make comparisons and extrapolations with the data from samples within those 8 mapping units.

Sample I_{ME}

This 32-map (6%) sample was obtained by randomly selecting 2 NWI maps from each of the 16 1:100,000 map units (Fig. 1). Previous migration studies documented that >50% of whooping crane stopover roosts were in wetlands <1 ha and that wetlands as small as 0.01 ha were used (Howe 1987, 1989). Therefore, any wetland large enough (≥0.04 ha) to appear on a NWI map was evaluated. Wetlands were subtotaled on each 256-ha (640-acre) section and totaled for the entire NWI map.

All wetlands were categorized by roost type (open, emergent, or riverine) and by size (<1 or ≥1 ha). I used a Wainer Land Area and Slope Indicator to estimate wetland size. Streams were difficult to categorize by size. Therefore the entire reach of a stream in a 1.6-km^2 (1-mi^2) section was considered 1 wetland; a stream that crossed more than 1 section was recorded as a separate wetland in each section. This portion of my evaluation was equivalent to Armbruster's (1990) Suitability Index of Wetland Size (SIWS) (Fig. 2A).

A Suitability Index of Horizontal Visibility (SIHV) was then determined for each wetland. A wetland <25 m from obstructions ≥1 m height, such as shrubs, trees, or steep slopes along ≥5° of shoreline, is considered unsuitable for crane roosting (Armbruster 1990) and was assigned SIHV of 0 (Fig. 2B). The aerial photographs used to delineate wetlands by the USFWS were not readily available, so they could not be used to check for visual obstructions. Soil

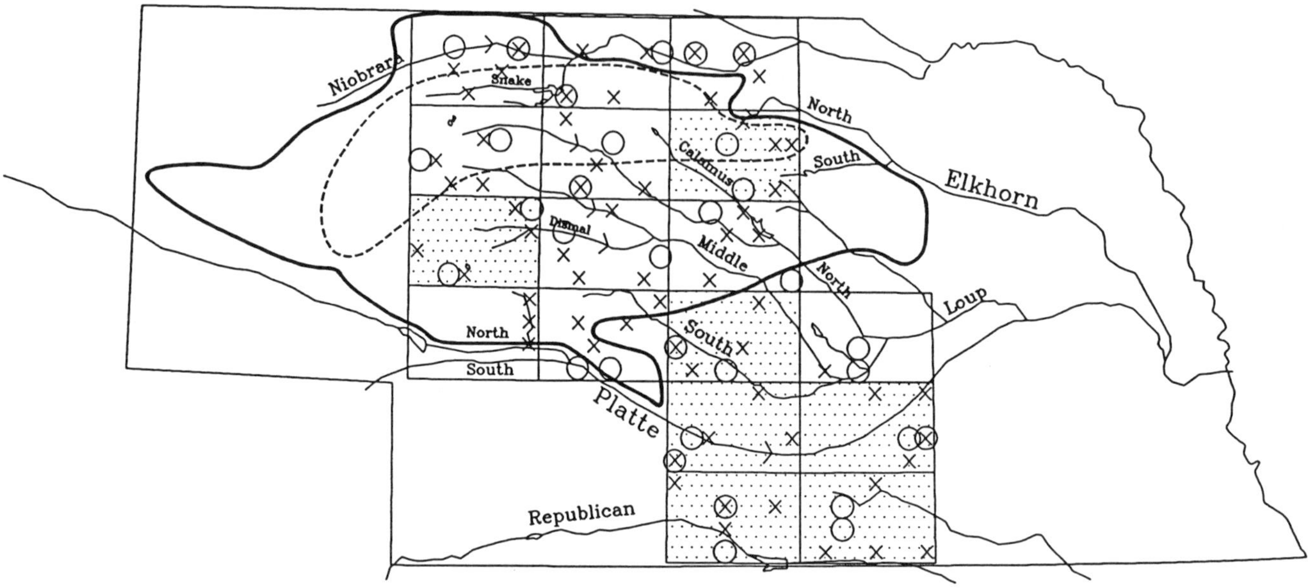

Fig. 1. Rivers and sample National Wetlands Inventory maps in central Nebraska. Sample I maps are X's and Sample II maps are O's. All 17 1:100,000 map units are shown; shaded 1:100,000 map units are those where the total numbers of wetlands by type were known. The dark line encircles the Sandhills; the light line encircles the area containing many natural lakes. Arrows (>) on major rivers indicate point below which there are consistently high SIHV's.

Conservation Service county aerial photographs (Stahlecker 1992) were also not available for all Nebraska counties in the sample, so I used standard USGS topographic maps to look for potential visual obstructions. I used contour lines on NWI maps to estimate slope; 2 or more 3-m contour lines within 30 m of the shoreline for ≥180° portion of the wetland also resulted in an SIHV of 0 and deletion from further consideration.

Those wetlands meeting the SIHV criteria were then evaluated for disturbance factors (Table 2) (Armbruster 1990). If at least 0.04 ha of a wetland was beyond the range of all disturbance factors shown, and measured on the NWI map, it had Passed Map Examination (PME).

Sample I_{GT}

I randomly selected 3 sections (2.56 km^2) from each NWI map in Sample I_{ME} (n = 96) to ground truth. During October 1992, I visited 86 sample sections (I was denied, or could not find, access to 10 sections). For each wetland an SIHV was estimated and disturbance factors evaluated. If SIHV and disturbance criteria were met, I determined a Suitability Index of Water Depth (SIWD) by either measuring or estimating water depth. An SIWD of 1 was assigned if water was <30 cm deep at 20 m from shore; otherwise SIWD was 0 (Armbruster 1990). If at least 1 wetland in a section passed all 3 tests, the section had Passed Ground Truthing (PGT). PGT results were compared directly with PME results quantified in the same sections during Sample I_{ME}.

Sample II_{ME}

The study in Oklahoma had shown that so few small wetlands (<1 ha) were suitable roosts that a second sample only of larger (≥1 ha) wetlands was taken (Stahlecker 1992). In Nebraska I again randomly selected 4 NWI maps from each 1:100,000 unit (Fig. 1), a total of 64 maps (12%), to sample wetlands ≥1 ha. I followed the same procedures outlined in Sample I_{ME} to eliminate unsuitable wetlands.

Sample II_{GT}

During October 1992 I visited 55 open, 82 emergent, and 233 riverine wetlands ≥1 ha in a stratified random sample of the total counted in Sample II_{ME}. No more than 3 wetlands of the open and emergent roost types were visited in each NWI map in Sample II_{ME}; all riverine sections were visited because riverine habitat was much less common. Each wetland was evaluated following the same procedure as outlined for Sample I_{GT}. Since many large wetlands were seasonally dry during ground testing, I also recorded whether a wetland was likely to be suitable during the drier autumn period rather than the wetter spring period when aerial photographs for

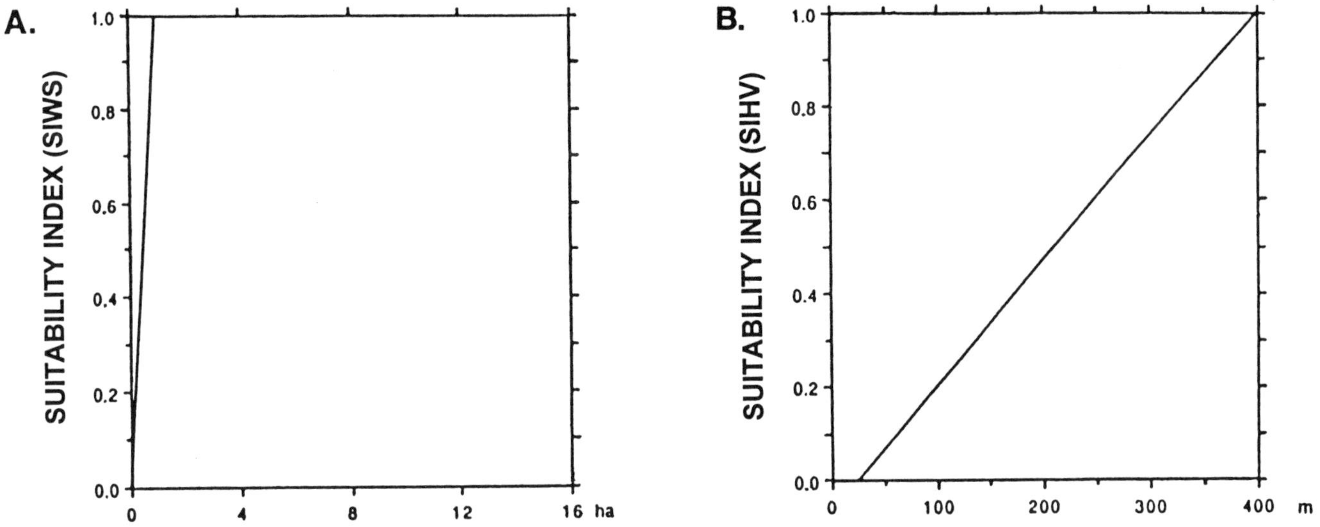

Fig. 2. Predicted relationships between (A) wetland size and (B) horizontal visibility and roosting suitability for non-traditional (migration) stopover sites (from Armbruster 1990).

NWI evaluation are generally taken.

Analysis

I categorized wetlands by roost type, size, and suitability criteria for each NWI map in Samples I_{ME} and II_{ME} and for each section for Sample I_{GT}, then converted to wetlands/100 km^2 and wetlands/km^2, respectively. I used regression analysis and paired *t*-tests to evaluate the relationships between variables within Samples I_{ME} and II_{ME} and a binary response model to test the ability of PME wetlands to predict PGT wetlands in Sample I_{GT}.

RESULTS

Sample I_{ME}

More than 10,500 wetlands were evaluated on NWI maps. Many (62%) met SIHV or disturbance suitability criteria. Emergent wetlands accounted for 79% of all wetlands in the sample and 81% passed map examination. Five percent of 2,172 open wetlands and 54% of 87 riverine wetlands passed map exam (Table 3).

The average number of potential roosts/100 km^2 was 209 (range = 12–1,582), and the mean number of PME roosts was 135/100 km^2 (range = 4–1,482). Potential roosts did not decline from north to south ($F = 0.51$, $P = 0.481$), PME roosts <1 ha did, though not significantly ($F = 3.16$, $P = 0.086$), but the north-south decline in numbers of PME roosts ≥1 ha was significant ($F = 7.45$, $P < 0.001$).

Potential open wetland roosts increased significantly from north to south ($F = 41.53$, $P < 0.001$), but PME open wetlands did not ($F = 0.07$, $P = 0.800$). Though both potential and PME emergent ($F = 2.57$, $P = 0.120$; $F = 3.24$, $P = 0.082$; respectively) and riverine roosts ($F = 2.13$, $P = 0.155$; $F = 2.20$, $P = 0.149$; respectively) declined southward, none of the declines was significant.

If emergent wetlands are excluded, mean potential crane roosts/100 km^2 were 45 (range = 0–192) and increased significantly to the southeast ($F = 40.94$, $P < 0.001$).

Table 2. Disturbance types and their estimated zones of influence (m) on whooping crane selection of stopover roosts. From Armbruster and Farmer (1981) as adapted by Armbruster (1990).

Disturbance type	Width of affected area (m)[a]
Windmill[b]	100
Private road	100
Power line	100
Gravel road	200
Rural dwelling	200
Recreation area	200
Paved road	400
Railroad	400
Bridge	400
Urban dwellings	800
Commercial development	800

[a] Width of a band on each side of a linear feature, or radius of a circle around a point.

[b] Added as a result of this study.

Table 3. Percentages of central Nebraska wetlands that passed or failed map and/or field examination.

Sample	Wetland type	*n*	Passed examination		Failed examination			
			< 1 ha	≥1 ha	Visibility	Disturbance	Slope/depth	Misidentified
I_{ME}	Open	2,172	3	2	+	2	93	
	Emergent	8,263	78	3	+	13	6	
	Riverine	87		54	32	14	0	
	Totals	10,522	62	3	+	11	24	
I_T	Open	120	5	3	0	4	88	
	Emergent	426	64	3	2	20	11	
	Riverine	6		66	17	0	17	
	Totals	552	51	3	1	17	28	
I_{GT}	Open	110	12	6	19	22	41	
	Emergent	382	59	2	8	27	4	
	Riverine	6		66	17	0	17	
	Totals	498	48	4	10	26	12	
II_{ME}	Open	190		58	0	5	37	
	Emergent	895		94	1	4	1	
	Riverine	233		41	54		5	
	Totals	1,294		80	9	4	7	
II_{GT}	Open	55		58	13	16	9	4
	Emergent	82		69	7	15	2	8
	Riverine	233		34	48	10	8	
	Totals	370		45	34	12	6	3

Because a high percentage of open wetlands failed map exam, an average of only 2.8 PME open roosts/100 km^2 (range = 0–9.5) was in the study area and no directional trend occurred (F = 2.51, P = 0.123).

Sample I_{GT}

The percentages of PME wetlands/km^2 in the 86 randomly selected 2.56-km^2 sections in Sample I tested (I_T) appeared to be representative of Sample I_{ME}. Fifty-four percent of the combined roost types of Sample I_T passed map exam, and 52% of Sample I_{GT} passed ground truthing (Table 3). The percentage of PGT open wetlands was 18%, up from 8% PME open wetlands; PGT emergent wetlands declined to 61% from 67% PME emergent wetlands, and PGT and PME riverine wetlands were constant at 66%.

The purpose of ground truthing was to test the predictive value of map examination by asking the question "Does a 2.56-km^2 section have at least 1 suitable crane roost?" A binary response model produced a 2 × 2 contingency table (Table 4) which showed a high association between map and ground examinations (χ^2 = 11.26, df = 1, P = 0.001). NWI map review was best as a predictor of unsuitability because it was correct 35 of 48 times (73%) but also correctly predicted suitability 24 of 38 times (63%).

Sample II_{ME}

Potential roosts ≥1 ha examined in this 64-map sample totaled 1,294 ($\bar{x}$ = 12.8/100 km^2; range = 0–92). The majority were emergent wetlands (65%); riverine wetlands comprised 21% of the sample, and open wetlands were least common (14%). Open wetlands increased significantly toward the south (F = 5.20, df = 63, P = 0.026); emergent wetlands decreased in the same direction (F = 5.71, P = 0.020) and riverine wetlands showed no pattern (F = 0.53, P = 0.468). In total, potential roosts ≥1 ha decreased toward the southeast, but the trend was just short of significant (F = 3.06, P = 0.086).

Table 4. A 2 × 2 contingency table of Sample I_T and I_{GT} sections containing (Yes) or not containing (No) suitable whooping crane roosts.

		Map Examination (I_T)		
		No	Yes	Total
Ground truthing	No	35	13	48
	Yes	14	24	38
	Total	49	37	86

Table 5. Roost suitability and average SIHV of major rivers and their tributaries, Sample II_{GT}.

River	No. of NWI maps	No. of sections[a]	Suitable sections[a] No.	%	$\bar{x}$ SIHV
Niobrara	5	42	16	38	0.92
Snake[b]	1	8	0	0	0.00
North Loup	4	28	27	96	0.57
Calamus	1	9	6	67	0.50
Middle Loup	3	22	13	59	0.67
South Loup	2	13	0	0	0.00
North Platte	1	19	1	05	0.10
Birdwood	3	18	10	56	0.80
Platte	3	28	6	21	0.17
Republican	4	46	0	0	0.00

[a] A section is 2.56 km^2 (1 mi^2).
[b] When the name of a stream is indented, the stream is a tributary of river listed above it.

PME wetlands averaged 10.3/100 km^2 (range = 0–92). Most PME roosts were emergent (81%) wetlands because they predominated within the sample and most (94%) passed map exam. Total PME wetlands ≥1 ha decreased significantly (F = 6.21, P = 0.015) from northwest to southeast, primarily because PME emergent wetlands ≥1 ha decreased significantly (F = 5.82, P = 0.019).

Sample II_{GT}

Comparable percentages of open (58% PME, 58% PGT) and riverine (41% PME, 34% PGT) wetlands ≥1 ha passed both map examination and ground truthing. The percentage of emergent wetlands (69%) passing ground truthing, however, was considerably less than that (94%) passing map examination (Table 3). Most of 32 open (91%) and all 79 riverine (100%) PGT roosts were likely to have shallow standing water during autumn migration, but only 23% of 56 PGT emergent roosts met the same criteria.

Open roosts ≥1 ha were of significantly higher quality in the Sandhills than in Farmlands. Ninety-six percent of 24 open roosts checked in the Sandhills passed ground truthing with a mean SIHV of 0.95, but only 23% of 31 Farmland open roosts passed ground truthing and their mean SIHV was only 0.21 (t = 12.94, P < 0.001).

More emergent roosts in the Sandhills, 84% of 43, passed ground truthing with a higher mean SIHV (0.84) than outside the Sandhills (67% of 37; SIHV $\bar{x}$ = 0.67), but the difference was not significant (t = 0.097, P = 0.340). However, if SIHV was averaged over all roosts visited, whether suitable or not, the SIHV means were lower (Sandhills = 0.70, Farmlands = 0.38) and the difference was significant (t = 3.47, P < 0.000).

Central Nebraska is drained by 6 major rivers and their tributaries (Table 5, Fig. 1). These rivers flow generally west to east, perpendicular to the whooping crane migration corridor. Sample II included 25 maps with potential riverine roosting habitat. Within Sample II, 5 rivers had no or few sections with suitable roosting habitat and low mean SIHV's, and 5 had moderate to high suitability rates and mean SIHV's (Table 5).

Comparisons of Samples

The USFWS has totaled wetlands within 8 of the 17 1:100,000 maps in the study area (Fig. 1); consequently, it was possible to compare my randomly selected samples with actual wetland totals. I grouped wetlands into the 3 roost types (open, emergent, riverine), calculated the percentage for each roost category, and compared these percentages for Sample I_{ME}, Sample I_{GT}, and the actual wetlands available within each 1:100,000 map unit with a series of 3 × 3 contingency tables (Table 6). In 4 of the 1:100,000 map units, the samples were representative of the actual wetlands present. In the 4 units for which samples were not representative, the smallest sample, I_{GT}, was the aberrant sample.

Samples I_{ME} and II_{ME} tabulated PME wetlands ≥1 ha, so results could be tested with 2-sample t-tests. In comparisons of open (t = 1.34, P = 0.18), emergent (t = 1.52, P = 0.13), riverine (t = 0.2, P = 0.84), and total PME (t = 1.47, P = 0.15) roosts, none was significantly different. This similarity indicates that both samples were from the same population of wetlands capable of passing map examination.

Table 6. Results of χ^2 analysis of percentages of 3 potential roost types in Samples I_{ME} and I_{GT} with the actual wetland totals within 8 1:100,000 map units in the central Nebraska study area.

Map unit	Geographic region	χ^2	P	Samples represent actual?	Aberrant sample
Valentine NW	Sandhills	3.39	0.495	Yes	none
O'Neill SW	Sandhills	6.89	0.142	Yes	none
N. Platte NW	Sandhills	19.50	0.000	No	I_{GT}
Broken Bow SW	Farmlands	4.67	0.323	Yes	none
Grand Island NW	Farmlands	60.26	0.000	No	I_{GT}
Grand Island NE	Farmlands	12.18	0.016	No	I_{GT}
Grand Island SW	Farmlands	6.93	0.140	Yes	none
Grand Island SE	Farmlands	17.79	0.001	No	I_{GT}

DISCUSSION

Evaluation of Technique

In Oklahoma, NWI map examination overestimated suitable roost sites, primarily because it was difficult to ascertain the presence and height of woody vegetation blocking horizontal visibility and because most mapped wetlands were small, steep-sided, dammed stock ponds that did not have adequate shallows to meet minimum criteria (Stahlecker 1992). In Nebraska, vision blocking vegetation adjacent to potential roost sites was much less common, though impossible to locate from NWI and difficult to locate from topographic maps. However, I was more familiar with Nebraska landscapes than I had been with Oklahoma landscapes when I initiated respective map reviews. That experience made analysis of SIHV much easier than it had been in Oklahoma. The reduced visibility problems in Nebraska were ultimately confirmed by ground truthing; 84% of wetlands ≥0.04 ha checked in Oklahoma (Stahlecker 1992) were rejected for SIHV, but only 10% were rejected in Nebraska (Table 3).

Narrow stock ponds sampled in Oklahoma never had suitable shallows for crane roosting, so I rejected similar impoundments during review of Nebraska NWI maps. However, during ground truthing I found that some impoundments had suitable shallows, apparently because Nebraska's sandier soils are more erodible and filled inlets more quickly. In the 4 southernmost 1:100,000 map units there were 9 sample sections that had no PME wetlands that in fact had PGT wetlands. However, there were also 10 sample sections that map review correctly predicted did not have any suitable roosts. Though the proportion of impoundments that have suitable roosting shallows might be predicted, the suitability of a particular wetland cannot be predicted from NWI maps.

Sampling 2 of 32 NWI maps/1:100,000 mapping unit appears to have been representative of the proportion of available wetland types, but the ground truthing sample of 3 sections/NWI map was not representative in one-half of the tests conducted (Table 6). The important aspect of ground truthing, however, was to test the predictive value of map examination, no matter what wetland types were present, and it appeared to do so successfully.

Stopover Roost Availability in Nebraska

The large number and wide distribution of wetlands within the whooping crane migration corridor in Nebraska suggest that this population should have multiple options of roost sites during migration through the state. Much higher proportions of wetlands of all sizes passed map and ground testing in Nebraska (Table 3) than in Oklahoma (Stahlecker 1992). The average number of PME wetlands ≥1 ha/100 km^2 was 3–6 times higher in Nebraska (12.8/100 km^2) than was estimated for Oklahoma (2–4/100 km^2, Stahlecker 1992).

Emergent wetlands were 78% of all wetlands tested during map examination. Tall emergents, such as cattails (*Typha* sp.) and bulrushes (*Scirpus* sp.) are not attractive to roosting cranes because they interfere with horizontal visibility. However, 80–97% of the emergent wetlands in the 8 1:100,000 map units totaled by USFWS were classified as temporary (PEMA) or seasonal (PEMC); similar rates were likely for the other 9 map units in the study. Throughout Nebraska, but particularly in the Sandhills, these ephemeral wetlands are cut for hay late in the summer. They are most likely to have standing water in spring after snowmelt and therefore are most available to northbound whooping cranes. However, 77% of all PGT emergent wetlands ≥1 ha (n = 56) examined in October 1992 did not have standing water and therefore would not attract southbound cranes. Therefore, whooping cranes in autumn have an average of 4.8 PME roosts ≥1 ha/100 km^2, a rate more comparable to that predicted for Oklahoma, regardless of season. An even higher rate of unsuitable emergent wetlands <1 ha is likely in autumn.

Suitable roost sites also varied in quality. The Sandhill region had palustrine wetlands ≥1 ha with significantly higher SIHV ratings than the remainder of the state. Though it could not be tested in the same manner, palustrine wetlands <1 ha likely showed a similar difference. Man-made palustrine wetlands, especially those <5 ha, do not have the same high qualities as crane roosts as their natural counterparts.

Though not statistically confirmed in this sample, natural open wetlands, especially those ≥1 ha, in the Sandhill region are not randomly or uniformly distributed. The majority of these highly attractive roost sites are concentrated in an arc that includes the higher elevations of the Sandhills and the headwaters of its major streams (Fig. 1). Fortunately, adjacent lower elevation areas were crossed by major eastward flowing rivers and/or areas where man-made stock ponds were common.

Nebraska's eastward flowing rivers are well spaced north-to-south and could provide an abundance of roost site possibilities for migrant whooping cranes. The relatively unbridled Niobrara, North Loup, and Middle Loup rivers in the northern two-thirds of Nebraska had high SIHV values (Table 5) once they had attained flows that provided active wide and sandy channels. The South Loup River is much smaller and never attains flows needed to create suitable roosting channels. The southernmost of Nebraska's rivers, the Platte and Republican rivers, have major mainstem dams and diversions. The impact of control and diversion on channel activity in the Platte (Currier et al. 1985) and its primary tributaries (McDonald and Sidle 1992) are well documented. The Platte provides wide sandy channels, likely to have high SIHV's, only after some of the diverted water is returned below Lexington. Only 1 of 3 sample maps encompassing the Platte fell within this region. The Republican River, a considerably smaller river than the Platte, has been so controlled that its narrow channel, usually less than 30 m wide, is tree-lined and unsuitable for roosting cranes.

The minimum SIHV for this study was 0.1; this required a minimum channel width of 100 m for a river segment to pass map and/or ground examination. Whooping cranes have been seen roosting within Platte River channels as narrow as 52 m (SIHV = 0.0), though most roost sites were >150 m (SIHV >0.32) (Faanes et al. 1992). Faanes et al. (1992) found that whooping cranes selected channels that were >150 m wide and were wet for >80% of that width. Though whooping cranes will use channels meeting minimum criteria, they prefer higher quality roost sites. The approximate location where each Nebraska river becomes consistently attractive to migrant whooping cranes is indicated in Figure 1.

CONCLUSIONS AND RECOMMENDATIONS

NWI map and ground review of wetlands in Nebraska indicates that there are sufficient numbers of potential roosts that meet minimum criteria for migrating whooping cranes (Armbruster 1990), though quality varies extensively. Potential roosts are widely distributed. Riverine roosts meeting minimum standards (this study, Stahlecker 1992) and roosts that whooping cranes have chosen to use (Faanes et al. 1992) are quite different. Numerous sightings of migrant whooping cranes have been documented by state and federal agencies in Nebraska. Consequently, the next logical step appears to be a comparative study of these 2 data sets.

NWI map review was a better predictor of roost site unsuitability (0.73) than of roost site suitability (0.63). Both prediction rates fall within an acceptable range, though higher rates would be more useful. Predictive values from this study may be applicable to South Dakota, North Dakota, and Montana portions of the whooping crane migration corridor, where terrain, rainfall, and the general lack of woody vegetation are similar.

LITERATURE CITED

ARMBRUSTER, M. J. 1990. Characterization of habitat used by whooping cranes during migration. U.S. Fish Wildl. Serv. Biol. Rep. 90(4). 16pp.

______, and A. H. FARMER. 1982. Draft sandhill crane habitat suitability index model. Pages 136-143 *in* J. C. Lewis, ed. Proc. 1981 crane workshop. Natl. Audubon Soc., Tavernier, Fla.

COWARDIN, L. W., V. CARTER, F. G. GOLET, and E. T. LAROE. 1979. Classification of wetlands and deepwater habitats of the United States. U.S. Fish Wildl. Serv. FWS/OBS-79/31. 103pp.

CURRIER, P. J., G. R. LINGLE, and J. G. VANDERWALKER. 1985. Migratory bird habitat on the Platte and North Platte Rivers in Nebraska. Platte River Whooping Crane Maintenance Trust, Grand Island, Nebr. 155pp.

FAANES, C. A. 1992. Factors influencing the future of whooping crane habitat on the Platte River in Nebraska. Pages 101-109 *in* D. A. Wood, ed. Proc. 1988 North Am. crane workshop. Florida Game Fresh Water Fish Comm. Nongame Wildl. Program Tech. Rep. 12.

______, D. H. JOHNSON, and G. R. LINGLE. 1992. Characteristics of whooping crane roost sites in the Platte River. Proc. North Am. Crane Workshop 6:90-94.

HOWE, M. A. 1987. Habitat use by migrating whooping cranes in the Aransas-Wood Buffalo corridor. Pages 303-311 *in* J. C. Lewis, ed. Proc. 1985 crane workshop. Platte River Whooping Crane Maintenance Trust, Grand Island, Nebr.

______. 1989. Migration of radio-marked whooping cranes from the Aransas-Wood Buffalo population: patterns of habitat use, behavior, and survival. U.S. Fish Wildl. Serv. Fish Wildl. Tech. Rep. 21. 33pp.

JOHNSGARD, P., and R. REDFIELD. 1977. Sixty-five years of whooping crane records in Nebraska. Nebr. Bird Rev. 45:54-56.

KRAPU, G. L., K. J. REINECKE, and C. R. FRITH. 1982. Sandhill cranes and the Platte River. Trans. North Am. Wildl. Nat. Resour. Conf. 47:542-552.

KUYT, E. 1992. Aerial radio-tracking of whooping cranes migrating between Wood Buffalo National Park and Aransas National Wildlife Refuge, 1981-84. Can. Wildl. Serv. Occas. Pap. 74. 50pp.

LINGLE, G. R. 1982. Mormon Island Crane Meadows–protecting habitat for cranes along the Platte River, Nebraska. Pages 17-21 *in* J. C. Lewis, ed. Proc. 1981 crane workshop. Natl. Audubon Soc., Tavernier, Fla.

______. 1987. Status of whooping crane migration habitat within the Great Plains of North America. Pages 331-340 *in* J. C. Lewis, ed. Proc. 1985 crane workshop. Platte River Whooping Crane Maintenance Trust, Grand Island, Nebr.

______, P. J. CURRIER, and K. L. LINGLE. 1984. Physical characteristics of a whooping crane roost on the Platte River, Nebraska. Prairie Nat. 16:39-44.

McDONALD, P. M., and J. G. SIDLE. 1992. Habitat changes above and below water projects on the North Platte and South Platte Rivers in Nebraska. Prairie Nat. 24:149-158.

MELVIN, S. M., and S. A. TEMPLE. 1982. Migration ecology of sandhill cranes: a review. Pages 73-87 *in* J. C. Lewis, ed. Proc. 1981 crane workshop. Natl. Audubon Soc., Tavernier, Fla.

MONTANARI, J. H., and J. E. TOWNSEND. 1977. Status of the National Wetlands Inventory. Trans. North Am. Wildl. Nat. Resour. Conf. 42:66-72.

STAHLECKER, D. W. 1992. Using National Wetland Inventory maps to quantify whooping crane stopover habitat in Oklahoma. Proc. North Am. Crane Workshop 6:62-68.

SWENK, M. 1933. The present status of the whooping crane. Nebr. Bird Rev. 1:111-129.

U.S. FISH AND WILDLIFE SERVICE. 1986. Whooping crane recovery plan. U.S. Fish Wildl. Serv., Albuquerque, N.M. 283pp.

WARD, J. P., and S. H. ANDERSON. 1987. Roost site use versus preference by two migrating whooping cranes. Pages 283-288 *in* J. C. Lewis, ed. Proc. 1985 crane workshop. Platte River Whooping Crane Maintenance Trust, Grand Island, Nebr.

WOODY VEGETATION EXPANSION AND CONTINUING DECLINES IN OPEN CHANNEL HABITAT ON THE PLATTE RIVER IN NEBRASKA

PAUL J. CURRIER, Platte River Whooping Crane Maintenance Trust, 2550 Diers Ave., Suite H, Grand Island, NE 68803, USA

Abstract: Aerial videography in 1988 and 1994 of the same selected segments of the Big Bend reach of the Platte River in central Nebraska was compared. During this 6-year period, channel area declined 4 to 41% (median 26%). In river segments where the channel had already narrowed significantly and at sites where channel habitat is actively managed declines were less (17–18%). These substantial changes occurred during a time that included a relatively low flow period (1990–92). Flows during the summer germination and establishment period (mid-May through August) for cottonwood (*Populus deltoides*) and willow (*Salix* spp.), the primary woody species, were less than 17 m^3/second 64% of the time during the period and less than 3 m^3/second 25% of the time in 1990 and 1991. Such low flows allowed extensive exposure of the riverbed where germination could occur. Peak scouring and ice flows were also low, remaining below 80 m^3/second throughout 1990–92 except for a few days in May 1991. Because the expanding vegetation recorded in this study had grown to an advanced stage, it most likely became permanently established. Furthermore, high flows in the 340–450 m^3/second range, which occurred in June and July 1995, were effective in removing only a fraction of the newly developed growth.

PROC. NORTH AM. CRANE WORKSHOP 7:141-152

Key words: aerial videography, cottonwood, crane roosting, flow management, *Grus americana*, *Grus canadensis*, habitat declines, open channel, Platte River, *Populus deltoides*, *Salix* spp., sandhill crane, whooping crane, willow, woody vegetation.

Each spring, more than 400,000 sandhill cranes (*Grus canadensis*) stage on the Big Bend reach of the Platte River in central Nebraska for a month or more while en route to nesting grounds in Canada, Alaska, and Siberia. At night the cranes roost in shallow, open channels of the river (Sidle et al. 1993). Daytime hours are spent feeding on waste grain in croplands, and resting and foraging in wet meadows near the river channel. Endangered whooping cranes (*G. americana*) also migrate through the central Platte each spring and fall, but they normally stay for only 1 or 2 days. Critical habitat has been designated for the whooping crane along the Platte between Lexington and Shelton (43 Federal Register 94:20938-20942, 1978). Open channel roost habitat for sandhill cranes and whooping cranes on the Platte River has been declining (Currier et al. 1985, Sidle et al. 1989, McDonald and Sidle 1992, Faanes and LeValley 1993) for decades. As peak and mean annual flows have been reduced, the channel has narrowed and become vegetated with woodland and herbaceous islands. Diversions for irrigation and power generation have been primarily responsible for the reduced flows that allowed woodland expansion and island development in the river channel (Williams 1978, Eschner et al. 1981, Currier 1982, Currier et al. 1985, Peake et al. 1985, Johnson 1994). Flow depletions, particularly reductions in peak or pulse flows, have effectively narrowed the streambed and allowed woody vegetation development on former active channels.

Sidle et al. (1989) concluded that 80% or more of the 1860's (pre-development) floodplain channel area had filled with trees along extensive reaches of the Platte and North Platte rivers. Cranes had virtually abandoned some river segments, particularly between North Platte and the J-2 return near Overton (Currier et al. 1985, Faanes and LeValley 1993). Substantial channel narrowing had also occurred in the downstream habitat reaches between Overton and Chapman (Currier et al. 1985, Sidle et al. 1989, Johnson 1994). However, the remaining open-channel areas within this reach (particularly between Kearney and Chapman) continued to provide important migratory habitat for cranes (U.S. Fish and Wildlife Service 1981, Krapu et al. 1982, Krapu et al. 1984, Currier et al. 1985) (Fig. 1). Open channels and mid-river sandbars also provide habitat for other migratory birds, including summer-nesting least terns (*Sterna antillarum*) and piping plovers (*Charadrius melodus*), waterfowl, shorebirds, bald eagles (*Haliaeetus leucocephalus*), and American white pelicans (*Pelecanus erythrorhynchus*) (Currier et al. 1985, Ziewitz et al. 1992).

Faanes and LeValley (1993) documented a significant shift in sandhill crane populations in response to these habitat changes. In 1957, 60% of the crane population was centered in the Lexington to Kearney reach, but by the late 1980's only 5% of the population was using that area. Downstream in the Kearney to Chapman segment, however, crane use increased from 9% to 80% of the population between 1957 and 1989. Shifts in whooping crane use of the river also followed these patterns (Faanes and LeValley 1993). The highest concentrations of sandhill cranes on the river are now between Wood River and Chapman (708 cranes/km on average), where the widest unobstructed channels remain (mean maximum width of 216 m) and there are abundant wet meadows (108 ha/river km). A combination of wet meadow loss and declines in open river channels has led to

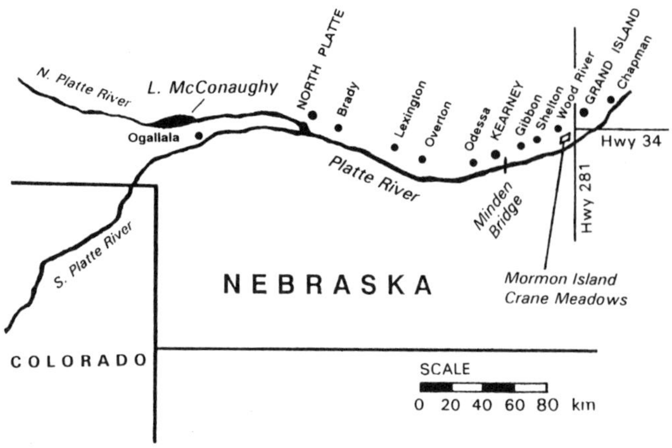

Fig. 1. Big Bend reach of the Platte River in central Nebraska. River segments examined in the study were near Gibbon and Shelton, and in a 24-km reach between Wood River and Chapman.

these shifts in crane distribution, and suitable roosting habitat is currently limited on the river (Faanes and LeValley 1993, Sidle et al. 1993).

Channel Stability and Narrowing

Declines in open channel habitat and shifts in crane populations have long been recognized; however, there has been debate about the specific causes of woodland expansion, and whether such expansion has reached a "steady-state" or "quasi-equilibrium" (Lyons and Randle 1988, CNPPID and NPPD 1990, Currier et al. 1990, Johnson 1994). In such an equilibrium, sediment inputs and outputs in a given river reach would be nearly equal, and overall channel widths would be expected to remain relatively stable in a condition of self-regulation.

Channel narrowing appears to have reached such an equilibrium in the upper stretches of the river between Brady and Overton (Lyons and Randle 1988). In the Big Bend reach between Overton and Grand Island, however, even though sand loads appeared to be in a quasi-equilibrium, width adjustments continued through 1983, and "future adjustments in channel width" were thought possible in the downstream portion of this reach (Lyons and Randle 1988). Additional narrowing was predicted where the widest channels remain. Although CNPPID and NPPD (1990) and Johnson (1994) agreed that extensive channel narrowing occurred as a result of water development, they suggested that the last major episodes of narrowing occurred in the 1960's. In the near term, they predicted that channel and woodland area would remain in a steady state. Currier et al. (1990) provided an alternative view that substantial channel narrowing was continuing in some reaches, noting that CNPPID's and NPPD's own studies (CNPPID and NPPD 1990) indicated considerable woody seedling recruitment and survival had occurred in recent years. Furthermore, the studies by CNPPID and NPPD (1990) and Johnson (1994) excluded a 48-km downstream reach of the river (Wood River to Chapman) which had wide channels and was used extensively by roosting cranes.

Cottonwood and Willow Establishment

Cottonwood and willow trees establish on the Platte and in other riverine systems under a narrow set of environmental conditions (Moss 1938; McVaugh 1947; Kapustka 1972; McLeod and McPherson 1973; Currier 1982, 1986). Weather and flow patterns play a major role in establishment because they affect the length of the growing season, the timing of seed release, the distribution of seeds, and the availability of suitable germination and establishment sites. Along the Platte, the short-lived seeds of cottonwood and willow are released, germinate, and establish seedlings in mid-May through July, mostly on the exposed sandy riverbed as spring flows recede. When seeds land or are washed onto exposed, generally unvegetated sandbars, germination can occur followed by seedling establishment if the site remains moist for 2–3 weeks (Currier 1982, O'Brien and Currier 1987, Johnson 1994).

Cottonwood and willow are pioneering species that are well adapted to the variability of the environment on floodplains. They produce huge numbers of seeds (up to 650,000 per tree, Kapustka 1972), time their seed dispersal to coincide with the availability of suitable conditions for establishment, have high germinability (80–90%), and grow very rapidly. These species can reproduce vegetatively (particularly after physical damage), and their seedlings are tolerant of flooding, drought, sedimentation, soil saturation, and low soil fertility (Currier 1982, Johnson 1994). During their first year, mid-summer fluctuating flows can bury or scour seedlings from the riverbed, and drought can eliminate seedlings through desiccation. Generally, the earlier in the season the Platte riverbed is exposed, and the longer the riverbed remains at low flows, the greater the potential for extensive seedling germination, survival, and establishment (Currier 1982, Johnson 1994). Earlier riverbed exposure allows seedlings to grow large enough to resist moderate flow fluctuations and droughts. Cooler, wetter growing seasons and limited mid-summer flow fluctuations provide the best conditions for survival. However, subsequent high flow events during winter and late spring can remove substantial numbers of seedlings (Johnson 1994). Beyond the first year, seedlings become more and more resistant to removal by flow events (Currier et al. 1990, Johnson 1994).

During a high water period in the 1970's and mid-1980's, vegetation expansion slowed on the river (Currier et al. 1990, McDonald and Sidle 1992, Johnson 1994). However, widespread woody seedling establishment was noted during a drought in the late 1980's and early 1990's, particularly on the main channel of the river in the Wood River to Chapman reach (Currier et al. 1990). The extent of this open-channel vegetation expansion was examined in this paper for the period 1988–94 in selected segments of the river channel. Hydrologic conditions under which these changes occurred and the potential for scouring flows and mechanical clearing to remove these newly vegetated sites was investigated.

The author appreciates the generous contributions to this study by J. Sidle, who piloted the flights, and the U.S. Fish and Wildlife Service, Grand Island, Nebraska, for providing the camera and computer equipment needed to acquire and print the aerial video images. Thanks also to J. Sidle, G. Lingle, D. Carlson, B. Goldowitz, J. O'Brien, and J. Echeverria for their critical reviews of this manuscript. B. Goldowitz and B. Harvey provided valuable assistance with statistical analyses.

METHODS

Aerial videography taken on 5 July 1988 and 10 June 1994 was used to compare woodland development and island stabilization on the Platte River channel by methodologies described in Sidle and Ziewitz (1990). Depending upon the camera lens in use, videos were flown at 500–800 m above the ground surface. The videos from both years had approximately the same ground coverage. Although river discharge was not identical during the 2 periods, flows were similar enough and low enough that major islands and well established vegetation (cottonwoods and willows at heights of 1.5–3.0 m) were clearly distinguishable during both 1988 and 1994. Flows at Overton were 27 m^3/second and 25 m^3/second in 1988 and 1994, respectively, when the videography was taken, while flows at Grand Island were 49 m^3/second and 38 m^3/second in the 2 years. Videography was taken in both years following relatively high spring flows which precluded the presence of widespread annual vegetation during the sampling periods.

Differences in perennial vegetation and channel area were noted by comparing the percentages of these 2 cover types in the 2 years. Percentage channel area changes were statistically analyzed with a 2-tailed *t*-test to compare differences in mechanically cleared and uncleared sites and to examine area changes in relation to the extent of channel narrowing that had occurred prior to 1988. River flows during the 1988 to 1994 period were examined to investigate potential causes for vegetation expansion. Changes in channel width were assessed in addition to channel area because cranes generally avoid roosting near islands or banks with vegetation a meter or more in height (U.S. Fish and Wildlife 1981, Krapu et al. 1984). As a measure of changes in view from a crane roost, the width of the widest unobstructed channel was measured on the video images along 20 cross-river transects at each site (i.e., the single, widest channel bounded by banks or vegetated islands >1 m in height was measured on each transect). An index of unobstructed width was constructed by expressing the width of these widest channels as a percentage of the total open channel width present in 1988 (i.e., the sum of all channels on a transect). Percentage widths were averaged for each site and compared between the 2 time periods.

Study Segments

Eleven 0.8-km river segments, located between Kearney and Chapman, were examined. Two sites were selected near Gibbon and Shelton, but the majority were located in a downstream, 24-km reach of the river from west of Highway 281 (south of Grand Island) to the Highway 34 bridge east of Grand Island. Study locations were chosen primarily because of their high use by roosting cranes, a lack of disturbance, and the availability of comparable video images from the 2 time periods. The Trust and other groups have attempted to physically maintain open channel habitat in a number of areas including Audubon's Rowe Sanctuary (Minden to Gibbon bridges) and at Mormon Island Crane Meadows (upstream of the Highway 281 bridge). Sampling was avoided in these river reaches because of the disturbance from clearing, but 2 sites near Mormon Island (river miles 177 and 175) were chosen in order to examine the difference between "natural" vegetation and island development and progression on a managed site. Heavily vegetated river segments were also avoided because they had already extensively narrowed and, in some cases, cranes had abandoned or reduced their use of such areas (Faanes and LeValley 1993). A segment near Shelton (river mile 196) and another near Highway 34 (river mile 169), where substantial channel narrowing had already taken place, were included in the analysis.

RESULTS

Substantial island and woody vegetation expansion occurred on the Platte between 1988 and 1994 (Figs. 2–4). Newly formed islands were generally colonized by trees in the same year; consequently, by 1994 the majority (≥90%) were dominated by even-aged stands of 1.5- to 3.0-m-tall cottonwoods and willows (Fig. 2). Although some woody

Fig. 2. Recent (1988–94) island development in the Platte River near Grand Island, Nebraska (river miles 173 and 172). Photos (January 1995) are upstream (top) and downstream (bottom) of the South Locust bridge.

seedlings most likely developed annually between 1988 and the spring of 1994, many of these seedlings did not survive over the long-term (Johnson 1994). A major episode of successful recruitment, however, was noted during the summer of 1990 (Currier et al. 1990), and to a lesser extent in the summer of 1991 (P. Currier, pers. observ.). These seedlings were able to survive the summer and the following winter, stabilized exposed sandbars, and eventually allowed new islands to develop.

None of the channels examined increased in channel area or width between 1988 and 1994. The percentage cover of open, active channels, and vegetated islands and banks in the 2 sampling periods is summarized in Table 1. In comparison to 1988, open channel declines generally ranged from 16% to 41% in the segments studied, with a median decline of

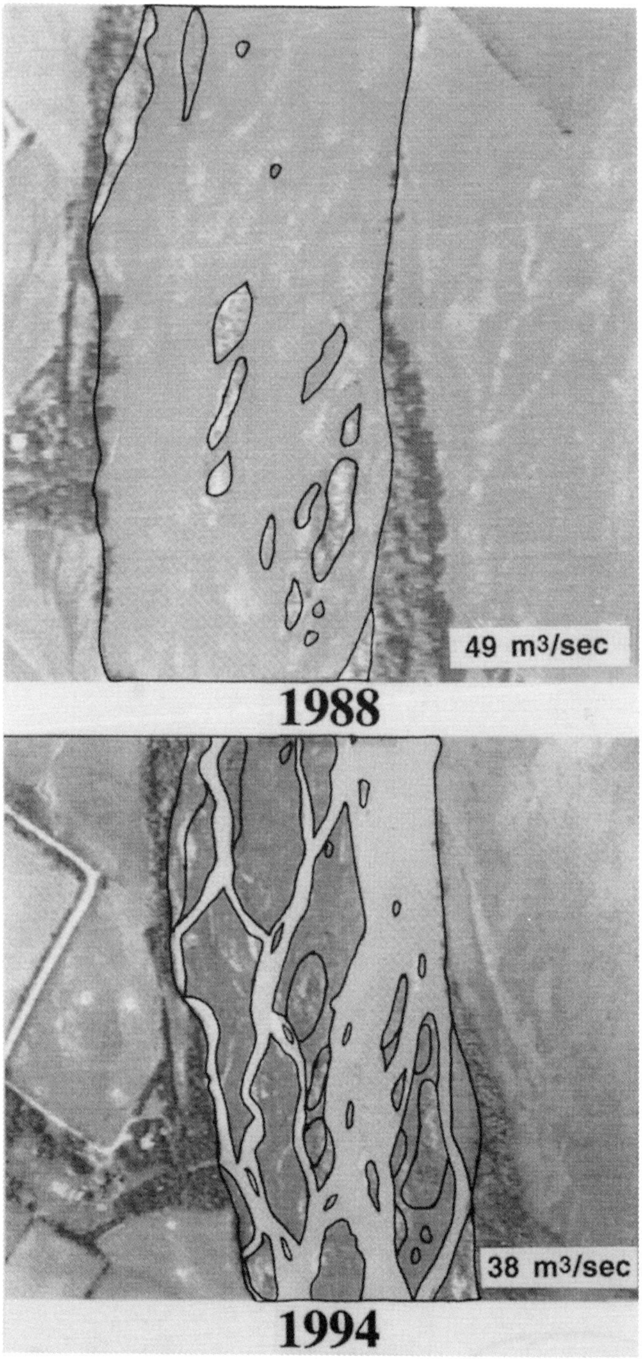

Fig. 3. Aerial videography comparisons of the river channel at river mile 171.5, 1.6 km below I-80 Bridge, near Grand Island, Nebraska. Open channel area declined by 41% in this segment during the 1988–94 period. Island subdivisions marked on the 1994 image represent portions of the channel that were vegetated in 1988. Flows measured at Grand Island.

26%. One exception was the most downstream segment near the Highway 34 Bridge (river mile 168), which showed a

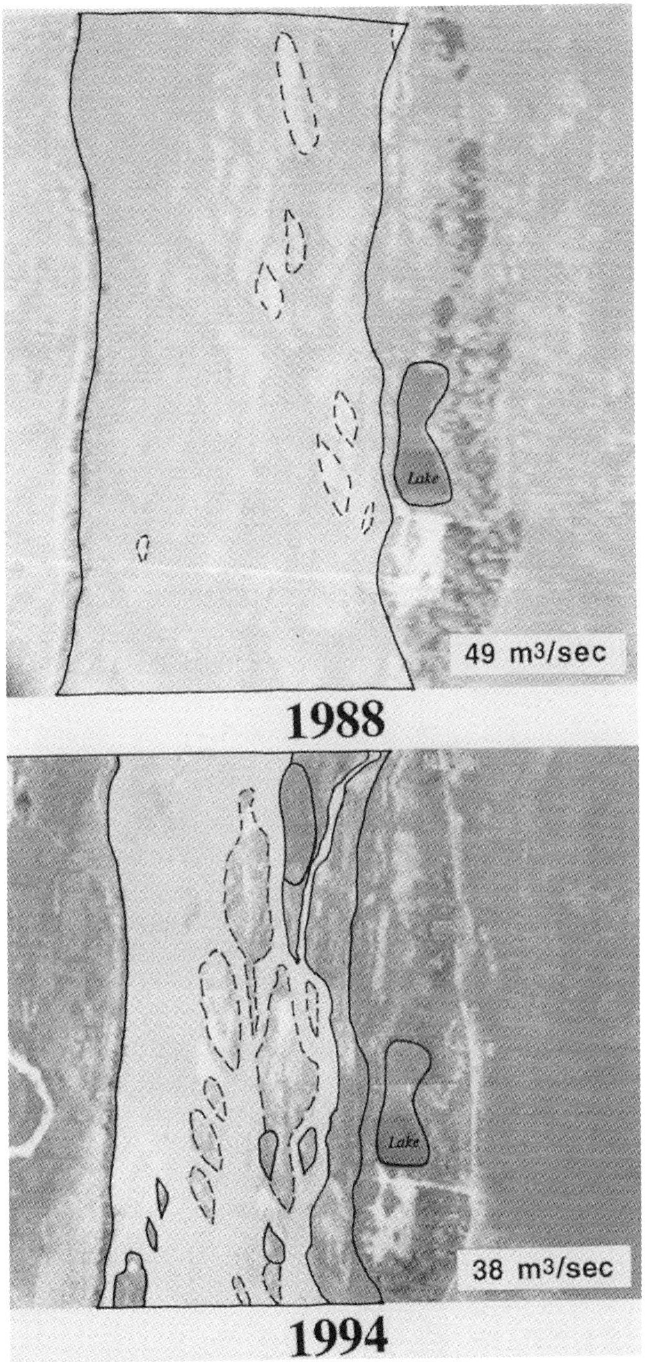

Fig. 4. Aerial videography at Mormon Island Crane Meadows, Nebraska, river mile 177. Dashed lines indicate cleared islands remaining in part of the active channel; solid lines indicate stabilized areas. Channel area declined 17.7% between 1988 and 1994. Flows measured at Grand Island.

relatively small decline of only 4%.

In the segments located 1.6 km above Highway 34 (river mile 169) and near Shelton (river mile 196.7), a substantial portion of the river channel (28% and 30%, respectively) had already narrowed prior to 1988. The 16.8% mean decline in channel area in these segments (river miles 169 and 196.7, $n = 2$) was significantly lower than in the other segments ($\bar{x} = 27.6\%$, $n = 9$) ($t = -2.911$, $P = 0.0195$).

At sites where the channel had been managed by clearing woody growth (river miles 175 and 177), declines in channel area were also lower, but still averaged 19.6% ($n = 2$) and were not statistically different from the remaining unmanaged sites ($\bar{x} = 29.0\%$, $n = 6$) ($t = -1.609$, $P = 0.1588$) (Table 1). Three sites were omitted from the analysis because of the artificial constriction of the channel at mile 168 and substantial prior narrowing in segments 169 and 196.7. At Mormon Island Crane Meadows (river mile 177), the entire width of the channel was cleared and disked on a nearly annual basis between 1982 and 1995. The majority of channel loss at the site was due to island extension along the northern riverbank (Fig. 4). Potential channel losses without clearing were estimated at 34-36%, or about 15% greater than with clearing (Table 1).

Maximum unobstructed channel width (i.e., visibility across the channel) also declined in most segments as vegetation expanded (Table 2). An index of unobstructed width, a measure of the quality of habitat for roosting cranes, was expressed as a percentage of the total channel width in 1988. Except for the Highway 34 Bridge segment (river mile 168), where channel losses were minimal, and the Shelton segment (river mile 196.7), where substantial narrowing had previously occurred, unobstructed width declines were greater than channel area losses. Even though the extent of woody island development was in some cases relatively small, the presence of woody vegetation on islands in the middle of wide-channel areas significantly reduced visibility from the channel. In the Gibbon reach (river mile 201), for example, channel losses were only half (26%) of the 51% decline in unobstructed width. In much narrower reaches, such as Shelton, unobstructed width declines were actually less than overall channel area declines (a loss of 6% compared to 17%).

DISCUSSION

Vegetation Expansion

The channel changes observed between 1988 and 1994 varied depending upon segment location, river management, and the extent of previous channel narrowing. The smallest change, a 4% decline in channel area at the Highway 34 Bridge (river mile 168) occurred in an area where the channel was constricted by concrete rip-rap along the banks. In areas where mechanical clearing was used, channel

Table 1. Percentage of open channel and vegetated island habitat area in selected segments of the Platte River between Kearney and Grand Island, Nebraska, 1988 and 1994. Each segment represents approximately 0.8 km of river channel.

Segment	Open channel		Vegetated islands		% change in open channel 1988-94
	1988	1994	1988	1994	
Highway 34 Bridge (mile 168)	80.6	76.5	19.4	23.5	-4.1
1.6 km above Highway 34 (mile 169)	72.5	56.1	27.5	43.9	-16.4[a]
1.6 km below I-80 (mile 171.5)	91.4	50.1	8.6	49.9	-41.3
I-80 (mile 172.6)	95.5	72.1	4.5	27.9	-23.4
South Locust St. (mile 173)	98.7	71.8	1.3	28.2	-26.9
Hannon's Bridge (mile 174.3)	96.4	60.9	3.6	39.1	-35.5
Highway 281 (mile 175.3)	95.8	74.4[b]	4.2	25.6	-21.4[c]
(without clearing)		(59.4)	(40.6)		(-36.4)
Mormon Island (mile 177)	97.0	79.3[b]	3.0	20.7	-17.7[c]
(without clearing)		(63.4)	(36.6)		(-33.6)
1.6 mile above Shelton (mile 196.7)	70.1	52.9	29.9	47.1[d]	-17.2[a]
1.6 km below Gibbon (mile 201)	97.5	71.8	2.5	28.2	-25.7
Kearney Hike/Bike Trail (mile 209.8)	85.9	64.8	14.1	35.2	-21.1

[a] Segments with substantial narrowing prior to 1988.
[b] Includes cleared areas remaining in active channel.
[c] Channel changes adjusted for clearing.
[d] Includes 0.2% loss (erosion) of islands.

narrowing was also less severe and averaged about 15% less than at uncleared sites. Clearing and disking destabilized many of the islands at Mormon Island (river mile 177), but a number of permanent islands still formed at this site during the 1988-94 period (Fig. 4).

At Shelton and 1.6 km above Highway 34, much of the channel had already narrowed prior to 1988. Additional channel losses averaged 17% in these segments and were statistically lower than the declines at the remaining sites (Table 1). Because large areas of the channel had already filled with vegetated islands, fewer areas were available for additional recruitment. Island extension and attachment to neighboring islands (Eschner et al. 1981) were responsible for much of the additional decline. In more open segments, the majority of channel narrowing occurred as new mid-channel islands (Fig. 3).

The growth forms of cottonwood and willow make them particularly adapted to the gradual sedimentation process that occurs when river islands build. With each layer of sediment deposition, these species are able to form adventitious roots and continue to grow, effectively trapping sediment and stabilizing islands. As flows over-top the islands, additional sediment is deposited, aided by the increased surface roughness of the vegetation (Hupp 1992). Without moderate scouring flows to remove young seedlings and islands, the sedimentation process eventually leads to the development of permanent river features.

The expansion of woody vegetation during the 1988-94 period was primarily the result of low summer flows and relatively low pulse flows during the following winters and springs (Fig. 5). Low summer flows exposed extensive areas of the riverbed where successful recruitment could occur, while subsequent winter and spring flows were insufficient to provide the scouring, ice-flows, or riverbed "restructuring" needed to remove the previous summer's seedlings. From 1988 to 1994, flows during the cottonwood and willow seedling germination and establishment period (mid-May through August) were less than 17 m^3/second 64% of the time. In 1990 and 1991, summer flows were even lower, with discharges less than 3 m^3/second occurring 25% of the time.

To illustrate the potential for recruitment in high and low flow periods, I compared flows during the seedling germination period, 1983-87, when little recruitment was observed, and 1988-92, when more than 50% of the active riverbed was covered with new seedlings (Currier et al. 1990, Johnson 1994, P. Currier, unpubl. data). During the germination period, there were more than 3 times as many days in 1988-92 (436 days) when flows were less than 23 m^3/second than in the 1983-87 period (130 days). At flows of 0-11 m^3/second, the difference was even greater, more than 8 times the number of low flow days occurring in the 1988-92 period. Increased exposure of the riverbed in 1988-92 provided ample opportunity for seedling germination and establishment.

The large recruitment of seedlings in 1990 and 1991

Table 2. Index of unobstructed channel width in selected segments of the Platte River, Nebraska, in 1988 and 1994. Numbers represent the maximum unobstructed width of the river as a percentage of total channel width in 1988. Measurements from 20 transects at each river segment were averaged to arrive at the index values.

Segment	Percent $\bar{x}$ max. unobstructed channel width 1988	Percent $\bar{x}$ max. unobstructed channel width 1994	% change in open width (1988-94)
Highway 34 Bridge (mile 168)	65	65	0
1.6 km above Highway 34 (mile 169)	56	39	-17
1.6 km below I-80 (mile 171.5)	77	33	-44
I-80 (mile 172.6)	94	51	-43
South Locust St. (mile 173)	96	59	-37
Hannon's Bridge (mile 174.3)	82	35	-47
Highway 281 (mile 175.3)	91[a]	53	-38
Mormon Island (mile 177)	100[a]	76	-24
1.6 km above Shelton (mile 196.7)	28	22[b]	-6
1.6 km below Gibbon (mile 201)	90	39	-51
Kearney Hike/Bike Trail (mile 209.8)	63	39	-24

[a] Includes cleared areas remaining in active channel.
[b] Includes 0.2% loss (erosion) of islands.

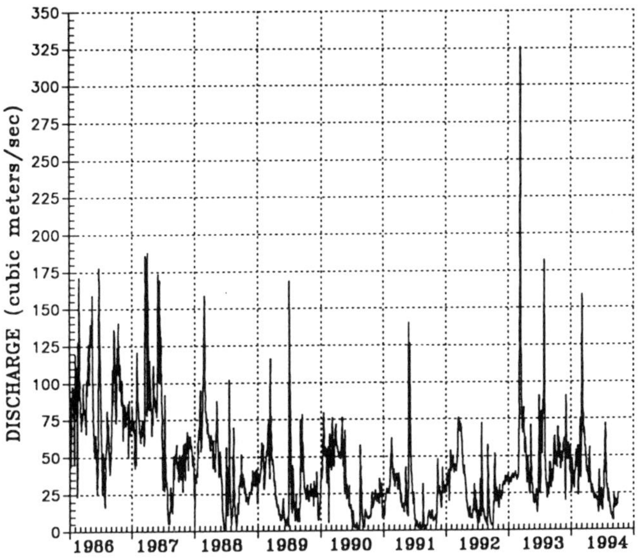

Fig. 5. Hydrograph of the Platte River at Grand Island, Nebraska, 1986-94. Extensive woody seedling recruitment and establishment occurred in mid-summer 1990 and 1991 during low flows. Higher flows in 1993-94 were not sufficient to remove much of the new vegetative growth.

(Currier 1990, pers. observ.), combined with reduced winter and spring mortality, most likely accounted for the majority of the vegetation expansion observed in this study. Throughout the 1990-92 period when widespread seedling recruitment was occurring, winter ice-flows and spring scouring flows remained below 79 m^3/second, except for a short period in May 1991 when flows reached nearly 142 m^3/second. As a result, little removal of seedlings occurred during the winter and spring of these years, a period when seedling mortality is normally relatively high (Johnson 1994).

In contrast, during 1988, 1989, and 1992, much of the riverbed was exposed quite early (April-May), providing an opportunity for widespread recruitment, but mid-summer fluctuating flows most likely eliminated many seedlings (Johnson 1994). In addition, spring flows of 221 to 326 m^3/second in 1993, resulted in an estimated loss of 20-25% of the seedlings and young trees that had developed during the previous 3 years in the Wood River to Chapman reach (P. Currier, pers. observ.). However, 75% or more of the new growth remained. As time has advanced, these remaining woody seedlings have grown, further stabilizing islands, and making it less likely that they will be removed except by exceptionally high flow events (Currier et al. 1990, Johnson 1994).

This interpretation of vegetation expansion is consistent with detailed woody seedling demography studies conducted by Johnson (1994) on various segments of the Platte. He observed increased recruitment and survival of tree seedlings during the 1988-92 drought, and concluded that this had led to the establishment of relatively permanent vegetation in areas of the channel that had been unvegetated in 1985-88. Johnson also found an inverse relationship between mean June flow and seedling recruitment, suggesting that extensive seedling development can occur whenever mean June flows are less than 79 m^3/second. Low flows below 79 m^3/second provided ample areas for colonization, and also allowed for rapid seedling growth and potentially higher seedling survival.

High summer flows, on the other hand, covered the unvegetated riverbed, and generally precluded seedling recruitment or were responsible for eroding young germinants (Johnson 1994). Of course, the extent of riverbed exposure is a function of the low flows that are averaged in the June mean. The pattern of June flows (i.e., when peaks occur and how rapidly they decline) is an important determinant of recruitment as well. Johnson (1994) also identified a combination of winter ice cover and relatively high base flows as a seedling removal mechanism. Highest seedling survival occurred when flows were less than 75-85 m^3/second. Likewise, summertime peak flows of 125-225 m^3/second were quite effective in scouring young seedlings, but they were generally ineffective in removing 1-year-old or older trees. Finally, high spring peaks (170-225 m^3/sec) were also found effective in removing the previous year's seedlings, but flows in this range do not occur with great regularity (Johnson 1994).

Table 3 shows the June and July means, and winter-spring peak flows that occurred in 1988-94. Woody tree seedling response in these years was generally consistent with the flow ranges identified by Johnson (1994). Based on Johnson's (1994) criteria, seedling recruitment should have been high in all the years between 1988 and 1993, as June mean flows were less than 79 m^3/second. However, at the end of June in 1989, flows were quite high (128 m^3/sec average during the last 5 days of the month), and undoubtedly most seedling establishment was precluded. In addition, mean July flows may also be an indictor of widespread seedling recruitment and establishment. For instance, during 1990 and 1991, low mean flows in July signaled an extended opportunity for recruitment in 1990 and 1991 (longest extended exposure of the riverbed). Mean July flows in other years were considerably below the 125–225 m^3/second level Johnson (1994) identified for mid-summer removal of young seedlings. These flows, however, were sufficient to substantially inundate the riverbed, limiting opportunities for seedling establishment and survival (e.g., flows of 22–81 m^3/sec, including nearly bank-to-bank inundation in 1993). Winter peak flows in 1988, 1989, 1992, and 1993 also exceeded Johnson's (1994) 75–85 m^3/second criteria associated with higher mortality. The lower winter flows in 1990 and 1991 (62 m^3/sec and 76 m^3/sec, respectively), on the other hand, limited extensive overwinter losses and allowed substantial seedling survival into the next growing season.

Table 3. Mean June, mean July, and following winter flows (maximum pulse or ice flows) at Grand Island, Nebraska, during the 1988-94 period when extensive woody vegetation expansion occurred in downstream reaches of the Platte River channel.

Seedling year	$\bar{x}$ June flow (m^3/sec)	$\bar{x}$ July flow (m^3/sec)	Winter max. flow (m^3/sec) (after Jan–Mar)
1988	15	32	116
1989	27[a]	34	79
1990	19[b]	4.1	62
1991	39[c]	3.7	76
1992	16[d]	22	325
1993	31	81	218

[a] 128 m^3/sec last 5 days.
[b] 14 m^3/sec last 15 days.
[c] 18 m^3/sec last 15 days.
[d] 14 m^3/sec last 15 days.

Long-term Trends in Channel Stability

The 16% to 41% channel reductions identified in this study represent a continuation of long-term trends in vegetation expansion and channel narrowing. The river appears to be continuing its upstream to downstream (west to east) progression from a wide, open-channel braided form towards a series of highly anabranched channels running between stabilized wooded islands (Eschner et al. 1981, O'Brien and Currier 1987, Sidle et al. 1989, Johnson 1994). Although the present analysis was primarily limited to the downstream portion of the Big Bend reach (Wood River to Chapman), it demonstrates that rapid vegetation expansion is continuing in the widest remaining channels of the river. Until recently, these braided segments had maintained 60% to 70% of their 1938 widths (Peake et al. 1985). Even in already highly vegetated and narrowed segments of the river channel, continued narrowing may occur, although most likely at a reduced rate.

The relatively large reductions in active channel between 1988 and 1994 support the view that the river has not yet reached a steady-state equilibrium. This is consistent with Lyons and Randle's (1988) findings that additional channel adjustments were possible in the Odessa to Grand Island reach. Although sediment loads in this reach appear to be in balance, channel morphology has most likely not adjusted completely, particularly in the downstream Wood River to Chapman reach. Furthermore, it is questionable whether upstream and downstream sediment loads are actually in balance. Because Lyons and Randle (1988) used the same discharge rating curves at both the upstream and downstream stations, their "balanced" sediment loads may simply reflect the rating curves, rather than an indication of the true sediment supply.

CNPPID and NPPD's (1990) and Johnson's (1994) conclusions that channel narrowing has essentially ceased and that the river is now in a stable-channel equilibrium are not supported by the channel changes identified in this study. Johnson (1994), as well, recognized additional permanent vegetation development on the river during the 1988–94 drought. Because the CNPPID and NPPD (1990) and Johnson (1994) studies were begun in the mid-1980's, they primarily reflect the limited channel narrowing that occurred during several high flow periods in the 1970's and 1980's. As a result, they concluded that major changes in channel adjustments had ceased. Not until a low flow drought occurred in the late 1980's and early 1990's did another major episode of vegetation expansion occur. Their studies also ignored the area of most rapid channel change (downstream of Wood River) highlighted in this paper.

CNPPID and NPPD (1990) and Johnson (1994) did predict that some additional changes in channel vegetation and narrowing would occur in the steady-state condition.

Under their model, the channel would tend to expand and contract in response to annual fluctuations in flows, but in the long run the prediction was that overall width of the channel would remain relatively steady. This interpretation has some validity in relation to river segments that have already narrowed significantly (e.g., channels in the upstream Brady-Cozad reach that average less than 25% of their historical width). Such channels have adjusted to the increased vegetation by maintaining a more U-shaped or canal-like cross-section. The steady-state theory does not seem to apply, however, to the remaining wide, flat, and braided channels in the river. The extent of recent vegetation expansion in these reaches, the compressed time period over which it occurred (a 2–3 year episode of narrowing), and the advanced stage of growth (some trees >5 years old) suggest that major permanent adjustments in channel width could still occur.

Some of the 1988–94 vegetative growth could eventually be removed by high flows. Flows in the spring of 1993, for example, were high enough (325 m^3/sec) to remove 20–25% of the 2- to 3-year-old seedlings that were present (P. Currier, pers. observ.). High flows on the North Platte River (McDonald and Sidle 1992) and on the Platte (Carlson 1987, Johnson 1994) have reportedly resulted in channel widening. Johnson (1994) noted that highly narrowed channels on the Platte (≤30% of historic width) increased by 4% to 8% as a result of high flows in 1983–84 (565 to 700 m^3/sec), but that widening on broader channels was limited to only 1–2%. A 10–15% increase in channel width was reported by Carlson (1987), however, for the same flow event. Based on these observations during 1983–84 and during a recent high flow in 1995, it seems unlikely that flows will eliminate all of the new 1988–94 vegetative growth. The infrequency of flood events (flows in 1983 were a 40-year high; those in 1995 were the highest in 12 years), and the extent and rapid growth of the newly established vegetation suggest that the long-term trend on the central Platte is still a net loss of open channel habitat.

As Faanes and LeValley (1993) found, many severely narrowed segments have already been abandoned by cranes because they no longer provide suitable roosting habitat. Although cranes prefer wide, shallow, braided channels (Sidle et al. 1993), they are also known to roost in narrower channels. On average, however, a minimum unobstructed channel width of 150 m has been identified for sandhill cranes (U.S. Fish and Wildlife Service 1981; Krapu et al. 1982, 1984; Currier et al. 1985), whereas whooping cranes strongly prefer much wider channels of >300 m (U.S. Bureau of Reclamation 1993). To protect the remaining crane habitat on the Platte, therefore, maintaining channel widths in the 150 to 300 m range is important.

Implications for Instream Flow Management

The data presented here suggest the need for both summer base flows and channel-forming pulse flows in maintaining the channel in the Big Bend reach. Because the Platte has a relatively flat, shallow channel, substrate exposure varies a great deal from high to low flow (Fig. 6). The channel depicted in Figure 6 represents a typical cross-channel profile based on Instream Flow Incremental Methodology (IFIM) data for Mormon Island (river mile 177) (U.S. Fish and Wildlife Service 1990). At high flows (57 m^3/sec to bankfull flows of 340 to 450 m^3/sec), ≥95% of the channel is underwater, leaving virtually no available sites for seedling establishment. But at moderate flows (23–28 m^3/sec), 25% to 34% of the riverbed is exposed, and at low flows (6–11 m^3/sec), 70–80% of the channel is exposed. Maintaining a base flow of 23–28 m^3/second in the channel during the seed germination and establishment period (mid-May through August) would therefore greatly limit opportunities for additional seedling recruitment on the remaining unvegetated riverbed. Summer base flows would also provide important ecological benefits for fish and aquatic organisms.

Some seedling establishment, however, will undoubtedly continue even with improved base flows. Seedlings and saplings that become established can be dislodged and removed through careful management of spring and early summer pulse flows. Bowman and Carlson (1994) identified the need to time such flows to coincide with historic fluctuations in February–March and May–June. At a minimum, these pulses should reach a magnitude of 75 to 225 m^3/second as identified by Johnson (1994). However, Bowman and Carlson (1994) and O'Brien (1995) defended the need for even higher flow events (340–450 m^3/sec) on a periodic basis (return period of 2.5–3 years) for seedling removal, channel maintenance, and a variety of other purposes, including groundwater recharge and nutrient redistribution. Flows in June and July 1995 of 340–450 m^3/second were quite effective in restructuring the bars in unvegetated areas of the channel and where clearing and disking had been done. However, because flows of this magnitude occur infrequently (the last occurred 12 years ago), much of the vegetative growth since 1988 (3- to 5-year-old saplings) was not removed (Fig. 7). Preliminary estimates suggest that 15–25% of the new growth was scoured, but that most of the islands vegetated since 1988 were simply "thinned" by this event.

Implications for Habitat Management

The channel changes noted in this paper will affect the long-term needs of sandhill cranes and whooping cranes that depend on open channel habitat. Such alterations could also

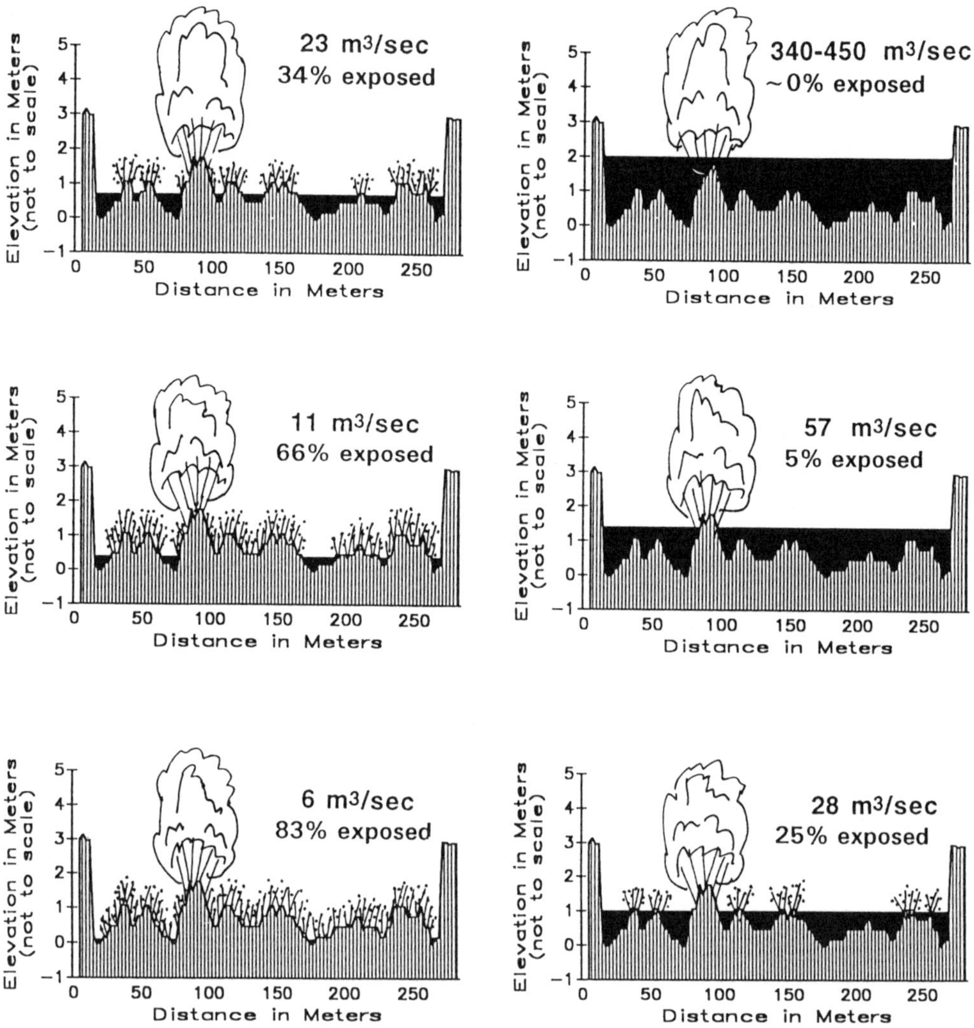

Fig. 6. Proportion of a typical 244-m riverbed cross-section that is exposed at various flows, Platte River, Nebraska. At flows less than 23 m^3/sec, extensive areas are exposed where cottonwood and willow seedling recruitment and establishment can occur.

change the distribution and abundance of waterfowl, shorebirds, least terns and piping plovers, and other species that use the Platte (Currier et al. 1985, Sidle et al. 1990, U.S. Department of Interior 1990). Over the past 10 years, the Platte River Whooping Crane Maintenance Trust, the National Audubon Society, and The Nature Conservancy have actively maintained approximately 27 km of open river channel by mechanical clearing, mowing, and disking on a nearly annual basis (Currier 1984, 1991). These treatments have been an expensive and difficult undertaking, but in conjunction with high pulse flows, they have been at least moderately successful in maintaining wide, open-channel areas on the river (Currier 1991, Currier et al. 1995). Although clearing destabilizes islands and river bars, it is the instream flows that flood and over-top them that are responsible for restructuring the riverbed and maintaining its braided character. Clearing is not a substitute for flows. Through a combination of pulse flows, base flows, and selective clearing, it is hoped that open-channel habitat can be maintained for the thousands of cranes, waterfowl, and other species that use the Platte.

SUMMARY

Analysis of aerial videography documented a trend towards continued narrowing of river channels in some of the remaining major crane roost areas in the Big Bend reach of the Platte. The most rapid expansion of vegetation occurred in a 24-km reach of the river between Wood River and Chapman, but declines in channel area were also observed in

Fig. 7. Woody vegetation development (1988-94) remaining after the 340-450 m^3/sec high water flows in June-July 1995, Platte River, Nebraska. Photos (October 1995) are at river mile 174 near Hannon's Bridge (top) and at river mile 175 near Highway 281 (bottom). In the bottom photo, the channel has been managed and cleared at the right, but uncleared on the left.

open-channel habitat near Minden and Gibbon, and in areas where clearing (Mormon Island Crane Meadows and near Highway 281) and substantial channel narrowing (near Shelton and Highway 34) have occurred. Extensive low flow events in the mid-summer germination and development period for cottonwood and willow, as well as reduced winter and spring peak flows, appear to have allowed vegetation expansion. This interpretation is consistent with the demography studies conducted by Johnson (1994), but the presence of substantial, permanently established vegetation during the 1988-94 period is counter to Johnson's and CNPPID and NPPD's (1990) conclusion that the river has reached a steady-state and that significant additional narrowing will not occur. Instead, the data presented here suggest that the river is not in equilibrium, and that further channel adjustments are possible under the existing water regime. Unless water management is undertaken to maintain fluctuating spring and summer pulse flows (75 to 225 m^3/sec annual minimum; 340 to 450 m^3/sec on a 2.5- to 3-year return period) and mid-summer base flows (23 to 28 m^3/sec range), channel narrowing trends will likely continue. Although physical management of riverbed vegetation and islands can help to sustain open sites for cranes and other species, this type of management alone is not sufficient to maintain the habitat. A combination of land and water management will be required to preserve valuable wildlife resources on the Platte.

LITERATURE CITED

BOWMAN, D., and D. CARLSON. 1994. Pulse flow requirements for the central Platte River. Unpubl. Rep., U.S. Fish Wildl. Serv., Grand Island, Nebr. 11pp.

CARLSON, M. 1987. Channel scouring capabilities of high flows in the Platte River. Unpubl. Rep., U.S. Fish Wildl. Serv., Grand Island, Nebr. 12pp.

CNPPID and NPPD (Central Nebraska Public Power and Irrigation District and Nebraska Public Power District). 1990. Joint response to Federal Energy Regulatory Commission, deficiency notice, 7 December 1984. Washington, D.C. 225pp.

CURRIER, P. J. 1982. The floodplain vegetation of the Platte River: phytosociology, forest development, and seedling establishment. Ph.D. Thesis, Iowa State Univ., Ames. 332pp.

______. 1984. Woody vegetation clearing on the Platte River aids restoration of sandhill crane roosting habitat (Nebraska). Restor. Manage. Notes 2:38.

______. 1986. Drought tolerance and scouring of riverbed cottonwood (*Populus deltoides* Marsh.) seedlings on the Loup River in Nebraska. Unpubl. Rep., Platte River Whooping Crane Maintenance Trust, Grand Island, Nebr. 20pp.

______. 1991. Reclamation of crane roosting habitat on the Platte River and restoration of riverine wetlands. Pages 403-407 *in* J. Harris, ed. Proc. 1987 int. crane workshop. Int. Crane Found., Baraboo, Wis.

______, B. S. GOLDOWITZ, G. R. LINGLE, and J. G. VANDERWALKER. 1990. Preliminary comments on the deficiency response of the CNPPID and NPPD for Project Nos. 1417 and 1835. Platte River Whooping Crane Maintenance Trust, Grand Island, Nebr. 122pp.

______, B. S. GOLDOWITZ, T. A. EMERTON, and G. R. LINGLE. 1995. Comments on the Nebraska Public Power District's May 9, 1995 comments on recent agency filings. Platte River Whooping Crane Maintenance Trust, Grand Island, Nebr. 67pp.

______, G. R. LINGLE, and J. G. VANDERWALKER. 1985. Migratory bird habitat on the Platte and North Platte Rivers in Nebraska. Platte River Whooping Crane Maintenance Trust, Grand Island, Nebr. 177pp.

ESCHNER, T., R. HADLEY, and K. CROWLEY. 1981.

Hydrologic and morphologic changes in the Platte River Basin: a historical perspective. Unpubl. Rep. 81-1125, U.S. Geol. Surv., Denver, Colo. 57pp.

FAANES, C. A., and M. J. LEVALLEY. 1993. Is the distribution of sandhill cranes on the Platte River changing? Great Plains Res. (August):297-304.

HUPP, C. R. 1992. Stream-grade variation and riparian forest ecology along Passage Creek, Virginia. Bull. Torr. Bot. Club 109:488-499.

JOHNSON, W. C. 1994. Woodland expansion in the Platte River, Nebraska: patterns and causes. Ecol. Monogr. 64:45-84.

KAPUSTKA, L. A. 1972. Germination and establishment of *Populus deltoides* in eastern Nebraska. M.S. Thesis, Univ. Nebraska, Lincoln. 72pp.

KRAPU, G. L., K. J. REINECKE, and C. R. FRITH. 1982. Sandhill cranes and the Platte River. Trans. North Am. Wildl. Nat. Resour. Conf. 47:542-552.

______, D. E. FACEY, E. K. FRITZELL, and D. H. JOHNSON. 1984. Habitat use by migrant sandhill cranes in Nebraska. J. Wildl. Manage. 48:407-14.

LYONS, J., and T. RANDLE. 1988. Platte River channel characteristics in the Big Bend Reach–Prairie Bend project. Unpubl. Rep., U.S. Bur. Reclamation, Sedimentation and River Hydraulics Sect., Denver, Colo. 28pp.

MCDONALD, P., and J. SIDLE. 1992. Habitat changes above and below water projects on the North Platte and South Platte Rivers in Nebraska. Prairie Nat. 24:149-158.

MCLEOD, K. W., and J. J. MCPHERSON. 1973. Factors limiting the distribution of *Salix nigra*. Bull. Torr. Bot. Club. 100:102-110.

MCVAUGH, R. 1947. Establishment of vegetation on sandflats along the Hudson River, New York. Ecology 28:189-194.

MOSS, E. H. 1938. Longevity of seed and establishment of seedlings in species of *Populus*. Bot. Gaz. 99:529-542.

O'BRIEN, J. S., and P. J. CURRIER. 1987. Platte River channel morphology and riparian vegetation changes in the Big Bend Reach and minimum streamflow criteria for channel maintenance. Unpubl. Rep., Platte River Whooping Crane Maintenance Trust, Grand Island, Nebr. 47pp.

______. 1995. Comments regarding pulse flow recommendations and channel morphology of the central Platte in response to the Nebraska Public Power District, May 9, 1995, filing on new evidence to the FERC. FLO Engineering, Breckenridge, Colo. 17pp.

PEAKE, J., M. PETERSON, and M. LAUSTRUP. 1985. Interpretation of vegetation encroachment and flow relationships in the Platte River by use of remote sensing techniques. Unpubl. Rep. to Nebr. Water Resour. Cent., Remote Sensing Applications Lab., Univ. Nebraska, Omaha. 36pp.

SIDLE, J. G., and J. W. ZIEWITZ. 1990. Use of aerial videography in wildlife habitat studies. Wildl. Soc. Bull. 18:56-62.

______, E. D. MILLER, and P. J. CURRIER. 1989. Changing habitats in the Platte River valley of Nebraska. Prairie Nat. 21:91-104.

______, C. A. FAANES, and W. G. JOBMAN. 1990. Occurrence of American white pelicans along the Platte River, Nebraska. Prairie Nat. 22:165-170.

______, H. G. NAGEL, R. CLARK, C. GILBERT, D. STUART, K. WILLBURN, and M. ORR. 1993. Aerial thermal infrared imaging of sandhill cranes on the Platte River, Nebraska. Remote Sens. Environ. 43:333-341.

U.S. BUREAU OF RECLAMATION. 1993. Platte River habitat conservation plan, draft report of the management alternatives workgroup, Platte River management joint study. U.S. Bur. Reclamation, Mills, Wyo. 131pp.

U.S. DEPARTMENT OF THE INTERIOR. 1990. Comments on the 1990 applications of CNPPID and NPPD and 10J recommendations for projects nos. 1417 and 1835 before the Federal Energy Regulatory Commission (November 1990). U.S. Solicitor's Off., Dep. Interior, Denver, Colo. 60pp.

U.S. FISH AND WILDLIFE SERVICE. 1981. Platte River ecology study. Unpubl. Rep., Northern Prairie Wildl. Res. Cent., Jamestown, N.D. 186pp.

______. 1990. Final report of the biology work group, Platte River management joint study. U.S. Fish Wildl. Serv., Denver, Colo. 131pp.

WILLIAMS, G. P. 1978. The case of the shrinking channels–North Platte and Platte Rivers in Nebraska. U.S. Geol. Surv. Circ. 781. 48pp.

ZIEWITZ, J. W., J. G. SIDLE, and J. J. DINAN. 1992. Habitat conservation for nesting least terns and piping plovers along the Platte River, Nebraska. Prairie Nat. 24:1-20.

HARTS BASIN, COLORADO: A SPRING MIGRATION STOPOVER AREA FOR ROCKY MOUNTAIN SANDHILL CRANES AND WHOOPING CRANES

MELVIN K. PETERSON, 1348 Highway 65, Eckert, CO 81418, USA
RODERICK C. DREWIEN, Hornocker Wildlife Research Institute, University of Idaho, P.O. Box 3246, Moscow, ID 83843, USA

Abstract: Greater sandhill cranes (*Grus canadensis tabida*) from the Rocky Mountain Population (RMP) and whooping cranes (*G. americana*) reared by RMP foster parents were monitored during 1980–95 at a traditional spring migration stopover at Harts Basin in western Colorado. Counts during March and April 1986–93 showed that a mean of 10,656 RMP sandhill cranes stopped annually for 1 or more nights en route from their spring staging area in the San Luis Valley, Colorado, to summer grounds. More than 96% stopped between 11 March and 10 April. Count data indicated that >50% of the RMP stopped annually at Harts Basin during spring migration. Ninety-three whooping crane sightings were recorded during 16 years ($\bar{x}$ = 5.8, annual range 2–12) with peak migrations (77.4%) during 1–10 April. Most whooping cranes (88.2%) migrated in April through Harts Basin during the latter half of the sandhill crane migration period. Managing human disturbance near roost sites at Fruit Growers Reservoir between 10 March and 20 April would help ensure continued long-term use of the area by migrants.

PROC. NORTH AM. CRANE WORKSHOP 7:153-156

Key words: Colorado, greater sandhill crane, *Grus americana*, *Grus canadensis tabida*, Harts Basin, Rocky Mountain Population, spring migration stopover, whooping crane.

Greater sandhill cranes from the RMP, and whooping cranes raised by cross-fostering to selected RMP breeding pairs (Drewien and Bizeau 1978), migrate annually to their traditional spring and fall staging area in the San Luis Valley, in southcentral Colorado, where they spend 2–6 weeks each season (Drewien and Bizeau 1974, 1978; Drewien et al. 1987, 1995). The primary RMP spring and fall migration route passes through the Gunnison River Valley in western Colorado (Drewien and Bizeau 1974, 1981); no cranes from other populations are known to migrate through this region. Crane migrations through western Colorado have been documented for 100 years (Cooke 1897:62; Keyser 1902:315; Bailey and Niedrach 1965:301; Drewien and Bizeau 1974, 1981), although no traditional stopover areas were identified.

Recent studies by H. E. Heusser (Grand Junction, Colo., pers. commun.) during 1985–87 and 1990–91 revealed that RMP sandhill cranes and occasionally whooping cranes stopped during migration, mainly during spring, in the Grand Valley in western Colorado. In addition, W. A. Davis (*in* Drewien and Bizeau 1974:730) reported more than 400 sandhill cranes near Eckert, Colorado (Harts Basin), in October 1970. J. D. Vela (Eckert, Colo., pers. commun.) noted that sandhill cranes have stopped annually at Harts Basin since 1948. Since the 1960's, M. K. Peterson observed cranes stopping at Fruit Growers Reservoir in Harts Basin, especially during spring, and initiated spring monitoring surveys in 1980. In this paper we report (1) numbers and migration chronology of whooping cranes observed at Harts Basin, Colorado, during spring 1980–95, (2) numbers and migration chronology of RMP sandhill cranes visiting Harts Basin during spring 1986–93, and (3) sightings of whooping cranes during fall migration.

We thank K. D. Peterson for assisting with surveys and S. R. Hemphill, E. J. Horn, and J. Vanko for reporting sightings of cranes. J. D. Vela kindly allowed access to his property near Fruit Growers Reservoir to observe cranes. We thank the U.S. Forest Service, Grand Mesa, Uncompahgre, and Gunnison National Forests, for allowing time for M. Peterson to conduct some surveys while an employee. We thank W. M. Brown, J. C. Lewis, R. E. Shea, T. V. Stehn and an anonymous reviewer for commenting on the manuscript, and R. J. Schneider for preparing a map.

STUDY AREA AND METHODS

Harts Basin and Fruit Growers Reservoir (reservoir) are 2 km east-southeast of Eckert, Delta County, in west-central Colorado (Fig. 1). Grand Mesa, about 15 km north of Harts Basin, is one of the largest flat-top mountains in the world, and, with elevations exceeding 3,300 m, dominates the landscape. The reservoir, at an elevation of 1,672 m, was built in 1898 for irrigation but washed out in 1937 and was reconstructed in 1939 by the Bureau of Reclamation. Water diversions from Surface and Current Creeks, tributaries of the Gunnison River, and natural runoff are the primary water sources for the reservoir. When full, the reservoir covers 188 ha, is 3.2 km long, and has a maximum width of 1.2 km.

Migrating cranes primarily used shallow water areas along the north and northwest reservoir shoreline for night roosts. They frequented meadows near these shores and foraged in harvested cornfields, hayfields, and cattle pastures within 1 km of the reservoir.

Cranes stopping at Harts Basin during spring migrations were monitored annually for 4–7 days weekly over 6–7 weeks during March and April, 1980–95. Occurrences of whooping cranes were recorded during 1980–95 and numbers

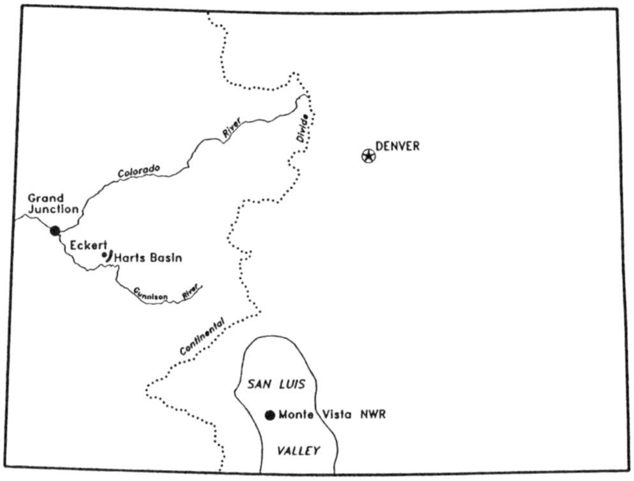

COLORADO

Fig. 1. Important spring migration areas in Colorado, including Harts Basin, Grand Valley (Grand Junction), and the San Luis Valley, used by greater sandhill cranes and whooping cranes of the Rocky Mountain Population.

of sandhill cranes were recorded during 1986–93. Annual estimates of total sandhill crane numbers using Harts Basin were obtained by counting small flocks (≤100–150), estimating numbers in large flocks, and summarizing daily totals by 10-day periods (1- to 11-day period, 21–31 Mar). Most cranes stopped only 1 night, therefore counts were excluded for days with inclement weather when cranes failed to migrate. This minimized duplicating counts of the same birds. Presence of marked cranes was also noted to identify individuals and to help interpret count data and turnover rates. To assess the proportion of the RMP using Harts Basin, we compared our count data with March population counts obtained by the U.S. Fish and Wildlife Service in the San Luis Valley during 1984–95 (Benning et al. 1997). Although few cranes stopped during fall migrations, surveys were conducted when whooping cranes were reported at Harts Basin.

RESULTS

Sandhill Crane Migrations

Counts during 8 spring migrations (1986–93) revealed a mean of 10,656 RMP sandhill cranes (SD = 650, annual range 9,469–11,532) stopped at Harts Basin annually (Table 1). The earliest and latest sightings were on 10 February 1990 (6 cranes) and 17 April 1987, respectively. However, in 1982, 7 cranes were present on 2 May and other reports revealed small numbers stopped occasionally after mid-April, indicating that dates of last observations (Table 1) were probably minimal for some years. The mean peak migration period was 21–31 March and included 45.3% of all cranes reported; 96.2% were tallied between 11 March and 10 April (Table 1). During March, an average of 60.3% (annual range 21.5–83.8%) of the migration arrived or passed through Harts Basin. All annual peak count periods occurred during 21–31 March or 1–10 April (Table 1).

Observations of marked and unmarked RMP cranes indicated that approximately 90% remained only 1 night. Most migrants arrived in late afternoon or evening, roosted overnight at the reservoir, foraged in nearby fields and pastures the following morning, and departed during mid- to late morning. During periods of inclement weather, cranes remained at Harts Basin for 2 or more days and resumed migration when weather improved. Marked RMP cranes (n = 51) observed at Harts Basin were from breeding areas in southeastern Idaho (Grays Lake and the upper Blackfoot River Valley in Bonneville and Caribou Counties) and from western Wyoming (Yellowstone National Park, Bear and Green River Valleys) (Drewien and Bizeau 1974, Drewien et al. 1987).

The mean spring (early to mid-March) population estimate, obtained from 5 comparable aerial surveys and 1 ground count for the RMP in the San Luis Valley, Colorado, during 1984–95 was 19,223 (Benning et al. 1997). We recorded a mean of 10,656 cranes, which indicated that ≥50% of the RMP stopped at Harts Basin annually. However, our counts are minimal because surveys were not conducted on all days during migrations and some cranes were probably missed.

Observations of Whooping Cranes

Ninety-three whooping crane sightings were recorded in Harts Basin during spring migrations over 16 years (Table 2); the annual mean was 5.8 (SD = 2.9, annual range 2–12). Of these sightings, 73 cranes were observed on the ground and 20 were migrating over the area. Of 73 cranes stopping, 66 (90.4%) remained for 1 night, 6 for 2 nights, and 1 for 3 nights. The earliest and latest observations were on 10 March 1983 and 2 May 1982, respectively. More than 77% of all observations occurred between 1–10 April, and 88.2% were during 1–20 April (Table 2).

Twenty-nine different whooping cranes, identified by colored leg bands, stopped at Harts Basin, representing 65.9% of the 44 known to be alive during 1 or more spring migrations during 1980–95 (R. Drewien, unpubl. data). Twenty individuals were observed for multiple springs, including 1 that was observed during 8 of 14 years and another during 6 of 10 years. Eight individuals stopped for 2

Table 1. Numbers and migration chronology of greater sandhill cranes from the Rocky Mountain Population recorded at Harts Basin, Delta County, Colorado, during spring migrations, 1986–93. Annual peak counts are in bold.

	March			April		Total	First	Last
Year	≤10	11–20	21–31	1–10	11–20	cranes	obs.	obs.
1986		367	**9,295**	1,520	350	11,532	13 Mar	13 Apr
1987		911	1,341	**6,718**	1,481	10,451	15 Mar	17 Apr
1988		952	**6,275**	3,525	38	10,790	14 Mar	15 Apr
1989	65	1975	**6,625**	1,710	275	10,650	7 Mar	13 Apr
1990	136[a]	1,825	**4,130**	4,023	278	10,392	10 Feb	15 Apr
1991	26	2,462	**5,350**	3,495	115	11,448	9 Mar	15 Apr
1992		751	2,678	**6,000**	40	9,469	13 Mar	12 Apr
1993	70	3,230	2,960	**3,916**	339	10,515	6 Mar	13 Apr
$\bar{x}$	37	1,559	4,832	3,863	365	10,656	7 Mar	14 Apr
%	0.3	14.6	45.3	36.3	3.4	99.9		

[a] The earliest observation of sandhill cranes; 11 were recorded in February, including 6 on 10 Feb and 5 on 23 Feb.

years, 7 for 3 years, 2 for 4 years, and 1 for 5 years.

We obtained 6 observations of whooping cranes (3 adults, 3 juveniles) stopping during fall migration; 5 stopped during periods of inclement weather (15–16 Oct 1984). All sightings were in October, including 5 in 1984 and 1 in 1987. None remained longer than 1 night and all were traveling with sandhill cranes. The 3 adults had stopped during prior spring migrations at Harts Basin.

Table 2. Numbers and migration chronology of whooping cranes from the Rocky Mountain Population recorded at Harts Basin, Delta County, Colorado, during spring migrations, 1980–95.

	March			April			May	
Year	1–10	11–20	21–31	1–10	11–20	21–30		Total
1980					2			2
1981				2				2
1982			1	2	1		1	5
1983	1			3	2			6
1984				8				8
1985				9				9
1986			5	5				10
1987		1		9	2			12
1988				5	1			6
1989			1	3				4
1990				6				6
1991			1	4				5
1992				8				8
1993				3				3
1994				2	2			4
1995				3				3
Total	1	1	8	72	10	0	1	93
%	1.1	1.1	8.6	77.4	10.8	0.0	1.1	100.1

DISCUSSION

Observations during 16 spring migrations showed that Harts Basin was used annually by migrating RMP sandhill and whooping cranes. However, observations during fall revealed sporadic use, although cranes were observed and heard migrating through the area, primarily during late September and October. We are uncertain why migrants infrequently stopped during fall yet made extensive use of the area during spring.

Harts Basin is 225 km northwest of the staging area at Monte Vista National Wildlife Refuge and vicinity in the San Luis Valley (Fig. 1). Radio tracking studies of migrating RMP cranes during 1979–82 (Drewien and Bizeau 1981; R. Drewien, unpubl. data) showed that they were flying at 3,415–3,750 m when crossing the Continental Divide between the Gunnison River drainage and the San Luis Valley. Flying at such high altitudes must be strenuous and tiring. A 225-km, high altitude flight during spring may tire many migrants and encourage them to stop at Harts Basin. Few other water areas with shallow roost sites and attractive adjacent habitat exist in the area.

Migration chronology data from Harts Basin showed that peak movements of RMP sandhill and whooping cranes were predictable among years and that most whooping cranes migrated during the latter half of the sandhill crane migration period. Most sandhill crane movements (81.6%) occurred during 21 March–10 April; most whooping cranes (77.4%) migrated through Harts Basin during 1–10 April. In March, only 10.8% of whooping crane sightings were recorded, whereas an average of 60.3% of all sandhill cranes had passed through or arrived in Harts Basin.

Repeated annual spring use by >50% of the RMP

revealed that Harts Basin is a traditional stopover, even though most cranes used the area for only 1 night. This use pattern does not conform to the definition of a "traditional stopover" (Melvin and Temple 1982:75-76) because cranes did not remain for extended periods nor were abundant food sources available. We prefer the term "staging area" for fall premigration and migratory locations used annually over extended periods by a population or major population segment. "Traditional stopover" better defines those sites such as Harts Basin or the Grand Valley (H. E. Heusser, pers. commun.), Colorado, the Rio Grande and Rio Chama Valleys in northcentral New Mexico (Stahlecker 1992), and the Green River Valley in northeastern Utah (Drewien and Bizeau 1974, 1981) that receive repeated annual migration use for short time periods by RMP cranes.

Cranes using Harts Basin are a 1-day flight from the RMP spring staging area in the San Luis Valley where they have spent 2–6 weeks foraging and resting prior to resuming migration to summer areas. The primary attractions at Harts Basin as an overnight stopover are adequate roosting sites, adjacent feeding areas, and minimal human disturbance. Long-term use of Fruit Growers Reservoir and Harts Basin by migrating cranes can be ensured only if human disturbance is managed near roost sites during early morning and late evening hours between 10 March and 20 April. During the 1980's, a local conservation group proposed establishing nature trails along the reservoir shoreline. Such human activities during the spring migratory period could greatly reduce or eliminate crane use of this major stopover area.

LITERATURE CITED

BAILEY, A. M., and R. J. NIEDRACH. 1965. Birds of Colorado. Vol. 1. Denver Mus. Nat. Hist., Denver, Colo. 454pp.

BENNING, D. S., R. C. DREWIEN, D. H. JOHNSON, W. M. BROWN, and E. L. BOEKER. 1997. Spring population estimates of Rocky Mountain greater sandhill cranes in Colorado. Proc. North Am. Crane Workshop 7:165-172.

COOKE, W. W. 1897. Birds of Colorado. State Agric. Coll. Exp. Stn. Bull. 37, Tech. Ser. 2. 143pp.

DREWIEN, R. C., and E. G. BIZEAU. 1974. Status and distribution of greater sandhill cranes in the Rocky Mountains. J. Wildl. Manage. 38:720-742.

______, and ______. 1978. Cross-fostering whooping cranes to sandhill crane foster parents. Pages 201-222 *in* S. A. Temple, ed. Endangered birds: management techniques for preserving threatened species. Univ. Wisconsin Press, Madison.

______, and ______. 1981. Use of radiotelemetry to study movements of juvenile whooping cranes. Pages 130-134 *in* J. C. Lewis and H. Masatomi, eds. Crane research around the world. Int. Crane Found., Baraboo, Wis.

______, W. M. BROWN, and W. L. KENDALL. 1995. Recruitment in Rocky Mountain greater sandhill cranes and comparison with other crane populations. J. Wildl. Manage. 59:339-356.

______, W. M. BROWN, and J. D. VARLEY. 1987. The greater sandhill crane in Yellowstone National Park: a preliminary survey. Pages 27-38 *in* J. C. Lewis, ed. Proc. 1985 crane workshop. Platte River Whooping Crane Maintenance Trust, Grand Island, Nebr.

KEYSER, L. S. 1902. Birds of the Rockies. A. C. McClurg and Co., Chicago, Ill. 335pp.

MELVIN, S. M., and S. A. TEMPLE. 1982. Migration ecology of sandhill cranes: a review. Pages 73-87 *in* J. C. Lewis, ed. Proc. 1981 crane workshop. Natl. Audubon Soc., Tavernier, Fla.

STAHLECKER, D. W. 1992. Crane migration in northern New Mexico. Pages 1-12 *in* D. A. Wood, ed. Proc. 1988 North Am. crane workshop. Florida Game Fresh Water Fish Comm. Nongame Wildl. Program Tech. Rep. 12.

SANDHILL CRANE USE OF THE MID-PECOS VALLEY OF EASTERN NEW MEXICO

JAMES B. MONTGOMERY, JR., Department of Biology, New Mexico Military Institute, Roswell, NM 88201, USA

Abstract: Sandhill cranes (*Grus canadensis*) were monitored during winters 1989–94 in the mid-Pecos River Valley of eastern New Mexico to determine how many used this region following a population decline in the 1980's. Numbers annually increased to a late October/early November peak ($\bar{x}$ = 7,017), fluctuated around 6,000 through December, declined in January ($\bar{x}$ = 3,531), and again peaked in February/early March ($\bar{x}$ = 5,737). Annual crane use-days (weekly counts × 7) varied from 500,545 to 1,007,673 ($\bar{x}$ = 760,606). Cranes foraged mainly in corn and sorghum stubble and alfalfa fields. Crop averages (ha) during 1990–92 were alfalfa 21,449, corn 2,462 (cut green for ensilage), sorghum 169 (down from 5,666 ha in 1970), and other crops 6,441. Starting in 1991, Bitter Lake National Wildlife Refuge (NWR) provided 43–55 ha of sorghum and corn. The amount of available food appeared to limit the number of wintering cranes. Hunting disturbances caused approximately 50% of the cranes to leave the valley on 1 occasion, but usually cranes responded by moving from off-refuge sites to protected roosts at Bitter Lake NWR.

PROC. NORTH AM. CRANE WORKSHOP 7:157-164

Key words: abundance, *Grus canadensis*, hunting, New Mexico, Pecos Valley, population trend, sandhill crane, winter food.

The Pecos Valley of eastern New Mexico has been recognized as a migration route and wintering area for sandhill cranes since the 1940's. Walkinshaw (1949) reported 5,800 cranes in December 1942, 11,000 in March 1943, and 8,000 in November 1943 at Bitter Lake NWR near Roswell in Chaves County. In December 1960, Boeker et al. (1961) counted 10,229–16,706 at 4 roost sites in Chaves County. Except for Christmas bird count (CBC) data, there were no published records in Chaves County for the next 3 decades. Annual peak numbers of cranes at Bitter Lake NWR, however, were recorded in annual refuge reports. Peak numbers averaged 38,583 during the 1970's and 13,842 during the 1980's, indicating a population decline had occurred in the Pecos Valley. This study documents the numbers of sandhill cranes using the mid-Pecos Valley in Chaves County during 1989–94, identifies the principal factor affecting the numbers of cranes, and assesses crop-lands preferred by cranes. This information provides insight for managing crane populations and provides a baseline to compare future crane numbers in eastern New Mexico.

This study was a volunteer research project at Bitter Lake NWR. L. Marlatt, former Refuge Manager, suggested the project and provided access to refuge annual reports; W. R. Radke, current Refuge Manager, continued support for the study. M. Peckinpaugh, refuge volunteer, and refuge staff members M. L. Harper, R. Larrañaga, S. R. Najera, M. E. Rosenbaum, and T. A. Rossignol made weekly crane counts on the refuge. M. L. Burfield helped prepare the final manuscript. I appreciate financial support from a Roy F. Ward Professorship in Natural Science funded by the Kerr Foundation of Oklahoma City, Oklahoma, and administered by New Mexico Military Institute. I thank W. R. Radke for his advice and T. V. Stehn, R. C. Drewien, and an anonymous reviewer for their help and comments on drafts of this manuscript.

STUDY AREA

The study area included agricultural lands surrounding Roswell in southeast New Mexico (Fig. 1). In the Pecos Valley, most farming was confined to 4 areas of irrigated croplands near the communities of Fort Sumner, Roswell, Artesia, and Carlsbad (Fig. 1). Intervening shortgrass prairie separated these areas from each other and from farmlands in west Texas and eastern New Mexico near Muleshoe NWR. The north-to-south orientation of the Pecos River with islands of cropland provided a migration corridor through southeast New Mexico, termed the Pecos Valley Subflyway (Ligon 1961).

Sandhill cranes migrated through and wintered along this corridor, some moving into Mexico (Ligon 1961). The most important winter area was near Roswell in the mid-Pecos Valley where suitable roost sites existed along the Pecos River and food was available in nearby farmlands (Fig. 2).

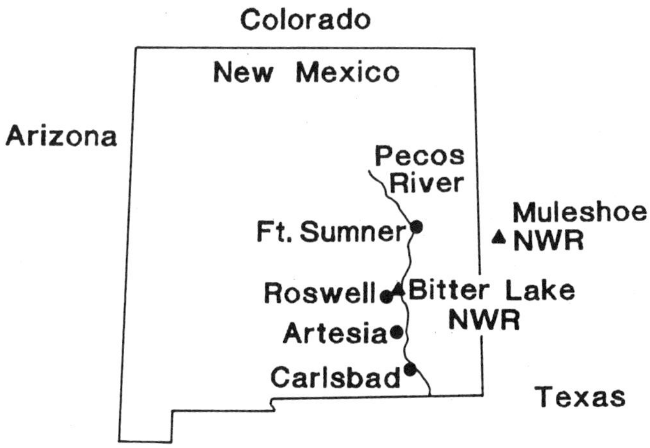

Fig. 1. Location of Roswell and Bitter Lake National Wildlife Refuge in southeast New Mexico.

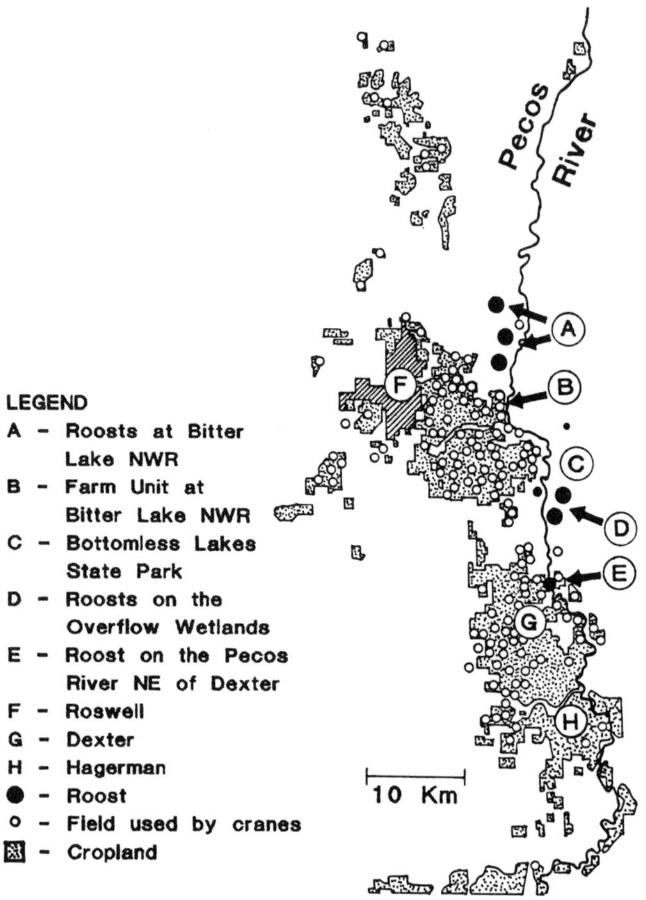

Fig. 2. Areas used by sandhill cranes in the mid-Pecos Valley, Chaves County, New Mexico, during October 1989–March 1995. Map after U.S. Department of Agriculture, Soil Conservation Service (1981).

Table 1. Annual peak numbers of sandhill cranes (peak) and numbers of sandhill cranes recorded during annual Christmas bird counts (CBC) at Bitter Lake National Wildlife Refuge and the hectares of corn, sorghum, and alfalfa grown in Chaves County, New Mexico, 1969–88.

	No. of cranes		Crops (ha)		
Year	Peak	CBC	Corn	Sorghum	Alfalfa
1969	66,000	9,200	2,226	5,463	15,378
1970	67,000	11,350	2,428	5,666	15,783
1971	23,500	10,512	2,023	3,642	15,662
1972	36,300	4,900	1,355	2,995	15,985
1973	30,300	9,282	1,408	3,076	16,795
1974	12,300	2,400	1,578	2,428	16,107
1975	37,300	9,064	2,185	4,856	16,188
1976	44,500	17,940	2,428	2,914	17,402
1977	70,000	33,090	1,862	1,255	17,807
1978	10,630	352	2,185	2,873	18,009
1979	54,000	6,076	2,550	2,833	18,211
1980	36,200	14,370	2,833	2,428	17,200
1981	18,000	10,226	1,700	1,841	17,402
1982	21,215	8,330	2,550	2,428	20,235
1983	9,005	9,005	1,174	931	20,235
1984	13,900	7,320	1,416	850	20,056
1985	12,000	3,730	1,700	1,255	20,235
1986	11,066	2,500	1,700	1,133	19,425
1987	5,400	2,432	1,072	809	20,235
1988	6,000	4,510	1,740	405	21,044

Approximately 39,580 ha of irrigated cropland occurred in Chaves County (see Methods) and 97% was within 25 km of the Pecos River (Fig. 2). Principal crops (30,310 ha) were alfalfa, cotton, corn, chile peppers, wheat, barley, and sorghum. Only a small portion of the corn crop was harvested for grain; 99% was cut while green for ensilage. Approximately 2,425 ha were in orchards, dairies, and vegetable farms, and the remaining 6,845 ha were either fallow or pasture. Non-croplands were mostly grassland. The average annual rainfall was 27.9 cm (Hodson et al. 1980).

Bitter Lake NWR, 10 km east of Roswell, provided 3 protected roosts and food crops (Fig. 2). Crops, including 40–55 ha of sorghum and corn, were grown for cranes and waterfowl on a 201-ha irrigated farm unit acquired by the refuge in 1988, and millet was occasionally grown on an additional 20 ha of poorly irrigated land. Another major roost was the Overflow Wetlands near Bottomless Lakes State Park on the east side of the Pecos River, 18 km southeast of Roswell (Fig. 2). This roost was important because it was 10–20 km closer than refuge roosts to farmlands in the southern part of Chaves County. The Overflow Wetlands roost, however, was not protected from human disturbance as most of the area was public land administered by the Bureau of Land Management or the state, and crane and waterfowl hunting occurred annually from October to January. The Pecos River, 5 km northeast of the town of Dexter (Fig. 2), and a temporary pond near the Overflow Wetlands were used by smaller numbers (<1,000) of cranes. These lesser roosts were used mainly during winter and spring when water levels were suitable and when cranes moved from the Overflow Wetlands because of disturbance from hunting. A temporary pond north of Bottomless Lakes State Park was also used by ±2,000 cranes for 2 weeks during January 1992.

METHODS

Trends in crane numbers using the Pecos Valley during 20 years (1969–88) preceding this study were identified by regression analyses of annual peak numbers recorded at Bitter Lake NWR (U.S. Fish and Wildlife Service, unpubl. data,

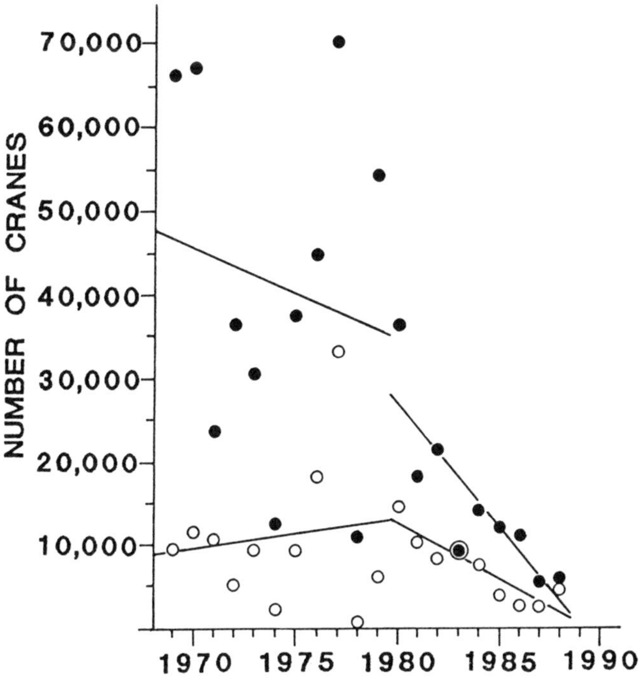

Fig. 3. Trends in annual peak sandhill crane numbers (solid dots) and in numbers of sandhill cranes recorded during annual Christmas birds counts (open circles) at Bitter Lake National Wildlife Refuge, New Mexico, 1969-88. Statistics are included in text.

Annu. Narrative Reps., Roswell, N.M.) and of crane numbers included in annual Christmas bird counts centered on the refuge (National Audubon Society 1970, 1971-89). Relevant agricultural cropland trends in Chaves County during these same years were identified by regression analyses of annual hectares of corn, sorghum, and alfalfa reported annually for New Mexico (U.S. Department of Agriculture 1969-88). Correlation coefficients between annual hectares of these crops and the annual peak and Christmas bird count crane numbers were calculated to assess associations.

Crane numbers were determined weekly by simultaneous dawn counts of roosts at Bitter Lake NWR and the Overflow Wetlands, along with counts obtained during the previous evening along the Pecos River near Dexter or at the temporary pond near the Overflow Wetlands. Weekly counts were usually made by 1 to 4 observers, depending on the number of occupied roosts. When weather conditions prevented accurate censuses, counts were repeated the following day or estimates were used.

Estimates of annual crane use, expressed as crane use-days, were made by multiplying crane numbers times the number of days in the count period to provide insight into the crane carrying capacity of the mid-Pecos Valley.

Observations of daily and monthly patterns of crane activity and field use between October 1990 through March 1993 were obtained from vantage points, by following flying cranes, and by scanning fields while driving through farmlands. It was not possible to randomize searches and observations because the county road net did not accommodate a random search pattern and some fields could not be observed from a public road. Instead, I concentrated searches in areas where cranes were known or suspected to be foraging.

When cranes were located in a field, individuals were counted or their number estimated and the field was classified as (1) alfalfa, (2) stubble (including disked fields with surface plant material), (3) sprouting green field, (4) plowed, (5) grass/prairie, or (6) wet area. Count data were also recorded by time of day, including (1) "early AM" observations made within 2 hours following sunrise, (2) "late AM" observations made during the remaining morning hours, and (3) "PM" observations made after noon, mainly ≤3 hours of sunset. Data were converted into percentages for comparisons. Data collections were not randomized because the relative proportions of alfalfa, stubble, and plowed fields changed weekly as farmers worked fields and statistical analyses were not conducted.

RESULTS

Crane Populations and Crop Trends: 1969-79 and 1980-88

Crane counts, both annual peak numbers and CBC's, were analyzed separately for the period 1969-79 and 1980-88 (Table 1, Fig. 3). Peak numbers did not decline during the 1970's (t = 0.529, df = 9, $P > 0.05$, Y =

Table 2. Relationships of annual peak numbers (peak) and Christmas bird count numbers (CBC) of sandhill cranes recorded at Bitter Lake National Wildlife Refuge with annual hectares of sorghum, corn, and alfalfa in Chaves County, New Mexico.

Crop	Years	Correlation coefficients[a] Peak	CBC
Sorghum	1969-79	0.275	-0.304
	1980-88	0.879**	0.689*
Corn	1969-79	0.416	0.088
	1980-88	0.865**	0.596[b]
Alfalfa	1969-79	−0.014	0.226
	1980-88	−0.666*	−0.582[b]

[a] $*P < 0.05, **P < 0.01$, df = 7.

[b] The critical value for $P = 0.05$, df = 7, is 0.666.

Table 3. Summary of weekly counts of sandhill cranes and annual crane use-days in the mid-Pecos Valley, New Mexico, 1989–94.

Count period	1989	1990	1991	1992	1993	1994
16 Sep–22 Sep	0	0	0	11	16	0
23 Sep–29 Sep	0	17	18	7	20	1
30 Sep– 6 Oct	11	451	139	394	130	74
7 Oct–13 Oct	955	1,692	1,101	2,056	589	4,300
14 Oct–20 Oct	1,920	6,056	3,908	3,295	774	1,420
21 Oct–27 Oct	1,802	5,737	4,653	3,215	1,558	2,550
28 Oct– 3 Nov	5,635	5,599	8,026	6,762	6,578	5,230
4 Nov–10 Nov	4,830	6,046	5,412	6,373	7,925	5,130
11 Nov–17 Nov	5,367	6,608	6,743	6,676	6,999	5,963
18 Nov–24 Nov	4,170	6,052	2,924	6,747	8,882	6,188
25 Nov– 1 Dec	4,830	5,274	4,022	5,726	7,900	6,892
2 Dec– 8 Dec	3,535	6,282	4,536	6,888	8,892	8,193
9 Dec–15 Dec	2,556	6,864	6,143	6,197	7,544	7,880
16 Dec–22 Dec	3,042	7,288	5,909	5,751	10,008	3,959
23 Dec–29 Dec	2,800	5,865	5,021	7,337	9,500	5,509
30 Dec– 5 Jan	3,137	5,750	3,500	5,582	5,530	6,230
6 Jan–12 Jan	3,568	6,600	3,271	5,755	5,432	3,458
13 Jan–19 Jan	3,575	5,309	3,344	6,182	7,109	3,327
20 Jan–26 Jan	3,197	6,163	2,771	2,473	4,785	3,463
27 Jan– 2 Feb	2,878	6,052	3,214	3,213	5,233	3,050
3 Feb– 9 Feb	4,800	6,915	5,103	2,682	6,218	3,154
10 Feb–16 Feb	3,862	5,881	4,349	2,298	7,360	3,550
17 Feb–23 Feb	2,661	4,721	4,825	2,943	6,687	3,640
24 Feb– 2 Mar	1,900	3,241	7,163	2,815	6,301	3,080
3 Mar– 9 Mar	300	2,384	3,729	2,034	7,778	3,950
10 Mar–16 Mar	22	3,857	2,845	2,150	3,084	1,390
17 Mar–23 Mar	86	935	1,262	289	941	633
24 Mar–30 Mar	49	51	138	0	148	104
31 Mar– 6 Apr	0	140	0	0	24	40
7 Apr–13 Apr	0	0	0	0	23	45
14 Apr–20 Apr	0	0	0	0	9	0
Crane-use days	500, 545	894,750	728,483	740,957	1,007,673	691,229

47,761 − 1,114*X*), but numbers declined (t = 4.092, df = 7, $P < 0.01$, Y = 29,413 − 2,932*X*) from 1980 to 1988 (Fig. 3). Cranes counted during Christmas bird counts showed no significant trend during the 1970's (t = 0.115, df = 9, $P >$ 0.05, Y = 8,798 + 309*X*), but numbers declined significantly during the 1980's (t = 5.754, df = 7, $P < 0.01$, Y = 13,582 − 1,329*X*). Changes in agriculture crops occurred in Chaves County during the 1970's and 1980's (Table 1). Sorghum production declined (t = 7.136, df = 18, $P <$ 0.01) from 5,463 ha and 5,666 ha in 1969 and 1970 to 405 ha in 1988. Simultaneously, alfalfa production increased (t = 9.364, df = 18, $P < 0.01$) from 15,378 ha in 1969 to >20,000 ha in the 1980's. The annual size of the corn crop (20-yr $\bar{x}$ = 1,908 ha), however, did not change (t = 1.370, df = 18, $P > 0.05$). An average of 11.1% of the corn crop was harvested for grain during 1969–75 but increased to 50% during 1976–79 and then fell to 6.6% during 1980–88. Most corn was cut for ensilage and, along with alfalfa hay, was used to support an expanding dairy industry.

Correlations between annual peak crane numbers and hectares of sorghum and corn were not significant during the 1970's but were significant during the 1980's (Table 2); peak numbers also showed a significant negative correlation (P = 0.05) with hectares of alfalfa during the 1980's. Cranes counted during CBC's also showed a similar, but weaker, relationship between numbers and hectares of corn and sorghum (Table 2).

Crane Population Trends: 1989–94

Cranes arrived in the Roswell area in late September/early October; numbers peaked in late October/early November and remained high through December (Tables 3 and 4). Although year-to-year variation occurred within each

Table 4. Means of weekly counts of sandhill cranes at Bitter Lake National Wildlife Refuge and off-Refuge sites in the mid-Pecos Valley, New Mexico, 1989–94.

Count period	Bitter Lake NWR		Off-Refuge site		Total	
	$\bar{x}$	SE	$\bar{x}$	SE	$\bar{x}$	SE
23 Sep–29 Sep	10	3	1	1	11	4
30 Sep– 6 Oct	140	50	60	23	200	73
7 Oct–13 Oct	1,208	554	574	128	1,782	548
14 Oct–20 Oct	1,473	676	1,422	294	2,895	792
21 Oct–27 Oct	1,721	431	1,532	339	3,253	673
28 Oct– 3 Nov	4,012	281	2,293	497	6,305	422
4 Nov–10 Nov	3,712	631	2,241	426	5,953	458
11 Nov–17 Nov	4,144	596	2,249	426	6,393	249
18 Nov–24 Nov	4,596	1,002	1,231	451	5,827	847
25 Nov– 1 Dec	4,603	805	1,171	520	5,774	577
2 Dec– 8 Dec	5,929	935	459	146	6,388	844
9 Dec–15 Dec	5,400	1,000	797	263	6,197	782
16 Dec–22 Dec	5,377	1,277	616	324	5,993	1,012
23 Dec–29 Dec	5,509	915	496	164	6,005	922
30 Dec– 5 Jan	4,395	671	560	194	4,955	529
6 Jan–12 Jan	3,883	612	798	135	4,681	581
13 Jan–13 Jan	3,324	848	1,484	335	4,808	666
20 Jan–26 Jan	1,811	687	1,998	252	3,809	573
27 Jan– 2 Feb	2,639	727	1,301	377	3,940	551
3 Feb– 9 Feb	3,364	664	1,448	308	4,812	678
10 Feb–16 Feb	3,282	776	1,268	367	4,550	736
17 Feb–17 Feb	3,144	708	1,102	277	4,246	608
24 Feb– 2 Mar	3,325	915	758	246	4,083	866
3 Mar– 9 Mar	2,620	1,109	743	470	3,363	1,034
10 Mar–16 Mar	1,826	553	399	183	2,225	558
17 Mar–17 Mar	518	205	173	96	691	181
24 Mar–30 Mar	32	22	50	24	82	24
31 Mar– 6 Apr	0	0	34	22	34	22

weekly count period, the 6-year means show approximately 6,000 cranes remained in the mid-Pecos Valley during the 9 count periods from 28 October to 29 December (Table 4). Average numbers then declined during January and rose again in February. Mean maximum (7,300) and minimum (3,531) winter counts (December and January, Table 3) differed significantly (t = 4.326, df = 5, P < 0.01) but mean fall (7,017) and spring (5,737) migration peaks (Table 3) were similar (t = 1.418, df = 5, P > 0.05).

The timing and magnitude of winter declines in crane numbers varied (Table 3). Numbers declined during December 1989, but no noticeable decline occurred during the 1990 migration and winter season. During the remaining 4 years when sorghum and corn were available on the refuge farm unit, steep declines (≥40–60%) coincided with depletion of these foods in late December and early January. For example, in November 1994 portions of the refuge corn and sorghum plots were made available to birds by periodic cuttings. In early December, when most cranes and geese in the Roswell area were feeding at the refuge farm, the remaining crops were left uncut in an effort to disperse the wintering goose flock. Crane numbers also subsequently declined from 7,880 on 9 December to 3,959 on 16 December. The remaining crop was then cut and crane numbers rose to 6,600 by 31 December, but the food was quickly consumed and numbers declined to 3,458 by 7 January.

Declines in crane numbers, with 1 observed exception, did not coincide with the opening of crane (mid-October) or duck (mid-November) hunting seasons, although declines in crane use of off-refuge roosts did (Table 4, Fig. 4). During mid-October, crane roosting was evenly divided between roosts at Bitter Lake NWR and off-refuge sites. However, beginning in November and coinciding with disturbances by both crane and waterfowl hunters, roosting declined on off-refuge sites because most cranes moved to the protected refuge roosts. Total crane numbers were not affected, but 1

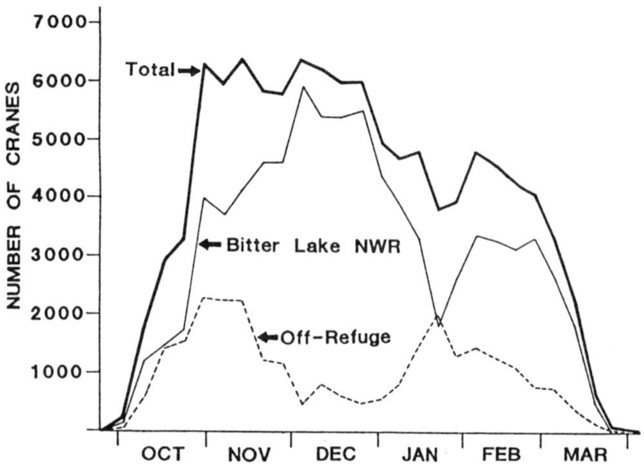

Fig. 4. Mean numbers of sandhill cranes in the mid-Pecos Valley, New Mexico, during the fall and spring migration and winter seasons, 1989–94. Means are from weekly counts at Bitter lake National Wildlife Refuge and off-Refuge sites in Chaves County.

incident of roost disturbance (of 38 recorded) caused a mass departure (>3,000 cranes, approximately 50% of total) from Chaves County in 1991 (Table 5). Total crane numbers, however, increased again by more than 3,000 during the next 4 weeks.

Field Use

The mean numbers of hectares by crop type in Chaves County during 1990–92 were alfalfa 21,449, cotton 4,789, corn 2,462 (99% cut for ensilage), chile peppers 1,079, wheat 573, and sorghum (all kinds) 169 (U.S. Department of Agriculture 1990–92). Counts of cranes observed in various fields (Table 6) showed that stubble (both corn and sorghum) received the most use. After leaving roosts during early morning hours, cranes either flew directly to stubble fields or, more commonly, gathered in an alfalfa field and then flew to stubble; a similar gathering in alfalfa was observed in afternoons prior to their return to the roost. A seasonal difference in early AM movements was observed. During October, most cranes (75%) flew directly from the roost to stubble. In November, 41.5% flew directly to stubble and in December 82.8% stopped first in alfalfa fields.

Incidental observations also confirmed that cranes were foraging in these crop types. During the 1990 winter when most cranes roosted on Bitter Lake and fed in sorghum fields, droppings at the roost were the characteristic color of sorghum grain (L. Marlatt, U.S. Fish and Wildlife Service, Roswell, N.M., pers. commun.). Examination of the esophagi of 3 cranes killed during 1990 revealed 2 were filled with corn kernels and 1 contained corn kernels, 16 alfalfa leaflets, and a grasshopper.

Table 5. Weekly counts of sandhill cranes at Bitter Lake National Wildlife Refuge and off-Refuge sites in the mid-Pecos Valley, New Mexico, before and after the start of the 1991 duck hunting season. The season opened on 16 November.

Date	Bitter Lake NWR	Off-Refuge sites	Total
8 Nov	2,802	3,330	5,412
15 Nov	3,511	3,232	6,743
22 Nov	2,814	110	2,924
29 Nov	3,879	143	4,022
6 Dec	4,006	530	4,536
14 Dec	5,587	556	6,143

Two other cropland types were used to a lesser extent for foraging. Sprouting green fields were used when the seeds were germinating and young plants were emerging, but cranes vacated these fields once plants were established. Flock sizes in these fields varied ($\bar{x}$ = 228, SE = 64.5, range 3–2,000, n = 35). Recently plowed fields were used by smaller flocks ($\bar{x}$ = 48, SE = 7.3, range 3–275, n = 67) for several days until the soil dried.

Cranes were not observed in cotton, chile peppers, or other crops except when these fields were losing or had lost their identity by disking or plowing. Crane observations in these latter fields were included in the "plowed" category.

The preceding observations were obtained in private farmland where food for wildlife was considered incidental to normal farming operations. Bitter Lake NWR began growing wildlife crops in 1989 on a 201-ha farm unit. Coinciding with a 55-ha planting of sorghum in 1991, cranes began using the refuge farm. During 1991 and subsequent years, they foraged mostly in sorghum and corn.

Although cranes foraged in alfalfa, especially during morning hours, they also used alfalfa fields for diurnal resting sites. Cranes also rested near small ponds and marsh areas not used for roosting, particularly in October and November (Table 6). Cranes occasionally used prairie during early morning after leaving roosts, but they usually flew to agricultural fields within the hour.

Daily Activity

The typical daily activity pattern was a mass morning flight from the roost to the farmland and a return in small groups from mid-morning through the afternoon or a mass return to the roost at sunset. Smaller afternoon flights from resting areas and roosts to farmlands also occurred. After departing roosts in the morning, cranes usually congregated in the first alfalfa fields they encountered before flying on to

Table 6. Distribution (%) of sandhill cranes in different field types, mid-Pecos Valley, New Mexico, October 1990–March 1993[a].

Month and time of day[b]	% of observations by field type[c,d]						
	No.	ALF	STB	SGF	PLW	G/P	WET
October							
Early AM	19,149	23.5	75.0	0.0	0.0	1.5	0.0
Late AM	3,938	17.3	82.7	0.0	0.0	0.0	0.0
PM	9,103	31.0	11.2	1.9	0.0	0.6	55.3
November							
Early AM	19,689	44.6	41.5	6.6	0.7	6.6	0.0
Late AM	7,982	43.2	52.1	4.7	0.0	0.0	0.0
PM	17,520	18.7	36.1	7.7	0.0	6.3	31.2
December							
Early AM	7,925	82.8	13.4	3.8	0.0	0.0	0.0
Late AM	5,038	34.7	46.5	18.8	0.0	0.0	0.0
PM	9,371	74.3	25.1	0.0	0.0	0.6	0.0
January							
Early AM	6,805	77.2	18.5	0.6	3.7	0.0	0.0
Late AM	4,575	11.0	84.1	4.1	0.8	0.0	0.0
PM	17,015	25.0	56.1	0.0	1.9	0.0	17.0
February							
Early Am	11,754	77.4	17.7	1.2	3.7	0.0	0.0
Late AM	10,731	12.6	62.7	20.2	4.4	0.0	0.1
PM	19,011	48.8	37.4	1.6	5.3	0.0	6.9
March							
Early AM	12,380	69.1	12.9	0.0	3.3	14.7	0.0
Late AM	5,824	38.7	59.6	0.0	1.7	0.0	0.0
PM	11,024	65.4	26.1	6.4	0.2	0.2	1.7
All months combined							
Early AM	77,702	55.0	36.7	2.3	1.6	4.4	0.0
Late AM	38,088	26.2	62.5	9.7	1.6	0.0	0.0
PM	83,044	40.7	35.2	3.1	1.6	1.5	17.9

[a] Bitter Lake National Wildlife farm not included.

[b] Early AM = ≤2 hours after sunrise, late AM = remaining morning hours, PM = all afternoon hours, mostly ≤3 hours before sunset.

[c] ALF = alfalfa, STB = corn or sorghum stubble, SGF = sprouting green field (germinating wheat, alfalfa, or barley), PLW = plowed field (no stalks visible), G/P = grass or prairie, and WET = wet area.

[d] No cranes were observed in cotton, chile peppers, or other field crops.

stubble fields to feed. Before returning to the roost in late afternoon, cranes often congregated in alfalfa fields or wet areas. During January, however, more cranes remained in stubble fields until sunset (Table 6).

Foraging cranes were not randomly distributed, but generally concentrated in a few fields. For example, in December 1990 and January 1991, more than 90% of the cranes in Chaves County used 1 roost (at Bitter Lake NWR) and flew 22 km northwest to feed in 168 ha of sorghum fields that were left undisked for livestock foraging. Similarly, during spring 1993, up to 5,000 cranes foraged daily in a 45-ha corn field grown for grain and left undisked for livestock. Most cranes from a roost foraged as a single flock in 1 area for several days to weeks and then, in response to food depletion or disturbances, moved to another area. Locations of fields used by cranes are shown in Figure 2. Some fields were used throughout the study, but most were used only 1 or 2 years and were fallow, planted in other crops, or converted to other uses in other years.

DISCUSSION

Crane numbers in the mid-Pecos Valley during 1989–94 rose to an initial migratory peak in late October/early November, averaged ±6,000 through December, and then declined significantly. This decline appeared to be related to available food, and food was apparently the principal factor limiting winter population levels. During the 1991–94 seasons, crane numbers declined sharply when food on the refuge farm was depleted. In contrast, a winter decline did not occur during the 1990 season when most cranes used an apparently rich food source northwest of the refuge. Annual peak numbers at Bitter Lake NWR were also correlated with hectares of corn and sorghum during the 1980's when both crane numbers and amount of sorghum grown annually were decreasing.

Hunting also affected numbers and distribution during 1989–94, but not as severely as reported by Boeker et al. (1961). They observed a 96% decline in the crane population in Chaves County within 2 weeks after the start of the first crane hunting season. Lovvorn and Kirkpatrick (1981) summarized studies also showing population shifts and roost abandonment following the opening of waterfowl seasons in various regions of the United States. Although 1 hunting-caused mass departure was recorded in Chaves County, crane numbers generally remained high throughout crane and waterfowl seasons. Cranes usually responded to hunting disturbances by moving to protected roosts at Bitter Lake NWR.

Observations of field use, esophageal contents, and droppings at the roost noted in my study were consistent with other studies in New Mexico and west Texas. Boeker at al. (1961) reported that cranes in the Roswell area mainly ate sorghum and green alfalfa. Walker and Schemnitz (1985) found corn, sorghum, and alfalfa were the principal foods in central and southwestern New Mexico. They noted that grasshoppers also were important in the diet and concluded that most were consumed when cranes were foraging in alfalfa.

Iverson et al. (1985*a*) observed large increases in crane numbers in west Texas between January and mid-February.

Similar increases occurred in early February in the mid-Pecos Valley. Iverson et al. (1985*b*) also reported that cranes landed in milo (sorghum) or cotton stubble soon after leaving roosts and later concentrated in milo stubble farther from roosts. During afternoons, large flocks gathered in cotton stubble near roosts. A similar pattern was observed in this study, but alfalfa fields (not present in west Texas) and wet areas replaced cotton stubble as gathering areas in the Pecos Valley.

MANAGEMENT IMPLICATIONS

How many cranes can the mid-Pecos Valley support with current food resources? The amount of off-refuge food varied annually but was sufficient in 1990 to support 5,000–6,000 cranes throughout most of the winter. Bitter Lake NWR provided a more consistent food supply, but the amount was insufficient to sustain recent wintering numbers beyond mid-January.

The number of crane use-days gives a measure of the amount of use the mid-Pecos Valley receives during a winter season and, therefore, provides an indirect measure of available food resources. The 6-year average was 760,606 crane use-days. If agriculture production remains constant, food resources should be sufficient to maintain approximately 4,700 cranes from 7 October through 16 March (161 days) or larger numbers for shorter periods. Because the number of cranes annually exceeds this level, a winter decline is inevitable unless more food is provided as in 1990.

Food provided by Bitter Lake NWR since 1991 has attracted the majority of cranes remaining in the Roswell area through December. If the management goal is to increase crane numbers wintering in the mid-Pecos Valley beyond December, then additional food crops are needed. Agricultural trends on private lands, however, have been toward smaller sorghum crops and earlier disking and plowing of corn stubble. It is also unlikely that food production will be increased at Bitter Lake NWR due to budget restrictions, limited availability of suitable cropland, and a possible reduction in allocated water. Or, if the management goal is to maintain a smaller, but constant, number throughout the winter, then food grown on the refuge should be rationed by carefully monitored cuttings to encourage part of the population to winter elsewhere. However, if the goal is to maintain a larger number of cranes in the Roswell area until the end of the hunting season, the current practice of making all food on the refuge available in November and December should remain unchanged.

LITERATURE CITED

BOEKER, E. L., J. W. ALDRICH, and W. S. HUEY, 1961. Study of experimental sandhill crane hunting season in New Mexico during January 1961. U.S. Fish Wildl. Serv. Spec. Sci. Rep. Wildl. 63. 24pp.

HODSON, M. V., T. E. CALHOUN, C. L. CHASTAIN, L. W. HACKER, W. G. HENDERSON, and C. R. SEAGRAVES. 1980. Soil survey of Chaves County, New Mexico, southern part. U.S. Dep. Agric., Soil Conserv. Serv., Albuquerque, N.M. 143pp.

IVERSON, G. C., P. A. VOHS, and T. C. TACHA. 1985*a*. Distribution and abundance of sandhill cranes wintering in western Texas. J. Wildl. Manage. 49:250-255.

______, ______, and ______. 1985*b*. Habitat use by sandhill cranes wintering in western Texas. J. Wildl. Manage. 49:1074-1083.

LIGON, J. S. 1961. New Mexico birds and where to find them. Univ. New Mexico Press, Albuquerque. 360pp.

LOVVORN, J. S., and C. M. KIRKPATRICK. 1981. Roosting behavior and habitat of migrant greater sandhill cranes. J. Wildl. Manage. 45:842-857.

NATIONAL AUDUBON SOCIETY. 1970. 70th Christmas bird count, #771, Roswell, N.M. Audubon Field Notes 24:410.

______. 1971-89. 71st-89th Christmas bird count [nos. variable], Roswell, N.M. Am. Birds, vols. 25-43.

U.S. DEPARTMENT OF AGRICULTURE. 1969-88, 1990-92. New Mexico agricultural statistics, editions of 1969 through 1988 and 1990 through 1992. New Mexico Crop and Livestock Rep. Serv., Las Cruces, N.M.

______, SOIL CONSERVATION SERVICE. 1981. Important farmlands, Chaves County, New Mexico (map). Sheets 1 and 2. U.S. Dep. Agric., Albuquerque, N.M.

WALKER, D. L., and S. D. SCHEMNITZ. 1987. Food habits of sandhill cranes in relation to agriculture in central and southwestern New Mexico. Pages 201-213 *in* J. C. Lewis, ed., Proc. 1985 crane workshop. Platte River Whooping Crane Maintenance Trust, Grand Island, Nebr.

WALKINSHAW, L. H. 1949. The sandhill cranes. Cranbrook Inst. Sci. Bull. 29. 202pp.

SPRING POPULATION ESTIMATES OF ROCKY MOUNTAIN GREATER SANDHILL CRANES IN COLORADO

DOUGLAS S. BENNING, U.S. Fish and Wildlife Service, Office of Migratory Bird Management, P.O. Box 25486, DFC, Denver, CO 80225, USA
RODERICK C. DREWIEN, Hornocker Wildlife Research Institute, University of Idaho, P.O. Box 3246, Moscow, ID 83843, USA
DOUGLAS H. JOHNSON, Northern Prairie Science Center, U.S. Geological Survey–Biological Resources Division, 8711 37th Street SE, Jamestown, ND 58401, USA
WENDY M. BROWN,[1] Hornocker Wildlife Research Institute, University of Idaho, P.O. Box 3246, Moscow, ID 83843, USA
ERWIN L. BOEKER, 8219 W. Virginia, Lakewood, CO 80226, USA

Abstract: We developed an aerial survey to estimate the Rocky Mountain Population (RMP) of greater sandhill cranes (*Grus canadensis tabida*) (greaters) annually during March 1984–95 in the San Luis Valley, Colorado (SLV). We flew aerial transects covering a 21,220-ha contiguous area. Aerial photographs of crane flocks were taken to correct for estimation errors. We employed ground surveys to correct for presence of lesser sandhill cranes (*G. c. canadensis*) (lessers), which mixed with RMP cranes. To adjust for RMP cranes not in the survey area, we made ground counts in traditional use areas outside the SLV. During 1984–95, 5 comparable aerial surveys and 1 ground count yielded estimates of the RMP ranging from 16,220 to 21,496 ($\bar{x}$ = 19,223). Counts indicated that the population was stable. Cloud cover, partial snow cover, and lack of background color contrast in certain habitat types reduced visibility of cranes. Increases in non-RMP cranes mixed with RMP flocks are jeopardizing the accuracy of counts and assessment of population trend. An alternative census method may be needed.

PROC. NORTH AM. CRANE WORKSHOP 7:165-172

Key words: aerial count, Colorado, greater sandhill crane, *Grus canadensis tabida*, population estimate, Rocky Mountain Population, San Luis Valley.

Greater sandhill cranes from the RMP were estimated annually in March 1984–95 at their spring migration staging area in the SLV. Although the population is dispersed during most of the year, virtually the entire RMP assembles annually in early to mid-March in the SLV, providing an opportunity to assess its abundance (Drewien and Bizeau 1974). Aerial estimates were used to assess population size and help arrive at annual harvest rates for limited quota permit hunts currently held at selected locations in Arizona, Montana, New Mexico, Utah, and Wyoming (Smith 1991). Due to low recruitment rates and delayed sexual maturity, knowledge of changes in population numbers is needed to manage hunted crane populations (Drewien et al. 1995). The annual March survey is a joint effort of the U.S. Fish and Wildlife Service (FWS) and the Pacific and Central Flyway Councils. In this paper we report (1) March population estimates of RMP cranes during 1984–95 in the SLV, (2) survey methodology, and (3) factors influencing annual estimates.

We thank the following Fish and Wildlife Service personnel for assisting with various aspects of the study: S. S. Berlinger, J. E. Cornely, J. P. Cossick, R. L. Croft, D. Falzone (volunteer), R. J. Garcia, J. A. Harr, J. W. Haskins, L. M. Jones, J. D. Kauffeld, R. E. Kirby, A. E. Morkill, M. T. Nail, G. C. Rennick, R. W. Schnaderbeck, J. M. Steiner, M. B. Suthers, and J. P. Taylor. We appreciate reviews of the manuscript by J. E. Austin, J. C. Bartonek, W. E. Newton, D. E. Sharp, R. E. Shea, and J. F. Voelzer. We thank R. J. Schneider for preparing a map and E. A. Bublitz for typing the manuscript. Surveys were funded by the FWS.

STUDY AREAS

The SLV in south-central Colorado is a high-elevation (2,280 to 2,380 m) mountain valley about 170 × 32–80 km. The valley is intensively farmed, chiefly for barley, wheat, potatoes, and alfalfa, and mainly irrigated by pivot sprinklers. Native vegetation is usually limited to areas too saline or too moist to cultivate, riparian corridors, state wildlife management areas, and the Alamosa and Monte Vista National Wildlife Refuges (NWR). Detailed descriptions of the SLV and sandhill crane habitats are provided by Kauffeld (1982, 1987), Brown and Drewien (1995), and Drewien et al. (1995).

METHODS

Aerial Survey Design

An aerial survey design similar to that described by Ferguson et al. (1979) and Benning and Johnson (1987) was employed for this survey. A fixed-wing single-engine aircraft

---

[1]Present address: U.S. Fish and Wildlife Service, Endangered Species, P.O. Box 1306, Albuquerque, NM 87103, USA.

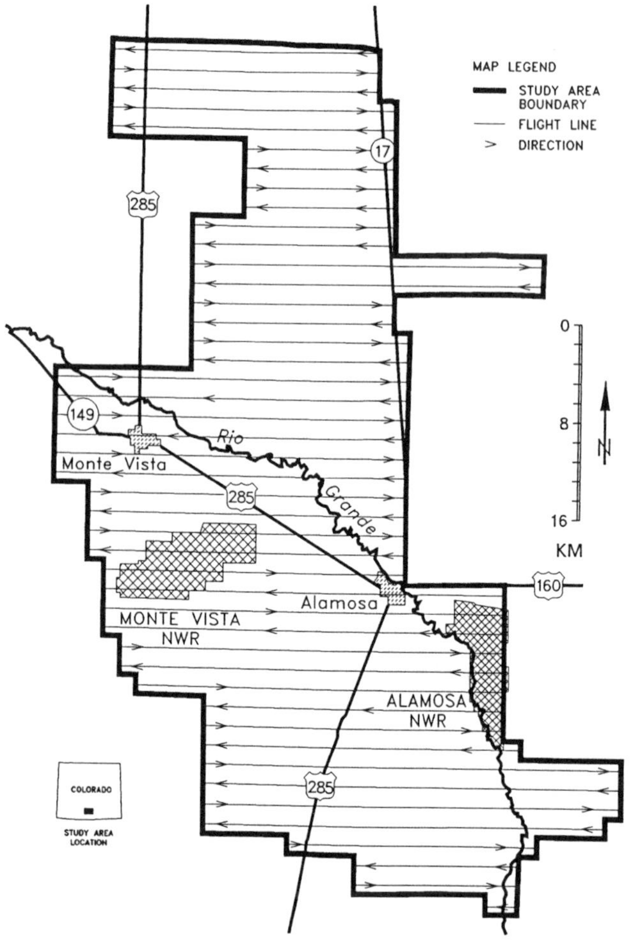

Fig. 1. Transect locations used for March aerial surveys of the Rocky Mountain Population of greater sandhill cranes in the San Luis Valley, Colorado, 1984–95.

(Cessna 185) flown at ~160 km/h ground speed and at an altitude of ~120 m above ground level to optimize observer visibility was used for surveys. The aircraft was equipped in 1995 with a Trimble 2000T Global Position System (GPS) to facilitate accurate navigation of transects and precise identification of individual segments (use of brand names does not imply product endorsement by authors or their employers). An observer (right side) and the pilot (left side) estimated numbers of cranes in flocks within 840 m of the aircraft; each flock was identified by transect and segment. Transect start and end times, pertinent weather and habitat conditions, and soaring cranes were also recorded. Surveys were initiated at ~0730 hours (MST) and ended at ~1730 hours, and were completed in 1 day in most years. In 1984–86 and 1988, 2 consecutive days were required because of inclement weather.

The aerial survey included 45 east-west transects totaling 1,263 km, which varied in length from 1.6 to 40 km, were 1.6 km wide, and covered 21,220 ha (Fig. 1). Section lines were used as transect boundaries and generally were defined by roads, trails, or fence lines; quarter-section lines were used as transect center lines. Each transect was subdivided into 1.6-km segments identified by topographical features or the GPS.

Surveys were scheduled annually in early to mid-March when peak numbers of RMP cranes were in the SLV. Annual RMP migration chronology was determined by monitoring crane numbers at principal wintering sites in the middle Rio Grande Valley, New Mexico, including Bosque del Apache NWR (Drewien and Bizeau 1974; Drewien et al. 1987, 1995). Typically, cranes migrated from winter areas after mid-February (Drewien and Bizeau 1974). Aerial surveys in the SLV were scheduled when winter areas in the Rio Grande Valley were vacated (<1,000 cranes remained) and few cranes had migrated from the SLV. Transects were flown in sequences to minimize the effects of crane movements, and transect order was selected to optimize crane observability.

Aerial surveys were flown in all years except 1992, when deep snow (30–40 cm) resulted in large numbers of cranes, Canada geese (*Branta canadensis*), and ducks crowding into dense flocks at 6 major and 7 minor feeding sites. Cranes were counted during a coordinated morning ground survey at all 13 feeding sites.

Photo Correction for Estimating Bias

Flocks were photographed from the back seat of the aircraft to correct for observer estimation bias (incorrectly estimating numbers) or accuracy error (Bayliss and Yeomans 1990). Single or multiple photographs were taken to compare with pilot and observer flock estimates. The photographer also recorded cranes missed by pilot or observer. We used a Canon 35-mm EOS 630 with 35–105mm zoom lens and a shutter speed of 1/250, and Kodak 100 ASA color print film. We maximized crane image size with the zoom lens, while encompassing the entire flock.

Flock photographs were taken opportunistically when adequate light conditions prevailed. Circling was sometimes required to enhance imagery, but flock estimates were recorded before flight paths were altered. To obtain adequate sample sizes, we completed additional flock photography the day following the survey. We attempted to photograph 40 flocks annually for both pilot and observer.

Photographs were enlarged to 20 × 30 cm in glossy format. Where multiple photographs were taken to encompass large flocks, areas of overlap were eliminated to avoid double counting. Each crane was counted with an electronic pressure-activated stylus. Counts were audited by a second person to ensure that overlap lines were appropriately depicted, all cranes were counted, and other species (primarily geese) were not included.

Subspecies Composition

We sampled flocks by ground surveys and classified cranes as greaters or lessers to assess proportions of RMP and non-RMP cranes; however, field appraisals are not adequate to separate Canadian (intermediate-size sandhill cranes (*G. c. rowani*) (Canadians) and greaters (Drewien and Bizeau 1974, Drewien et al. 1995). Surveys were conducted within 3 days of the aerial count. In 1995, we also obtained an estimate of the proportion of lessers and Canadians mixed in RMP flocks by collecting mid-toe track measurements. Track measurement data were used to adjust raw crane population counts and arrive at an estimate of the number of RMP cranes. Buller and Boeker (1965) and Buller (1967) used mid-toe track measurements to separate lessers (≤95 mm) and greaters (>95 mm). Drewien and Bizeau (1974:735) found that mid-toe measurements for RMP greaters averaged 107.2 mm (SD = 5.7, range 94-123 mm) and that 99% were >95 mm. Guthery (1975) reported that Canadians averaged 96.1 mm (SD = 3.7, range 89-103 mm) and lessers averaged 86.6 mm (range 78-92mm). These data showed that mid-toe track measurements >103 mm were greaters and those ≤95 mm were lessers and the smaller one-half of the Canadians. However, measurements of 96-103 mm overlapped greaters and Canadians and subspecies could not be separated.

Air-ground Comparison Counts for Visibility Bias

Air-ground comparison counts were obtained in 1994-95 to assess visibility bias (cranes not seen by pilot or observer). Selected fields were counted by ground observers within 5 minutes of the aircraft overflight.

Data Analysis

Four potential sources of error influence annual population estimates: (1) absence of some RMP cranes from the survey area, (2) presence of non-RMP cranes in the survey area, (3) inaccurate counts (estimation bias), and (4) inability to observe all cranes (visibility bias). Most wildlife counts also have statistical sampling error; however, the entire crane habitat in the SLV was included in our transects, thus eliminating sampling error.

We minimized the first source of error by scheduling annual counts when virtually all RMP cranes were in the SLV and having cooperators monitor RMP winter and migration areas south and north of the SLV to record cranes present during survey dates. During early surveys, we occasionally found a few cranes outside the original survey area, but we enlarged the area to encompass all utilized crane habitat in the SLV.

To overcome the second source of error, we estimated p, the fraction of cranes in the surveyed area that were RMP cranes based upon data collected during ground surveys. Subspecies composition data were not collected in 1984-86, so for these years we assumed that the same proportions existed as determined in 1987. We applied a binomial variance to the estimates:

$$V(\hat{p})=\hat{p}(1-\hat{p})/n,$$

where $\hat{p}$ is the estimated fraction of the total number of cranes (n) that are greaters.

We accounted for the third source of error (estimation bias) by photographing samples of various-sized flocks and comparing visual estimates to counts from aerial photographs. Sample means and variances for the corrections were calculated. We computed ratio estimates (c) and their variances ($V(c)$) for the pilot and observer separately (Cochran 1977: 6.11).

The total population (T) of RMP sandhill cranes was then estimated by

$$\hat{T}=\hat{p}(\hat{N}_1+\hat{N}_2),$$

where $\hat{N}_1$ and $\hat{N}_2$ estimate number of cranes in the areas surveyed by the pilot and observer, respectively. The fraction of the counted cranes that were greaters (p) was assumed to be the same for both pilot and observer and was estimated by $\hat{p}$. Accounting for the errors in estimating flock sizes, we get

$$\hat{T}=\hat{p}(\bar{c}_1X_1+\bar{c}_2X_2).$$

If estimation errors and the fraction of cranes that were greaters were independent (we assumed so), the estimated variance of this quantity was

$$V(\hat{T})=V(\hat{p})(\bar{c}_1X_1+\bar{c}_2X_2)^2+\hat{p}^2V(\bar{c}_1x_1+\bar{c}_2X_2)$$

$$=V(\hat{p})(\hat{N}_1+\hat{N}_2)^2+\hat{p}^2[V(\bar{c}_1)X_1^2+V(\bar{c}_2)X_2^2]$$

$$=\hat{N}^2\hat{p}(1-\hat{p})/n+\hat{p}^2[V(\bar{c}_1)x_1^2+V(\bar{c}_2)X_2^2],$$

where $\hat{N}=\hat{N}_1+\hat{N}_2$ was the total estimated population size.

We attempted to assess the fourth source of error (visibility bias) by air-ground comparisons in 1994-95. Corrections for visibility bias were not used because of

Table 1. March counts of the Rocky Mountain Population of greater sandhill cranes in the San Luis Valley, Colorado, 1984–95.

Year	Day	Raw aerial estimate	Adjusted for estimation bias: Estimate	SE	Crew estimation error: %	n[b]	Proportion of lessers: %	n	Adjusted to remove lessers[a]: Estimate	SE	RMP cranes outside survey area	Aerial survey conditions	Population index
1984	14–15	10,962	14,488	918	−24.3	42	6.4[c]		13,562	859	550 (N.M. 500) (Colo. 50)	Poor	14,112
1985	13–14	18,393	21,773	1,403	−15.5	75	6.4[c]		20,382	1,314	0	Good	20,382
1986	8–9	14,031	NA[d]		NA	8	6.4[c]		13,135	17	(Wyo.) 20	Poor	13,155
1987	14	13,561	15,661	778	−13.4	21	6.4	5,056	14,660	729	0	Poor	14,660
1988	3–4	17,510	NA		NA	2	6.5	7,617	16,381	21	(Ut.) 22	Poor	16,403
1989	11	17,302	18,389	963	−5.9	50	7.5	6,795	17,004	891	0	Good	17,004
1990	14	20,851	24,593	1,341	−15.2	83	13.7	8,298	21,221	1,157	(Colo.) 275	Good	21,496
1991	9	19,990	18,405	685	+8.6	69	12.8	9,406	16,045	598	(Colo.) 175	Good	16,220
1992	14	23,516[e]	NA		NA	0	15.0	5,998	19,999	28	(Colo.) 9	ground[e]	20,008
1993	11	17,576	NA		NA	0	6.3	5,698	16,478	18	1,260 (Colo. 70) (N.M. 1,190)	Poor	17,738
1994	6	17,229	16,036	884	+7.4	33	6.1	7,353	15,063	831	203 (N.M. 137) (Colo. 66)	Fair	15,226
1995	7	25,276	23,390	1,215	+8.1	70	12.6	7,448	20,229	1,052	0	Good	20,229
$\bar{x}$ (12 yr)	10	18,016	19,092	1,023	−6.8	38	9.6[f]	7,074	17,013	626	210		17,223
$\bar{x}$ (6 yr)[g]	11	20,362	21,310	1,121	−3.9	69	11.3	7,589	19,147	840	76	Good	19,223

[a] Calculated after removing estimation bias.
[b] Number of flocks in photographic sample.
[c] Adjusted with 1987 value because no surveys to determine proportion of lessers were conducted in 1984–86.
[d] NA = No adjustment made (sample too small).
[e] Ground counts on feeding sites (n = 13).
[f] Years 1984–86 not included.
[g] Includes 6 years considered comparable (1985, 1989, 1990, 1991, 1992, and 1995).

insufficient data in those years.

Annual aerial survey conditions were judged to be good, fair, or poor. Good conditions had light winds (<17 km/h), minimal cloud cover (<10%), little or no snow cover, few (<500) or no cranes known to be outside the SLV, and an adequate sample (≥40) of usable flock photographs to correct for estimation bias. Poor survey conditions were characterized by weather factors that adversely influenced crane visibility and quality of counts: ≥10% cloud cover, high winds (≥24 km/h), partial or complete snow cover, large (≥500) numbers of RMP cranes outside the SLV, and few or no suitable flock photographs to correct for estimation bias. Fair conditions were intermediate between good and poor.

RESULTS

Twelve population estimates (1 ground and 11 aerial) were obtained during 1984–95. All surveys were conducted 3–15 March. After adjustment for estimation bias, presence of lessers, and RMP cranes outside the survey area, population estimates for the 12 years varied from 13,155 (1986) to 21,496 (1990) and averaged 17,223 (Table 1). During the 11 aerial surveys, conditions were considered good during 5, fair during 1, and poor during 5. In addition, we judged that the 1992 ground count provided a good population estimate. The RMP mean estimate during 6 good surveys was 19,223, range 16,220 to 21,496 (Table 1), and based on these 6 estimates the population appeared stable (Fig. 2).

Flock photography to correct for estimation bias was considered adequate during 8 aerial surveys, with 21 to 83 flocks photographed annually (Table 1). Over-estimates of crane numbers occurred in 3 years and under-estimates in 5 years. During the 5 years with comparable surveys, annual crew (pilot and observer) estimation errors ranged from −15.5% to +8.6% with a mean of −3.9% (Table 1). The best estimate (closest to the number of cranes counted from photographs) was −5.9% in 1989.

Correction rates for estimation errors for the pilot and

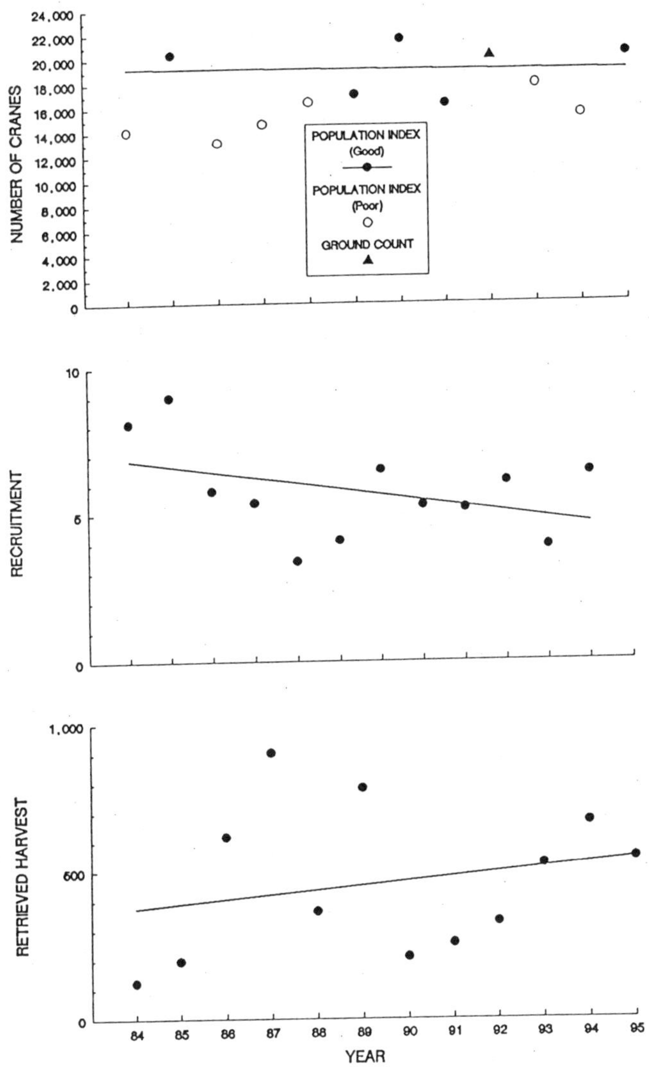

Fig. 2. Estimates of March crane population, recruitment, and retrieved harvest of the Rocky Mountain Population of greater sandhill cranes in the San Luis Valley, Colorado, 1984–95.

observer were averaged for years with similar condition classes prevailing during surveys (Table 2). The pilot tended to underestimate flocks consistently, and more so under good survey conditions. The observer, conversely, had no overall bias when conditions were good, but underestimated flock sizes markedly under poor survey conditions. Fair survey conditions occurred only during 1994 (Table 1), limiting interpretation of these correction rates.

During most years, flock photography samples were increased by obtaining additional photographs and flock size estimates the day following counts. Combined estimation error for both the pilot and observer over all years showed no significant difference (P = 0.63) during on-survey vs. off-survey days.

Table 2. Mean correction rates for estimation error of pilot and observer for sandhill crane numbers recorded in flocks during aerial counts, March 1984–95 (except 1992–93), San Luis Valley, Colorado. Correction rates (rate × flock estimate) were assessed from cranes counted on aerial photographs taken of sample crane flocks.

	Pilot			Observer		
Survey condition	*n* years/ *n* flocks	$\bar{x}$ corr. rate	SD	*n* years/ *n* flocks	$\bar{x}$ corr. rate	SD
Good	5/176	1.19	0.13	5/171	1.00	0.14
Fair	1/18	1.07		1/15	0.87	
Poor	4/39	1.06	0.15	2/12	1.37	0.29

Field appraisals during 1987–95 showed that the lesser subspecies was present annually and ranged from 6.1% (1994) to 15.0% (1992) and averaged 9.6% of cranes surveyed (Table 1). A March 1978 survey found 3.4% lessers (R. C. Drewien, unpubl. data). The proportion of non-RMP cranes during March in the SLV has been increasing (Fig. 3).

Mid-toe track measurements (n = 2,519) collected in 1995 confirmed the presence of both lessers and Canadians. Estimates of proportions of subspecies from mid-toe measurement data indicated that 26.6% were lessers and approximately 50% of the Canadian subspecies (≤95 mm), 47.2% were RMP greaters (>103 mm), and 26.2% overlapped between RMP greaters and Canadians (96–103 mm). Adjusting the 1995 count for only greaters (>103 mm) yielded an RMP estimate of 11,936, whereas removing the proportion of lessers and those Canadians with mid-toe measurements ≤95mm (26.6%) yielded an estimate of 18,557. The high estimate included an unknown number of Canadians whose mid-toe measurements overlapped with greaters.

During 1994–95, we obtained air-ground comparison counts along 28 km of transects to assess visibility bias (Table 3). These data indicated that aerial estimates were lower than ground counts in greasewood (*Sarcobatus vermiculatus*), wet meadows, barley stubble, and chiseled stubble fields; aerial estimates were similar or higher than ground counts in bare, disked fields, wheat stubble, and alkali flats. Results indicated that visibility bias was a definite factor in 4 of 22 samples, but may have affected others. Where visibility bias was the only identifiable source of error in a given sample, few (<100) cranes were involved.

We monitored crane soaring activities during aerial counts, but soaring was rare. Cranes that were noted by the photographer, but missed by the pilot or observer, were also

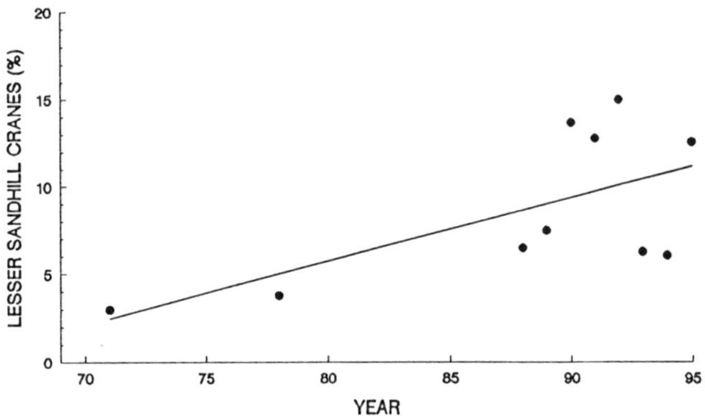

Fig. 3. Proportion of lesser sandhill cranes mixed with Rocky Mountain greater sandhill cranes during March, San Luis Valley, Colorado, 1971–95.

infrequent and involved only small groups, generally <20 birds.

DISCUSSION

The original survey plan called for complete coverage of sandhill crane habitat within the SLV to define distribution patterns and develop a survey design to estimate population size. Once crane distribution was defined, we planned to stratify the survey area and determine the optimal sampling rate within each stratum. During early surveys, high-use areas were identified; however, substantial distributional changes occurred between years and we found numbers often shifted because of (1) changes in availability and distribution of surface water used as overnight roosts and mid-day loafing areas and (2) changes in cropping patterns and tillage practices of preferred food crops. Consequently, we continued total coverage of crane habitats in all years.

All surveys were conducted during 3–15 March when peak numbers of the RMP were judged to be in the SLV. Surveys were not conducted after mid-March because migrants started leaving the SLV in large numbers during most years (Peterson and Drewien 1997). Although we accounted for RMP cranes on winter sites and stopover areas north of the SLV, several migration stopover sites in the Rio Grande and Rio Chama Valleys in north-central New Mexico (Stahlecker 1992) were not surveyed and in some years may have harbored some RMP cranes during our count period. A major spring migration stopover where ≥50% of the RMP stopped annually at Harts Basin near Eckert, Colorado, 225 km northwest of Monte Vista NWR (Peterson and Drewien 1997), was monitored each spring (1985–95) until our surveys were completed. Within the SLV, the aerial crew checked potential crane habitat outside the transect area. On several occasions small flocks (<50) were found, and they were included in the population estimate; the survey area was expanded to incorporate those habitats.

To control known or suspected visibility biases, we developed criteria for acceptable survey conditions. We found that cloud cover reduced or eliminated crane shadows and contributed to visibility errors during aerial surveys, as did Ferguson et al. (1979) and Benning and Johnson (1987). We also found that partial snow cover formed mottled patterns which reduced observability of cranes standing near snow patches. Excessive cloud cover and partial snow cover also resulted in smaller photo samples to correct for estimation bias because these conditions yielded many photographs that were unsuitable for counting cranes. Primarily because of weather and related visibility factors, we judged that 6 surveys were conducted under sub-optimal conditions (Bayliss and Giles 1985:692) and excluded them from our population estimates.

Preliminary data from air-ground comparison counts in 1994–95 indicated that sandhill cranes were also more difficult to see in certain habitat types because of lack of background contrast. Visibility in some riparian habitats may also have been poor because of tree and shrub cover. We believe that observers adequately scanned transect strips in most habitat types; counts obtained from aerial photographs and air-ground comparison surveys did not indicate that transect width created a problem. We intend to obtain additional information to better address crane visibility biases by different habitat types. Under-counting cranes in several

Table 3. Air-ground comparison counts on 28 km of aerial transects during March surveys of the Rocky Mountain Population of greater sandhill cranes in the San Luis Valley, Colorado, 1994 and 1995.

Cover type	Sample (km^2)	Air count	Ground count	Correction (ground/air)
Native cover				
Greasewood-grass complex	3.9	45	144	3.20
Alkali flat	1.3	20	4	0.20
Wet meadow-wetland complex	2.6	270	553	2.05
Subtotal	7.8	335	701	2.09
Cropland				
Alfalfa	3.9	225	220	0.98
Stubble (barley/wheat)	7.8	693	1,299	1.87
Chiselled stubble (barley/wheat)	7.8	1,935	2,108	1.09
Bare soil	1.3	70	81	1.16
Subtotal	20.7	2,923	3,708	1.27
Total	28.5	3,258	4,409	1.35

habitat types would reduce population estimates.

To minimize annual variation in estimation error, we used the same pilot/observer for all surveys and the same observer for 11 of 12 surveys. Experienced observers tend to reduce estimation error with increased knowledge of their individual error rates from previous surveys (Erwin 1982, Bayliss and Yeomans 1990). Inaccurate estimates, in part, resulted from the short time available to count cranes as the aircraft passed overhead. We used aerial photography to correct for estimation bias (Benning and Johnson 1987) and to improve accuracy of our estimates, and recommend the continued use of this technique.

Some consistency of estimation error by an observer for a given set of survey conditions was evident (Table 2). Under either good or poor survey conditions, the pilot was fairly consistent (SD = 0.13/0.15) in underestimating flock sizes (correction rates = 1.19/1.06). The observer under good conditions had no estimation bias, but under poor conditions underestimated flock size appreciably (correction rate = 1.37), and the error was not consistent among years (SD = 0.29), although photography samples were small over 2 years. These findings suggest that for the pilot under good or poor conditions and the observer under good conditions, it might be possible to use data from earlier years with similar conditions in combination with data from the current year to provide an estimator that is more accurate than the current-year value. For example, survey conditions in 1996 were good. The correction rate for estimation by the pilot in 1996 could be combined with the average value (1.19) for the earliest 5 years with good survey conditions. Such combinations, in which the 2 values are weighted by their standard deviations, are an application of the empirical Bayes methodology (e.g., see Morris [1983] for description of theory and Johnson [1989] for an application to wildlife surveys).

A possible improvement in analysis of survey results would involve use of finite population corrections in estimates of the correction rate. The number of flocks photographed to estimate rates was an appreciable fraction of the total surveyed population. Thus, use of the finite population correction (Cochran 1977) would decrease estimated variance by that fraction. For example, in a year in which photographs were taken of 40% of the flocks that the pilot observed, the variance of the estimated correction rate would be reduced by 40%. This revision could be explored for future analyses.

In years when poor survey conditions prevailed, conditions for determining correction rates for estimation errors were also poor. In essence, we were obtaining a poor estimator and were unable to improve it through photographic corrections. Inclement weather or otherwise poor conditions for a survey preclude accurate results. The limited time frame (±7 days) to conduct March surveys when the population peaks precluded extended delays waiting for improved survey conditions. Managers should not make decisions about the RMP based upon a single year's count, especially when the data are poor. Only with several years of high-quality data can any apparent trends in population size be realized. A poor estimate for any year is not really better than having no estimate.

Subspecies composition data from field appraisals and mid-toe track measurements confirmed that all 3 migratory subspecies occurred in the SLV during March surveys. The proportion of non-RMP cranes (lessers and Canadians) has increased substantially since 1971 and this increase adversely influences accuracy of current RMP estimates. Although we can adjust for the presence of lessers, we cannot identify all Canadians by field appraisals or mid-toe track samples.

Mid-toe track measurements in 1995 suggested that a minimum of 26.5% of the population consisted of non-RMP cranes (≤95 mm) and that another 26.2% had measurements that overlapped (96–103 mm) between Canadians and greaters. In contrast, similar data collected in March 1971 showed only 3% of the population were non-RMP cranes, although 30% had overlapping measurements between Canadians and greaters (Drewien and Bizeau 1974:732). Ancillary information suggests that the actual percentage of Canadians mixed with the RMP was probably much lower, possibly <25% of the population segment with overlapping mid-toe track measurements. Data collected at hunter check stations in the principal RMP winter range in western New Mexico since 1982 and southeastern Arizona since 1981 showed that Canadians averaged 7–13% of the harvest in most hunts (C. G. Schmitt and P. M. Smith, pers. commun., *in* Drewien et al. 1995:351, Schmitt and Hale 1997) . Mid-toe track measurements can be used to adjust RMP estimates to obtain a minimum (% track samples >103 mm) and maximum (% track samples >95 mm) range, but it must be recognized that an unknown proportion of track samples measuring 96–103 mm was Canadian subspecies. However, it is not known if the proportion of Canadians represented in the harvest on RMP winter sites was similar to those occurring in the SLV. Recognizing these limitations, we believe March aerial counts can provide estimates of the RMP if suitable survey conditions prevail and adequate samples of mid-toe track measurements are collected.

The Pacific and Central Flyways Management Plan for the RMP calls for assessing the population status annually during March in the SLV and every 5 years on fall pre-migration staging areas (and the SLV) (Smith 1991:42). Coordinated air and ground counts at pre-migration staging areas in Colorado, Idaho, Montana, Utah, and Wyoming in September provide a minimal RMP estimate that is not contaminated by non-RMP cranes. Drawbacks to September counts are (1) cranes are widespread and occupy >40 areas in 5 states, (2) some cranes may not have arrived at staging areas while others may have started migration to the SLV, and (3) considerable effort is required to coordinate this count, which involves 5 state agencies, the FWS, and other interested parties. September counts were made in 1987 (17,481), 1992 (19,297), and 1995 (16,028) and confirm that the RMP contained >16,000–19,000 cranes during this period.

The March population index during 1984–95 indicated a stable population, whereas the retrieved harvest has increased (Sharp 1995), and annual recruitment has decreased (Drewien et al. 1995, R. C. Drewien, unpubl. data) (Fig. 2). Further, long-term (25 yr) count data at 2 of the largest RMP premigration staging areas in Idaho (Teton Basin, Grays Lake) show that fall populations have declined >50% during 1984–95 (R. C. Drewien and W. M. Brown, unpubl. data). The above data suggest that the population may be declining. Increasing numbers of Canadian cranes during March surveys could be masking a population decline. Due to conflicting data on population trend, we recommend against increasing harvest of RMP cranes until (1) population recruitment improves ≥the long-term average (1986–95 $\bar{x}$ = 5.2%, SD = 1.1, range 3.4–6.5; 1972–95 $\bar{x}$ = 7.8, SD = 2.6, range 3.4–12.0) over a 3-year period, (2) counts from hunted premigration staging areas in Montana, Utah, and Wyoming are reviewed and show a stable or increasing population trend over a 3-year period, and (3) RMP counts in the SLV or elsewhere demonstrate an increase. We also recommend that other census methods be considered, including September pre-migration staging area counts and an October count in the SLV. These counts should be evaluated to determine if they could provide a more reliable estimate of the RMP.

LITERATURE CITED

BAYLISS, P., and J. GILES. 1985. Factors affecting the visibility of kangaroos counted during aerial surveys. J. Wildl. Manage. 49:686-692.

______, and K. M. YEOMANS. 1990. Use of low-level aerial photographs to correct bias in aerial survey estimates of magpie goose and whistling duck density in the Northern Territory. Aust. Wildl. Res. 17:1-10.

BENNING, D. S., and D. H. JOHNSON. 1987. Recent improvements to sandhill crane surveys in Nebraska's central Platte Valley. Pages 10-16 *in* J. C. Lewis, ed. Proc. 1985 crane workshop. Platte River Whooping Crane Maintenance Trust, Grand Island, Nebr.

BROWN, W. M., and R. C. DREWIEN. 1995. Evaluation of two power line markers to reduce crane and waterfowl collision mortality. Wildl. Soc. Bull. 23:217-227.

BULLER, R. J. 1967. Sandhill crane study in the Central Flyway. U.S. Fish Wildl. Serv. Spec. Sci. Rep. Wildl. 113. 17pp.

______, and E. L. BOEKER. 1965. Coordinated sandhill crane study in the Central Flyway. Trans. North Am. Wildl. Nat. Resour. Conf. 30:100-112.

COCHRAN, W. G. 1977. Sampling techniques. Third ed. John Wiley and Sons, New York, N.Y. 428pp.

DREWIEN, R. C., and E. G. BIZEAU. 1974. Status and distribution of greater sandhill cranes in the Rocky Mountains. J. Wildl. Manage. 38:720-742.

______, W. M. BROWN, and W. L. KENDALL. 1995. Recruitment in Rocky Mountain greater sandhill cranes and comparison with other crane populations. J. Wildl. Manage. 59:339-356.

______, W. H. MULLINS, and E. G. BIZEAU. 1987. Winter and spring distribution of greater sandhill cranes from south-central Idaho. Pages 89-92 *in* J. C. Lewis, ed. Proc. 1985 crane workshop. Platte River Whooping Crane Maintenance Trust, Grand Island, Nebr.

ERWIN, R. M. 1982. Observer variability in estimating numbers: an experiment. J. Field Ornithol. 53:159-167.

FERGUSON, E. L., D. S. GILMER, D. H. JOHNSON, N. LYMAN, and D. S. BENNING. 1979. Experimental surveys of sandhill cranes in Nebraska. Pages 41-49 *in* J. C. Lewis, ed. Proc. 1978 crane workshop. Colorado State Univ. Printing Serv., Fort Collins.

GUTHERY, F. S. 1975. Footprint measurements of Canadian sandhill cranes. J. Wildl. Manage. 39:447-448.

KAUFFELD, J. D. 1982. Management of migratory crane habitat on Alamosa and Monte Vista National Wildlife Refuges. Pages 117-121 *in* J. C. Lewis, ed. Proc. 1981 crane workshop. Natl. Audubon Soc., Tavernier, Fla.

______. 1987. An avian cholera epizootic among sandhill cranes in Colorado. Pages 145-148 *in* J. C. Lewis, ed. Proc. 1985 crane workshop. Platte River Whooping Crane Maintenance Trust, Grand Island, Nebr.

JOHNSON, D. H. 1989. An empirical Bayes approach to analyzing recurring animal surveys. Ecology 70:945-952.

MORRIS, C. N. 1983. Parametric empirical Bayes inference: theory and applications (with discussion). J. Am. Stat. Assn. 78:47-65.

PETERSON, M. K., and R. C. DREWIEN. 1997. Harts Basin, Colorado: a spring migration stopover area for Rocky Mountain sandhill cranes and whooping cranes. Proc. North Am. Crane Workshop 7:153-156.

SCHMITT, C. G., and B. HALE. 1997. Sandhill crane hunts in the Rio Grande Valley and southwest New Mexico. Proc. North Am. Crane Workshop 7:219-231.

SHARP, D. E. 1995. Status and harvests of the Mid-continent and Rocky Mountain Populations of sandhill cranes. Unpubl. Rep., U.S. Fish Wildl. Serv., Golden, Colo. 8pp.

SMITH, P., chairman. 1991. Management plan of the Pacific and Central flyways for the Rocky Mountain Population of greater sandhill cranes. Pac. Flyway Stud. Comm. and Cent. Flyway Tech. Comm., U.S. Fish Wildl. Serv., Portland, Oreg. 55pp.

STAHLECKER, D. W. 1992. Crane migration in northern New Mexico. Pages 1-12 *in* D. A. Wood, ed. Proc. 1988 North Am. crane workshop. Florida Game Fresh Water Fish Comm. Nongame Wildl. Program Tech. Rep. 12.

MODELING THE POPULATION DYNAMICS OF GULF COAST SANDHILL CRANES

DOUGLAS H. JOHNSON, Northern Prairie Science Center, U.S. Geological Survey–Biological Resources Division, Jamestown, ND 58401, USA

WILLIAM L. KENDALL,[1] Office of Migratory Bird Management, U.S. Fish and Wildlife Service, Laurel, MD 20708, USA

Abstract: The Mid-continent Population of sandhill cranes (*Grus canadensis*) has a large geographic range, contains nearly 500,000 birds, and is hunted in much of its range. The population includes 3 subspecies; the numbers of 2 of these are uncertain, and they should be afforded protection from hunting that would be detrimental to their population. The 2 subspecies of concern tend to concentrate in the eastern part of the Great Plains during fall and spring and to winter along the Gulf Coast in Texas. This paper uses the limited information available about the Gulf Coast subpopulation in a model. We included in the model 5 input parameters: population size, annual survival rate in absence of hunting, the number of birds taken by hunters, the extent of additivity of hunting mortality, and recruitment rate, measured as the fraction of juveniles in the winter population. Using 3 widely ranging estimates of each parameter, we examined the general behavior of the simulated population. Realistic population projections occurred with medium (60,000) or large (166,000) population sizes, low (2,000) or moderate (4,000) harvests, and recruitment rates of 0.07 and 0.11. All values of survival in the absence of hunting and additivity of hunting yielded some realistic projections. Results of modeling suggest that the variables warranting closer monitoring are population size and recruitment rate.

PROC. NORTH AM. CRANE WORKSHOP 7:173-180

Key words: *Grus canadensis*, Gulf Coast subpopulation, Mid-continent, population model, sandhill crane, Texas.

The sandhill crane has one of the most widespread breeding distributions of any crane species; populations range from Siberia, Alaska, western and southern Canada, much of the 48 contiguous states, and into Cuba (Johnsgard 1983). Many populations are disjunct, and ornithologists have distinguished several subspecies on the basis of geographical distribution, morphology, plumage, and behavioral characteristics (Walkinshaw 1973). The Cuban (*G. c. nesiotes*), Florida (*G. c. pratensis*), and Mississippi (*G. c. pulla*) subspecies are nonmigratory, have relatively few members, and are not legally hunted. The other 3 subspecies, greater (*G. c. tabida*), Canadian (*G. c. rowani*), and lesser (*G. c. canadensis*), are migratory and do encounter legal sport and, in some areas, subsistence hunting (Tacha et al. 1994). Greaters comprise 5 identified populations (Drewien and Lewis 1987, Drewien et al. 1995): the Eastern, which is increasing in number and is not hunted; 3 Western populations, which vary in number from 2,000 to about 20,000 and 1 of which is hunted; and the Texas Coastal or Prairie population, part of the Mid-continent group, which is of uncertain size and is hunted. Texas Coastal or Prairie greaters migrate along similar paths and have overlapping winter ranges with Canadians and lessers. The entirety of the Mid-continent Population–about 500,000 birds (Sharp and Vogel 1992, Sharp 1995)–is surveyed each spring in Nebraska (Ferguson et al. 1979, Benning and Johnson 1987), but the racial composition is not known. For purposes here, the Mid-continent Population has been divided into 2 geographical subpopulations, the Western and the Gulf Coast (Tacha et al. 1984, 1986).

Hunting of sandhill cranes was terminated by the Migratory Bird Treaty, ratified in 1916. Legal hunting resumed in the 1960's, and the species has become an important game bird, especially in North Dakota and Texas (Lewis 1977). About 6,600 hunters in the continental U.S. spend 18,000 days each year in pursuit of cranes and harvest about 15,500 cranes annually. Sport or subsistence hunting also takes place in Alaska, Canada, and Mexico. These large birds are of special interest to non-hunters also. Outside the breeding season, sandhill cranes concentrate in large flocks, which often attract large numbers of serious and casual birdwatchers (Lingle 1992).

The advent of legal hunting aroused concerns; Sherwood (1971) and Miller (1974) made dire forecasts about the plight of sandhill crane populations under harvest. Unlike most hunted birds, cranes have a long lifespan and correspondingly low reproductive potential, and their populations cannot rebound quickly from any major decline. Concerns about the effects of hunting led federal and state agencies to active research and monitoring in the 1970's and to a model of sandhill crane population dynamics that incorporated information then available (Johnson 1979). Later, the Central Flyway Waterfowl Council (1981) adopted a management plan for the Mid-continent Population of cranes (all 3 subspecies combined).

The plan, as well as the model of Johnson (1979), treated the Mid-continent Population as if it were homogeneous. It has been long recognized that the hunted population consists

---

[1]Present address: Patuxent Wildlife Research Center, U.S. Geological Survey–Biological Resources Division, 11510 American Holly Drive, Laurel, MD 20708-4017, USA.

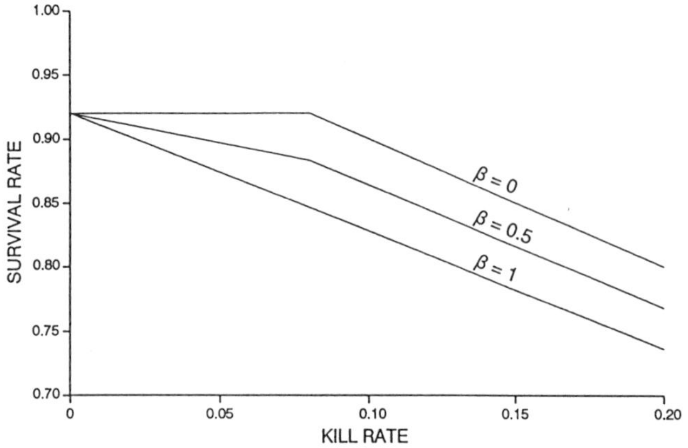

Fig. 1. Survival rate of adult sandhill cranes (S_A) as a function of the kill rate (k) for 3 levels of additivity of hunting mortality (β).

of all 3 migratory subspecies (Johnson and Stewart 1973). Although the lessers are relatively plentiful, the status of the Canadians is uncertain, and the greaters were sufficiently uncommon to be listed as a rare subspecies as recently as 1968 (Committee on Rare and Endangered Wildlife Species 1968). An objective of regulation-setting was to concentrate the harvest primarily on the most abundant subspecies, the lesser.

Within the Mid-continent, greater and Canadian sandhill cranes tend to be distributed farther east than most lessers. This feature has been identified among fall migrants in the northern Great Plains (Johnson and Stewart 1973), wintering cranes in Texas (Guthery and Lewis 1979), and spring migrants in Nebraska (Lewis 1979). Cranes that migrate along the eastern part of the Great Plains and winter along the Gulf Coast of Texas probably include most greater and Canadian cranes in the Mid-continent Population (Guthery and Lewis 1979, Melvin and Temple 1980, Gaines and Warren 1984, Tacha et al. 1984). Accordingly, proper management of that subpopulation would ensure adequate protection of the 2 less-common subspecies (Ad Hoc Mid-continent Sandhill Crane Subcommittee 1994). The objective of work presented in this paper was to model the population dynamics of the Gulf Coast subpopulation of sandhill cranes to gain a better understanding of the population and to identify needs for further research and monitoring.

Some key parameters in the model are estimated poorly if at all. In addition, the model ignores the age structure of the population. Therefore, the model should be used only as a rough approximation to reality. Modeling has several related purposes, including (1) making predictions based on an understanding of the system, which can usefully guide management of the system; (2) identifying important variables in the system, which can be helpful for directing monitoring of the system; and (3) identifying gaps in our understanding of the system, which can be valuable for guiding research. For the system we are examining, the population dynamics of Gulf Coast sandhill cranes, we are beginning to identify gaps in understanding. Modeling may also be useful for shaping a more effective monitoring program.

We appreciate comments on this manuscript by J. E. Austin, R. C. Drewien, R. R. George, F. S. Guthery, S. C. Kohn, M. E. O'Meilia, J. A. Roberson, D. W. Stahlecker, and G. W. Smith.

THE MODEL

The model incorporates the following 5 components for which estimates of parameters must be provided:

SPRING: the population size at the start of a year,
S_0: annual survival rate in absence of hunting,
KILL: the number of birds taken by hunters,
β: extent of additivity of hunting mortality, and
r: the recruitment rate, measured as the fraction of juvenile cranes in the winter population.

The following 4 components are determined internally within the model, according to the prescribed equations:

RATIO: the ratio of juveniles to adults in the winter population; RATIO $= r/(1 - r)$.
FALL: the population size in fall, determined as FALL $=$ SPRING$(\sqrt{S_0}(1 + \text{RATIO}))$, under the assumption that mortality for the half-year from spring to fall occurs at the same rate as during the other half-year, and that RATIO recruits are added to the population for each adult.
k: the fraction of birds in the fall population that are killed by hunters, determined as $k = \text{KILL}/\text{FALL}$.
S_A: the survival rate of adults, determined as $S_A = S_0(1-\beta k) + (\beta-1)\times[k-(1-S_0)]\times \mathrm{I}[k>(1-S_0)]$, where I($Z$) is an indicator function: I(Z) $= 1$ if Z is true, I(Z) $= 0$ if Z is false.

The last equation follows from a survival model (Burnham et al. 1984) that allows hunting mortality to be completely additive to natural mortality (if $\beta = 1$), completely compensated for by natural mortality (if $\beta = 0$), or partially compensated for (if $0 < \beta < 1$). Compensation, either complete or partial, can apply only to mortality that would have occurred in the absence of hunting (i.e., for $k \leq 1-S_0$). Figure 1 illustrates this survival model for 3 values of β.

We executed the model under a variety of situations representing various combinations of parameter values.

Table 1. Estimates of the size of the Gulf Coast subpopulation of sandhill cranes.

Estimate	Source	Comment
13,528	Guthery (1972)	For 1970–71
7,000–12,550	Guthery (1972)	For 1971–72
23,000	Lewis et al. (1977)	
21,960	Guthery and Lewis (1979)	"Conservative estimate"
25,000	Guthery and Lewis (1979)	As cited by Iverson et al. (1985)
50,000–70,000	R. R. George, pers. commun.	Cited by Tacha et al. (1984)
60,000	R. R. George, pers. commun.	Cited by Iverson et al. (1985)
166,000	Muehl (1994)	Cited by Tacha et al. (1994); SE = 81,000
~ 30,000	Tacha et al. (1994)	"Stable" long-term trend

Because some parameters are so poorly known, we used values reflecting the extreme estimates. This effort does not provide model predictions sufficiently accurate to use for management (purpose 1 listed above). It should, however, indicate whether a particular parameter is strongly influential in the model and therefore merits additional attention (purposes 2 and 3 above). We allowed the model to simulate a 25-year period and looked at its general behavior; i.e., did the simulated population become extinct, grow explosively, or remain within realistic bounds? From those results, we identified combinations of parameter values that generated realistic population scenarios.

DATA

Spring Population Size

The Mid-continent Population of sandhill cranes is systematically surveyed each spring in Nebraska (Benning and Johnson 1987), but members of the Gulf Coast subpopulation mingle with the other (Western) subpopulation and cannot be separately counted. Therefore, we used estimates for the wintering Gulf Coast subpopulation, after harvest (Table 1). Most estimates, unfortunately, are subjective. Some natural mortality occurs between late winter and spring, but we assume it is negligible. Estimates of the size of the Gulf Coast subpopulation have varied widely, from 7,000–12,550 birds (Guthery 1972) to 166,000 birds (Muehl 1994). For the model, we used 2 extreme values, 22,000 and 166,000, and 1 intermediate value, 60,000.

Annual Survival in the Absence of Hunting (S_0)

There have been no banding programs or other marking studies that provided useful estimates of the survival rate of cranes in this subpopulation. We had no choice but to use estimates obtained indirectly or from other populations (Table 2). We executed the model with values reflecting the range of most estimates: S_0 = 0.88, 0.92, 0.96.

Number of Birds Taken by Hunters (KILL)

Sandhill cranes are legal game in much of western North America. Cranes in the Mid-continent Population are harvested in Alaska, Saskatchewan, Manitoba, North Dakota, South Dakota, Kansas, New Mexico, Oklahoma, Texas, and Mexico, but most harvest occurs in North Dakota and Texas (Sharp and Vogel 1992, Sharp 1995). The legal sport harvest of sandhill cranes in the United States portion of the Mid-continent is the best known variable used in the model, because a federal permit has been required to hunt cranes in the Central Flyway (Sharp and Vogel 1992). Permit-holders form a sampling universe from which 30–100% of them have been randomly selected each year to receive questionnaires

Table 2. Estimates of annual survival rate of sandhill cranes in the absence of hunting.

Estimate	Source	Comment
0.93–0.96	Johnson (1979)	Indirect, from model
0.82–0.94	Bennett and Bennett (1990)	Florida sandhill cranes
0.88	Tacha et al. (1994)	Eastern and Florida greaters
~0.95	Tacha et al. (1994)	Not evident from cited source
0.91–0.95	Drewien et al. (1995)	Indirect, Rocky Mountain greaters

Table 3. Percentages of juvenile sandhill cranes in fall or winter Gulf Coast populations.

Estimate (%)	Source	Comment
~ 14.5	Guthery (1972)	"Randomly" collected cranes in Texas
10.8, 17.1	Tacha and Vohs (1984)	Oklahoma
21	Tacha and Vohs (1984)	Eastern North Dakota
~ 20	Tacha and Vohs (1984)	Eastern North Dakota, western Oklahoma; cited in Tacha et al. (1984:1032)
18	Tacha et al. (1994:88)	Gulf Coast; not evident how arrived at from sources cited
18.2	Tacha et al. (1986:81)	Eastern Texas, but only 11.7% in bag

about their crane-hunting activity and success. Summaries of hunter kill (harvest plus unretrieved birds) are published annually by the U.S. Fish and Wildlife Service (e.g., Sorenson and Reeves 1976, Martin 1995).

For this analysis, we needed estimates for the Gulf Coast subpopulation. For Texas, we included the total kill in the 29 coastal counties defined by Guthery and Lewis (1979). The average for the 1989–90 through 1994–95 hunting seasons was 1,797. It is not known which counties in North Dakota contribute to the Gulf Coast subpopulation, but presumably they are mostly eastern counties. The 5-year (1990–94) average kill in eastern counties (Rolette, Pierce, Wells, Stutsman and counties eastward) was 1,727. The average for the next westward tier of counties (here termed central North Dakota) was 2,590. Kansas instituted a hunting season in 1993; most harvest occurs in the central part of the state. Most of the cranes that have been checked were Canadian or greater subspecies, and likely affiliated with the Gulf Coast subpopulation. Estimated harvest in Kansas was 655 in 1993 and 821 in 1994 ($\bar{x}$ = 738). Substantial numbers of cranes are harvested in Oklahoma, but most are in western counties and probably not associated with the Gulf Coast subpopulation.

For the model runs, we used KILL values of 2,000, 4,000, and 7,000. The lowest value corresponds roughly to the harvest only in coastal Texas. The high value incorporates also all the harvest in eastern and central North Dakota and a harvest in Kansas equalling the average during 1993–94.

Additivity of Hunting Mortality (β)

The extent to which changes in natural mortality compensate for hunting mortality is very difficult to estimate, even for well-studied and heavily hunted species such as the mallard (*Anas platyrhynchos*) (Burnham et al. 1984, Smith and Reynolds 1992). For the sandhill crane, we can only speculate about β. Because the species is highly territorial on its breeding grounds, it seems plausible that mortality there might be density-dependent, but the extent of breeding-ground mortality is not known. For purposes of modeling, we used the extreme values of β as well as an intermediate value: β = 0, 0.5, 1.

Recruitment Rate (*r*)

This parameter ideally should be measured before hunting commences, because young birds are more susceptible to hunting (Johnson 1979, Tacha and Vohs 1984); disproportionate harvest of juveniles would reduce their percentage in the population. We had several estimates available (Table 3), but some of them simply repeated or summarized earlier ones. Guthery (1972) reported a value of 14.5%, based on cranes assumed to have been collected randomly; if juvenile cranes were more vulnerable, this estimate would be biased high. The high values (18–21%) reported by Tacha and coworkers have been questioned because they exceed values recorded from a variety of studies of sandhill cranes, many of them conducted for several years (Drewien et al. 1995). Relevant values range from 2.4 to 14.3% (Table 4). Tacha's values are in fact greater than those reported for any other species of crane (Johnsgard 1983:38). For our modeling effort, we used recruitment rates of 7%, 11%, and 20%, again representing the range of

Table 4. Percentages of juveniles in sandhill crane populations other than Gulf Coast populations (Drewien et al. 1995).

Estimate (%)	Population
9.2–14.3	Eastern greaters
6.5–13.4	Rocky Mountain greaters
3.1– 8.1	Lower Colorado River Valley greaters
4.5– 6.3	Central Valley greaters
3.8– 9.0	Pacific Coast lessers
4.8–11.3	Mid-continent lessers and Canadians
9.7–11.2	Western subpopulation, Mid-continent lessers and Canadians
5.0–11.2	Florida sandhill cranes
2.4	Mississippi sandhill cranes

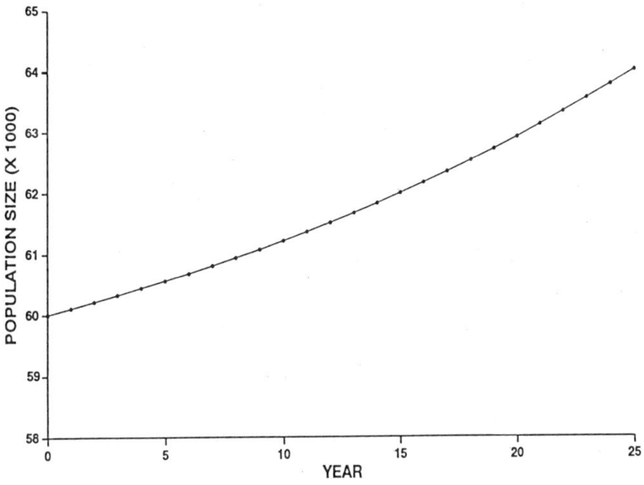

Fig. 2. Example of simulated sandhill crane population projection, based on input parameter values: SPRING = 60,000, r = 0.11, S_0 = 0.92, β = 0.5, and KILL = 4,000.

available estimates.

RESULTS

We executed the model with $3^5 = 243$ combinations of parameter values, and for 25 years under each combination. An example of such a simulation is shown in Fig. 2, which is based on the medial values of each input parameter: SPRING = 60,000, r = 0.11, S_0 = 0.92, β = 0.5, and KILL = 4,000. From its initial size of 60,000, the population increased to 63,987 after 25 years. The projection was very smooth, because the model included no stochastic variation. The approximate stability of the simulated population provided encouraging support for those parameter values.

We considered as potentially realistic those combinations of parameter values for which the simulated population did not decline by more than one-half, or more than double in size, during the 25 years. Only 37 of the 243 combinations met that criterion (Appendix A). Most of these involved the larger 2 SPRING population sizes (60,000 and 166,000); most populations starting at 22,000 collapsed (Table 5). The highest level of harvest (KILL = 7,000) resulted in relatively few realistic scenarios; most of them led to extinction. Conversely, realistic simulated populations were generated almost equally often for the 3 levels of additivity of hunting mortality. Recruitment levels of r = 0.07 and r = 0.11 commonly produced realistic results; the high value, r = 0.20, never did, but usually led to an exploding population. The intermediate value of survival in the absence of hunting (S_0 = 0.92) yielded reasonable projections more frequently than did either extreme value.

Table 5. Results of 25-year simulations of sandhill crane populations for various values of input parameters.

Parameter	Value	Frequency of populations that Collapsed	Remained stable	Exploded
SPRING	22,000	62	1	18
	60,000	34	13	34
	166,000	15	23	43
KILL	2,000	24	16	41
	4,000	36	13	32
	7,000	51	8	22
β	0	27	14	40
	0.5	39	12	30
	1	45	11	25
r	0.07	60	17	4
	0.11	38	20	23
	0.20	13	0	68
S_0	0.88	51	9	21
	0.92	34	17	30
	0.96	26	11	44

If a more stringent criterion were adopted–that the final simulated population after 25 years remain within 20% of the initial value–only 3 combinations of parameter values would meet it. All of these involved the intermediate initial population size, SPRING = 60,000, and relatively low harvest rates.

DISCUSSION

The Gulf Coast subpopulation is too poorly understood at this time to model accurately. Modeling does, however, allow us to put together the best available information. Although the product will not be immediately suitable for fine-tuning management, we hope it can contribute to decisions about further studies and monitoring plans.

In that regard, we evaluated how frequently changes in the values of 5 input parameters caused qualitative changes in the behavior of the modeled population. We recognize that the influence of an input variable as measured here reflects not only its mathematical importance in the model, but also the range of values we used as input. Changes in additivity of hunting mortality (β) produced smaller output changes than any other input variable (Table 5). This finding is fortunate, as β is the most difficult parameter to estimate. Annual survival rate in the absence of hunting (S_0) also was not strongly influential and, again, its value is not well known. The number of cranes harvested (KILL) was the next least influential parameter in the model. Harvest by county is estimated each year, so if the affiliation of each county to the

Gulf Coast subpopulation were better understood, the harvest affecting that subpopulation would be known. The 2 most influential variables were population size (SPRING) and recruitment rate (r). We used very disparate values of SPRING in the model, reflecting divergent estimates that have been made. Fortunately, a systematic survey of the wintering population was recently concluded; a preliminary analysis of results suggests that the population in 1995–96 contained about 50,000 cranes (F. S. Guthery, Texas A&M University, Kingsville, pers. commun.; J. A. Roberson, Texas Parks and Wildlife Department, Austin, pers. commun.). Input values of recruitment rate also represented extreme variation, notably the high value of $r = 0.20$. That value led to no realistic population projections (Table 5), so it seems incompatible with the other estimated parameters and is well outside the range of values estimated for other populations (Table 4).

Research and monitoring are unlikely to provide accurate estimates of β or S_0 in the near future. Crane harvest, by county, is well estimated, but the association of each county with the Gulf Coast subpopulation has not been established, except, of course, for the Texas wintering area. Fortunately, measurements of hunter-shot cranes in North Dakota and Kansas are regularly taken; this information warrants further analysis to identify the geographical distribution of harvest by subspecies, although it is not known how well the measured sample of cranes reflects the actual harvest. A project to refine the subspecies classification procedure is also being planned.

The population size, a critically important parameter in our modeling, lends itself to improved estimation. The recently completed survey (F. S. Guthery, pers. commun.; J. A. Roberson, pers. commun.) may offer a more accurate estimate than is currently available. That improved accuracy will reduce some of the uncertainty in the modeling results. Recent estimates from adjacent Tamaulipas, Mexico, suggest that a minimum of about 2,500 Gulf Coast cranes winter in the area (Drewien et al. 1996).

Likewise, the other highly influential parameter, recruitment rate, can be estimated with greater accuracy by intensive monitoring. Care needs to be taken because of potential biases in sampling procedures and field determination of age (Drewien et al. 1995). Also, recruitment rate can vary dramatically from year to year, so it should be monitored for a number of years.

The model presented here helps to identify some important features of population dynamics, but is too crude for management purposes. It ignores important features such as age-dependent reproduction and differential vulnerability to harvest. Therefore, population surveys and age composition assessments should be conducted on an annual, or at least periodic, basis. In addition, uncertainty about how harvest is allocated between the Gulf Coast and Western subpopulations should be reduced. This kind of information could contribute to the development of better predictive models.

Assessment of the status of the greater sandhill crane, as a subspecies, is more difficult. Field assessments of the subspecies composition of the harvest (Kendall et al. 1997) can be used to estimate the kill of each subspecies. Population counts of greaters can be accomplished only on their breeding grounds. Absent that information, the subspecies composition of the harvest from the wintering grounds could be used in conjunction with the aerial count to approximate their numbers if we can assume that the distributions of harvest pressure and crane abundance across the wintering area are equivalent, and that, in any given location, the relative vulnerability is the same for all subspecies.

LITERATURE CITED

AD HOC MID-CONTINENT SANDHILL CRANE SUBCOMMITTEE. 1994. Meeting summary. Unpubl. Rep., Cent. Manage. Unit Tech. Comm., Golden, Colo. 16pp.

BENNETT, A. J., and L. A. BENNETT. 1990. Survival rates and mortality factors of Florida sandhill cranes in Georgia. North Am. Bird Bander 15:85-88.

BENNING, D. S., and D. H. JOHNSON. 1987. Recent improvements to sandhill crane surveys in Nebraska's central Platte valley. Pages 10-16 *in* J. C. Lewis, ed. Proc. 1985 crane workshop. Platte River Whooping Crane Maintenance Trust, Grand Island, Nebr.

BURNHAM, K. P., G. C. WHITE, and D. R. ANDERSON. 1984. Estimating the effect of hunting on annual survival rates of adult mallards. J. Wildl. Manage. 48:350-361.

CENTRAL FLYWAY WATERFOWL COUNCIL. 1981. Management plan for Mid-continent sandhill cranes. Cent. Flyway Waterfowl Counc., Denver, Colo. 21pp.

COMMITTEE ON RARE AND ENDANGERED WILDLIFE SPECIES. 1968. Rare and endangered fish and wildlife of the United States. U.S. Fish Wildl. Serv. Resour. Publ. 34. [various pagination].

DREWIEN, R. C., W. M. BROWN, and D. S. BENNING. 1996. Distribution and abundance of sandhill cranes in Mexico. J. Wildl. Manage. 60:270-285.

______, W. M. BROWN, and W. L. KENDALL. 1995. Recruitment in Rocky Mountain greater sandhill cranes and comparison with other crane populations. J. Wildl. Manage. 59:339-356.

______, and J. C. LEWIS. 1987. Status and distribution of cranes in North America. Pages 469-477 *in* G. W. Archibald and R. F. Pasquier, eds. Proc. 1983 int. crane workshop. Int. Crane Found., Baraboo, Wis.

FERGUSON, E. L., D. S. GILMER, D. H. JOHNSON, N. LYMAN, and D. S. BENNING. 1979. Experimental surveys of sandhill cranes in Nebraska. Pages 41-49 *in* J. C. Lewis, ed. Proc. 1978 crane workshop. Colorado State Univ. Printing

Serv., Fort Collins.

GAINES, G. D., and R. J. WARREN. 1984. Genetics and morphology of sandhill crane populations in Texas. J. Wildl. Manage. 48:1387-1393.

GUTHERY, F. S. 1972. Food habits, habitat, distribution, numbers, and subspecies of sandhill cranes wintering in southern Texas. M.S. Thesis, Texas A&M Univ., College Station. 81pp.

______, and J. C. LEWIS. 1979. Sandhill cranes in coastal counties of southern Texas: taxonomy, distribution, and populations. Pages 121-128 *in* J. C. Lewis, ed. Proc. 1978 crane workshop. Colorado State Univ. Printing Serv., Fort Collins.

IVERSON, G. C., P. A. VOHS, and T. C. TACHA. 1985. Distribution and abundance of sandhill cranes wintering in western Texas. J. Wildl. Manage. 49:250-255.

JOHNSGARD, P. A. 1983. Cranes of the world. Indiana Univ. Press, Bloomington. 257pp.

JOHNSON, D. H. 1979. Modeling sandhill crane population dynamics. U.S. Fish Wildl. Serv. Spec. Sci. Rep. 222. 10pp.

______, and R. E. STEWART. 1973. Racial composition of migrant populations of sandhill cranes in the northern plains states. Wilson Bull. 85:148-162.

KENDALL, W. L., D. H. JOHNSON, and S. C. KOHN. 1997. Subspecies composition of sandhill crane harvest in North Dakota, 1968-94. Proc. North Am. Crane Workshop 7:201-208.

LEWIS, J. C. 1979. Taxonomy, food, and feeding habitat of sandhill cranes, Platte Valley, Nebraska. Pages 21-28 *in* J. C. Lewis, ed. Proc. 1978 crane workshop. Colorado State Univ. Printing Serv., Fort Collins.

______, chairman. 1977. Sandhill crane (*Grus canadensis*). Pages 4-43 *in* G. C. Sanderson, ed. Management of migratory shore and upland game birds in North America. Int. Assoc. Fish Wildl. Agencies, Washington, D.C.

LINGLE, G. R. 1992. History and economic impact of crane-watching in central Nebraska. Proc. North Am. Crane Workshop 6:33-37.

MARTIN, E. M. 1995. Sandhill crane harvest and hunter activity in the Central Flyway during the 1994-95 hunting season. Unpubl. Rep., U.S. Fish Wildl. Serv., Off. Migr. Bird Manage., Laurel, Md. 12pp.

MELVIN, S. M., and S. A. TEMPLE. 1980. Migration ecology and wintering grounds of sandhill cranes from the interlake region of Manitoba. Unpubl. Rep., Dep. Wildl. Ecol., Univ. Wisconsin, Madison. 60pp.

MILLER, R. S. 1974. The programmed extinction of the sandhill crane. Nat. Hist. 83(2):62-69.

MUEHL, G. T. 1994. Distribution and abundance of water birds and wetlands in coastal Texas. M.S. Thesis, Texas A&M Univ.-Kingsville, Kingsville. 130pp.

SHARP, D. E. 1995. Status and harvests of the Mid-continent and Rocky Mountain Populations of sandhill cranes, 1995. Unpubl. Rep., U.S. Fish Wildl. Serv., Off. Migr. Bird Manage., Golden, Colo. 22pp.

______, and W. O. VOGEL. 1992. Population status, hunting regulations, hunting activity, and harvests of mid-continent sandhill cranes. Proc. North Am. Crane Workshop 6:24-32.

SHERWOOD, G. 1971. If it's big and flies-shoot it! Audubon Mag. 73:72-99.

SMITH, G. W., and R. E. REYNOLDS. 1992. Hunting and mallard survival, 1979-88. J. Wildl. Manage. 56:306-316.

SORENSON, M. F., and H. M. REEVES. 1976. Sandhill crane harvest and hunter activity in the Central Flyway during the 1975-76 hunting season. Unpubl. Rep., U.S. Fish Wildl. Serv., Off. Migr. Bird Manage., Laurel, Md. 9pp.

TACHA, T. C., D. E. HALEY, and R. R. GEORGE. 1986. Population and harvest characteristics of sandhill cranes in southern Texas. J. Wildl. Manage. 50:80-83.

______, S. A. NESBITT, and P. A. VOHS. 1994. Sandhill crane. Pages 77-94 *in* T. C. Tacha and C. E. Braun, eds. Migratory shore and upland game bird management in North America. Int. Assoc. Fish Wildl. Agencies, Washington, D.C.

______, and P. A. VOHS. 1984. Some population parameters of sandhill cranes from mid-continental North America. J. Wildl. Manage. 48:89-98.

______, ______, and G. C. IVERSON. 1984. Migration routes of sandhill cranes from mid-continental North America. J. Wildl. Manage. 48:1028-1033.

WALKINSHAW, L. H. 1973. Cranes of the world. Winchester Press, New York, N.Y. 370pp.

Appendix A. Combinations of parameter values for which the simulated sandhill crane population neither declined by more than one-half nor more than doubled during 25 years.

Population size					
Initial	Final	KILL	r	S_0	β
22,000	16,585	2,000	0.11	0.88	0.0
60,000	45,790	2,000	0.07	0.92	0.0
60,000	95,895	2,000	0.07	0.96	0.5
60,000	59,094[a]	2,000	0.07	0.96	1.0
60,000	45,234	2,000	0.11	0.88	0.0
60,000	100,711	2,000	0.11	0.92	0.5
60,000	63,987[a]	2,000	0.11	0.92	1.0
60,000	45,240	4,000	0.07	0.92	0.0
60,000	89,593	4,000	0.07	0.96	0.0
60,000	45,234	4,000	0.11	0.88	0.0
60,000	63,987[a]	4,000	0.11	0.92	0.5
60,000	117,340	4,000	0.11	0.96	1.0
60,000	76,591	7,000	0.11	0.92	0.0
60,000	98,246	7,000	0.11	0.96	0.0
166,000	126,686	2,000	0.07	0.92	0.0
166,000	105,561	2,000	0.07	0.92	0.5
166,000	84,435	2,000	0.07	0.92	1.0
166,000	330,325	2,000	0.07	0.96	0.5
166,000	293,524	2,000	0.07	0.96	1.0
166,000	125,148	2,000	0.11	0.88	0.0
166,000	104,602	2,000	0.11	0.88	0.5
166,000	84,056	2,000	0.11	0.88	1.0
166,000	306,789	2,000	0.11	0.92	1.0
166,000	126,686	4,000	0.07	0.92	0.0
166,000	84,435	4,000	0.07	0.92	0.5
166,000	293,524	4,000	0.07	0.96	0.5
166,000	219,923	4,000	0.07	0.96	1.0
166,000	125,148	4,000	0.11	0.88	0.0
166,000	84,056	4,000	0.11	0.88	0.5
166,000	306,789	4,000	0.11	0.92	0.5
166,000	233,341	4,000	0.11	0.92	1.0
166,000	126,686	7,000	0.07	0.92	0.0
166,000	238,318	7,000	0.07	0.96	0.5
166,000	109,520	7,000	0.07	0.96	1.0
166,000	125,148	7,000	0.11	0.88	0.0
166,000	251,703	7,000	0.11	0.92	0.5
166,000	123,170	7,000	0.11	0.92	1.0

[a] Final population within 20% of initial population size.

COMPUTER SIMULATIONS OF POSSIBLE FUTURES FOR TWO FLOCKS OF WHOOPING CRANES

CLAIRE M. MIRANDE, International Crane Foundation, E-11376 Shady Lane Road, Baraboo, WI 53913, USA
JOHN R. CANNON, Program in Sustainable Development and Conservation Biology, University of Maryland, College Park, MD 20742, USA
KIMBERLY AGZIGIAN, Program in Sustainable Development and Conservation Biology, University of Maryland, College Park, MD 20742, USA
ROXANNE E. BOGART, Program in Sustainable Development and Conservation Biology, University of Maryland, College Park, MD 20742, USA
SARAH CHRISTIANSEN, Program in Sustainable Development and Conservation Biology, University of Maryland, College Park, MD 20742, USA
JASON DUBOW, Program in Sustainable Development and Conservation Biology, University of Maryland, College Park, MD 20742, USA
A. KATYA FERNANDEZ, Program in Sustainable Development and Conservation Biology, University of Maryland, College Park, MD 20742, USA
DUSTIN K. HOWARTH, Program in Sustainable Development and Conservation Biology, University of Maryland, College Park, MD 20742, USA
CLAUDIA JONES, Program in Sustainable Development and Conservation Biology, University of Maryland, College Park, MD 20742, USA
KATHERINE G. MUNSON, Program in Sustainable Development and Conservation Biology, University of Maryland, College Park, MD 20742, USA
SONAL I. PANDYA, Program in Sustainable Development and Conservation Biology, University of Maryland, College Park, MD 20742, USA
GINA SEDAGHATKISH, Program in Sustainable Development and Conservation Biology, University of Maryland, College Park, MD 20742, USA
KEVIN L. SKERL, Program in Sustainable Development and Conservation Biology, University of Maryland, College Park, MD 20742, USA
SUSAN A. STENQUIST, Program in Sustainable Development and Conservation Biology, University of Maryland, College Park, MD 20742, USA
JENNIFER WHEELER, Program in Sustainable Development and Conservation Biology, University of Maryland, College Park, MD 20742, USA

Abstract: We conducted computer simulations using the program VORTEX (version 7) to project population sizes, growth rates, genetic diversity, and probabilities of extinction over the next 100 years for 2 flocks of whooping cranes (*Grus americana*), the Aransas/Wood Buffalo population and the experimental Florida population. Standard runs based on best estimates of demographic, genetic, and environmental parameter values were used as a baseline to which several alternative scenarios were compared. Results generally supported the conclusion of the earlier Population Viability Assessment (Mirande et al. 1991) that the Aransas/Wood Buffalo population will continue to grow steadily with less than a 1% probability of extinction. It was noted, however, that a combination of negative factors such as shrinking habitat and increased probabilities of catastrophes accompanied by increased mortality rates could put this population at risk. Results for the Florida population were less optimistic. The standard run produced a population growth rate (r) of only 0.0026 for the next 100 years, and this shifted down to −0.0001 over a 200-year time frame. Adult mortality in this flock would have to be about 20% lower than the predicted value (10%) in order to raise growth rates to above r = 0.02. Amount and duration of supplementation of the Florida flock had minimal impacts on the long-term growth rate of the flock. It is the enduring rates of mortality, breeding, and disease risk that will have major effects on this population. For example, if disease risks turn out to be greater than the best-estimate scenario, this population could face a relatively high risk of extinction (17%). The formula for success in Florida is lower adult mortality, lower age of first breeding, lower disease risk, and higher productivity than the best-guess estimates. Fortunately, there are some potential management interventions (e.g., predator control, vaccines and health monitoring, selective introductions to balance the sex ratio of the flock) that may be able to push the odds in favor of success.

PROC. NORTH AM. CRANE WORKSHOP 7:181-200

Key Words: computer, *Grus americana*, population viability analysis, whooping crane.

In 1970 the whooping crane was listed as an endangered species by the United States Fish and Wildlife Service (USFWS). This species is the most endangered of the world's 15 crane species. The USFWS designated critical habitat in 1978 and has developed recovery plans detailing management goals and strategies for whooping crane conservation. To provide protection for the migratory population across international borders, the Canadian Wildlife Service (CWS) and the USFWS jointly implement whooping crane recovery efforts through the Memorandum of Understanding on the Conservation of the Whooping Crane (USFWS, NBS, CWS, and PC 1995).

One captive and 3 wild populations of whooping cranes currently exist, all of which originate from the same (<14) founding birds (Mirande et al. 1991). The captive population consists of more than 100 birds divided among 4 facilities. The only remaining natural wild population migrates between Wood Buffalo National Park in the Northwest Territories of Canada and Aransas National Wildlife Refuge in Texas (hereafter AWP). In the fall of 1995, 155 birds of the AWP were counted on their wintering grounds in Texas. A second wild population is an experimental flock introduced to the Kissimmee Prairie, Florida, beginning in 1993 (hereafter FP). In December 1995, this flock had a population of 18 birds ranging from 1 to 3 years of age. A group of 3 wild whooping cranes that migrates annually between Gray's Lake National Wildlife Refuge in Idaho and Bosque del Apache National Wildlife Refuge in New Mexico is the remnant of an experiment to establish a migratory population by cross-fostering by sandhill cranes (*Grus canadensis*). The experiment was terminated in 1988 due to lack of breeding and a high mortality rate within the whooping crane flock. Failure to breed was attributed to sexual imprinting on the foster species (Mahan and Simmers 1992, USFWS 1994).

The goal of the U.S. Whooping Crane Recovery Plan of 1994 is to downlist the species to "threatened" by the year 2020. The maintenance for at least 10 years of 40 breeding pairs in the AWP, 25 breeding pairs in the FP, and 25 breeding pairs in an additional proposed wild flock would allow downlisting.

Interest in the plight of the whooping crane has provided the rare opportunity to conduct long-term population monitoring and research in genetics, captive-breeding, and reintroduction. Due to the efforts of biologists working with the last remaining natural population (AWP), 57 years of demographic data on whooping cranes are available. These data (Appendix A) allow extensive analyses of population viability to be conducted under a variety of potential scenarios.

This report presents whooping crane population viability assessments (PVA's) for both the AWP and the FP using VORTEX version 7. VORTEX is a "Monte Carlo simulation of the effects of deterministic forces as well as demographic, environmental, and genetic stochastic events on wildlife populations" (Lacy et al. 1995:4). A PVA is just one of many tools that can be used to guide management decisions and research priorities for populations of species of concern. It should be understood that this tool is only as credible as the data that are used, and that long-term population studies are a prerequisite to receiving reliable PVA results.

The objectives of this study were to (1) update the 1991 PVA of the AWP by making adjustments to specific demographic and environmental parameters (these adjustments were made based on new population data obtained from 1991 to 1995 and other more recent estimates of environmental variability and human-induced threats to habitat); (2) update the 1991 PVA of the FP by incorporating population data obtained from 1993 to 1995 and information on environmental variability and human-induced threats to habitat; (3) identify how specific variations in demographic, genetic, and environmental parameters affect population size, extinction probability, rate of growth, and genetic diversity in both populations; and (4) determine how potential management strategies may impact the populations and assist in meeting the USFWS downlisting goal by 2020.

We are extremely grateful for the generous assistance provided by R. C. Lacy, T. Stehn, S. Nesbitt, J. C. Lewis, S. Fisher, G. Gee, C. Bocetti, J. Hatfield, J. Langenberg, C. Brand, and M. Spalding. Reviews by J. C. Lewis, R. C. Lacy, T. Stehn, and D. Stahlecker improved the original manuscript.

METHODS

VORTEX Input Parameters for the AWP

The set of standard parameters developed for the AWP is shown in Table 1. These values are our best estimates based on the best available data, expert opinion, current literature, and existing conditions and management technology. These values provide a baseline to which other simulations can be compared. Variations from standard values were tested to explore the response of the population to variation in these parameters. Those variations that proved most interesting or sensitive were explored further through combination variations. The AWP viability assessment results table presented later in the report itemizes the parameter(s) varied for each AWP simulation. These parameters are discussed below.

Using the inputs for the standard run, and both an estimated actual age distribution and a stable age distribution, we conducted 2 "check runs" to verify the accuracy of the input parameters. The population growth during the past 20 years was modeled with the actual 1975 population size of 49 individuals as the starting population. The estimated actual age distribution of the 49 individuals was calculated by using actual annual survival rates for the years prior to 1975 (see Appendix B for an example of this type of age distribution estimation).

For the purposes of all AWP simulations, "birth" was defined as the arrival of migrating chicks at Aransas National Wildlife Refuge in the fall because data exist on these arrivals since 1938–39. In a similar manner, "death" was defined as failure to return to Aransas the following year. Because of these conventions, values for reproduction are not fecundity rates for cranes and should not be viewed as such.

For the standard run 1,000 simulations were run for 100 years. For all variations, 100 simulations were run for 100 years, unless otherwise noted.

Inbreeding Depression.—Inbreeding depression is possible because the population experienced a bottleneck in 1941. The heterosis model was chosen since there is not much information about recessive lethal alleles in the population. The value "3 lethal equivalents" was chosen because that value represents an approximate average number for many large vertebrate species, such as mammals, which have been well-studied for inbreeding (Lacy et al. 1995).

Age of First Breeding.—A standard age of first breeding of 4 years was used for both males and females. This value was established from the estimated current age distribution (Appendix B) by determining that a breeding age of 4 was necessary to account for the actual number of nests present in Wood Buffalo National Park (WBNP) in 1995. A variation

Table 1. Input parameters for standard run of VORTEX simulation of Aransas/Wood Buffalo whooping crane population.

Parameter	Input value	Reference/rationale
Number of simulations	1,000	
Number of years	100	
Reporting interval	10	
Number of populations	1	
Inbreeding depression	N	Mirande et al. (1991)
Environmental variance (EV) correlation	Y	Edwards et al. (1994)
Number of catastrophe types	3	Mirande et al. (1991), consensus of expert estimates
Monogamous or polygamous	M	Ehrlich et al. (1988)
Female breeding age	4	Based on current number of breeding pairs and estimated age distribution
Male breeding age	4	Based on current number of breeding pairs and estimated age distribution
Maximum age	40	Based on actual annual survival rates (see Appendix B)
Sex ratio	0.5	Edwards et al. (1994)
Maximum brood size	1	Assumes continuing egg removals
Density dependent breeding	N	Edwards et al. (1994)
Percent brood size	0=52.4% 1=47.6%	Calculations based on breeding data in Edwards et al. (1994) applied to 20-year data set (see Appendix C)
EV-reproduction	10	Calculations based on 20-year data set (see Appendix C) Square root of annual variance in recruitment for the last 20 years minus expected stochastic variance (see Lacy et al. 1995)
Female mortality: age 0	8%	Weighted mean mortality for last 20 yrs (see Appendix C)
EV in mortality	3	Square root of variance in annual mortality for the last 20 years minus expected stochastic variance (see Lacy et al. 1995)
Female mortality: all ages above 0	8%	Weighted mean mortality for last 20 yrs (see Appendix C)
EV in mortality	3	Square root of variance in annual mortality for the last 20 years minus expected stochastic variance (see Lacy et al. 1995)
Male mortality: age 0	8%	Weighted mean mortality for last 20 yrs (see Appendix C)
EV in mortality	3	Square root of variance in annual mortality for the last 20 years minus expected stochastic variance (see Lacy et al. 1995)
Male mortality: all ages above 0	8%	Weighted mean mortality for last 20 yrs (see Appendix C)
EV in Mortality	3	Square root of variance in annual mortality for the last 20 years minus expected stochastic variance (see Lacy et al. 1995)
P of catastrophe 1 (chemical spill)	2%	T. Stehn (pers. commun.)
Severity-reproduction	1.0	Impact reflected in reduced survival
Severity-survival	0.9	T. Stehn (pers. commun.)
P of catastrophe 2 (hurricane)	0.33%	Mirande et al. (1991), T. Stehn (pers. commun.)
Severity-reproduction	1.0	Impact reflected in reduced survival
Severity-survival	0.5	Mirande et al. (1991), consensus of expert estimates
P of catastrophe 3 (disease)	5%	J. Langenberg, C. Brand, and T. Stehn (pers. commun.)
Severity-reproduction	1.0	Impact reflected in reduced survival
Severity-survival	0.9	J. Langenberg, C. Brand, T. Stehn (pers. commun.)
All male breeders?	1	Mirande et al. (1991)
Stable age distribution?	N	Number of known breeders too large for stable age distribution assumption
Initial female ages: age 1 to age 40	Age 1 to age 40[a]	All of the values are based on estimated actual age distribution derived from actual annual survival rates (see Appendix B)
Initial male ages: age 1 to age 40	Age 1 to age 40[b]	All of the values are based on estimated actual age distribution dervied from actual annual survival rates (see Appendix B)
Carrying capacity (K)	500	T. Stehn (pers. commun.)
EV in K	0	T. Stehn (pers. commun.)
Trend in K	N	T. Stehn (pers. commun.)
Harvest	N	Illegal
Supplement	N	Consensus of U.S. and Canadian Recovery Teams

[a] 3,7,6,3,4,6,5,6,5,4,3,1,2,0,1,1,0,1,2,1,1,0,1,0,1,0,1,1,0,0,0,0,0,0,1,0,0,0,0,1 years.
[b] 4,7,6,3,4,6,5,5,5,3,3,1,1,0,1,1,1,2,1,0,0,0,0,1,0,1,0,0,0,1,1,0,0,0,0,0,0,0,0,0 years.

was also conducted using 5 as the age of first breeding.

Maximum Age.—Maximum age was determined from the age distribution calculated from annual survival rates over the last 50 years (Appendix B). Variations on maximum age were used to determine how reduced longevity, and thus productivity, affected population growth (*r*) and the final population size.

Twinning/Productivity.—Because some natural breeding pairs produce twins, twinning was tested to compare against the standard run that reflects egg removal and a maximum brood size of 1. For years when data were available on the number of known breeding pairs at WBNP (Edwards et al. 1994), the average percentage of the total flock that were potential breeders was calculated to be 49%. This value was then used to determine the total number of breeders and breeding pairs (i.e., one-half the number of breeders) in the 57-year data set (804 pairs) and the 20-year data set (502 pairs).

Since no twins arrived at Aransas during the 20-year data set period, the percentage of successful nests during this period was determined as the total number of chicks arriving during the 20 years divided by the total number of pairs of breeders. This was the value for brood size 1 during the 20-year period (47.6%). Brood size 0 was simply 1.0 minus 0.476, or 52.4%. Similar calculations were obtained for the 57-year data set, no twins assumed (a false assumption for the past, but a possible scenario for the future if egg removals in WBNP were to continue indefinitely).

To obtain brood sizes for scenarios that included twinning, we analyzed data from the 29-year period of Aransas records before egg removals began (i.e., 1938–39 through 1966–67) (T. Stehn, USFWS, pers. commun.). During this period, nesting success (again, as indicated by Aransas arrivals) was as follows: brood size 0 = 41.7%, brood size 1 = 50.8%, and brood size 2 = 7.5%. These values account for the 131 chicks (101 singletons and 15 pairs of twins) that were produced by 199 pairs of breeders (0.5 × 49% of the 815 total birds) during the 29-year period. Twinning was also run for 1,000 simulations to more adequately test the impacts of having brood sizes of 2. These latter brood size percentages constitute our best estimates of broods that would be produced under conditions of no removal or manipulation (i.e., nest switching) of eggs.

To test for the effect of twinning with all other variables (e.g., number of breeding birds) held constant, we conducted a simulation in which the brood size 0 of 52.4% from the standard run was used and the other 47.6% that was brood size 1 in the standard run was divided into brood size 1 of 41.5% and brood size 2 of 6.1%. These latter percentages were calculated by using the historic twinning ratio derived from the data from 1938–39 through 1966–67. This simulation was run for 1,000 iterations just like the standard run. Additional sensitivity runs were conducted setting the twinning percentages at 2% and 10% (i.e., in the low twinning run, 0=52.4%, 1=45.6%, and 2=2%; in the high twinning run, 0=52.4%, 1=37.6%, and 2=10%). The environmental variation (EV) value for reproduction was calculated by taking the square root of the variance in the annual number of chicks per female minus the expected stochastic variance (Lacy et al. 1995).

Mortality.—The mortality values used for the AWP simulations were weighted means from AWP data for the last 20 years (Appendix C) or for the last 57 years (Appendix A). The effects of higher and lower annual mortality were explored because, potentially, mortality can be managed to some extent (e.g., mortality due to powerlines can be decreased by management practices). The EV for mortality was calculated by taking the square root of the variance in annual mortality minus the expected stochastic variance (Lacy et al. 1995).

Catastrophes.—The major catastrophes potentially affecting the AWP are disease, hurricanes, and chemical spills. The hurricane season usually occurs when the flock is in Canada (Mirande et. al 1991). Thus, there is a low probability that a hurricane will affect the survival of the population.

Several experts have estimated the upper range of the probability of disease outbreak to be about 5% annually (J. Langenberg, C. Brand, and T. Stehn, pers. commun.). Chemical spills from barges traveling along the Gulf Intracoastal Waterway (GIWW) may pose the greatest threat to the survival of this population (USFWS 1994). Chemical spills are also the catastrophes over which people have the most potential for control through GIWW traffic restrictions. To explore this catastrophe further, we varied the severity and frequency of chemical spills. Note that in all these variations the severity of reproductive impacts was considered to be reflected in the severity of survival impacts and thus was kept at a value of 1.0 (no impact).

Stable Age Distribution.—A simulation with a stable age distribution was performed to compare to the results obtained from use of the estimated actual age distribution.

Carrying Capacity.—The standard value for carrying capacity (*K*) was based on expert opinion (T. Stehn, pers. commun.). Carrying capacity was varied due to uncertainty regarding what constitutes acceptable habitat and how existing habitat may be altered in the future. The reduction of freshwater flow into the habitat surrounding Aransas is projected to cause a decline in the blue crab (*Callinectes sapidus*) population, a major food source for the cranes (see Trends in *K*, below) (T. Stehn, pers. commun.). Higher carrying capacities were included as variations to explore the effect of increased habitat on population growth.

Trends in Carrying Capacity.—Trends in carrying capacity were modeled to reflect the impact of reduced freshwater inflow into the Aransas area caused by upstream diversions by humans. Discussions with experts indicated that reduced freshwater inflow is expected to reduce the blue crab population by about 8% over the next 50 years. This reduction could increase crane mortality rates, particularly for juveniles. Experts estimate that mortality could increase to the maximum that has ever occurred over the past 20 years (T. Stehn, pers. commun.). A possible relationship between crab population numbers and whooping crane nesting the following spring has been observed. For example, the crab population was low in the winter of 1993-94 (Chavez-Ramirez 1996), and the following spring only 61% of pairs nested. In 1994-95, a year of high blue crab population, 96% of available pairs nested. In addition to impacting mortality and reproduction, reduced inflows could lower overall carrying capacity of the habitat.

Supplementation.—Supplementation was modeled to determine any potential positive impacts on population growth rate and size.

Combination Variations.—Combinations of reduced carrying capacity and increased catastrophe probability and severity were modeled to reflect possible worst case scenarios for the AWP.

Simulation Based on the 57-year Data Set.—Data from 1938-39 to 1994-1995 (i.e., all 57 years as opposed to the last 20 years) were investigated due to their higher variability. A set of parameters based on the 57-year data set was derived in a similar fashion as that for the standard run (refer to Table 1 and above discussions). Simulations using the 57-year data were run both with catastrophes and without catastrophes. It can be argued that the 57-year mortality data set extends over a sufficiently long period of time to include natural catastrophes. For example, the existence of 7 outliers in the mortality data could be viewed as caused by catastrophes (Mirande et al. 1991). Following this argument, adding catastrophes to the simulation might in fact double their projected impact. On the other hand, if catastrophe risks are increasing over time (e.g., chemical spills), it can be argued that the 57-year data set underestimates the probable impacts of catastrophes during the next 100 years. To allow for the possible validity of both arguments, we conducted simulations both with and without catastrophes included. In addition, variations in mortality were made to the 57-year-based parameter set following the same reasoning used to vary the standard run (described above).

VORTEX Input Parameters for the FP

The parameters for the FP standard run are presented in Table 2. These parameter values reflect the professional knowledge and data available at the time the simulations were run. Parameters are based upon the 57 years of study of the AWP, as well as expert knowledge and observations by crane biologists at Kissimmee Prairie, Florida (hereafter Kissimmee). The FP viability assessment results table presented later in the report itemizes the parameter(s) varied for each FP simulation. These parameters are discussed below.

Inbreeding Depression.—As discussed for the AWP, a simulation with inbreeding depression (we used a heterosis model with 3.0 lethal equivalents) was performed.

Age of First Breeding.—An age of first breeding of 5 years was selected as the standard value for the FP (in comparison to 4 for the AWP) because observations of captive behavior indicate that introduced birds are generally expected to begin reproduction later than wild birds. Age of first breeding was increased to 7 under the assumption that the effect of captive rearing was even more pronounced. However, it was also decreased to 4, based on the uncertainty regarding future crane behavior at Kissimmee (e.g., a pair of introduced 3-year-old birds built a nest in 1995).

Maximum Age.—The standard maximum age of 40 years was based on the findings for the AWP. This parameter was decreased to 30 years and to 20 years to gauge its effect on population viability.

Brood Size/Productivity.—The calculations of the standard run values for percentages for each brood size were based on the historical productivity data from the AWP prior to egg removals. These calculations were described above in the AWP section of the paper. To explore the effects of varied productivity on population viability, we varied productivity by ±10%. The 10% increase in productivity scenario had brood-size values as follows: 0 = 35.7%, 1 = 56.0%, and 2 = 8.3%. The 10% decrease in productivity scenario had the following brood-size values: 0 = 47.7%, 1 = 45.6%, and 2 = 6.7%. The reproduction EV value (10%) was the value calculated for the AWP standard run.

Mortality.—The chick mortality value (26.7%) and associated EV value (18.4%) were based on AWP data from 1976 through 1989 (Mirande et al. 1991). Comparable chick mortality rates for non-migratory Florida sandhill cranes (*G. c. pratensis*) range from 12.8% (Mirande et al. 1991) to 41% (Nesbitt 1992, Seal et al. 1992). The selected standard mortality value (10%) and EV (7%) for older birds were based on the 57-year data set from the AWP (see Appendix A) and on data from non-migratory Florida sandhill cranes (Mirande et al. 1991). To explore the effects of mortality, we varied adult mortality values (increased to 12% and decreased to 8%). In addition, a simulation was conducted with both chick mortality and adult mortality set at 10%, and a simulation was conducted with chick mortality at 10% and

Table 2. Input parameters for standard run of VORTEX simulation of Florida whooping crane population.

Parameter	Input value	Reference/rationale
Number of simulations	1,000	
Number of years	100	
Reporting interval	5	
Number of populations	1	
Inbreeding depression	N	Consensus of captive breeders
Environmental variance (EV) correlation	Y	Edwards et al. (1994)
Number of catastrophe types	2	Consensus of expert estimates
Monogamous or polygamous	M	Ehrlich et al. (1988)
Female breeding age	5	Consensus of captive breeders
Male breeding age	5	Consensus of captive breeders
Maximum age	40	Based on actual annual survival rates for Aransas/Wood Buffalo population (AWP) (see Appendix B)
Sex ratio	0.5	Edwards et al. (1994)
Maximum brood size	2	Assumes twinning at rates of AWP prior to egg removals
Density dependent breeding	N	Edwards et al. (1994)
Percent brood size	0=41.7% 1=50.8% 2=7.5%	Calculations based on AWP data prior to egg removals
EV-reproduction	10	Value calculated for AWP standard run
Female mortality: age 0	26.7%	Mirande et al. (1991)
EV in mortality	18.4	Mirande et al. (1991)
Female mortality: all ages above 0	10%	Based on approximate rate for AWP adults using 57-year data and non-migratory Florida sandhill crane data (Mirande et al. 1991)
EV in mortality	7	Calculated from AWP data for last 57 years (see Lacy et al. 1995)
Male mortality: age 0	26.7%	Mirande et al. (1991)
EV in mortality	18.4	Mirande et al. (1991)
Male mortality: all ages above 0	10%	Based on approximate rate for AWP adults using 57-year data and non-migratory Florida sandhill crane data (Mirande et al. 1991)
EV in mortality	7	Calculated from AWP data for last 57 years (see Lacy et al. 1995)
P of catastrophe 1 (disease)	10%	Average of expert estimates
Severity-reproduction	1.0	Impact reflected in reduced survival
Severity-survival	0.75	Average of expert estimates
P of catastrophe 2 (hurricane)	0.33%	S. Nesbitt (pers. commun.)
Severity-reproduction	1.0	Impact reflected in reduced survival
Severity-survival	0.9	S. Nesbitt (pers. commun.)
All male breeders?	Y	Mirande et al. (1991)
Stable age distribution?	N	Current population does not constitute a stable age distribution
Initial female ages: ages 1, 2, 3	5, 5, 1	Values are based on known ages of live birds at Kissimmee as of 20 Sep 1995
Initial male ages: ages 1, 2, 3	9, 1, 3	Values are based on known ages of live birds at Kissimmee as of 20 Sep 1995
Carrying capacity (K)	300	S. Nesbitt (pers. commun.)
EV in K	0	S. Nesbitt (pers. commun.)
Trend in K	N	S. Nesbitt (pers. commun.)
Harvest	N	Illegal
Supplement	Y	U.S. Fish and Wildlife Service (1994)
First year of supplementation	1	All of the values are based on a conservative estimate of birds available for supplementation (see text)
Last year of supplementation	20	
Supplementation interval	1	
Females age 1	15	
Females above age 1	0	
Males age 1	15	
Males above age 1	0	

adult mortality at 8%. For all FP simulations (including the standard), the non-chick mortality EV value used was 7%, calculated from the 57-year data on the AWP. The 57-year data set was utilized because these data include the maximum variation found in the initially small AWP.

Catastrophes.—Disease outbreaks are expected to have a significant impact on mortality of the FP. Experts are particularly concerned about eastern equine encephalitis (EEE). Introduced birds are currently inoculated against this disease, but it is unknown whether the inoculations will be effective over the long run, or to what extent the disease may affect the progeny of the introduced birds. Expert opinion varies greatly about the extent to which disease outbreaks may occur. Because of this uncertainty, the probability of disease outbreak was varied from 10% in the standard run to 20% (increasing probability assumed) and to 5% (an optimistic value). Severity of a disease catastrophe was varied from the standard of 25% mortality to a lower value of 10% mortality.

Because Kissimmee is approximately 100 km inland, the probability of hurricanes of force 4 or 5 at this location is estimated at only 0.33% (S. Nesbitt, pers. commun.). It was not considered realistic to run a variation which decreased the chances of a hurricane hitting, because weather patterns appear to be increasing in severity (Suplee 1995). Therefore, a simulation in which hurricane probability was doubled (to 0.66%) was conducted. For comparison purposes, a simulation was also run without any catastrophes included.

Age Distribution.—The standard values for age distribution were based upon the current actual distribution of introduced birds at Kissimmee. A simulation incorporating the stable age distribution was also conducted.

Carrying Capacity.—The standard carrying capacity (K = 300) was estimated based upon expert opinion and studies at Aransas and Kissimmee. It is unknown whether the birds will move into habitat currently believed to be unsuitable. Likewise, it is unknown how human pressures will affect K. Four additional simulations were performed, increasing K to 500, 800, and 1,000, and decreasing K to 150.

Supplementation.—The standard scenario assumes that 40 birds (20 males and 20 females) will be available for introduction into Kissimmee every year for the next 20 years. An immediate (within 30 days) 25% mortality rate was assumed for these 40 birds (so the values entered into VORTEX were actually 15 birds of each sex). Two simulations were performed in which the length of supplementation was shortened (to 10 years and 5 years) and 1 simulation was performed in which more birds were successfully introduced (20 birds of each sex entered into VORTEX rather than 15). Finally, a simulation was conducted in which 20 birds per year were introduced for 10 years. These variations were selected in order to explore the effects and possible management options regarding supplementation.

Combination Scenarios.—The effects of mortality variations were examined in combination with variations in age of first breeding, productivity, catastrophic disease probability, and supplementation length and numbers. The values were selected to create scenarios that reflected the "best" mortality level in combination with "best values" of the other parameters, the "best" mortality level with the "worst values" of the other parameters, and 2 levels of higher mortality (12% and 15%) combined with the "best values" of the other parameters, based on the descriptions of these parameters given earlier.

General Model Parameters

Because "it is impossible to simulate the complete range of complex processes and dynamics typical of a wild population" (Lacy 1993:56), some general assumptions are implicit in the VORTEX model. To the degree that these biological and mathematical assumptions are unrealistic in terms of the actual ecology and biology of whooping cranes, they will add uncertainty to the model's outputs. The most notable of these assumptions are described below.

Individuals in the founder population (the initial population specified for VORTEX simulations) and any supplemental animals are assumed to be unrelated. This assumption is not valid for the current world population of whooping cranes, all of which are derived from less than 14 (probably 6–8) founder individuals. This assumption results in an underestimate of inbreeding depression in the model simulations, if inbreeding depression is selected as a parameter.

When the inbreeding depression option is selected, VORTEX models inbreeding depression as reduced survival in the first year of life only. This assumption also results in an underestimate of inbreeding depression. Moreover, for each modeling year, breeding partners are randomly reshuffled by VORTEX. This assumption is not valid for whooping cranes, which generally mate for life. This assumption results in increased genetic diversity in a population, and thus underestimates inbreeding depression, if the inbreeding depression parameter is selected.

Most inputs into VORTEX cannot be varied during the modeling interval (except randomly within the specified range of environmental variation). That is, values for mortality, productivity, and catastrophes are assumed to be fixed during the 100- or 200-year simulations. (An exception is carrying capacity, K, for which an increasing or decreasing trend can be entered). Averages derived from existing data were typically used for the fixed model inputs; however, this approach does not account for steadily increasing or decreasing trends. For example, data for the AWP show that both mortality and reproduction have been decreasing over time. As for the FP, trends in demographic data are not unexpected as the number of experienced birds hatched in the wild increases or inoculations for disease lose their effectiveness. Finally, economic or environmental trends that may affect the probability or severity of catastrophes cannot be accounted for directly. The use of fixed values will underestimate or overestimate population sizes depending on the trends and the values used. Therefore, variations in these parameters must be tested in separate experimental runs.

VORTEX assumes that the effects of a catastrophe are felt only in the year in which the catastrophe occurs. This assumption most likely is inappropriate for some catastrophes which may befall whooping cranes. For example, hurricanes and major chemical spills may have lasting effects (e.g., effects on food supply, habitat, reproduction) which apply to more than 1 year.

Finally, in VORTEX, survival is assumed to be density independent up to the point that carrying capacity, K, is reached. Thus, there is no negative feedback mechanism on survival as the population size increases towards K. Based on ecological principles, this type of population growth is unlikely for whooping cranes. If mortality increases as population size approaches K (e.g., at the wintering grounds) then VORTEX overestimates population sizes for later years and underestimates the time until K is reached.

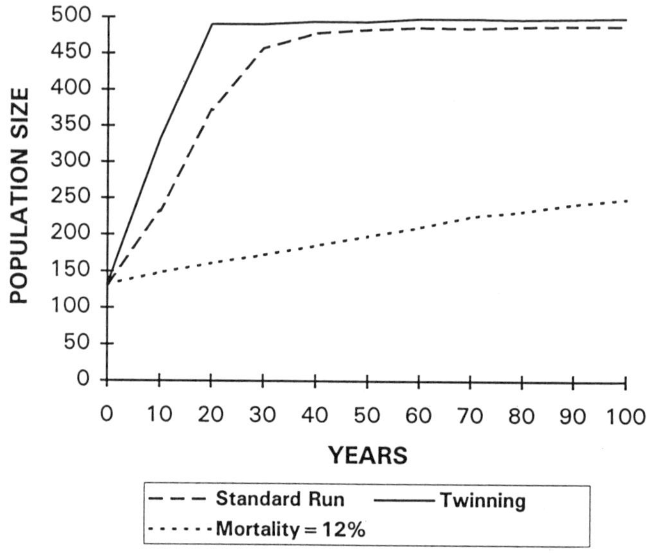

Fig. 1. VORTEX simulation showing effects of twinning and high mortality on the future of the Aransas/Wood Buffalo whooping crane population.

RESULTS AND DISCUSSION

Analysis of AWP VORTEX Simulation Results

This discussion focuses on the most notable results of the VORTEX analyses for the AWP. The results of all runs appear in Table 3.

Check Runs.—Prior to carrying out the population viability assessment for the AWP, we ran 2 retrospective simulations from 1975 to 1995 to confirm the accuracy of the standard inputs. These runs used the AWP 1975 population size of 49 cranes and the standard inputs, simulating for 20 years to the present. When compared against the actual AWP of spring 1995 (132 cranes), the outcomes of the retrospective runs were remarkably accurate. Use of the stable age distribution and the actual age distribution in the simulations resulted in population sizes of 131.9 and 129.3, respectively. With the accuracy of the standard inputs established, the population viability analysis was carried out.

Standard Run.—The standard run, with an estimated carrying capacity of 500, resulted in a mean population size of 487.8 after 100 years. As illustrated in Figure 1, use of the standard set of parameters leads to a projection of rapid growth for the next 30 years, at which time the population approaches the established carrying capacity of the habitat. It is the carrying capacity "lid" that causes the growth rate to level off at this point. The average rate of growth for the population before bumping into the carrying capacity lid was $r = 0.0498$, with a 0% probability of extinction over the 100 years of the simulation. These outcomes indicate that, if the standard parameters are fairly accurate, the AWP will remain viable.

In the VORTEX program, 2 unique alleles are assigned to each individual animal that is present in the population at the beginning of a simulation. The last 2 columns of Table 3 present the data on loss of allelic diversity for each simulation. For the standard run, there were 131 animals in the starting population (see Appendix B); therefore, there were 262 unique alleles in the starting population. By the end of 100 years, there were only 71 of these unique alleles remaining in the population. This result suggests that loss of allelic diversity may be a potential problem for the AWP. The implications of this finding will be discussed later.

Single Parameter Variations of the Standard Run.—Stable Age Distribution: A stable age distribution was used in place of the estimated actual age distribution but yielded very similar results.

Inbreeding Depression: Introducing inbreeding depression into the population had no serious effects on the population's viability. However, given the species' susceptibility to disease, it is suspected that loss of genetic diversity is in actuality affecting this population in combination with genetic

Table 3. VORTEX simulation of the Aransas/Wood Buffalo whooping crane population over 100 years. Refer to Table 1 for parameter values in standard run.

Description[a]	Final population *N*	Final population SD	Intrinsic growth rate (*r*)	*P* of extinction	Expected hetero-zygosity	Observed hetero-zygosity	Original unique founder alleles	Final alleles[b]
20-year check runs (1975–95)								
Actual age distribution	129.32	48.19	0.0448	0	0.9697	0.9855	98	51.56
Stable age distribution	131.91	49.15	0.0456	0	0.9701	0.9855	98	51.99
Standard run (1,000 simulations)								
Actual age distribution	**487.79**	**35.99**	**0.0498**	**0**	**0.9746**	**0.9773**	**262**	**70.68**
Stable age distribution	481.87	47.11	0.0492	0	0.9750	0.9778	264	71.62
Inbreeding depression								
Heterosis; 3 lethal equivalents	489.04	35.09	0.0489	0	0.9746	0.9784	262	70.81
Age of first breeding								
M=5, F=5	487.23	27.57	0.0378	0	0.9756	0.9784	262	72.21
Maximum age								
M=30, F=30	485.06	35.78	0.0462	0	0.9726	0.9753	262	68.21
Brood size								
Decrease, 0=60%, 1=40%	479.62	46.37	0.0367	0	0.9741	0.9759	262	70.11
Decrease, 0=67%, 1=33%	402.73	113.67	0.0159	0.001	0.9644	0.9688	262	55.41
Twinning								
0=41.7%, 1=50.8%, 2=7.5%	499.22	10.74	0.0842	0	0.9735	0.9760	262	68.15
0=41.7%, 1=50.8%, 2=7.5% (1,000 sims.)	491.83	32.54	0.0830	0	0.9733	0.9763	262	67.38
0=52.4%, 1=41.5%, 2=6.1% (1,000 sims.)	491.10	31.35	0.0614	0	0.9735	0.9765	262	68.47
0=52.4%, 1=45.6%, 2=2%	482.04	45.85	0.0544	0	0.9746	0.9772	262	70.68
0=52.4%, 1=37.6%, 2=10%	495.62	21.73	0.0681	0	0.9729	0.9753	262	67.02
Mortality								
Mortality = 6%	490.14	37.23	0.0711	0	0.9764	0.9794	262	76.08
Mortality = 10%	465.62	59.17	0.0277	0	0.9688	0.9716	262	59.25
Mortality = 12%	249.02	139.96	0.0045	0.011	0.9351	0.9450	262	33.20
57-year, 10% (no catastrophes)	447.89	69.05	0.0411	0	0.9699	0.9732	262	61.12
57-year, 11% (no catastrophes)	422.94	92.74	0.0307	0	0.9641	0.9677	262	54.01
Catastrophes (vars. re: chemical spill)								
P = 5%, survivorship = 0.90, *K* = 500	486.53	35.01	0.0464	0	0.9737	0.9761	262	69.00
P = 5%, survivorship = 0.50, *K* = 500	315.10	176.94	0.0188	0	0.9364	0.9489	262	39.34
P = 5%, survivorship = 0.90, *K* = 300	288.58	26.70	0.0472	0	0.9648	0.9688	262	49.91
P = 5%, survivorship = 0.50, *K* = 300	192.89	101.89	0.0162	0.04	0.9273	0.9414	262	29.61
P = 2%, survivorship = 0.75, *K* = 500	477.92	52.13	0.0469	0	0.9732	0.9761	262	68.37
P = 2%, survivorship = 0.50, *K* = 500	445.35	98.47	0.0379	0	0.9672	0.9696	262	58.45
P = 2%, survivorship = 0.75, *K* = 300	290.14	22.17	0.0450	0	0.9633	0.9677	262	48.35
P = 2%, survivorship = 0.50, *K* = 300	255.89	68.59	0.0370	0	0.9563	0.9620	262	42.04
Carrying capacity								
K = 300	285.20	32.94	0.0492	0	0.9645	0.9690	262	49.87
K = 1,000	964.05	100.99	0.0489	0	0.9799	0.9813	262	92.27
Decrease of 0.16%/yr for 50 yrs, SD = 0	445.75	35.04	0.0509	0	0.9735	0.9775	262	68.24
Decrease of 1.00%/yr for 50 yrs, SD = 0	245.31	14.82	0.0500	0	0.9622	0.9677	262	46.55
Decrease of 0.16%/yr. for 50 yrs; SD=50	421.56	47.60	0.0490	0	0.9717	0.9743	262	64.05
Decrease of 1.00%/yr for 50 yrs, SD = 50	194.60	41.35	0.0501	0	0.9536	0.9586	262	38.80
Decr., 1.00%/yr; 50 yrs, mortality up 20%	231.13	23.53	0.0309	0	0.9581	0.9659	262	42.73
Decr., 1.00%/yr, 50 yrs; recruit. down 20%	229.89	35.50	0.0271	0	0.9606	0.9670	262	45.93
Supplementation								
5M + 5F/year, years 5–24	482.37	47.40	0.0495	0	0.9802	0.9834	262	91.15
57-year data								
With catastrophes	446.02	71.03	0.0405	0	0.9664	0.9684	262	58.56
No catastrophes	462.96	56.43	0.0522	0	0.9718	0.9749	262	65.43
Mortality = 12%	362.74	143.72	0.0180	0	0.9447	0.9539	262	42.78
Mortality = 6%	479.86	38.64	0.0875	0	0.9760	0.9775	262	75.17

[a] All scenarios were run for 100 simulations and for 100 years unless otherwise noted.

[b] Due to supplementation, the final alleles value has only relative meaning.

drift (e.g., loss of genetic diversity such as Major Histocompatibility Complex, MHC, genes).

Age of First Breeding: Increasing the age of first breeding from 4 to 5 years old decreased r by nearly 25% but did not have a significant impact on the viability of the population within the 100-year timeframe.

Maximum Age: The maximum age for the cranes was reduced from the standard 40 to 30, but this had little effect on the results. This is because cranes in the higher age classes are few in number and are therefore not making a large contribution to the flock productivity.

Brood Size/Productivity: The percent of brood size 0 was increased to 60 and then to 67 (with brood size 1 decreased to 40 and 33, respectively) to examine the effects of lower recruitment rates on the population. The first level of reduction in productivity had only minor impacts on the population (N = 479.62, r = 0.0367). The second level of reduced productivity led to a decrease in N (402.73), a large decrease in r (0.0159), and a 0.001 probability of extinction. These results indicate that the percentage of successful nests is an important parameter in maintaining the flock's growth. Therefore, if the cranes experience sustained decreased recruitment in the future (e.g., during drought years), management strategies must address the issue of productivity (see twinning below).

Twinning: In these variations, it was assumed that the flock is allowed to have twins instead of the current practice of removing surplus eggs from the nests of breeding pairs. For the first twinning variation, inputs were based on actual twinning data from 1938 to 1967 where the estimated brood size percentages were 0 = 41.7%, 1 = 50.8%, and 2 = 7.5%. As illustrated in Fig. 1, in this simulation the population reaches the set carrying capacity of 500 very quickly (about 20 years), and r shows a substantial increase from the standard run (0.0842 vs.0.0498). This simulation was repeated for 1,000 iterations with very similar results.

In a second series of twinning variations, the standard run value for brood size 0 (52.4%) was used so that the effect of twinning could be compared to the standard run with all other variables held constant. Under these conditions, when brood size 2 was set at 6.1% (the expected value based on the historic twinning ratio), the growth rate of the population was 0.0614 as compared to the standard run value of r = 0.0498. This result indicates that allowing twinning may result in a notable increase in the rate of population growth. Other parameters such as final population size and genetic diversity did not change appreciably from the standard run. The sensitivity runs, varying the brood size 2 percentage, had predictable effects (see Table 3). With low twinning (brood size 2 = 2%), the growth rate (r = 0.0544) was higher than the standard run but lower than the run with the best estimate of the twinning rate if twinning were allowed. With high twinning (brood size 2 = 10%), the growth rate was predictably higher (r = 0.0681). Again, in all twinning variations, parameters other than growth rate did not change appreciably from the standard run.

Mortality: Simulations were run varying the level of mortality from the standard 8% down to 6%, and up to 10% and 12% to determine the population's sensitivity to this parameter. A notable result from these simulations was that at 12% mortality, r = 0.0045, which was the lowest growth rate for any scenario in the AWP population viability assessment. This outcome indicates that, given the other standard parameters, mortality rates cannot be allowed to increase much beyond 12% if the flock is to remain viable. At 12% mortality, allelic diversity decreases much farther than in the standard run.

Catastrophes: The probability of a chemical spill catastrophe was increased from 2% to 5% with no remarkable changes in outcome. When the probability of a chemical spill is maintained at the standard (2%), but survivorship is decreased to 0.5, r decreases to 0.0379. When the 2 parameters are combined (an increase in catastrophe probability and a decrease in survivorship), the resultant population size decreases to 315, and r decreases still further to 0.0188. However, even in the latter case, r remains positive and the management goal of maintaining 40 breeding pairs is still achievable. (Also see the results of multiple parameter variations below). VORTEX assumes no long-term effects of catastrophes. That is, the effects on reproduction and mortality are felt only in the year in which the catastrophe occurs.

Carrying Capacity: A decrease in the carrying capacity from the standard of 500 down to 300 resulted in a final population size of 285.2. In this simulation, the average growth rate (0.0492) was almost exactly the same as in the standard run (0.0498). The major negative effect of the reduced carrying capacity (other than placing a cap on the population size) was the loss of allelic diversity (49.87 final alleles vs. 70.68 final alleles for the standard run). When the carrying capacity was increased to 1,000, the final population reached 964.05, there was no notable change in r from the standard run, and retained allelic diversity was maximized (final alleles = 92.27). [Note: the r calculated by VORTEX is based on population growth prior to truncation due to limited carrying capacity, so unless K becomes very small, r is not usually affected by changes in carrying capacity.]

Trend in Carrying Capacity: A series of simulations was conducted incorporating a trend of decreased carrying capacity over time (e.g., if the amount of fresh water decreases in the cranes' wintering grounds). None of the decreases in K modeled over time produced any noteworthy impacts on the viability of the flock. That is, although final population sizes decreased with the decreasing *K*, population growth rates (*r*) remained as high as with the standard run. There were, however, negative impacts on allelic diversity in the simulations that included a 1% per year decrease in carrying capacity for a period of 50 years. Predictions about carrying capacity and its trends are extremely hard to make because, over time, additional wintering habitat may become more or less suitable and the cranes may or may not expand into alternative areas from those they currently occupy.

Supplementation: Supplementing the AWP showed no notable differences in the population size, or in the rate of growth, by the end of 100 years when compared to the standard run. Also, the apparently higher level of preservation of genetic diversity (both heterozygosity and unique alleles) in this simulation is misleading because, as previously noted, VORTEX treats supplemental animals as if they were unrelated to the receiving population and, in the case of the AWP, this assumption is not true. In general, the results of the supplementation simulation support the current management decision not to supplement the AWP.

Multiple Parameter Variations of the Standard Run.—Based on the results of the standard run and the initial variations, combinations of parameters were varied within single simulations.

Catastrophe and Carrying Capacity: When the effects of an increased chemical spill catastrophe probability ($P = 5\%$) were combined with a lower carrying capacity ($K = 300$), the final population size dropped to 288.58 but *r* remained high (0.0472). Another variation on this scenario was simulated with the assumption of a greater impact on survivorship (value = 0.5) in addition to increased probability. This combination produced the following results: final population size dropped to 192.89, *r* decreased to 0.0162, there was a 4% probability of extinction, and final unique alleles dropped to 29.61. The combined effects of increased probability and severity of catastrophe and lowered carrying capacity clearly impact the population in terms of lowering both *N* and *r*, and decreasing genetic diversity. This result suggests that the flock is put at substantial risk if the impacts of catastrophes are greater than expected and if the carrying capacity of the wintering grounds is reduced.

Trend in Carrying Capacity and Increasing Mortality or Decreasing Recruitment: When increased mortality (by 20%) or decreased recruitment (by 20%) were modeled simultaneously with an annual 1% decrease in *K* (over 50 years), there were noticeable reductions in the growth rate of the population from the standard of $r = 0.0498$ to $r = 0.0309$ and $r = 0.0271$, respectively. In addition, both of these simulations resulted in notably decreased allelic diversity and heterozygosity. However, even with these negative effects, the population remained viable.

Simulations Based on the 57-year Data Set.—The standard run is based on the 20-year data set (1975–95) for the AWP. Simulations were run by substituting some of the standard inputs with values derived from the 57-year data. As noted earlier, there was a logical rationale for either excluding or including catastrophes in these simulations. Therefore, simulations were run both ways. The results of these runs were fairly consistent with those of the standard run. Without catastrophes, the growth rate with the 57-year data set was somewhat higher ($r = 0.0522$) than in the standard run. When catastrophes were included, the growth rate with the 57-year data set was somewhat lower ($r = 0.0405$) than the standard run. The impacts of varying mortalities, with the 57-year data set, were similar to those variations with the 20-year data set.

Observations on Heterozygosity.—A general goal for the management of endangered species is to maintain 90% of the original population's genetic diversity for 100 years (Soulé et al. 1986). All of the variations modeled in this analysis appear to meet this goal; however, the VORTEX program assumes that the 1995 population begins with 100% heterozygosity, whereas in actuality it begins with only approximately 87% heterozygosity (87% is the final value derived from the 1991 PVA, which included the effects of the 1941 bottleneck, see Mirande et al. 1991). The resultant values of observed heterozygosity were converted into values reflecting the number of autozygous alleles remaining in the population by calculating 1 minus the inbreeding coefficient (F). The range of values for F falls between 0.0665 and 0.0141. A value of 0 implies that there is no inbreeding; a value of 1 implies that the population is totally inbred. The results in this analysis remain very close to 0, indicating that the population is not significantly inbred. Based on this PVA, the genetic diversity of the AWP appears to be in good standing. VORTEX underestimates inbreeding however (see Methods). The simulations are based on the assumption that the AWP starts with 131 unique lines in year 1 rather than the less than 14 unique lines that survived the population bottleneck of 1941.

Observations on Allelic Diversity.—The founding population in this analysis-the 131 cranes of the AWP in the spring of 1995 (see Appendix B)-are assigned 262 unique founder alleles in the VORTEX program. All variations carried out on the population revealed great losses in the number of alleles over time. The parameter variations such as severe catastrophes and high mortality rates had particularly devastating effects on allelic diversity. Carrying capacity set at 1,000 as well as the supplementation scenario resulted in the greatest maintenance of unique alleles, but even these numbers were less than one-half of the founder alleles (and, as noted above, the supplementation simulation overestimates diversity because of the false assumption that the supplemental birds are not related to the receiving population). These low values for final alleles can be expected due to the genetic drift associated with small populations and the fact that small populations are more sensitive to stochastic variation and therefore more likely to lose individuals with unique alleles. The loss in allelic diversity may explain the high vulnerability of the species to disease (e.g., the EEE outbreak in the captive flock and tuberculosis in the Gray's Lake population and the AWP). [Note again: VORTEX assumes that the 131 birds in the starting population are unrelated; therefore, 262 unique alleles are possible at any 1 locus. In fact, if less than 14 unique lines survived the 1941 bottleneck, the maximum number of unique alleles at any 1 locus would be 28.]

Analysis of FP VORTEX Simulation Results

As with the AWP analysis, this discussion covers only those simulations which reveal notable outcomes. Table 4 presents the results of all simulations.

Standard Run.—As expected, the FP exhibits a steady increase in flock size during the years of supplementation. However, as soon as supplementation is terminated at year 20, the growth rate decreases, leaving the final population at one-half the designated carrying capacity (162 individuals, SD = 90.51), with an average post-supplementation growth rate of 0.0026, and with a 1.1% chance of extinction over 100 years. In light of these findings, another simulation was run using the standard inputs, but this time for 200 years, to gain more insight into the population trends. The 200-year simulation showed a 17% chance of extinction over those years and a final population size of 136 individual cranes, with the average r becoming negative (-0.0001). Based on the outcomes of these standard runs, the post-supplementation population growth rate is an issue of great concern. Much of the analysis using the following variations focuses on which parameters can be varied to achieve positive growth rates beyond the supplementation years.

Single Parameter Variations on the Standard Run.—Because the years of supplementation have positive values for r in every simulation (as would be expected), the discussion of results for each variation focuses on the years following supplementation.

Inbreeding Depression: Incorporating inbreeding depression into the model produces a negative growth rate for the population by the end of 100 years. The negative value, however, is only slightly below the standard run final r value, as are the resultant levels of genetic and allelic diversity. From this simulation, it appears that inbreeding depression produces mild effects on the population which can be

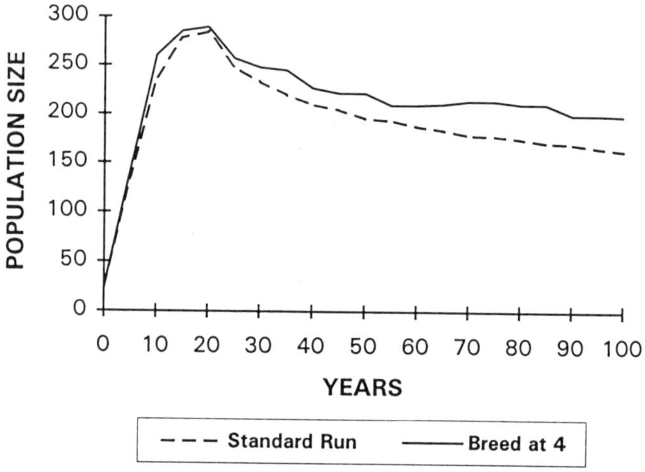

Fig. 2. VORTEX simulation showing effect of early breeding on the future of the Florida whooping crane population.

significant if growth rates are low.

Age of First Breeding: Lowering the age of first breeding from 5 to 4 years increases the growth rate of the population notably above the r of the standard run (0.0152 vs. 0.0026). On the other hand, raising the breeding age from 5 to 7 has a clearly detrimental effect on the growth rate (r = -0.0196). These results indicate that breeding age is a variable that can have significant impacts on the population (Fig. 2).

Maximum age: Reducing the maximum age from 40 to 30 to 20 had a slightly adverse impact on the number of individuals remaining in the population at the end of 100 years. A maximum age of 20 leaves the population with only 129 cranes. However, these variations did not notably impact the growth rates, or amounts of genetic or allelic diversity as compared to those of the standard run. For these reasons, maximum age was ruled out as a sensitive parameter.

Productivity: This parameter exhibited a great deal of sensitivity on the down side. Decreasing productivity by 10% dropped the growth rate to -0.0117 and the final population size to 108 individuals. Under these conditions, the flock had a 7% probability of going extinct. Alternatively, increasing the productivity by 10% produced negligible positive effects on the final population size and growth rate, suggesting that investing in techniques to increase productivity will not yield great rewards.

Mortality: Reducing adult mortality from 10% to 8% produced a notably positive growth rate (r = 0.0279). This was the highest growth rate exhibited in a single parameter variation. This scenario was re-run for 200 years to confirm the longevity of this growth rate, and similar results were achieved. On the other hand, increasing the adult mortality to 12% reduced the population's growth rate to r = -0.0223.

Table 4. VORTEX simulation of the Florida whooping crane population over 100 years (or 200 years as indicated). Refer to Table 2 for parameter values in standard run.

Description[a]	Final population *N*	Final population SD	Growth rate (*r*) With supplementation	Growth rate (*r*) Without supplementation	*P* of extinction	Expected heterozygosity	Observed heterozygosity	Final alleles[b]
Standard run (1,000 simulations)								
100-year run	**162.58**	**90.51**	**0.1529**	**0.0026**	**0.011**	**0.9406**	**0.9572**	**36.83**
200-year run	136.40	85.91	0.1502	-0.0001	0.17	0.8627	0.8769	14.99
Inbreeding depression								
Heterosis, 3 lethal equivalents	122.37	86.30	0.1539	-0.0053	0.02	0.9270	0.9517	32.21
Age of first breeding								
M=4, F=4	199.16	79.33	0.1658	0.0152	0	0.9519	0.9604	39.52
M=7, F=7	89.47	80.62	0.1362	-0.0196	0.11	0.9180	0.9500	27.60
Maximum age								
M=20, F=20	129.66	86.96	0.1530	-0.0081	0.03	0.9286	0.9457	29.05
M=30, F=30	150.07	91.79	0.1552	-0.0003	0.03	0.9363	0.9558	33.76
Productivity								
Low (decreased 10%)	108.87	75.65	0.1522	-0.0117	0.07	0.9258	0.9510	29.80
High (increased 10%)	169.53	85.95	0.1593	0.0067	0.01	0.9425	0.9557	37.26
Mortality								
Non-chick mortality = 8% (100 year run)	252.86	56.18	0.1683	0.0279	0	0.9679	0.9740	55.60
Non-chick mortality = 8% (200 year run)	255.67	58.89	0.1647	0.0279	0	0.9330	0.9396	27.81
Non-chick mortality =12%	65.74	66.40	0.1408	-0.0223	0.06	0.8812	0.9236	20.65
Chick mortality = 10%	231.63	70.26	0.1608	0.0212	0	0.9588	0.9689	44.27
Non-chick mortality = 8%; chick mort. = 10%	261.30	48.17	0.1717	0.0391	0	0.9659	0.9708	51.34
Catastrophes								
No catastrophes	256.61	42.94	0.1717	0.0327	0	0.9693	0.9752	56.45
Disease, *P* = 20%, survivorship = 0.75	45.00	46.10	0.1380	-0.0309	0.17	0.8463	0.9200	15.34
Disease, *P* = 5%, survivorship = 0.75	218.40	72.79	0.1655	0.0148	0	0.9588	0.9698	45.99
Disease, *P* = 10%, survivorship = 0.90	239.26	63.88	0.1620	0.0239	0	0.9651	0.9714	51.75
Hurricane, *P* = 0.66%, survivorship = 0.90	160.05	84.92	0.1518	-0.0018	0.06	0.9408	0.9561	35.48
Stable age distribution	165.43	86.42	0.1563	0.0009	0.02	0.9477	0.9627	36.74
Carrying capacity								
K=150	75.40	45.22	0.1913	-0.0037	0.04	0.8744	0.9118	17.30
K=500	270.86	139.94	0.1472	0.0030	0	0.9611	0.9749	58.57
K=800	419.61	230.99	0.1444	0.0047	0	0.9735	0.9765	81.05
K=1,000	510.85	301.41	0.1458	0.0052	0.01	0.9715	0.9807	94.62
Supplementation								
30 birds per year for 5 years	130.95	93.64	0.3449	-0.0029	0.09	0.8973	0.9180	23.15
20 birds per year for 10 years	155.51	92.64	0.1927	0.0019	0.04	0.9192	0.9373	27.65
30 birds per year for 10 years	151.84	89.42	0.2289	0.0016	0.02	0.9285	0.9480	30.88
30 birds per year for 20 years (standard run)	162.58	90.51	0.1529	0.0026	0.011	0.9406	0.9572	36.83
40 birds per year for 20 years	166.40	88.70	0.1770	0.0036	0	0.9464	0.9580	38.25
Combination scenario: best case								
Non-chick mortality = 8%, breeding age = 4, catastrophic disease probability = 5%, high productivity (+10%), suppl. with 40/yr	290.93	22.79	0.2218	0.0661	0	0.9691	0.9736	56.93
Combination scenario: low mortality with worst case for other parameters								
Non-chick mortality = 8%, breeding age = 7, catastrophic disease probability = 20%, low productivity (-10%), suppl.with 30 for 5 years	21.67	21.74	0.3381	-0.0405	0.54	0.7749	0.8778	9.20
Combination scenario: best case, but 12% adult mortality								
Non-chick mortality = 12%, breeding age = 4, catastrophic disease probability = 5%, high productivity (+10%), suppl. with 40/yr	230.52	66.74	0.1858	0.0201	0	0.9580	0.9642	43.14
Combination scenario: best case, but 15% adult mortality								
Non-chick mortality = 15%, breeding age = 4, catastrophic disease probability = 5%, high productivity (+10%), suppl. with 40/yr	101.97	85.33	0.1702	-0.0146	0.06	0.9088	0.9446	24.00

[a] All scenarios were run for 100 simulations and for 100 years unless otherwise noted.

[b] Due to supplementation,the final alleles value has only relative meaning.

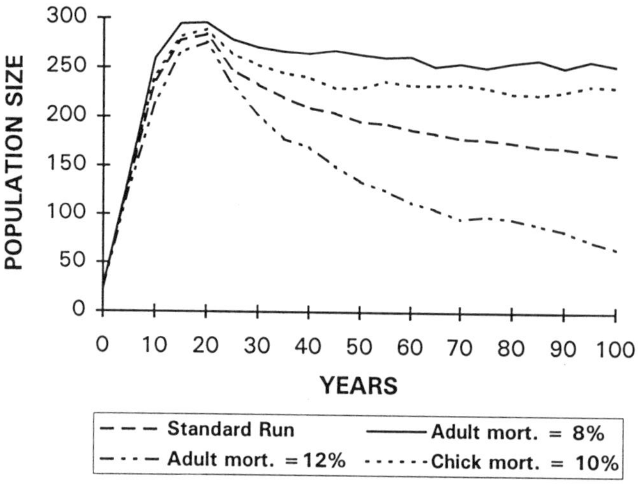

Fig. 3. VORTEX simulation showing effects of variation in mortality on the future of the Florida whooping crane population.

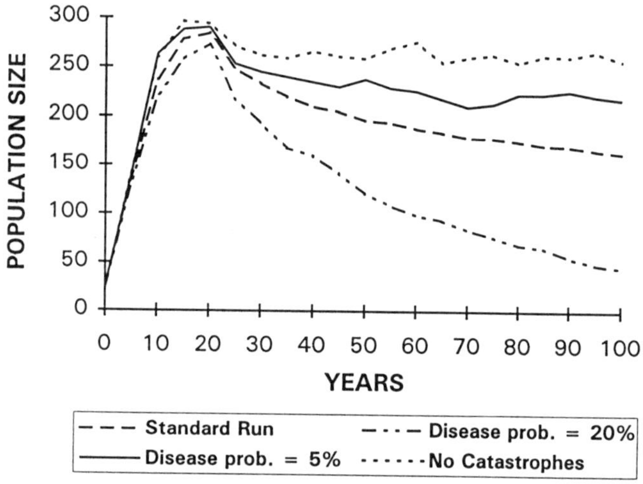

Fig. 4. VORTEX simulation showing effects of catastrophes on the future of the Florida whooping crane population.

Reducing chick mortality from 26.7% to 10% increased the growth rate from 0.0026 to 0.0212, and the final population size increased from 162.58 to 231.63. With chick mortality at 10% and adult mortality at 8%, the growth rate climbed to 0.0391, and the final population size was 261.30. From these results, it appears that chick and adult mortality rates are variables of the FP that should receive intensive monitoring and close management attention (Fig. 3).

Catastrophes: The probability and severity of disease in the FP are issues of great concern. When the probability of a disease catastrophe is increased from 10 to 20 times in 100 years, the flock has a 17% chance of extinction. Alternatively, decreasing the probability to 5% yields positive results as compared with the standard run (N = 218.40, r = 0.0148). Also, decreasing severity of catastrophic disease impact from 25% mortality to 10% mortality has an even more positive effect than reducing the probability of disease (N = 239.26, r = 0.0239). It appears that disease is a very significant variable for the FP. Hurricanes, on the other hand, do not have as severe an impact on the viability of the population when the probability of their occurrence is increased. When a simulation is conducted with no catastrophes included, the results are very positive for the FP: the final population size is 256.61 and the population growth rate is 0.0327 after supplementation has ceased (Fig. 4).

Stable Age Distribution: This variation was run to verify that inputting the actual age distribution of this population did not hinder the ability of VORTEX to model the population accurately. The results were very similar to those of the standard run.

Carrying Capacity: Varying the carrying capacity produced fairly predictable results. Decreasing K yielded a smaller final population with a negative growth rate, and increasing K yielded larger final populations with slightly higher growth rates than the standard run. It is interesting to note that, in all these scenarios, the population grew only to approximately one-half of the designated carrying capacity. This outcome was probably due to the large standard deviations in the distributions of final population sizes in these simulations. The large standard deviations, in themselves, constitute an index of the instability of final population size.

Supplementation: Variations in the supplementation scenario yielded negligible differences over the long run (Fig. 5). Varying supplementation to take place over longer or shorter periods of time had immediate effects on the population within the first 5 to 20 years but did not have notable positive or negative effects on the population by the end of 100 years. These results, and those of the adult mortality variations, suggest that if adult mortality is high, supplementation will not be helpful over the long run. However, if adult mortality is low, supplementing at low levels over 5, 10, or 20 years will be helpful in establishing a viable flock. Supplementation may also be an important insurance strategy during particular periods of high mortality or during periods when there are insufficient data to discern clear trends in other critical parameters. For example, supplementation may sustain the population during an initial period until adult mortality can be reduced.

Multiple Parameter Variations on the Standard Run.—Results from the single parameter variation simulations indicate that adult mortality rate was one of the most significant parameters for the FP. Therefore, the

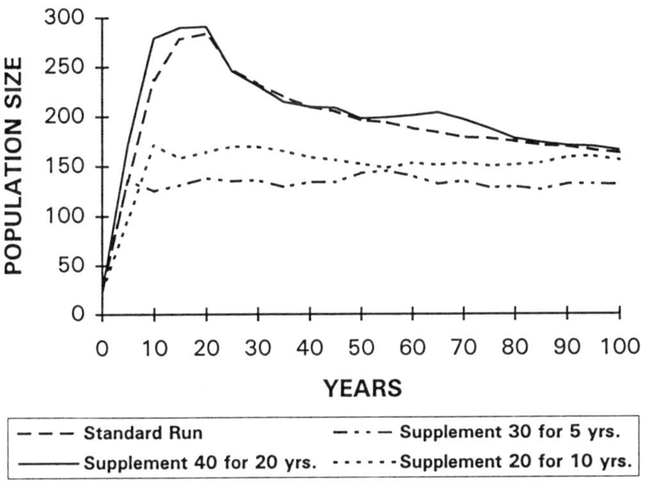

Fig. 5. VORTEX simulation showing effects of variation in supplementation rate on the future of the Florida whooping crane population.

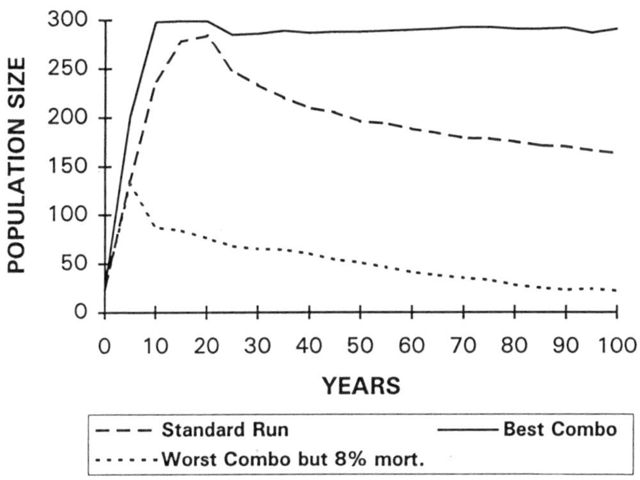

Fig. 6. VORTEX simulation showing effects of combination scenarios (see Table 4) on the future of the Florida whooping crane population.

multiple parameter variations were designed to test scenarios centering around variations in the adult mortality rate (Fig. 6). The "best case" scenario was to model adult mortality at 8% with an age of first breeding of 4, disease catastrophe probability at 5%, productivity increased by 10%, and supplementing 40 birds per year for 20 years. The impacts of these variations were strongly positive. The final population came within 10 cranes of the carrying capacity (K = 300). The growth rate following supplementation was 0.0661, and there was a 0% probability of extinction. This scenario was also modeled using all of the "best case" parameters except for adult mortality, which was set at 12%. In this simulation, the population remained viable over the 100-year time period with a final population size of 230.52 and an average growth rate of 0.0201. However, when mortality was increased to 15%, in combination with all the other "best case" parameters, the simulation resulted in a negative growth rate (-0.0146) with a 6% chance of extinction. The results of this series of simulations indicate that under certain conditions, the negative effects of high adult mortality rates can be overcome; however, there is a threshold at which point adult mortality rates dominate.

One final simulation was run which included a "best case" adult mortality rate (8%) and "worst case" values for the other 4 parameters listed above (i.e., age of first breeding = 7, disease catastrophe probability = 20%, productivity decreased by 10%, and supplementing 30 birds per year for 5 years). It is apparent from these results (N=21.67, r = -0.0405) that a low adult mortality rate is not enough to save the population if other important parameters are at values significantly worse than the expected values. The probability of extinction in this case was 54%.

CONCLUSIONS

AWP Viability Assessment

The 1991 PVA concluded that the "population is projected to have a very low (less than 1%) probability of extinction within the next 100 years," with a growth rate of 0.028 when surplus eggs are removed and a growth rate of 0.046 when they are not (Mirande et al. 1991:4,28). The findings of the present analysis, which took into account new data available since 1991, additional expert opinions on parameters, and new information on environmental scenarios, also indicate that the AWP has a very low probability of extinction with a growth rate of 0.0498 with continued egg removal and a growth rate of 0.0842 when twinning is allowed. If the selected standard parameters of this PVA are indeed accurate and will continue to be accurate in the coming years, the AWP should clearly meet the recovery goal of a minimum of 40 breeding pairs for 10 years or more before the year 2020 (Fig. 1).

In the simulations described here, the most notable positive effects on population growth were caused by twinning and by decreased mortality. If historic rates of twinning re-occur, discontinuing the current practice of egg removal could have a significant impact on the rate of population increase. If mortality rates are reduced, the population will also reap the benefits of a higher growth rate.

The most negative impacts on the population are realized when the severity and probability of a chemical spill catastrophe are increased, when the carrying capacity is lowered to 300, and when mortality rates increase or recruitment rates decrease. Under no conditions, however, did the population experience a negative growth rate, and only when mortality was increased to 12%, or when increased probability and severity of catastrophe were combined with decreased carrying capacity, did the results show a notable probability of extinction.

As they have in the past, human actions can strongly influence the future prospects of the AWP. Following are some actions that can keep the odds stacked in favor of AWP survival and growth: protecting against catastrophic losses, maintaining potential for expanded habitat, reducing mortality (e.g., marking powerlines and securing/restoring safe stop-over locations for migrating birds), and increasing recruitment (e.g., allowing twinning and managing the blue crab population at Aransas). If these actions can yield even more favorable parameters than were assumed in the best estimates used for this analysis, the AWP will reach carrying capacity more quickly, with less loss of heterozygosity and rare alleles, and the flock will have even better chances of maintaining a self-sustaining viable population in the future.

FP Viability Assessment

Two major questions concerning the potential success of the Florida reintroduction experiment are: (1) Can adult mortality from predation and disease be reduced? and (2) Will the captive-reared birds breed naturally in the wild? Concerning the first question, changing release sites to avoid concentrations of bobcats (*Lynx rufus*) has already produced a notable drop in early mortality (Nesbitt et al. 1997), and it is hoped that these gains will endure as the birds reach adulthood. Also, research is currently underway to assess the effects of pre-release inoculations against disease. (Inoculations may be only a stop-gap measure because chicks hatched in the wild will not be protected and titers in released birds may drop over time).

Concerning the second question, at this writing, there are two rays of hope already on the horizon: the pairing of 2-year-olds and the nesting attempt of 3-year-old whooping cranes during the spring of 1995 at Kissimmee (Nesbitt et al. 1997) and the natural breeding behavior that has been observed in released, captive-reared sandhill cranes (Duan 1994).

Even if the introduced whooping cranes breed successfully, the simulations reported here indicate that the Florida project has borderline prospects for long-term success (Fig. 2). If our standard ("best-estimate") parameters are close to correct, growth of the flock will be very slow (or even negative) after supplementation efforts have ceased. As Table 4 indicates, each of several key variables has the potential to "make or break" the Kissimmee experiment. In positive combination, these key variables will provide a strong base for success in Florida. If adult mortality is 1 or 2 percentage points lower than our best-estimate scenario, prospects are favorable as long as breeding (i.e., productivity and age of first breeding) and disease risk remain at or near best-estimate values. Even if adult mortality is at, or slightly above, the standard value, as long as breeding and disease risk are somewhat better than the standard, the outcome appears good. On the negative side, if mortality rates for young birds are higher than our best estimate, or if adult mortality approaches 12% annually, then all other key variables will have to be at better-than-standard values to avoid extinction of the experimental population.

The amount and duration of supplementation have only minimal effects on the prospects for success in Florida. It is the enduring rates of mortality, breeding, and disease risk that will determine the outcome of the experiment. Fortunately, there are some management actions that may have an influence on the values of these key variables. As noted above, careful selection of release sites can affect the mortality resulting from bobcat predation. Release site selection can also consider the proximity of other birds that may function as role models for introduced cranes in terms of where to roost safely and where to find food, for example. Ongoing research on vaccines and health monitoring may help to minimize the risk of a catastrophic disease outbreak, and selective releases of new birds can help to balance the sex ratio, increase the genetic diversity, and bolster the flock against periods of high adult mortality . All of these management actions have the potential to push the odds in favor of success. Thus, with careful monitoring of the key variables, and creative interventions to influence them positively, the Florida reintroduction experiment could provide a major boost to the prospects for long-term recovery and survival of the whooping crane.

LITERATURE CITED

CHAVEZ-RAMIREZ, F. 1996. Food availability, foraging ecology, and energetics of whooping cranes wintering in Texas. Ph.D. Thesis, Texas A&M Univ., College Station. 103pp.

DUAN, W. 1994. Social and reproductive behavior of isolation-reared, released sandhill cranes (*Grus canadensis*). M.S. Thesis, Ohio State Univ., Columbus. 177pp.

EDWARDS, R., S. BRECHTEL, R. BROMLEY, D. HJERTAAS, B. JOHNS. E. KUYT, J. LEWIS, N. MANNERS, R. STARDOM, and G. TARRY. 1994. National recovery plan for the whooping crane. Rep. 6. Recovery Nat. Endangered Wildl. Comm., Ottawa, Ont. 39pp.

EHRLICH, P. R., D. S. DOBKIN, and D. WHEYE. 1988. The birder's handbook. Simon and Schuster, New York, N.Y. 785pp.

LACY, R. C. 1993. Vortex: a computer simulation model for population viability analysis. Wildl. Res. 20:45-65.

______, K. A. HUGHES, and P. S. MILLER. 1995. Vortex version 7 user's manual. IUCN/SSC Conserv. Breeding Specialist Group, Apple Valley, Minn. 111pp.

MAHAN, T. A., and B. S. SIMMERS. 1992. Social preference of four cross-foster reared sandhill cranes. Proc. North Am. Crane Workshop 6:114-119.

MIRANDE, C., R. C. LACY, and U. S. SEAL, eds. 1991. Whooping crane conservation viability assessment. IUCN/SSC Captive Breeding Specialist Group, Apple Valley, Minn. [various pagination].

NESBITT, S. A. 1992. First reproductive success and individual productivity in sandhill cranes. J. Wildl. Manage. 56:573-577.

______, M. J. FOLK, M. G. SPALDING, J. A. SCHMIDT, S. T. SCHWIKERT, J. M. NICOLICH, M. WELLINGTON, J. C. LEWIS, and T. H. LOGAN. 1997. An experimental release of whooping cranes in Florida–the first three years. Proc. North Am. Crane Workshop 7:79-85.

SEAL, U. S., and S. HEREFORD, eds. 1992. Mississippi sandhill crane (*Grus canadensis pulla*) population and habitat viability assessment workshop. IUCN/SSC Conserv. Breeding Specialist Group, Apple Valley, Minn. 146pp.

SOULÉ, M., M. GILPIN, W. CONWAY, and T. FOOSE. 1986. The millennium ark: how long a voyage, how many staterooms, how many passengers? Zoo Biol. 5:101-114.

SUPLEE, C. 1995. Over century, U.S. weather turns less variable, more often violent. Page A13 *in* Washington Post (D.C.), 21 Sep 1995.

U.S. FISH AND WILDLIFE SERVICE, NATIONAL BIOLOGICAL SERVICE, CANADIAN WILDLIFE SERVICE, and PARKS CANADA (USFWS, NBS, CWS, and PC). 1995. Memorandum of understanding between the United States Fish and Wildlife Service and the National Biological Service of the United States Department of the Interior and the Canadian Wildlife Service of Environment Canada and Parks Canada of Canadian Heritage on conservation of the whooping crane. USFWS, NBS, Washington, D.C., and CWS, PC, Ottawa, Ont. 16pp.

U.S. FISH AND WILDLIFE SERVICE. 1994. Whooping crane recovery plan. U.S. Fish Wildl. Serv., Albuquerque, N.M. 92pp.

Appendix A. Numbers, mortality, and recruitment of whooping cranes in the Aransas/Wood Buffalo population. Data collected during 57 years at Aransas National Wildlife Refuge, Texas.

Year	Total cranes arriving at ANWR	Cranes not returning next fall	Mortality rate (non-chicks)	Adults arriving	Chicks arriving	Recruitment (chicks/ total cranes)
1938/39	18	3	0.16666667	14	4	0.22222222
1940	22	1	0.04545455	15	7	0.31818182
1941	26	12	0.46153846	21	5	0.19230769
1942	16	1	0.0625	14	2	0.125
1943	19	3	0.15789474	15	4	0.21052632
1944	21	6	0.28571429	16	5	0.23809524
1945	18	0	0	15	3	0.16666667
1946	22	0	0	18	4	0.18181818
1947	25	0	0	22	3	0.12
1948	31	4	0.12903226	25	6	0.19354839
1949	30	0	0	27	3	0.1
1950	34	8	0.23529412	30	4	0.11764706
1951	31	11	0.35483871	26	5	0.16129032
1952	25	6	0.24	20	5	0.2
1953	21	0	0	19	2	0.0952381
1954	24	3	0.125	21	3	0.125
1955	21	1	0.04761905	21	0	0
1956	28	6	0.21428571	20	8	0.28571429
1957	24	2	0.08333333	22	2	0.08333333
1958	26	3	0.11538462	22	4	0.15384615
1959	32	1	0.03125	23	9	0.28125
1960	33	3	0.09090909	31	2	0.06060606
1961	36	2	0.05555556	30	6	0.16666667
1962	39	7	0.17948718	34	5	0.12820513
1963	32	6	0.1875	32	0	0
1964	33	1	0.03030303	26	7	0.21212121
1965	42	6	0.14285714	32	10	0.23809524
1966	44	6	0.13636364	36	8	0.18181818
1967	43	4	0.09302326	38	5	0.11627907
1968	48	4	0.08333333	39	9	0.1875
1969	50	2	0.04	44	6	0.12
1970	56	5	0.08928571	48	8	0.14285714
1971	57	3	0.05263158	51	6	0.10526316
1972	59	13	0.22033898	54	5	0.08474576
1973	51	4	0.07843137	46	5	0.09803922
1974	49	2	0.04081633	47	2	0.04081633
1975	49	0	0	47	2	0.04081633
1976	57	0	0	49	8	0.14035088
1977	69	7	0.10144928	57	12	0.17391304
1978	72	4	0.05555556	62	10	0.13888889
1979	75	5	0.06666667	68	7	0.09333333
1980	76	4	0.05263158	70	6	0.07894737
1981	78	7	0.08974359	72	6	0.07692308
1982	73	6	0.08219178	71	2	0.02739726
1983	73	5	0.06849315	67	6	0.08219178
1984	75	4	0.05333333	68	7	0.09333333
1985	86	5	0.05813953	71	15	0.1744186
1986	97	8	0.08247423	81	16	0.16494845
1987	110	1	0.00909091	89	21	0.19090909
1988	134	15	0.1119403	109	25	0.18656716
1989	138	12	0.08695652	119	19	0.13768116
1990	146	13	0.0890411	126	20	0.1369863
1991	146	22	0.15068493	133	13	0.0890411
1992	132	11	0.08333333	124	8	0.06060606
1993	136	9	0.06617647	121	15	0.11029412
1994	143	18	0.12587413	127	16	0.11188811
1994/95	132[a]	?	?	125	7	0.0530303
Totals	3,151	295	5.71041907	2,870	413	7.81716469

Weighted mean mortality (295/3,151) = 0.09362107

Weighted mean recruitment (413/3,283)=0.12579957

[a] Not included in column total.

Appendix B. Estimated actual age distribution of the Aransas/Wood Buffalo whooping crane population in 1994-95.

Year	Age	VORTEX age class	No. of chicks	(A) Annual survival rate[a]	(B) Previous years' survival product[a,b]	(C) Cumulative survival (A × B)[a]	Live birds (C × No. of chicks)[a]	Rounding remainder[c]	Adjusted total[c]
1995	0	1	7	1.00	NA	1.00	7.00	0.00	7
1994	1	2	16	0.87	1.00	0.87	13.92	0.00	14
1993	2	3	15	0.93	0.87	0.81	12.14	-0.08	12
1992	3	4	8	0.92	0.81	0.74	5.96	0.06	6
1991	4	5	13	0.85	0.74	0.63	8.18	0.02	8
1990	5	6	20	0.91	0.63	0.58	11.47	0.20	12
1989	6	7	19	0.91	0.58	0.52	10.03	-0.33	10
1988	7	8	25	0.89	0.52	0.47	11.57	-0.30	11
1987	8	9	21	0.99	0.47	0.46	9.77	0.27	10
1986	9	10	16	0.92	0.46	0.42	6.77	0.04	7
1985	10	11	15	0.94	0.42	0.40	5.92	-0.19	6
1984	11	12	7	0.95	0.40	0.38	2.66	-0.27	2
1983	12	13	6	0.93	0.38	0.35	2.12	0.40	3
1982	13	14	2	0.92	0.35	0.32	0.64	-0.48	0
1981	14	15	6	0.91	0.32	0.30	1.75	0.16	2
1980	15	16	6	0.95	0.30	0.28	1.71	-0.09	2
1979	16	17	7	0.93	0.28	0.26	1.82	-0.38	1
1978	17	18	10	0.94	0.26	0.25	2.44	0.44	3
1977	18	19	12	0.90	0.25	0.22	2.70	-0.12	3
1976	19	20	8	1.00	0.22	0.22	1.76	-0.42	1
1975	20	21	2	1.00	0.22	0.22	0.44	0.34	1
1974	21	22	2	0.96	0.22	0.21	0.42	-0.22	0
1973	22	23	5	0.92	0.21	0.19	0.97	0.20	1
1972	23	24	5	0.78	0.19	0.15	0.74	0.17	1
1971	24	25	6	0.95	0.15	0.14	0.86	-0.09	1
1970	25	26	8	0.91	0.14	0.13	1.02	-0.23	1
1969	26	27	6	0.96	0.13	0.13	0.75	-0.21	1
1968	27	28	9	0.92	0.13	0.12	1.08	-0.46	1
1967	28	29	5	0.91	0.12	0.11	0.55	-0.48	0
1966	29	30	8	0.86	0.11	0.09	0.76	0.07	1
1965	30	31	10	0.86	0.09	0.08	0.77	-0.17	1
1964	31	32	7	0.97	0.08	0.08	0.54	-0.40	0
1963	32	33	0	0.81	0.08	0.06	0.00	0.14	0
1962	33	34	5	0.82	0.06	0.05	0.25	0.14	0
1961	34	35	6	0.94	0.05	0.05	0.28	0.39	1
1960	35	36	2	0.91	0.05	0.04	0.09	-0.33	0
1959	36	37	9	0.97	0.04	0.04	0.35	-0.24	0
1958	37	38	4	0.88	0.04	0.04	0.14	0.11	0
1957	38	39	2	0.92	0.04	0.03	0.07	0.25	0
1956	39	40	8	0.79	0.03	0.03	0.19	0.32	1
1955	40	41	0	0.95	0.03	0.03	0.00	-0.49	0
1954	41	42	3	0.88	0.03	0.03	0.08	-0.49	0
1953	42	43	2	1.00	0.03	0.03	0.06	-0.41	0
1952	43	44	5	0.76	0.03	0.02	0.11	-0.35	0
1951	44	45	5	0.65	0.02	0.01	0.07	-0.24	0
1950	45	46	4	0.76	0.01	0.01	0.03	-0.17	0
1949	46	47	3	1.00	0.01	0.01	0.03	-0.14	0
1948	47	48	6	0.87	0.01	0.01	0.05	-0.11	0
1947	48	49	3	1.00	0.01	0.01	0.03	-0.08	0
1946	49	50	4	1.00	0.01	0.01	0.04	-0.04	0
1945	50	51	3	1.00	0.01	0.01	0.03	-0.01	0

[a] Rounded.

[b] B calculated as follows: e.g., for 1993, (1.00×0.87)=0.87; for 1992, (1.00×0.87×0.93)=0.81; for 1991, (1.00×0.87×0.93×0.92)=0.74; etc.

[c] Rounding procedure suggested by R. C. Lacy (pers. commun.).

Appendix C. Mortality and recruitment in the Aransas/Wood Buffalo whooping crane population during 20 years, 1975–95.

Year	Total cranes arriving at ANWR	Cranes not returning next fall	Mortality rate (non-chicks)	Adults arriving	Chicks arriving	Recruitment (chicks/ total cranes)
1975	49	0	0	47[a]	2[a]	0.040816[a]
1976	57	0	0	49	8	0.140351
1977	69	7	0.101449	57	12	0.173913
1978	72	4	0.055556	62	10	0.138889
1979	75	5	0.066667	68	7	0.093333
1980	76	4	0.052632	70	6	0.078947
1981	78	7	0.089744	72	6	0.076923
1982	73	6	0.082192	71	2	0.027397
1983	73	5	0.068493	67	6	0.082192
1984	75	4	0.053333	68	7	0.093333
1985	86	5	0.05814	71	15	0.174419
1986	97	8	0.082474	81	16	0.164948
1987	110	1	0.009091	89	21	0.190909
1988	134	15	0.11194	109	25	0.186567
1989	138	12	0.086957	119	19	0.137681
1990	146	13	0.089041	126	20	0.136986
1991	146	22	0.150685	133	13	0.089041
1992	132	11	0.083333	124	8	0.060606
1993	136	9	0.066176	121	15	0.110294
1994	143	18	0.125874	127	16	0.111888
1995	132[a]	?[a]	?[a]	125	7	0.053030
20-year total	1,965	156	1.4338	1,809	239	2.3216
20-year mean		7.8	0.0717	90.45	11.95	0.1161
20-year SD		5.7149	0.0376	27.958	6.0537	0.0461
Weighted mean			0.0794			0.1167

[a] Not included in 20-year means and standard deviations.

SUBSPECIES COMPOSITION OF SANDHILL CRANE HARVEST IN NORTH DAKOTA, 1968-94

WILLIAM L. KENDALL,[1] Office of Migratory Bird Management, U.S. Fish and Wildlife Service, Laurel, MD 20708, USA
DOUGLAS H. JOHNSON, Northern Prairie Science Center, U.S. Geological Survey-Biological Resources Division, Jamestown, ND 58401, USA
STANLEY C. KOHN, North Dakota Game and Fish Department, Bismarck, ND 58501, USA

Abstract: North Dakota is a major fall staging area for the Mid-continent Population (MCP) of sandhill cranes (*Grus canadensis*), which is composed of 3 subspecies: the greater (*G. c. tabida*), Canadian (*G. c. rowani*), and lesser (*G. c. canadensis*). The number of cranes killed by hunters in North Dakota averaged 6,793 during the 1990-94 seasons, ranking second highest among crane-hunting states. The distribution of harvest among subspecies is important, due to concerns about the poorly known status of these subspecies, especially the greater. We estimated subspecies composition of the harvest in North Dakota by using morphometric data collected from field samples of birds harvested since 1968. Subspecies composition varied both spatially (across counties from east to west) and temporally (among 3 periods of distinct harvest regulations and within season). Lessers predominated in the west and Canadians and greaters in the east. For the 1990-94 period we estimated that mortality from hunting in North Dakota averaged at least 1,085 (18%) greaters, 2,138 (36%) Canadians, and 2,716 (46%) lessers.

PROC. NORTH AM. CRANE WORKSHOP 7:201-208

Key Words: *Grus canadensis*, hunting, Mid-continent Population, North Dakota, sandhill crane, subspecies.

The Mid-continent Population (MCP) of sandhill cranes is the largest in North America, numbering about 500,000 (Sharp and Vogel 1992, Sharp 1995). Although currently managed as a single population, 3 subspecies have been recognized (Walkinshaw 1973): greater, Canadian, and lesser. Lessers are the most abundant, and greaters the least abundant, but there are no programs in place to monitor subspecific status or trends. The MCP is subject to legal sport hunting along much of its migration path, resulting in an average estimated yearly kill of 28,098 for the 1990/91-1994/95 period in Canada, the United States, and Mexico (Sharp 1996). The distribution of harvest among the 3 subspecies is not known. Because the greater represents the smallest portion of the MCP, and was once listed as a rare subspecies (Committee on Rare and Endangered Wildlife Species 1968), there is a desire among wildlife managers that harvest pressure be applied primarily to the most plentiful lesser (59 Federal Register 67:16765, 1994; Central Management Unit Technical Committee 1995).

Within the MCP, subspecies are geographically separated mostly during their breeding season. In fall and spring staging areas and wintering areas, greaters and Canadians are generally distributed farther east than lessers (Johnson and Stewart 1973; Guthery and Lewis 1979; Lewis 1979; Melvin and Temple 1980, 1983; Gaines and Warren 1984; Tacha et al. 1984), but they intermix to varying degrees, making subspecific harvest management impractical at this time.

Tacha et al. (1984, 1994) proposed managing the MCP as 2 geographic subpopulations, (1) Western-wintering mostly in western Texas and consisting largely of lessers and very few greaters, and (2) Gulf Coast-wintering mostly along the Gulf Coast of Texas and consisting mostly of Canadians and greaters.

North Dakota is a major fall staging area for all 3 subspecies of the MCP. The state initiated sandhill crane hunting in 1968 (Sharp and Cornely 1997), and harvest (plus crippling loss) averaged 6,793 for the 1990-94 period, ranking the state second highest (after Texas) in the United States. Because of the concern for the greater subspecies, the state of North Dakota has coordinated the measurement of harvested sandhill cranes since 1968. In this paper we estimate the subspecies composition of the harvest in North Dakota by using data from the measured cranes. We also test for differences in subspecies composition among counties, across periods of years related to major changes in harvest regulations (Sharp and Cornely 1997), and within seasons.

We thank C. Schroeder (deceased) and M. A. Johnson of the North Dakota Game and Fish Department, and all the cooperators who measured cranes. We also thank R. C. Drewien, R. E. Shea, and G. W. Smith for helpful comments on previous drafts of the manuscript.

METHODS

State and federal biologists in North Dakota collected morphometric, sex, and age data from harvested cranes in the field. Because samples were collected opportunistically, sampling intensity varied widely among counties. Each crane was sexed internally and aged (immature or adult). Three

[1]Present address: Patuxent Wildlife Research Center, U.S. Geological Survey-Biological Resources Division, 11510 American Holly Drive, Laurel, MD 20708-4017, USA.

Table 1. Subspecies composition (%) of hunter-killed sandhill cranes collected in North Dakota during 1968–94. Percentage data for each county are pooled within time period and presented with corresponding total cranes (*n*) collected.

	1968–76				1977–82				1983–94			
County	*n*	Greater	Canadian	Lesser	*n*	Greater	Canadian	Lesser	*n*	Greater	Canadian	Lesser
Benson					19	37	63	0	15	27	67	6
Bottineau									11	45	55	0
Burleigh					1	100	0	0	393	7	26	67
Divide									3	0	0	100
Emmons									12	0	0	100
Kidder	78	15	82	3	242	24	71	5	280	18	39	43
McHenry					11	55	36	9	85	24	53	24
McLean	53	0	15	85	182	4	16	80	128	5	14	81
Mountrail									3	0	33	67
Phillips	4	0	0	100								
Pierce	12	8	92	0	233	34	62	4	372	22	72	6
Sheridan					2	0	0	100	52	29	27	44
Stutsman					11	18	82	0	87	44	48	8
Ward									21	0	10	90
Wells									16	38	56	6

length measurements were taken: wing chord, tarsus, and culmen (from posterior nares). Assignment of birds to subspecies was based on results of discriminant function analysis involving those measurements. The discriminant functions ultimately were based on a series of measurements of cranes collected on breeding areas, so that their subspecies were known (Johnson and Stewart 1973). Sample sizes were limited, especially for greaters (7 females and 11 males) and Canadians (3 females and 10 males); lessers were better represented with 17 females and 36 males.

Johnson and Stewart (1973) found that the 3 subspecies differed significantly in morphological measurements. From their data, discriminant functions involving tarsus length, culmen length (from posterior nares), and wing chord length were developed. Separate functions were developed for adult males and females; juveniles were not used. The discriminant function analysis assigned to each specimen a "probability" that it belonged to each of the 3 subspecies. We classified each specimen into the subspecies with highest associated probability.

We compared subspecies composition spatially among counties, and temporally both between and within seasons. For comparisons across years, we pooled data for each county into 3 time periods a priori, defined by major differences in harvest regulations: (1) 1968–76, season lengths of 30 days beginning in early November; (2) 1977–82, with 5- to 16-day seasons in early to mid-September; and (3) 1983–94, with 21- to 58-day seasons in much of the state (excluding the portion east of U.S. Hwy. 281, which was closed to hunting). There were minor regulatory changes within these periods, such as small changes in season length, creation and dissolution of zones, and addition of counties, which could be used to refine these time periods (Sharp and Cornely 1997) but would reduce samples in each period. We used likelihood ratio tests (LRT) between generalized logit models (Agresti 1990:307) to test for a period effect on subspecies composition and for an interaction between period and county effects. We used Pearson Chi-square tests for comparisons among counties within each period.

For within-year temporal comparisons we considered only the most recent period, 1983–94. For each county with sufficient sample size, we compared 4 periods: (1) 1–15 September, (2) 16–30 September, (3) 1–15 October, and (4) 16–31 October, by using Pearson Chi-square tests.

RESULTS

Table 1 and Figures 1–3 illustrate the subspecies composition of sampled cranes across counties for the 3 time periods described above. Subspecies composition differed among counties within each of the 3 periods (1968–76: χ^2 = 105, 4 df, $P < 0.001$; 1977–82: χ^2 = 443, 10 df, $P <$ 0.001; 1983–94: χ^2 = 541, 22 df, $P < 0.001$). In each period there is an apparent cline in subspecies composition from west to east, with lessers predominant in the west, and Canadians and greaters predominant in the east.

Because of the small number of counties sampled in the early years, we first compared the 1968–76 and 1977–82 periods based only on Kidder and McLean Counties (sample of ≥10 cranes in both periods), and found no period effect

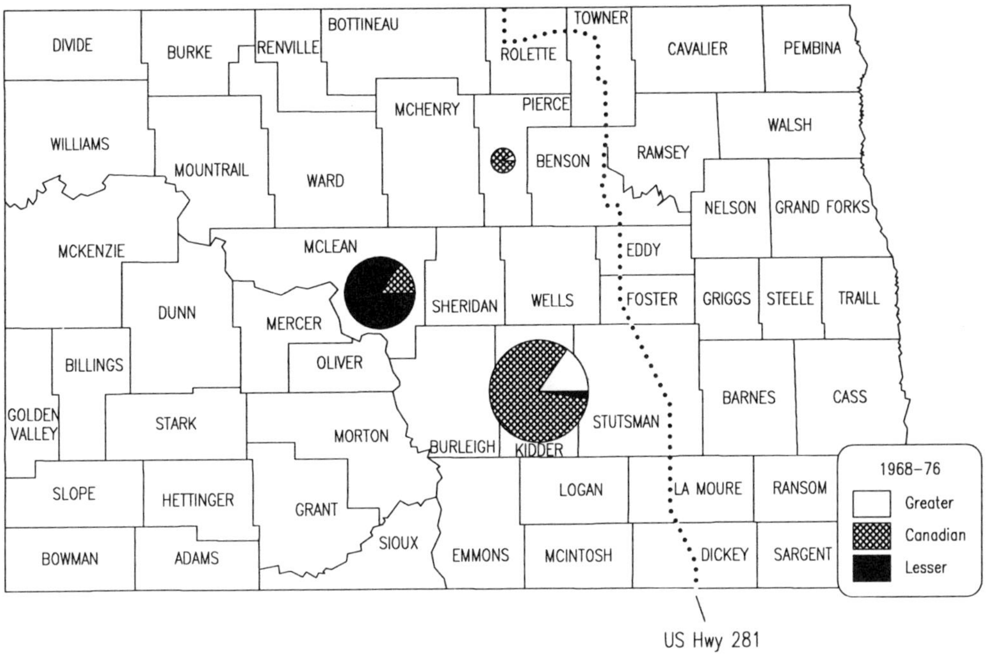

Fig. 1. Subspecies composition, by county, of harvested sandhill cranes collected in North Dakota, 1968–76.

(LRT χ^2 = 4.35, 2 df, P = 0.12). We compared periods 1977–82 and 1983–94 based on data from Benson, Kidder, McHenry, McLean, Pierce, and Stutsman Counties, and found both a period effect and an interaction between period and counties (LRT χ^2 = 67, 10 df, P < 0.001). Figures 2 and 3 indicate proportionally more lessers in eastern counties during 1983–94 than during 1977–82. This increase was not uniform from north to south, however, producing an interaction effect. For example, whereas a large proportional increase of lessers was indicated in Kidder County, there was

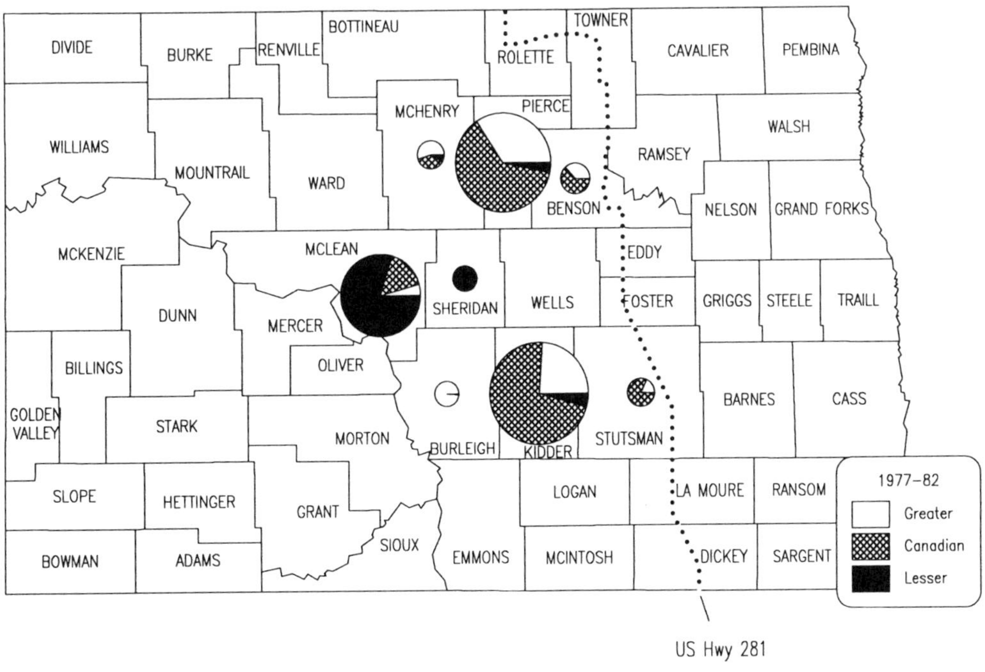

Fig. 2. Subspecies composition, by county, of harvested sandhill cranes collected in North Dakota, 1977–82.

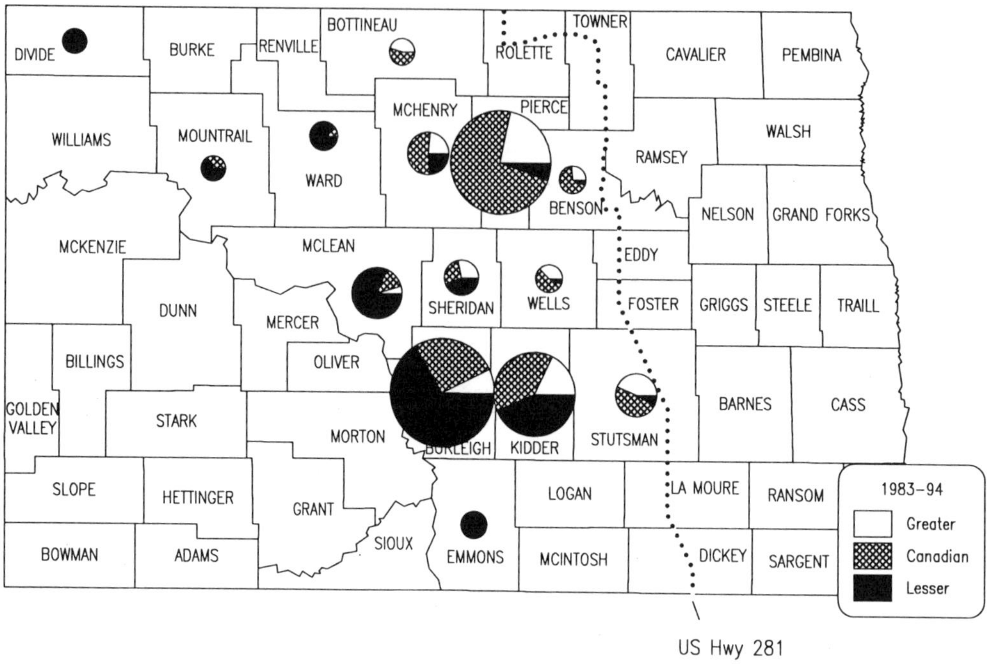

Fig. 3. Subspecies composition, by county, of harvested sandhill cranes collected in North Dakota, 1983–94.

a proportional increase in Canadians in Pierce County.

Table 2 contains county-specific subspecies composition by half-month intervals, pooled for the most recent multi-year period, 1983–94. For Burleigh, Kidder, and McHenry Counties the proportion of greaters and Canadians decreased from early September to mid-October, whereas the proportion of lessers increased for the same span of time. This pattern was statistically significant for Burleigh and Kidder Counties (χ^2 = 14, 4 df, P < 0.01 and χ^2 = 12, 4 df, P = 0.02, respectively), but not for McHenry County (χ^2 = 5.4, 4 df, P = 0.25). The data for Sheridan County also displayed this pattern for 16 September–15 October (χ^2 = 13, 2 df, P < 0.01). For McLean County, from 16 September to 15 October the proportion of greaters and Canadians increased, whereas the proportion of lessers decreased (χ^2 = 6.1, 2 df, P = 0.04). Subspecies composition was stable during 1 September–15 October for Pierce (χ^2 = 2.6, 4 df, P = 0.63) and Stutsman (χ^2 = 2.1, 4 df, P = 0.72) Counties. Sample sizes were small for 16–31 October in all counties.

We estimated a lower bound on the average annual subspecies-specific mortality due to hunting in North Dakota for the most recent 5 years, 1990–94. Figures 4 and 5, respectively, illustrate the distribution of subspecies composition and average total hunter kill for this period. Total average kill due to hunting in North Dakota for this period was 6,793 cranes (Novara and Handy 1991; Martin 1992, 1993, 1994, 1995). Table 3 lists the estimated kill per county, calculated by applying the subspecies composition of field-sampled cranes to the average kill for the county. For Benson and Ward Counties, where subspecies composition for 1990–94 was based on < 10 cranes, we used the subspecies composition for the 1983–94 period. We estimated an average kill of at least 1,085 greaters, 2,138 Canadians, and 2,716 lessers per year. Because some harvest occurred in unknown counties, and some cranes were taken in counties either not sampled for subspecies or insufficiently sampled, an annual average of 854 hunter-killed cranes was unaccounted for in this analysis.

DISCUSSION

We found differences in subspecies composition of harvest among counties within each of the 3 periods: 1968–76, 1977–82, and 1983–94. The cline in subspecies composition suggested in Figures 1-4, i.e., Canadians and greaters predominating in the east and lessers predominating in the west, is consistent with previous studies (Johnson and Stewart 1973, Melvin and Temple 1983). In addition, subspecies composition has changed over time within some counties, at least between the 1977–82 and 1983–94 periods. This is especially apparent in Kidder County, where the proportion of lessers in the sample of harvested cranes rose from 5% to 43%. When combined with increasing numbers of observations of cranes in areas east of U.S. Hwy. 281 (North Dakota Game and Fish Department, unpubl. data), where few were seen in the past and hunting is prohibited,

Table 2. Subspecies composition of hunter-killed sandhill cranes collected in North Dakota during 1983-94. For each county, data are pooled within 4 half-month periods. Only those counties for which ≥10 cranes were collected for ≥2 periods are represented.

		1-15 Sep		16-30 Sep		1-15 Oct		16-31 Oct	
County	Subspecies	*n*	(%)	*n*	(%)	*n*	(%)	*n*	(%)
Burleigh	Greater	10	(9)	18	(8)	1	(3)	0	(0)
	Canadian	40	(37)	60	(25)	3	(9)	0	(0)
	Lesser	58	(54)	158	(67)	29	(88)	12	(100)
	Total	108		236		33		12	
Kidder	Greater	14	(27)	25	(14)	5	(16)	5	(36)
	Canadian	25	(48)	63	(35)	12	(37)	8	(57)
	Lesser	13	(25)	93	(51)	15	(47)	1	(7)
	Total	52		181		32		14	
McHenry	Greater	5	(33)	12	(23)	3	(20)	0	(0)
	Canadian	9	(60)	29	(57)	6	(40)	1	(25)
	Lesser	1	(7)	10	(20)	6	(40)	3	(75)
	Total	15		51		15		4	
McLean	Greater	0	(0)	3	(4)	3	(8)	0	(0)
	Canadian	0	(0)	7	(9)	9	(24)	2	(18)
	Lesser	2	(100)	67	(87)	25	(68)	9	(80)
	Total	2		77		37		11	
Pierce	Greater	29	(21)	44	(22)	7	(26)	1	(50)
	Canadian	104	(74)	145	(71)	20	(74)	1	(50)
	Lesser	7	(5)	14	(7)	0	(0)	0	(0)
	Total	140		203		27		2	
Sheridan	Greater	2	(100)	5	(42)	8	(22)	0	(0)
	Canadian	0	(0)	7	(58)	7	(20)	0	(0)
	Lesser	0	(0)	0	(0)	21	(58)	2	(100)
	Total	2		12		36		2	
Stutsman	Greater	6	(50)	24	(41)	8	(53)	0	(0)
	Canadian	6	(50)	29	(49)	6	(40)	1	(100)
	Lesser	0	(0)	6	(10)	1	(7)	0	(0)
	Total	12		59		15		1	

this result indicates the possibility of a shift or expansion of the fall staging area to the east. This phenomenon, if it exists, could be caused by a variety of natural or man-made causes, including increased harvest pressure. Hunting seasons during 1983-94 were 58 days long in western counties and 21-58 days long in eastern counties, whereas season lengths during 1977-82 were 5-16 days. In addition, average crane kill by hunters during 1983-94 (6,344) was 91% higher than during 1977-82 (3,329) (Sharp and Cornely 1997).

For most counties sampled, we found that the proportion of lessers in the harvest increased from early September to mid-October, whereas the proportions of both greaters and Canadians decreased. This pattern is consistent with the belief that the larger subspecies migrate through earlier than the lessers (Johnson and Stewart 1973, Carlisle and Tacha 1983). Regardless of whether the larger subspecies leave the state earlier, large numbers of lessers arriving later could reduce the proportion of Canadians and greaters in the harvest. Conversely, for McLean County the proportion of Canadians and greaters in the harvest increased during 16 September-15 October. Movement between areas could cause either of these patterns. For Pierce and Stutsman counties, subspecies composition was relatively constant during 1 September-15 October, predominated by Canadians and greaters. This suggests that lessers simply have not made much use of those counties. Data from harvest after 15 October were too sparse to analyze.

For counties where subspecies composition varies within a season, the temporal distribution of samples should match that of the harvest, to minimize bias in estimates of subspecies composition. Because data were collected opportunistically, we do not know to what extent this was achieved.

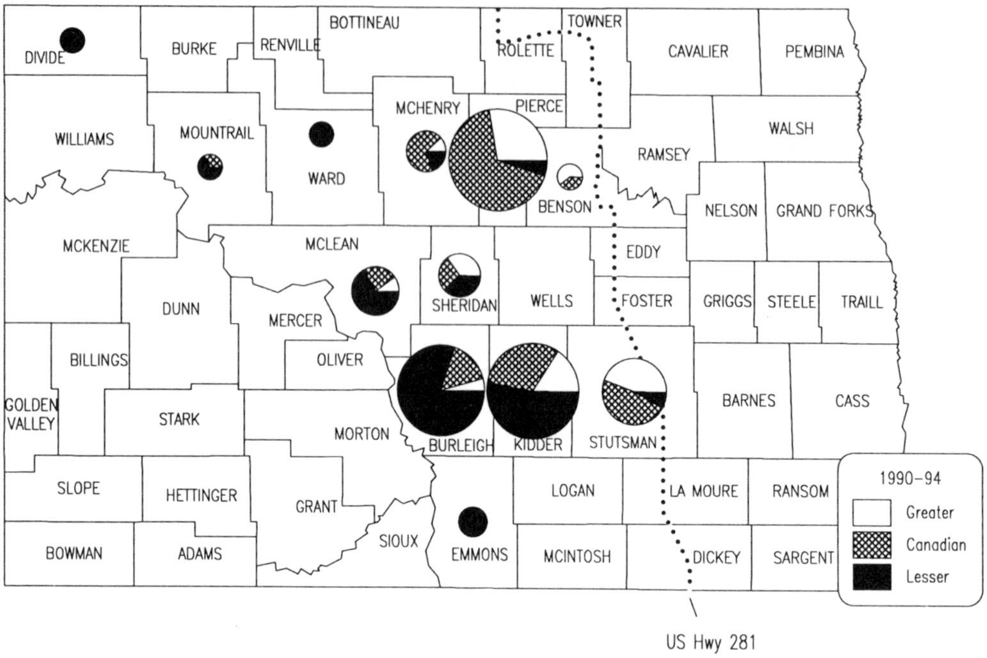

Fig. 4. Subspecies composition, by county, of harvested sandhill cranes collected in North Dakota, 1990–94.

The validity of the Canadian subspecies has been questioned (Stephen 1967), although the lesser and greater subspecies have long been recognized (Walkinshaw 1949, American Ornithologists' Union 1957). The Canadian subspecies was not formally described until 1965 (Walkinshaw 1965), although earlier references suggested its presence (e.g., Oberholser 1921, Walkinshaw 1949:64). Some authors (e.g., Tacha et al. 1985) have opined that subspecific designation, especially of Canadians, may be unwarranted, but that the various forms represent only a cline

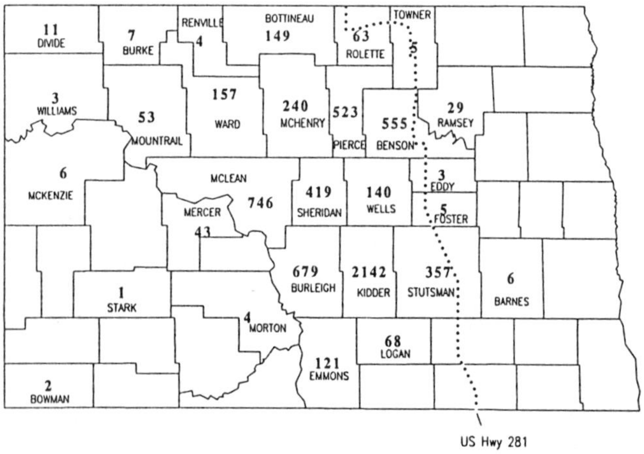

Fig. 5. Average mortality from hunting, by county, in North Dakota, 1990–94. An average of 248 cranes is unaccounted for here, due to unknown county of harvest.

in body size, from large birds (greaters) breeding in more southern locations, to small birds (lessers) breeding farthest north, and with intermediate birds (Canadians) breeding in between.

Currently available evidence suggests that the 3 subspecies should be distinguished, however. Johnson and Stewart (1973) found that each of 13 breeding-ground specimens of Canadians clearly differed from both lesser and greater populations on the basis of several morphometric measurements. Baldwin (1976) found that greater nestlings required an average of 9 hours to attain homeothermy, whereas Canadian and lesser nestlings were homeothermic at hatch. Baldwin (1977) noted differences in development as well: wings, height, tarsi, culmen, and midtoes of young greaters developed in proportion to overall body size, Canadian chicks showed priority for height, tarsus, and culmen development, and young lessers showed priority for wing development over body development. Gaines and Warren (1984) discovered that the presence or absence of protein P1 in the pancreas consistently distinguished 13 Canadians collected in Saskatchewan from 5 lessers collected in Alaska. They did not examine greaters for genetic distinction.

Johnson and Stewart (1973) noted, however, that their morphometric analysis could not exclude the possibility of a cline. Breeding specimens were not collected uniformly throughout the breeding range of sandhill cranes, and intermediate groups could have been missed. In addition, sandhill cranes have increased in number and range during

Table 3. Subspecies composition of hunter-killed sandhill cranes (%) and total kill from hunting in North Dakota, 1990–94. A county was included if the sample size of collected cranes was ≥10 for 1983–94. Subspecies composition of total hunter kill based on subspecies composition of collected cranes.

County	Sampled cranes				Average kill	Estimated kill		
	n	Greater	Canadian	Lesser		Greater	Canadian	Lesser
Benson	5	60	40	0	555	150[a]	372[a]	33[a]
Burleigh	138	6	15	79	679	41	102	536
Emmons	12	0	0	100	121	0	0	121
Kidder	145	16	31	53	2,142	343	664	1,135
McHenry	35	11	69	20	240	26	166	48
McLean	51	10	24	66	746	75	179	492
Pierce	158	28	67	5	523	146	351	26
Sheridan	40	35	28	37	419	147	117	155
Stutsman	86	44	48	8	357	157	171	29
Ward	2	0	0	100	157	0[a]	16[a]	141[a]

[a] Due to sample size of <10 cranes for 1990–94, estimated composition of harvest was based on 1983–94 collection (see Table 1).

the past 50 years, and it is possible that populations that once had distinct and non-overlapping breeding ranges now have expanded and possibly melded. Research to better assess the validity of the 3 subspecies and to develop improved classification schemes is now under way (D. H. Johnson and J. E. Austin, Northern Prairie Science Center, pers. commun.).

Because of the variation in subspecies composition of the harvest presented here, especially the extensive overlap of greaters and Canadians, we believe that subspecific harvest management in North Dakota would not be practical at this time. Tacha et al. (1994) suggested subpopulation-specific harvest management based on wintering populations. The U.S. Fish and Wildlife Service has stated its intent to manage the MCP based on 2 subpopulations (59 Federal Register 67:16765, 1994), and the Central Management Unit Technical Committee will consider revising its Mid-continent Population Management Plan to accommodate subpopulations in 1998 (Central Management Unit Technical Committee 1995).

Ideally, division of North Dakota into distinct geographic areas for each subpopulation would be based on banding or telemetry data relating fall staging areas to wintering areas. However, with the exception of limited studies by Melvin and Temple (1980) and Tacha et al. (1984), no such studies have been conducted and none are in progress. In the absence of such banding data, delineation of subpopulations in North Dakota based on the data presented here would be even more difficult, especially given that there is some movement between counties within a season. Nevertheless, the apparent cline in subspecies composition from east to west is consistent with the cline in the wintering grounds discussed by Tacha et al. (1994), and could lend itself to some kind of division.

The Central Flyway will be collecting measurements on harvested sandhill cranes for at least 2 more years before it considers changes to the crane management plan to account for subpopulations. We believe, however, that the results presented here could be useful in separating subpopulations. Managers should also consider that, although the entire MCP is apparently stable in size, the status and demographics of each subpopulation (especially the Gulf Coast subpopulation), and of the subspecies comprising those subpopulations (especially the greater), are poorly known or unknown. Modeling exercises such as that conducted for the Gulf Coast subpopulation by Johnson and Kendall (1997) could be helpful in setting reasonable bounds on harvest for each subpopulation, and for identifying additional demographic information needed to better manage these subpopulations.

LITERATURE CITED

AGRESTI, A. 1990. Categorical data analysis. John Wiley & Sons, New York, N.Y. 558pp.

AMERICAN ORNITHOLOGISTS' UNION. 1957. A.O.U. checklist of North American birds. Lord Baltimore Press, Baltimore, Md. 691pp.

BALDWIN, J. H. 1976. A study of thermoregulatory capacities of four subspecies of sandhill crane nestlings. Pages 291-295 *in* J. C. Lewis, ed. Proc. [1975] int. crane workshop. Oklahoma State Univ. Publ. Printing, Stillwater.

______. 1977. A comparative study of sandhill crane subspecies. Ph.D. Thesis, Univ. Wisconsin, Madison. 150pp.

CARLISLE, M. J., and T. C. TACHA. 1983. Fall migration of sandhill cranes in west-central North Dakota. J. Wildl. Manage. 47:818-821.

CENTRAL FLYWAY COUNCIL. 1993. Management plan for mid-continent sandhill cranes. Unpubl. Rep., Cent. Flyway

Counc., U.S. Fish Wildl. Serv., Golden, Colo. 21pp.

CENTRAL MANAGEMENT UNIT TECHNICAL COMMITTEE. 1995. Meeting summary: ad hoc Mid-continental sandhill crane subcommittee. Unpubl. Rep., Cent. Flyway Counc., U.S. Fish Wildl. Serv., Golden, Colo. 16pp.

COMMITTEE ON RARE AND ENDANGERED WILDLIFE SPECIES. 1968. Rare and endangered fish and wildlife of the United States. U.S. Fish Wildl. Serv. Resour. Publ. 34. [various pagination].

GAINES, G. D., and R. J. WARREN. 1984. Genetics and morphology of sandhill crane populations in Texas. J. Wildl. Manage. 48:1387-1393.

GUTHERY, F. S., and J. C. LEWIS. 1979. Sandhill cranes in coastal counties of southern Texas: taxonomy, distribution, and populations. Pages 121-128 *in* J. C. Lewis, ed. Proc. 1978 crane workshop. Colorado State Univ. Printing Serv., Fort Collins.

JOHNSON, D. H., and W. L. KENDALL. 1997. Modeling the population dynamics of Gulf Coast sandhill cranes. Proc. North Am. Crane Workshop 7:173-180.

______, and R. E. STEWART. 1973. Racial composition of migrant populations of sandhill cranes in the northern plains states. Wilson Bull. 85:148-162.

LEWIS, J. C. 1979. Taxonomy, food, and feeding habitat of sandhill cranes, Platte Valley, Nebraska. Pages 21-28 *in* J. C. Lewis, ed. Proc. 1978 crane workshop. Colorado State Univ. Printing Serv., Fort Collins.

MARTIN, E. M. 1992. Sandhill crane harvest and hunter activity in the Central Flyway during the 1991-92 hunting season. Unpubl. Rep., U.S. Fish Wildl. Serv., Laurel, Md. 12pp.

______. 1993. Sandhill crane harvest and hunter activity in the Central Flyway during the 1992-93 hunting season. Unpubl. Rep., U.S. Fish Wildl. Serv., Laurel, Md. 12pp.

______. 1994. Sandhill crane harvest and hunter activity in the Central Flyway during the 1993-94 hunting season. Unpubl. Rep., U.S. Fish Wildl. Serv., Laurel, Md. 12pp.

______. 1995. Sandhill crane harvest and hunter activity in the Central Flyway during the 1994-95 hunting season. Unpubl. Rep., U.S. Fish Wildl. Serv., Laurel, Md. 12pp.

MELVIN, S. M., and S. A. TEMPLE. 1980. Migration ecology and wintering grounds of sandhill cranes from the interlake region of Manitoba. Unpubl. Rep., Contract 14-16-009-78-050, U.S. Fish Wildl. Serv., Washington, D.C. 60pp.

______, and ______. 1983. Fall migration and mortality of Interlake, Manitoba sandhill cranes in North Dakota. J. Wildl. Manage. 47:805-817.

NOVARA, A. N., and M. HANDY. 1991. Sandhill crane harvest and hunter activity in the Central Flyway during the 1990-91 hunting season. Unpubl. Rep., U.S. Fish Wildl. Serv., Laurel, Md. 12pp.

OBERHOLSER, H. C. 1921. Notes on North American birds X. Auk 38:79-82.

SHARP, D. E. 1996. Status and harvests of the Mid-continent and Rocky Mountain Populations of sandhill cranes. Unpubl. Rep., U.S. Fish Wildl. Serv., Golden, Colo. 22pp.

______, and J. E. CORNELY. 1997. Summary of sandhill crane hunting seasons in North Dakota. Proc. North Am. Crane Workshop 7:209-218.

______, and W. O. VOGEL. 1992. Population status, hunting regulations, hunting activity, and harvests of Mid-continent sandhill cranes. Proc. North Am. Crane Workshop. 6:24-32.

STEPHEN, W. J. D. 1967. Bionomics of the sandhill crane. Can. Wildl. Serv. Rep. Ser. 2. 48pp.

TACHA, T. C., S. A. NESBITT, and P. A. VOHS. 1994. Sandhill crane. Pages 77-94 *in* T. C. Tacha and C. E. Braun, eds. Migratory shore and upland game bird management in North America. Int. Assoc. Fish Wildl. Agencies., Washington D.C.

______, P. A. VOHS, and G. C. IVERSON. 1984. Migration routes of sandhill cranes from mid-continental North America. J. Wildl. Manage. 48:1028-1033.

______, ______, and W. D. WARDE. 1985. Morphometric variation of sandhill cranes from mid-continental North America. J. Wildl. Manage. 49:246-250.

WALKINSHAW, L. H. 1949. The sandhill cranes. Cranbrook Inst. Sci. Bull. 29. 202pp.

______. 1965. A new sandhill crane from central Canada. Can. Field-Nat. 79:181-184.

______. 1973. Cranes of the world. Winchester Press, New York, N.Y. 370pp.

SUMMARY OF SANDHILL CRANE HUNTING SEASONS IN NORTH DAKOTA, 1968–94

DAVID E. SHARP, U.S. Fish and Wildlife Service, P.O. Box 25486, Denver Federal Center, Denver, CO 80225, USA
JOHN E. CORNELY, U.S. Fish and Wildlife Service, P.O. Box 25486, Denver Federal Center, Denver, CO 80225, USA

Abstract: The migratory Mid-continent Population (MCP), containing 3 subspecies, is the most numerous of all sandhill crane (*Grus canadensis*) populations. During fall the MCP uses major staging areas in the Canadian prairie provinces and northcentral United States. In North Dakota, sport hunting of the MCP resumed in 1968, after being closed during 1916–67. The resumption and expansion of crane hunting in North Dakota during 1968–94 followed a gradual pattern of implementation. Subspecies considerations, the presence of whooping cranes (*G. americana*), crop depredation complaints, and public reaction influenced the geographic and temporal expansion of seasons. Harvests gradually increased following each expansion and in 1993–94 reached near record levels (6,200–7,000), as seasons utilizing the full federal frameworks were implemented. Spring surveys indicate that the MCP is relatively stable, and current sport harvest levels appear appropriate to maintain current population size. A primary management concern is to maintain subspecies abundance, especially of the less numerous greater sandhill crane (*G. c. tabida*).

PROC. NORTH AM. CRANE WORKSHOP 7:209-218

Key Words: *Grus canadensis*, *Grus americana*, harvest, hunting, Mid-continent Population, North Dakota, sandhill cranes, whooping cranes.

Sandhill cranes are classified into 6 subspecies; 3 subspecies are sedentary and 3 are migratory (Walkinshaw 1949, 1965; Braun et al. 1975). For management purposes, sandhill cranes have been grouped into 9–10 populations (Drewien and Lewis 1987, Tacha et al. 1994). Nonmigratory populations are relatively small (<7,000–8,000) and are not hunted (Lewis 1977). Two migratory populations, the Rocky Mountain Population of greater sandhill cranes and the MCP, have carefully designed sport hunting programs with a goal to maintain stable population levels (Sharp 1995). The MCP, the largest hunted population, occurs in the Central Flyway during the fall and winter and is composed of 3 recognized subspecies: lesser (*G. c. canadensis*), Canadian (*G. c. rowani*), and greater (Braun et al. 1975, Tacha et al. 1994).

During the summer cranes of the MCP are distributed from western Ontario northwestward across central and northern Canada, Alaska, and into eastern Siberia (Lewis 1977). In September, cranes begin migration to traditional staging areas in the southern Canadian prairie provinces of Alberta, Saskatchewan, and Manitoba and in North Dakota (Buller 1967, Lewis 1977). Prairie staging areas provide opportunities to assimilate critical nutrients, primarily from small grain crops, necessary to complete fall migration and to prepare for winter (Madsen 1967, Sugden et al. 1988). In spring these habitats and the staging area along the Platte River Valley in Nebraska facilitate completion of nutrient acquisition to meet migration demands and to enhance body conditioning for nesting (Krapu et al. 1985, Iverson et al. 1987).

Sandhill crane hunting seasons (crane hunting) in North Dakota were closed for 52 years following the 1916 Convention for the Protection of Migratory Birds. In 1968 an experimental hunting season was initiated, and in 1975 the season was considered operational (Miller 1987). A Cooperative Management Plan for the MCP was prepared in 1981 (Central Flyway Council 1981) and was revised in 1993 (Central Flyway Council 1993). Since 1981 these plans have guided the regulation of annual crane harvests in the Central Flyway.

In North Dakota and the remainder of the Central Flyway, the resumption of crane hunting followed a gradual pattern of expansion. Crane hunting and expansion of seasons have come under intense scrutiny and controversy by anti-crane hunting groups and individuals (Miller 1974, Konrad 1977, Johnsgard 1982, Thompson and George 1987). However, this controversy also prompted studies of cranes (Miller 1987) and resulted in expansion and refinement of cooperative survey programs. In this paper we report (1) history and changes in sandhill crane hunting seasons in North Dakota, 1968–94, (2) harvests, (3) hunter activity and success, (4) population status of the MCP, and (5) impacts of hunting on whooping cranes.

We are indebted to the U.S. Fish and Wildlife Service (FWS), Office of Migratory Bird Management staff E. L. Martin, A. N. Novara, and M. F. Sorrenson, for conducting the special sandhill crane harvest surveys, and D. S. Benning, A. N. Novara, J. W. Solberg, and J. S. Walter for coordinating annual MCP spring surveys. We thank D. S. Benning, R. C. Drewien, D. H. Johnson, S. C. Kohn, and P. R. Schmidt for helpful reviews of this manuscript. We acknowledge M. A. Johnson, S. C. Kohn, and C. H. Schroeder for coordinating the annual distribution of federal crane hunting permits in North Dakota.

NORTH DAKOTA HUNTING SEASONS, 1968–94

Crane hunting seasons in North Dakota resumed in 1968. The presence of endangered whooping cranes, sandhill crane subspecies abundance and distribution patterns, crop depreda-

Table 1. Sandhill crane hunting seasons in North Dakota, 1968-94.

Year	Area	Bag/ possession	Season dates	Length (days)
1968		2/4	9 Nov- 8 Dec	30
1969		3/6	8 Nov- 7 Dec	30
1970		3/6	14 Nov-13 Dec	30
1971		3/6	13 Nov-12 Dec	30
1972		3/6	11 Nov-10 Dec	30
1973		3/6	10 Nov- 9 Dec	30
1974		3/6	9 Nov- 8 Dec	30
1975		3/6	8 Nov- 7 Dec	30
1976		3/6	6 Nov- 5 Dec	30
1977		3/6	7 Sep-11 Sep	5
1978		3/6	7 Sep-11 Sep	5
1979		3/6	7 Sep-11 Sep	5
1980	1	3/6	6 Sep-14 Sep	9
	2	3/6	6 Sep-10 Sep	5
1981	1	3/6	5 Sep-20 Sep	16
	2	3/6	5 Sep-13 Sep	9
1982	1	3/6	4 Sep-19 Sep	16
	2	3/6	4 Sep-12 Sep	9
1983	1	3/6	10 Sep- 6 Nov	58
	2	3/6	10 Sep-30 Sep	21
1984	1	3/6	8 Sep- 4 Nov	58
	2	3/6	8 Sep-28 Sep	21
1985	1	3/6	7 Sep- 3 Nov	58
	2	3/6	7 Sep-27 Sep	21
1986	1	3/6	6 Sep- 2 Nov	58
	2	3/6	6 Sep- 3 Oct	28
1987	1	3/6	5 Sep- 1 Nov	58
	2	3/6	5 Sep- 2 Oct	28
1988	1	3/6	10 Sep- 6 Nov	58
	2	3/6	10 Sep-30 Sep	21
1989	1	3/6	9 Sep- 5 Nov	58
	2	3/6	9 Sep-29 Sep	21
1990	1	3/6	8 Sep- 4 Nov	58
	2	3/6	8 Sep-14 Oct	37
1991	1	3/6	7 Sep- 3 Nov	58
	2	3/6	7 Sep-13 Oct	37
1992	1	3/6	5 Sep- 1 Nov	58
	2	3/6	5 Sep-11 Oct	37
1993		3/6	11 Sep- 7 Nov	58
1994		3/6	10 Sep- 6 Nov	58

tions, and harvest levels influenced season selections during subsequent expansions of hunting opportunities (Miller 1987). The following information summarizes changes in North Dakota crane seasons during 1968-94 (Table 1, Fig. 1).

1968

The first hunting season included a daily bag of 2 with a possession limit of 4 and was limited to Kidder and Stutsman Counties. The 30-day season was delayed until 9 November to avoid conflicts with migrating whooping cranes. A relatively small harvest resulted, because most cranes had migrated from the hunt area prior to the season. No conflicts with whooping cranes were reported.

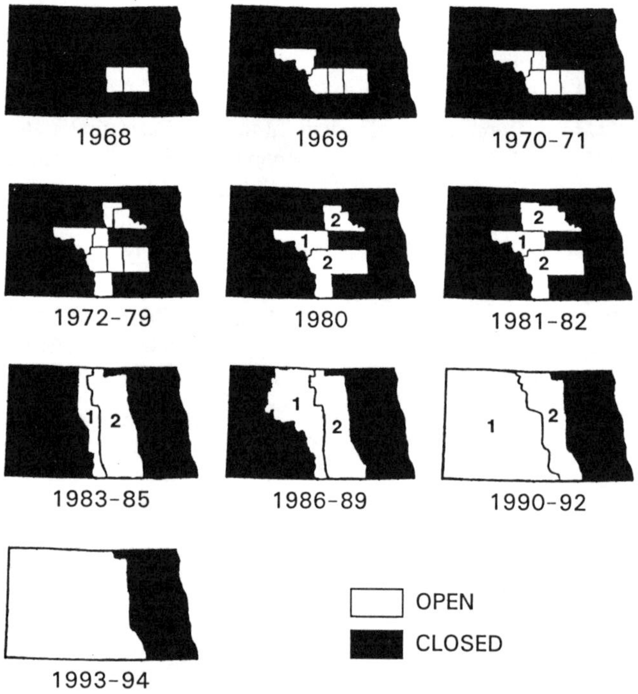

Fig. 1. Areas open to the hunting of sandhill cranes in North Dakota, 1968-94.

1969

Two additional counties were opened in 1969 (McLean and Burleigh), and these encompassed several additional crane concentration areas, which were probably used mainly by lessers and Canadians. A 30-day season was again held primarily in November to minimize conflicts with whooping cranes. Flyway-wide depredation concerns were addressed by increased bag and possession limits to 3 and 6, respectively. These bag and possession limits were followed during all subsequent seasons in North Dakota. Harvests were again relatively low and, because of historical knowledge of subspecies distributions during fall, likely contained a representation of all 3 subspecies (Johnson and Stewart 1973).

1970-71

In the third and fourth years, Sheridan County was

opened to hunting. A 30-day season was retained, but the opening date was delayed about 1 week, the 1970 and 1971 seasons beginning on 14 and 13 November, respectively. Harvests remained low but were reported to be increasing, as hunters learned more about crane behavior and improved their hunting skills.

1972-76

In years 5-9, Benson, Emmons, and Pierce Counties were opened to hunting. The season length remained 30 days, with opening dates of 6-11 November. Harvests during 1972-74 were believed to be low because of late opening dates. In 1975 a federal sandhill crane hunting permit was required for the first time, and a hunter survey indicated that 2,624 cranes were either retrieved or crippled and unretrieved (Fig. 2). In 1976 an early freeze-up occurred and most cranes departed before the 6 November opening date; consequently, only 84 cranes were retrieved or crippled. All seasons focused on the later migrating cohort of lessers; no conflicts with whooping cranes were reported.

1977-79

To increase hunting opportunity, a short, early season was offered and selected in favor of the later, longer season. A federal framework of 5 days in September was established for 1977-79. These seasons resulted in higher harvests (3,273-4,833). Again, no conflicts with whooping cranes were reported. In 1977, because of concern for increased harvest of the greater subspecies, hunter bag checks for subspecies composition were initiated by the North Dakota Game and Fish Department.

1980

To protect whooping cranes, reduce harvest of greaters, and increase hunter opportunity on the more numerous lessers, separate federal frameworks were established for 2 areas. Area 1 (McLean and Sheridan Counties) focused hunting on lessers and had more liberal September seasons of 9 days, while Area 2 (Kidder, Stutsman, Benson, Emmons, Pierce, and Burleigh counties) had reduced September seasons of 5 days. Statewide harvest declined to 2,688 (Fig. 2).

1981-82

In 1981 McHenry County (added to Area 2) was opened, representing the first area expansion since 1972. The season length was increased in both areas (from 9 to 16 days in Area

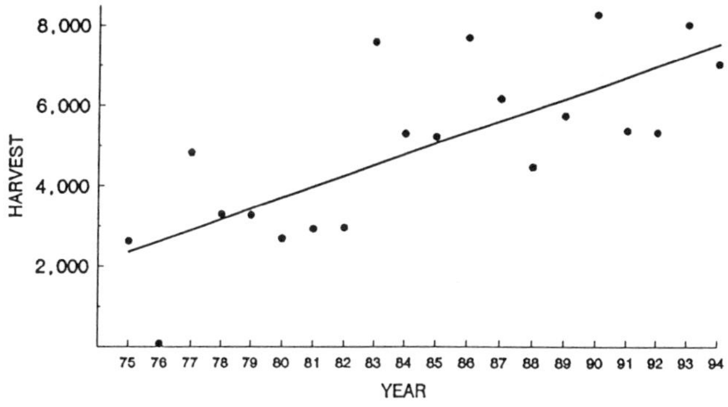

Fig. 2. Hunting mortality (retrieved and unretrieved) of sandhill cranes in North Dakota, 1975-94.

1 and from 5 to 9 days in Area 2). Harvest increased slightly but was $<3,000$ annually.

1983-85

The 1983 season marked the greatest change in the Central Flyway's crane hunting federal framework. The Central Flyway Council (1981) completed a management plan for MCP cranes, and frameworks for the 8 Central Flyway states had been gradually expanded since 1968. By 1985, annual flyway harvests reached 14,000, and North American harvests of 25,000 were believed to be appropriate for maintaining the current population status (Central Flyway Council, Recommendation 2, 18 March 1990). Therefore, instead of specific season descriptions, the FWS agreed to set a broad framework that allowed states flexibility in setting seasons to address local issues but maintain North American harvests near 25,000. Due to this more liberal framework, 12 additional counties were opened, totaling 21 open counties, that were grouped into 2 zones. The zones were oriented in a north-south direction, corresponding to known distributions and abundance of the 3 subspecies that were divided into the Western and Gulf Coast subpopulations (Tacha and Vohs 1984). In 1983, cranes were rarely found east of U.S. Highway 281 (Carlisle and Tacha 1983), which became the eastern boundary of Zone 2. A full federal framework of 58 days was selected in Zone 1, while a more restrictive 21-day season was selected in Zone 2. Seasons began in early September (Table 1) in both zones, but ended during early November in Zone 1 and during late September in Zone 2. As with earlier regulatory changes, the harvest more than doubled (150%) during the first year of expansion, from about 3,000 in 1982 to more than 7,500 in 1983. However, this increase was not maintained in the following 2 years, stabilizing at about 5,300 cranes during these seasons (Fig. 2). The relative proportion of the harvest of each subspecies was not known, but it was estimated that 13% (804/5,960) of the 1983 harvest was composed of greaters (S. C. Kohn, N.D. Game and Fish Dep., Bismarck, pers. commun.).

1986-89

In 1986, 3 more counties (Ward, McLean, and Renville) were opened to hunting. Most increases in harvest from the new areas were probably lesser and Canadian subspecies. The season was increased to 28 days in 1986-87, but decreased to 21 days in 1988-89. Harvests increased in 1986 to near 7,700 but declined in 1987 to about 6,200. In 1988 and 1989, harvests were ~4,500 and 5,700, respectively.

1990

The largest increase in harvest occurred in 1990 with the expansion of hunting seasons. Northwestern counties and other counties west of the Missouri River were opened, and the boundary between Zones 1 and 2 in Kidder County was changed; most of the county was placed in Zone 1. The harvest in Kidder County increased 128% from 1,192 to 2,721, and the statewide harvest increased 44% from 5,736 to 8,255.

1991-92

The federal sandhill crane hunting permit, first required in 1975, was issued free of charge to all Central Flyway crane hunters and facilitated compilation of a list of hunters and addresses for sampling purposes. In 1991 North Dakota established a $5 charge for the permit, and number of permits issued dropped 54% (7,268 to 3,353), active hunters fell 36% (from 2,684 to 1,684), and harvests decreased 32% (from 6,804 to 4,650). Reduced hunter participation and lower harvests were also reported in 1992.

1993-94

In 1993-94, zones were eliminated and the full federal framework was selected for crane hunting. The long-term impact of this change is unknown but initially resulted in higher harvests for those counties that were formerly in Zone 2. It may also have contributed to shifting cranes eastward into closed areas due to increased hunting activity (D. H. Johnson, Northern Prairie Sci. Cent., Jamestown, N.D., pers. commun.). In contrast to the 1990 expansion, any increases in harvest in the short- or long-term would likely involve higher numbers of Canadian and greater subspecies. Hunter numbers and harvests reached near record high levels during 1993-94 and effectively erased declines resulting

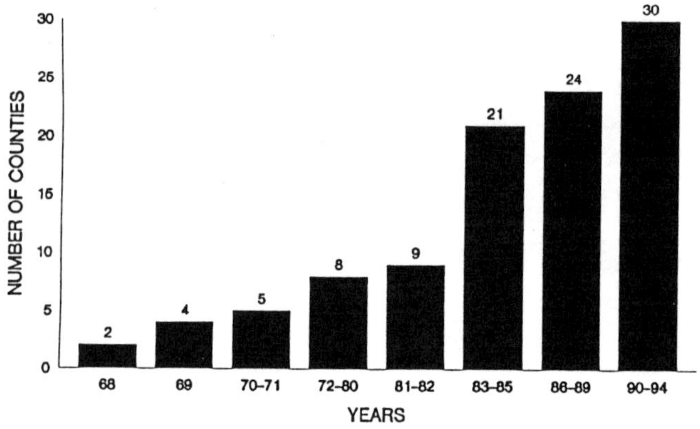

Fig. 3. Number of counties open to sandhill crane hunting in North Dakota, 1968-94.

from fee charges for hunting permits in 1991.

HARVESTS

Crane harvests in North Dakota have increased since modern day seasons resumed (Fig. 2). Most increases appeared to have resulted from opening additional counties to hunting (Fig. 3). In contrast, crane harvests in the remainder of the Central Flyway and in North America remained relatively stable since the current federal framework was established in 1983. Kidder County usually recorded the highest annual harvests, generally >1,000 or about 32% of the statewide harvest, followed by McLean and Pierce Counties. More than 50% of the state's harvests have consistently come from these 3 counties (Table 2).

Since 1975 the North Dakota and Central Flyway harvest has averaged about 21% and 60%, respectively, of the estimated North American MCP crane harvest (Table 3). North Dakota has maintained this proportion of harvest by expanding hunting opportunity, while the comparatively stable hunting opportunity in the remainder of the Central Flyway has resulted in a decline in the proportion of birds harvested in other states (Martin 1995).

Although the 1993 revision of the MCP Cooperative Management Plan eliminated the harvest threshold established in the 1981 plan, the FWS believed that past harvest levels should be maintained (59 Federal Register 67:16765, 1994). This decision was based on the belief that harvests were at an appropriate level for maintaining a stable population and were within established thresholds and near the long-term objective level. In addition, improved monitoring of the harvests and population status of MCP greaters were needed before any significant expansion of hunting occurred within the range of the Gulf Coast subpopulation (59 Federal Register 67:42477, 1994).

Table 2. Sandhill crane sport hunting harvests (retrieved) as reported by county in North Dakota, 1975–94.

County	1975	1976	1977	1978	1979	1980	1981	1982	1983	1984	1985	1986	1987	1988	1989	1990	1991	1992	1993	1994	%
Benson	118	5	500	287	339	361	228	119	305	159	265	554	511	228	328	448	316	486	638	489	8.6
Burleigh	153	5	91	147	165	81	145	95	678	491	543	565	240	381	324	412	562	460	611	875	9.1
Emmons	24	2	14	9	6	0	7	0	86	54	88	79	32	0	17	105	88	188	78	68	1.2
Kidder	513	3	1,320	939	867	643	528	520	1,996	1,246	1,320	1,556	1,831	1,290	1,192	2,721	1,401	1,149	2,115	1,515	31.8
McLean	1,020	18	264	182	450	259	526	845	2,003	1,002	600	783	533	344	615	535	548	439	878	897	16.4
Pierce	35	11	1,301	656	532	501	390	271	168	270	285	544	364	154	212	312	212	374	679	640	10.2
Stutsman	32	0	279	251	162	203	136	55	79	266	140	136	185	156	400	487	304	284	327	121	5.3
Sheridan	86	4	100	72	68	78	227	308	400	346	581	419	521	316	497	342	328	323	396	362	7.4
McHenry							84	99	264	225	205	189	230	132	144	156	120	87	275	378	3.3
Bottineau									6	82	78	108	103	125	58	213	118	168	99	0	1.5
Eddy									0	0	0	9	0	3	0	15	0	0	1	0	tr
Foster									0	0	0	0	0	0	6	3	2	0	0	16	tr
LaMoure									0	0	0	0	0	0	0	0	0	0	0	0	0.0
Logan									36	0	0	0	0	0	45	99	23	41	10	105	0.5
Mcintosh									0	0	0	0	0	0	0	0	0	0	0	0	0.0
Mercer									228	45	78	132	38	162	15	96	25	32	20	30	1.2
Ramsey									0	0	0	0	3	0	0	9	10	11	29	80	0.2
Rolette									0	10	6	3	0	0	24	66	32	81	97	15	0.4
Towner									0	0	0	0	0	0	12	9	0	11	6	0	tr
Ward									10	10	26	761	278	157	230	231	160	115	102	98	2.8
Wells									19	0	130	47	76	78	191	162	51	39	164	165	1.4
Burke												20	0	0	46	21	3	6	3	0	0.1
Mountrail												388	245	122	166	45	41	76	18	30	1.5
Renville												0	29	0	0	6	6	7	0	0	0.1
Bowman																0	0	5	0	0	tr
Divide																33	9	0	3	8	0.1
Dunn																3	1	0	0	0	tr
McKenzie																0	0	2	3	24	tr
Morton																12	0	0	1	0	tr
Stark																3	0	0	0	0	tr
Williams																0	0	0	12	0	tr
Unknown	141	4	209	134	144	114	124	157	193	161	186	270	115	157	134	215	218	223	413	319	4.7
Total	2,122	52	4,078	2,777	2,733	2,240	2,395	2,469	6,471	4,367	4,531	6,563	5,334	3,806	4,656	6,759	4,578	4,607	6,978	6,235	

During 1975–94 North Dakota's estimated crippling loss rates were generally higher (+2%) than those in the other Central Flyway states (Fig. 4). However, a declining trend in crippling rates is evident for North Dakota and the Central Flyway. Factors responsible for these trends are unknown, but the use of improved hunting techniques (e.g., decoys and calls) and enhanced understanding of crane behavior have likely contributed to improved hunter success and reduced crippling losses (Sharp and Vogel 1992).

In 1987 North Dakota implemented non-toxic shot shell requirements for crane hunting (M. A. Johnson, N.D. Game and Fish Dep., Bismarck, pers. commun.). Little public opposition to these regulations occurred, and a comparison of crippling loss rates for the pre- (1975–86) and post-implementation (1987–94) periods indicates no increase in crippling loss rates related to the use of non-toxic shot. Conversely, North Dakota's crippling rate declined from 18.0% to 15.3%; this followed similar declines of 14.6% to 11.8% in the remainder of the Central Flyway during the same period (Fig. 4).

HUNTER ACTIVITY AND SUCCESS

The number of federal crane hunting permits issued to North Dakota hunters increased during 1975–90 (Martin 1995). Beginning in 1991, North Dakota required a $5 fee for validation of the free federal permit. Subsequently, the number of permits issued declined, while the trend in total number of active hunters was unchanged (Fig. 5). This suggested that the additional fee had little effect on those hunters who previously hunted cranes, but eliminated casual crane hunters who acquired a free permit. It is not known if

Table 3. Annual sport hunting mortality (retrieved and unretrieved) of the Mid-continent Population of sandhill cranes, 1975-94.

Year	North Dakota				Central Flyway				North America		
	Retrieved	Unretrieved	Total	% N.A.[b]	Retrieved	Unretrieved	Total	% N.A.[b]	Retrieved	Unretrieved	Total
1975	2,122	502	2,624	11.5	9,497	1,885	11,382	49.8	18,272	4,568	22,840
1976	52	32	84	0.6	7,393	1,384	8,777	67.2	10,455	2,614	13,069
1977	4,078	755	4,833	27.9	12,251	2,013	14,164	81.7	13,874	3,468	17,342
1978	2,777	517	3,294	22.0	10,146	1,860	12,006	80.3	11,966	2,992	14,958
1979	2,733	540	3,273	16.7	10,379	1,655	12,029	61.4	15,684	3,921	19,605
1980	2,245	443	2,688	12.1	10,150	1,740	11,890	53.7	17,724	4,431	22,155
1981	2,395	533	2,928	16.2	10,130	1,780	11,910	66.0	14,431	3,608	18,039
1982	2,469	487	2,956	17.8	7,920	1,400	9,320	56.2	13,265	3,316	16,581
1983	6,471	1,114	7,585	32.1	12.960	2,220	15,180	64.3	18,880	4,720	23,600
1984	4,367	933	5,300	23.1	11,270	1,890	13,160	57.3	18,368	4,592	22,960
1985	4,531	690	5,221	20.5	12,580	1,860	14,440	56,8	20,331	5,083	25,414
1986	6,563	1,119	7,682	29.7	12,490	1,870	14,360	55.5	20,714	5,179	25,893
1987	5,334	832	6,166	24.5	12,770	1,710	14,480	57.6	20,119	5,030	25,149
1988	3,815	650	4,465	15.8	12,772	1,628	14,400	50.8	22,662	5,665	28,327
1989	4,656	1,080	5,736	22.3	13,639	2,081	15,720	61.2	20,543	5,136	25,679
1990	6,804	1,451	8,255	25.7	18,041	2,519	20,560	64.0	25,716	6,429	32,145
1991	4,580	785	5,365	20.3	13,079	1,768	14,847	56.3	21,098	5,274	26,372
1992	4,654	665	5,319	21.1	12,433	1,459	13,892	55.1	20,186	5,047	25,233
1993	6,985	1,023	8,008	27.3	18,005	2,252	20,257	69.2	23,429	5,857	29,286
1994	6,235	785	7,020	23.9	16,201	1,676	17,877	60.7	23,546	5,886	29,432
$\bar{x}$	4,193	747	4,940	21.3	12,200	1,832	14,033	60.5	18,563	4,641	23,204

[a] Percentage of total sport hunting mortality in North America.

this fee reduced recruitment of new crane hunters who previously acquired free permits and then participated in hunts opportunistically.

Numbers of active hunters declined during 1975-94 in North Dakota (Fig. 6) and in other Central Flyway states, but the decline was not as large as the drop in the sales of federal duck stamps. In the Central Flyway, duck populations declined, goose populations increased, and crane populations remained stable during 1985-94. The overall numbers of hunters declined but were probably greater for duck hunters than for crane or goose hunters. The downward trend in number of North Dakota crane hunters might have been larger without the expansion in hunter opportunity during 1975-94.

In North Dakota, during 1975-94, the numbers of days afield by active crane hunters increased from <2 to >3 days

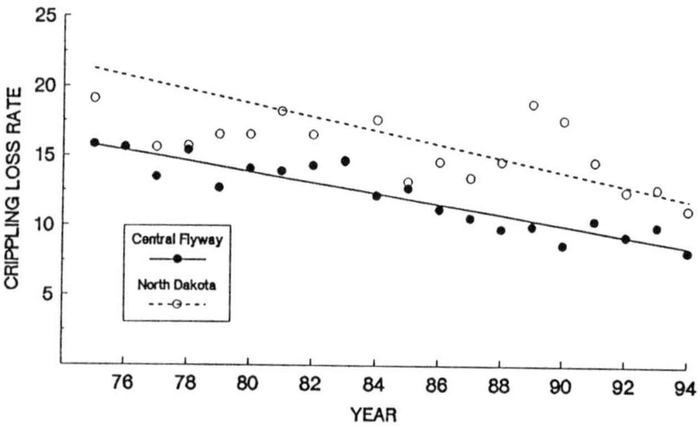

Fig. 4. Crippling loss rates (number crippled/total harvest) of sandhill cranes harvested in North Dakota and the remainder of the Central Flyway states, 1975-94.

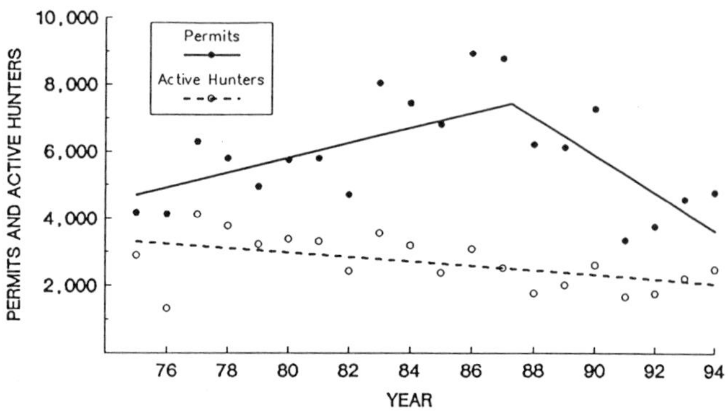

Fig. 5. Number of federal sandhill crane hunting permits and estimated active sandhill crane hunters in North Dakota, 1975-94.

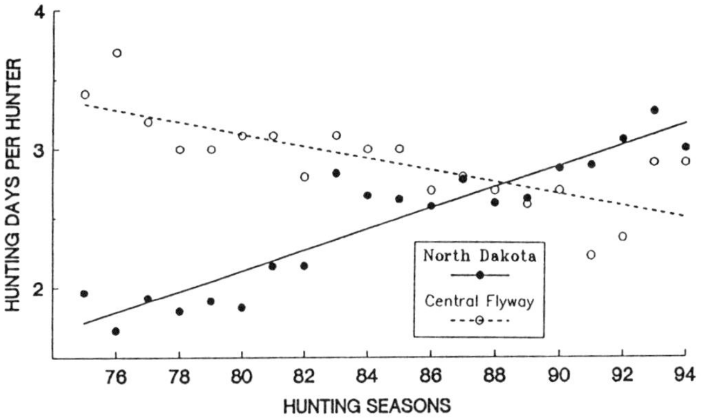

Fig. 6. Hunting days per active crane hunter in North Dakota and the Central Flyway, 1975-94.

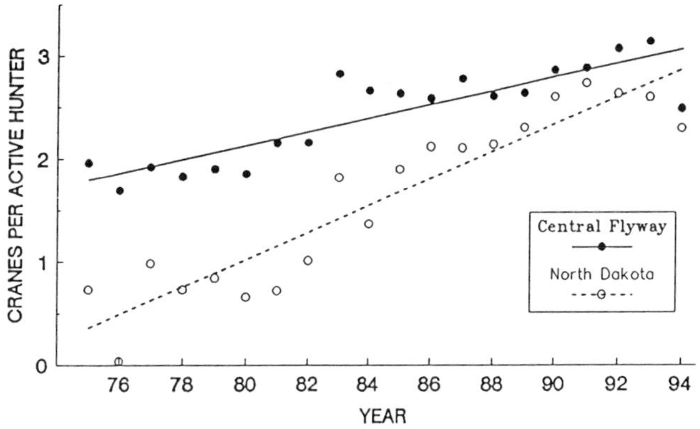

Fig. 7. Seasonal sandhill crane bag per active crane hunter in North Dakota and the Central Flyway, 1975-94.

annually (Fig. 6), but for Central Flyway hunters the number of days afield declined from ~3.5 to 2.5 days. During this period seasons were generally stable for most Central Flyway States but were greatly expanded in North Dakota.

A comparison of average seasonal crane bag of North Dakota hunters with other Central Flyway hunters indicates similar increasing trends during 1975-94 (Fig. 7). In 1994, Central Flyway hunters, including those in North Dakota, harvested >2 cranes per active adult hunter. This level of success has been maintained since 1987 and appears to have stabilized following increases from just over 1 crane per hunter in 1975.

POPULATION STATUS

The population dynamics of MCP cranes are poorly understood, but sandhill cranes are long-lived and have the lowest recruitment and highest survival of any game birds in North America (Lewis 1977, Drewien et al. 1995). In addition, MCP cranes may not breed until 5 years old (Tacha et al. 1994) and the potential for overharvest is high, although population declines may not be immediately evident. Crane harvest strategies must be conservative (Miller 1974, Lewis 1977, Johnson 1979, Drewien et al. 1995).

Numbers of MCP cranes are unknown. The spring migration period provides the best opportunity to conduct annual surveys (Benning and Johnson 1987) because the MCP is confined almost entirely to Nebraska's Central Platte River Valley. The 1993-95 3-year running average index for this area was 420,866, and was within the established population objective range of 343,000-465,000 (Walter 1995). In 1982 intensive photographic-transect spring surveys estimated the MCP at ~509,000 (Sharp and Vogel 1992). Subsequent annual indices (1983-95), derived from photo-correction of ocular-cruise surveys, indicate that the spring population was stable.

The size of the MCP during fall is related to the preceding spring population size, annual production, and summer mortality. Summer mortality was estimated at ~2% (Miller 1987) and average production was ~11% (Buller 1979, Drewien et al. 1995). During 1983-94 the estimated, annual fall flights probably ranged between 510,000 and 590,000 (Sharp and Vogel 1992). From a management perspective, the relatively constant spring populations observed during 1982-92 suggested that annual mortality, including sport harvest, has been at a level appropriate to stabilize the MCP at objective levels.

SUBSPECIES CONSIDERATIONS

The 1981 MCP Cooperative Management Plan (Central Flyway Council 1981) estimated that 65-70% of the MCP were lessers, 25-30% were Canadians, and 1-5% were greaters. Subsequent investigators have questioned this abundance of greaters and have estimated the population wintering on the Gulf Coast of Texas at ~5,000 greaters (C. D. Littlefield, Muleshoe, Tex., pers. commun.); others have estimated the subpopulation to be near 60,000 (Tacha et al. 1986, Muehl 1994). However, the current status of the Gulf Coast subpopulation is unknown, and its distribution patterns and abundance by subspecies and among states are poorly understood.

Before 1968 in the Central Flyway, subspecies were generally distributed according to size along a west-to-east gradient; lessers were most abundant along the western portion of the migration pathway, greaters along the eastern portion, and Canadians were primarily found in the central portions with interchange, especially Canadians and lessers, in most areas (Buller 1979). It is unknown if current fall distribution patterns are being influenced by hunting, but

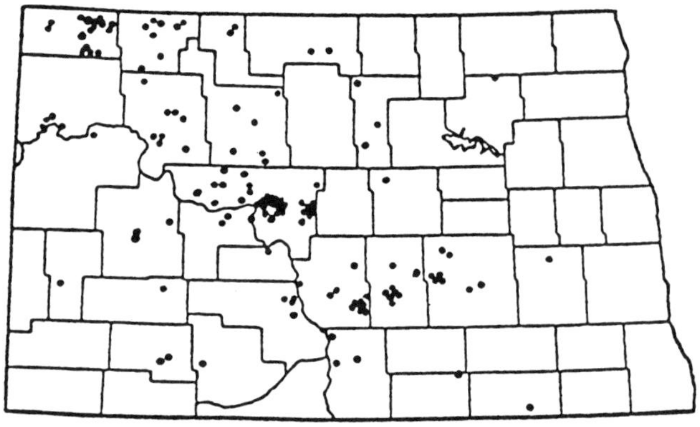

Fig. 8. Locations of 215 fall whooping crane sightings in North Dakota, 1955–93.

some changes were documented (Johnson 1976). Recently in North Dakota, cranes were observed east of U.S. Highway 281, but the relative abundance of each subspecies was unknown (S. C. Kohn, pers. commun.). Similar eastern directional shifts in winter distributions were also reported in Texas (R. R. George, Texas Parks and Wildl. Dep., Austin, pers. commun.).

In 1977 the North Dakota Game and Fish Department initiated annual crane bag checks and collected certain diagnostic morphometric measurements and data on sex of cranes. Discriminant function analysis was used to determine subspecific composition in the harvest (Johnson and Stewart 1973).

WHOOPING CRANES

The whooping crane is the only other species of crane native to North America. Only 22 individuals existed in 1941, and the species was designated as endangered in 1973 (U.S. Fish and Wildlife Service 1994). Recovery has been slow, and during winter 1992–93 a total of 136 wintered at Aransas National Wildlife Refuge, Texas. Historically, the species nested in North Dakota until 1915 (Konrad 1987).

Locations of 215 whooping crane fall migration sightings in North Dakota were recorded during 1955–93 (Fig. 8) and included at least 539 individuals. Observations occurred between 5 September and 9 November but mainly during October (70%), followed by September (22%), and November (8%). Most observations were reported from McLean County, followed by Divide, Burleigh, Kidder, Stutsman, and Mountrail. These observations indicated that during fall most whooping cranes moved southeasterly from Divide and Mountrail Counties in the northwest to McLean, Burleigh, Kidder, and Stutsman Counties in southcentral North Dakota.

Because of potential conflicts between whooping cranes and sandhill crane hunting (Thompson and George 1987), seasons in North Dakota were gradually expanded geographically and temporally (Konrad 1987). The FWS consults annually under Section 7 of the Endangered Species Act and develops a biological opinion for the migratory bird hunting program; potential conflicts between sandhill crane seasons and whooping cranes receive considerable attention. A contingency plan (U.S. Fish and Wildlife Service 1994) allows sandhill crane hunting to proceed, while providing for monitoring and protective actions, including temporary spot closures and public service announcements, if a whooping crane is discovered in a hunting area. Since sandhill crane seasons were resumed in 1968, no whooping crane has been accidentally shot by a sandhill crane hunter.

MANAGEMENT IMPLICATIONS

In North Dakota, seasons beginning in early September have higher annual harvests than those late seasons beginning in October–November. Whooping cranes are present during these early seasons, but the monitoring and protective actions have been adequate for their protection.

Concern over harvest levels of greaters in eastern North Dakota prompted annual collection of morphometric measurements since 1977. In 1980 the open portion of the state was divided into Area 1 (western), which had more liberal seasons, and harvest was focused on the more numerous lesser subspecies; and Area 2 (eastern), which had more restrictive seasons because all 3 subspecies were present. Data on subspecies are currently being analyzed and will be incorporated into the next revision of the MCP management plan.

In 1983 the FWS established a broad framework for crane hunting in the Central Flyway to allow states flexibility for setting seasons within the harvest threshold identified in the Cooperative Management Plan (Central Flyway Council 1981). federal frameworks for crane hunting have remained essentially unchanged for the MCP since 1983. The Central Flyway Council has recommended no change in hunting seasons until the MCP Cooperative Management Plan is revised in 2000.

Historical MCP distribution and traditional use of fall migration roost sites have been altered due to hunting pressure, drought, increases in acreages enrolled in the U.S. Department of Agriculture's Conservation Reserve Program, and any number of other possible influences. Complaints of depredations by cranes on agricultural crops have been used to help justify current population objectives, current hunting programs, and past expansions in federal hunting frameworks. However, these issues should be quantified and a comprehensive hunting strategy should be prepared that addresses these concerns throughout the Central Flyway.

In 1993 North Dakota recommended that the open area be expanded to counties east of U.S. Highway 281. The FWS denied this request until staff from the Office of Migratory Bird Management and the North Dakota Game and Fish Department could cooperatively assess the subspecies information gathered since 1977 and determine geographic and temporal harvest of greaters. The proposed change in hunting regulations upon the harvest of greaters in the Gulf Coast subpopulation should also be evaluated.

The MCP is currently at objective levels and its future remains favorable. However, significant management challenges remain, including (1) development of harvest strategies to maintain a stable population, (2) maintenance of current harvest distribution patterns, (3) maintenance of current subspecies abundance, and (4) assessment of the status and distribution of the numerically less abundant greater subspecies in North Dakota and throughout the Central Flyway. Data bases need much improvement and more attention should be directed to monitoring subspecies status for the future development and implementation of MCP hunting programs.

LITERATURE CITED

BENNING, D. S., and D. H. JOHNSON. 1987. Recent improvements to sandhill crane surveys in Nebraska's central Platte Valley. Pages 10-16 *in* J. C. Lewis, ed. Proc. 1985 crane workshop. Platte River Whooping Crane Maintenance Trust, Grand Island, Nebr.

BRAUN, C. E., R. C. DREWIEN, C. D. LITTLEFIELD, and L. H. WALKINSHAW. 1975. Conservation committee report on status of sandhill cranes. Wilson Bull. 87:297-302.

BULLER, R. J. 1967. Sandhill crane study in the Central Flyway. U.S. Fish Wildl. Serv. Spec. Sci. Rep. Wildl. 113. 17pp.

______. 1979. Lesser and Canadian sandhill crane populations, age structure, and harvest. U.S. Fish Wildl. Serv. Spec. Sci. Rep. 221. 10pp.

CARLISLE, M. J., and T. C. TACHA . 1983. Fall migration of sandhill cranes in west central North Dakota. J. Wildl. Manage. 47:818-821.

CENTRAL FLYWAY COUNCIL. 1981. Management plan for mid-continent sandhill cranes. Unpubl. Rep., Cent. Flyway Counc., Golden, Colo. 21pp.

______. 1993. Management plan for mid-continent sandhill cranes. Unpubl. Rep., Cent. Flyway Counc., U.S. Fish Wildl. Serv. Golden, Colo. 21pp.

DREWIEN, R. C., W. M. BROWN, and W. L. KENDALL. 1995. Recruitment in Rocky Mountain greater sandhill cranes and comparison with other crane populations. J. Wildl. Manage. 59:339-356.

______, and J. C. LEWIS. 1987. Status and distribution of cranes in North America. Pages 469-477 *in* J. C. Lewis, ed. Proc. 1985 crane workshop. Platte River Whooping Crane Maintenance Trust, Grand Island, Nebr.

IVERSON, G. C., P. A. VOHS, and T. C TACHA. 1987. Habitat use by mid-continent sandhill cranes during spring migration. J. Wildl. Manage. 51: 448-458.

KONRAD, P. M. 1987. Expanded sandhill crane hunting in the Dakotas and Oklahoma threatens endangered whooping cranes. Pages 69-77 *in* J. C. Lewis, ed. Proc. 1985 crane workshop. Platte River Whooping Crane Maintenance Trust, Grand Island, Nebr.

KRAPU, G. L., G. C. IVERSON, K. J. REINECKE, and C. M. BOISE. 1985. Fat deposition and usage by arctic nesting sandhill cranes during spring. Auk 102:362-368.

JOHNSGARD, P. A. 1982. Whooper recount. Nat. Hist. 91:70-75.

JOHNSON, D. H. 1976. The status of the sandhill crane in North Dakota. Pages 69-77 *in* J. C. Lewis, ed. Proc. [1975] int. crane workshop. Oklahoma State Univ. Publ. Printing, Stillwater.

______. 1979. Modeling sandhill crane population dynamics. U.S. Fish Wildl. Serv. Spec. Sci. Rep. 222. 10pp.

______, and R. E. STEWART. 1973. Racial composition of migrant populations of sandhill cranes in the northern plains states. Wilson. Bull. 85:148-162.

LEWIS, J. C., chairman. 1977. Sandhill crane (*Grus canadensis*). Pages 4-43 *in* G.C. Sanderson, ed. Management of migratory shore and upland game birds in North America. Int. Assoc. Fish Wildl. Agencies, Washington, D.C.

MADSEN, C. R. 1967. Food and habitat selection by fall migrant sandhill cranes in Kidder County, North Dakota. M.S. Thesis, Michigan State Univ., East Lansing. 60pp.

MARTIN, E. M. 1995. Sandhill crane harvest and hunter activity in the Central Flyway during the 1994-95 hunting season. Unpubl. Rep., U.S. Fish Wildl. Serv., Off. Migr. Bird Manage., Laurel, Md. 12pp.

MILLER, H. W. 1987. Hunting in the management of mid-continent sandhill cranes. Pages 30-46 *in* J. C. Lewis, ed. Proc. 1985 crane workshop. Platte River Whooping Crane Maintenance Trust, Grand Island, Nebr.

MILLER, R. S. 1974. The programmed extinction of the sandhill crane. Nat. Hist. 83:62-69.

MUEHL, G. T. 1994. Distribution and abundance of water birds and wetlands in coastal Texas. M.S. Thesis, Texas A&M Univ.-Kingsville, Kingsville. 130pp.

SHARP, D. E. 1995. Status and harvests of the Mid-continent and Rocky Mountain Populations of sandhill cranes. Unpubl. Rep., U.S. Fish Wildl. Serv., Golden, Colo. 8pp.

______, and W. O. VOGEL. 1992. Population status, hunting regulations, hunting activity, and harvests of mid-continent sandhill cranes. Proc. North Am. Crane Workshop 6:24-32.

SUGDEN, L. G., R. G. CLARK, E. J. WOODSWORTH, and H. GREENWOOD. 1988. Use of cereal fields by foraging sandhill cranes in Saskatchewan. J. Appl. Ecol. 25:111-124.

TACHA, T. C., D. E. HALEY, and R. R. GEORGE. 1986. Population and harvest characteristics of sandhill cranes in southern Texas. J. Wildl. Manage. 50:80-83.

______, S. A. NESBITT, and P. A. VOHS. 1994. Sandhill crane. Pages 76-94 *in* T. C. Tacha and C. E. Braun, eds. Migratory shore and upland game bird management in North America.

Int. Assoc. Fish Wildl. Agencies, Washington, D.C.

______, and P. A. VOHS. 1984. Some population parameters of sandhill cranes from mid-continental North America. J. Wildl. Manage. 48:89-98.

______, ______, and G. C. IVERSON. 1984. Migration routes of sandhill cranes from mid-continental North America. J. Wildl. Manage. 48:1028-1033.

THOMPSON, B. C., and R. R. GEORGE. 1987. Minimizing conflicts between migratory game bird hunters and whooping cranes in Texas. Pages 57-68 *in* J. C. Lewis, ed. Proc. 1985 crane workshop. Platte River Whooping Crane Maintenance Trust, Grand Island, Nebr.

U.S. FISH AND WILDLIFE SERVICE. 1994. Whooping crane recovery plan. U.S. Fish Wildl. Serv., Albuquerque, N.M. 92pp.

WALTER, J. S. 1995. Coordinated spring mid-continent sandhill crane survey. Unpubl. Rep., U.S. Fish Wildl. Serv., LaCrosse, Wis. 9pp.

WALKINSHAW. L. H. 1949. The sandhill cranes. Cranbrook Inst. Sci. Bull. 29. 202pp.

______. 1965. A new crane from central Canada. Can. Field-Nat. 79:181-184.

SANDHILL CRANE HUNTS IN THE RIO GRANDE VALLEY AND SOUTHWEST NEW MEXICO

C. GREGORY SCHMITT, New Mexico Department of Game and Fish, P.O. Box 25112, Santa Fe, NM 87504, USA
BARRY HALE, New Mexico Department of Game and Fish, P.O. Box 25112, Santa Fe, NM 87504, USA

Abstract: Annual limited permit hunting seasons for sandhill cranes (*Grus canadensis*) were initiated in southwestern New Mexico (SWNM) in 1982 and in the middle Rio Grande Valley (MRGV) in 1986. Within the MRGV and SWNM are wintering areas for 3 migratory subspecies, including the lesser sandhill crane (*G. c. canadensis*) (lessers), Canadian sandhill crane (*G. c. rowani*) (Canadians), and greater sandhill crane (*G. c. tabida*) (greaters). We summarized annual season dates, bag limits, requirements for permits, procedures for examination of crane specimens at check stations, number of hunters, estimated harvest, mean harvest per hunter, subspecific composition, sex ratios, and percent juveniles in the harvest by hunt area. Mean values for hunt parameters in the MRGV were: annual harvest 458.8 ± 272.0 (1 SD), number of hunters 309.7 ± 121.8, harvest per hunter 1.4 ± 0.8, crippling loss 20.4% ± 7.0, and subspecific composition during 9 years was 27.1% ± 23.1 lessers, 7.1% ± 3.5 Canadians, and 65.7% ± 20.6 greaters with juveniles averaging 13.4% ± 8.8 of the harvest. Mean values for hunt parameters in SWNM were: annual harvest 100.2 ± 45.9, number of hunters 108.8 ± 45.7, harvest per hunter 0.9 ± 0.3, crippling loss 41.9% ± 14.9, and mean subspecific composition for 9 years was 66.6% ± 18.2 lessers, 13.0% ± 13.6 Canadians, and 20.3% ± 10.6 greaters with juveniles representing a mean of 14.2% ± 7.1 of the harvest. Our data indicated an increased vulnerability of juvenile lessers and greaters to hunting as compared to adults. An analysis of variance revealed that 13 of 15 comparisons of subspecies-sex combinations (6) and measurements of morphological variables (5) were significantly different ($P < 0.001$). Only tail and wing chord measurements for male Canadians and female greaters did not differ significantly. Hunt procedures allowed us to successfully monitor annual crane harvest and adjust hunting seasons as necessary to remain within acceptable harvest rates of Rocky Mountain Population (RMP) cranes as established in Pacific and Central Flyway guidelines. Responsible management and monitoring of sandhill crane hunts can be accomplished by combining education programs and mandatory check stations where harvested cranes are examined.

PROC. NORTH AM. CRANE WORKSHOP 7:219-231

Key words: *Grus canadensis canadensis*, *Grus canadensis rowani*, *Grus canadensis tabida*, harvest, crane hunting seasons, New Mexico, Rio Grande Valley, sandhill crane, southwestern New Mexico.

Populations of sandhill cranes (cranes) wintering in the MRGV and portions of southwestern New Mexico have increased significantly during recent decades (Drewien and Bizeau 1974, Drewien et al. 1995). Lessers and Canadians are from the Western Subpopulation of the Mid-continent Population (Tacha et al. 1984) and greaters are from the RMP (Drewien and Bizeau 1974). Of the 3 migratory subspecies wintering in New Mexico, RMP greaters are most limited in numbers, and recent estimates revealed about 16,200–21,500 (Benning et al. 1997). Beginning in 1982, the New Mexico Department of Game and Fish (NMGF) established hunting seasons for cranes in portions of southwestern New Mexico and in 1986 in the MRGV. These seasons were in response to increased interest in crane hunting, increases in overall winter crane populations, and depredations to corn, chili peppers, and alfalfa. Hunts provided the opportunity to collect information on distribution, numbers, food habits, and subspecific composition of wintering cranes. In this paper we report (1) parameters of sandhill crane hunting seasons that were established in the MRGV and SWNM, (2) estimates of harvest and crippling losses, (3) subspecific, sex, and age composition of specimens examined at hunter check stations, and (4) recommendations concerning future crane hunting in New Mexico.

We thank J. L. Herring, J. M. Ward, and G. Downer with the NMGF, C. P. Lindberg, K. Waller, S. Miller, T. Collins, and students from Eastern New Mexico University and New Mexico State University for assistance with data collection at check stations. D. H. Sutcliffe, S. G. Gonzales, R. J. Salazar (NMGF), and L. Wiley-Kagen, U.S. Fish and Wildlife Service (USFWS) provided support with administrative assistance. We thank R. W. Dickerman, R. C. Drewien, and J. P. Hubbard for suggestions concerning critical aspects of identification of subspecies of sandhill cranes and for encouraging proper crane management. D. E. Cowley assisted with statistical analysis in comparing mensural data of sandhill cranes. We also thank R. C. Drewien and R. E. Shea for comments on previous drafts of the manuscript.

STUDY AREAS

The MRGV and portions of SWNM were regions where crane hunting seasons were established (Fig.1). In the MRGV, large numbers of cranes wintered along a 130-km strip from Los Lunas, Valencia County, south to the Bosque del Apache National Wildlife Refuge (NWR) in Socorro County. Smaller numbers of cranes wintered in the lower Rio Grande Valley floodplain (LRGV) between Caballo Reservoir

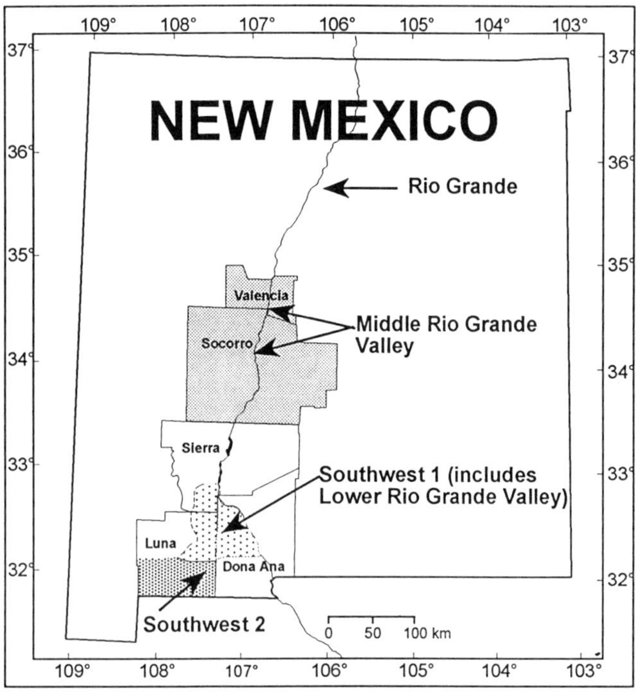

Fig. 1. Sandhill crane hunt boundaries during 1982–94 hunting seasons in the Middle Rio Grande Valley (Valencia and Socorro Counties), Southwest 1 (portions of Sierra, Doña, and Luna Counties including Lower Rio Grande Valley floodplain and the adjacent Uvas Valley), and Southwest 2 (Southern Luna County). Southwest New Mexico hunt boundaries include Southwest 1 and Southwest 2.

and the Hatch Valley in Sierra and Doña Ana Counties, the adjacent Uvas Valley in Sierra and Doña Ana Counties (Southwest Hunt Area 1), and agricultural areas near Deming and Columbus in Luna County (Southwest Hunt Area 2). In all areas, cranes concentrated in irrigated croplands. Drewien and Bizeau (1974), Walker and Schemnitz (1987), Taylor and Kirby (1990), and Drewien et al. (1995) described these wintering areas.

METHODS

Hunt Guidelines

All crane hunts in the MRGV and SWNM were cooperatively designed by personnel of the NMGF and USFWS. Operational plans were written and periodically revised and then agreed to by personnel of each agency and the Pacific and Central Flyway Councils to ensure adequate protection of whooping cranes (*G. americana*) and other protected species (Edens 1992, Lewis 1992). If necessary, hunts could be modified or canceled until such time that whooping cranes were not threatened. Operational plans also established a process to educate hunters on species identification and aspects of the whooping crane reintroduction experiment (Edens 1992).

The status of RMP greaters was a priority in designing hunts. To responsibly administer hunting of the RMP, representatives of state wildlife agencies in the Pacific and Central Flyways formulated a management plan acceptable to the USFWS that allocated the harvest allotment among states commensurate with the population status and provided guidelines for hunting seasons (Smith 1991, Sharp 1995). This plan also provided annual guidelines for hunting seasons and responsibilities of the respective state wildlife agencies that selected hunting seasons for RMP cranes (Smith 1991).

Hunt Structures

To accomplish agreements in Pacific and Central Flyway operational hunt plans (Smith 1991), NMGF structured hunts on a limited permit system to prevent overharvest of RMP greaters. Based on the annual harvest allotment of RMP greaters (Smith 1991, Sharp 1995), numbers of hunting permits were determined annually. Other information used to determine hunt structures included estimates of hunter participation and their success in harvesting cranes. Daily bag and possession limits were adjusted to maximize hunting opportunity and to reduce crop depredations. Permits were issued by random drawings held in September for the MRGV hunts, during 1982–84 and 1991–95 southwest hunts, and in November for remaining SWNM seasons.

Requirements for Hunters

Permittees were sent information packages containing (1) instructions on required hunter education classes, (2) information on season dates, hunt area boundaries, and daily and season harvest limits, (3) questionnaires requesting information on number of cranes bagged, number shot but not retrieved (crippled), number, locations, and descriptions on observations of color-marked or banded cranes, use of decoys, calls, type of hunt area, and an estimate of expenses per day, and (4) information on identification of whooping cranes and other protected avian species that could be mistakenly identified as a sandhill crane. Beginning in 1993, a toll-free telephone information system was established where hunters could access current information on the status of hunt periods, particularly hunt closures when harvest quotas were reached or spot closures due to the presence of whooping cranes.

Bird identification classes included information on the status and identification of whooping cranes and sandhill cranes within hunt areas. A presentation of color slides

Table 1. Combination of measurements used to identify subspecies of adult sandhill cranes wintering in New Mexico.

Subspecies	Measurements
Males	
Lesser	Posterior culmen[a] < 85 mm and tarsus[b] < 215 mm
Canadian	Posterior culmen ≥ 85 mm but < 99 mm and tarsus ≥ 215 mm and wing chord < 526 mm
Greater	Posterior culmen ≥ 99 mm and tarsus ≥ 215 mm or wing chord ≥ 526 mm and tarsus ≥ 215 mm
Females	
Lesser	Posterior culmen < 81 mm and tarsus < 210 mm
Canadian	Posterior culmen ≥ 81 mm and posterior culmen < 90 mm and tarsus ≥ 210 mm and wing chord < 520mm
Greater	Posterior culmen ≥ 90 mm and tarsus ≥ 210 mm or wing chord ≥ 520 mm and tarsus ≥ 210 mm

[a] Posterior culmen = culmen from posterior edge of nostril to tip of maxilla.
[b] Tarsus = diagonal of tarso-metatarsus (Baldwin et al. 1931).

showing these species, sometimes in mixed species groupings, and an examination to test prospective hunters' ability to distinguish a safe shooting situation from one that was not (e.g., a situation that might result in a species other than a sandhill crane being shot) were administered. Persons passing the test were issued certificates valid for 3 years before testing was required again for re-certification. Persons failing the test were ineligible to hunt cranes that particular year but were eligible to take an exam the following year. Details of testing procedures were provided by Edens (1992).

During 1982-83, 1983-84, and 1984-85 SWNM crane hunting seasons and 1994-95 MRGV crane hunting seasons, hunters were required to check in and out daily at check stations and submit completed questionnaires and/or provide information concerning their hunt. In all other hunt seasons, hunters were requested to return by mail completed questionnaires summarizing their hunting activities.

Check Stations

Cranes that were brought to check stations located near popular hunting areas were examined by NMGF biologists to determine sex (by examination of gonads), age, and subspecific identification. In the MRGV, check stations were located at Los Lunas, Bosque del Apache NWR, Belen area, San Antonio area, and at the Bernardo Wildlife Management Area. During the first season in 1986, 3 check stations were operational (Bosque del Apache NWR, Interstate 25 at north Belen exit, and at the junction of New Mexico Highways 47 and 6); in subsequent years, 1-2 check stations were operated. During 1986-89 seasons, check stations were open during weekends, and in 1990-95, stations were operated daily. For SWNM hunts, 1982-94, check stations were located near Deming and Hatch, and in 1994-95 a check station was operated only at Hatch. These check stations were operated daily. Annual projections of hunter participation, success, and crippling losses were made from data gathered at check stations and from questionnaires returned by hunters.

Age Determination

Specimens with red to reddish-orange colored papillose skin on their crowns and foreheads and having generally uniform gray plumage on the nape and secondary wing coverts were classified as adults. Cranes with fully feathered crowns/foreheads with brownish to various shades of gray feathers were classified as juveniles (Miller and Hatfield 1974, Lewis 1979). Drewien et al. (1995) noted juvenile greaters with adult-like head characteristics as early as October, and by late January as many as 54.4% juveniles had crowns that were partially red and up to 38.4% of juveniles had adult-like, red crown characteristics. Therefore, during our study, cranes with juvenal plumage on the nape (buff to brownish-tipped plumage) and secondary wing coverts (buff-colored plumage) were classified as juveniles even if they had adult-like heads.

Determination of Subspecies

Measurements (mm) of wing chord, posterior nares-culmen (posterior culmen), and tarsus (Baidwin et al. 1931) were recorded for each specimen. Beginning in 1987, additional measurements included anterior nares-culmen (anterior culmen), tail, and mass (g). Measurements (Table 1) on wing chord, tarsus, and culmen (culmen from posterior edge of nostril) from Walkinshaw (1965), Johnson and Stewart (1973), Aldrich (1979), Hubbard (1986), and Lockman et al. (1987) were used in comparisons with measurements of adult cranes of known sex that we examined. Adult cranes of known sex for which we had critical measurements (posterior culmen, wing chord, and tarsus) were compared with measurements listed in Table 1 and

Table 2. Characteristics of hunting seasons for sandhill cranes in the middle Rio Grande Valley and southwest New Mexico, 1982–83 to 1994–95.

Hunt area and year	Dates	No. of days	No. of permits	Daily limit	Season limit
Southwest[a]					
1982–83	30 Oct– 1 Nov	3	40	3	3
	4 Dec– 5 Dec	3	40	3	3
	15 Jan–17 Jan	3	40	3	3
1983–84	28 Oct–30 Oct	3	40	3	3
	16 Dec–18 Dec	3	40	3	3
	13 Jan–15 Jan	3	40	3	3
	15 Jan–17 Jan	3	40	3	3
1984–85	27 Oct–29 Oct	3	40	3	3
	15 Dec–17 Dec	3	40	3	3
	12 Jan–14 Jan	3	40	3	3
Southwest[b]					
1982–83	30 Oct– 1 Nov	3	40	3	3
	4 Dec– 5 Dec	3	40	3	3
	15 Jan–17 Jan	3	40	3	3
1983–84	28 Oct–30 Oct	3	40	3	3
	16 Dec–18 Dec	3	40	3	3
	13 Jan–15 Jan	3	40	3	3
1984–85	27 Oct–29 Oct	3	40	3	3
	15 Dec–17 Dec	3	40	3	3
	12 Jan–14 Jan	3	40	3	3
Southwest[c]					
1989	6 Jan– 8 Jan	3	350[f]	3	9
	13 Jan–15 Jan	3		3	9
	20 Jan–22 Jan	3		3	9
	27 Jan–29 Jan	3		3	9
1990	6 Jan– 8 Jan	3	350[f]	3	9
	12 Jan–14 Jan	3		3	9
	19 Jan–21 Jan	3		3	9
	26 Jan–28 Jan	3		3	9
1991	12 Jan–13 Jan	2	300[f]	3	3
	19 Jan–20 Jan	2		3	3
Southwest[c] (continued)					
1992	17 Jan–19 Jan	3	350[f]	3	3
	24 Jan–26 Jan	3		3	3
1993	22 Jan–24 Jan	3	250[f]	3	9
	29 Jan–31 Jan	3		3	9
1994	8 Jan– 9 Jan	2	100	3	6
1994–95	17 Dec–18 Dec	2	75	2	4
	14 Jan–15 Jan	2	75	2	4
Middle Rio Grande[d]					
1986	16 Oct–31 Oct	16	730	3	3
1987	17 Oct–24 Oct	8	500	3	9
	25 Oct–31 Oct	7	500	3	9
1988	15 Oct–22 Oct	8	365	3	9
	23 Oct–30 Oct	8	365	3	9
1989	18 Oct–24 Oct	7	365	3	9
	25 Oct–31 Oct	7	365	3	9
1990	26 Oct–28 Oct	3	300	3	3
1991	18 Oct–21 Oct	4	750[f]	3	3
	25 Oct–28 Oct	4		3	3
Middle Rio Grande[e]					
1992	16 Oct–18 Oct	3	450[f]	3	9
	23 Oct–25 Oct	3		3	9
1993–94	30 Oct–31 Oct	2	100[f]	3	6
	15 Jan–16 Jan	2		3	6
	4 Dec– 5 Dec	2	100[f]	3	6
	12 Jan–13 Jan	2		3	6
	18 Dec–19 Dec	2	100[f]	3	6
	25 Jan–26 Jan	2		3	6
1994–95	29 Oct–30 Oct	2	75	2	4
	10 Dec–11 Dec	2	75	2	4
	7 Jan– 8 Jan	2	75	2	4
	21 Jan–22 Jan	2	75	2	4

[a] Parts of Sierra, Doña Ana, and Luna Counties bounded by a line running south along I-25 from its junction with State Road 90 to intersection with I-10, along I-10 to U.S. 180, north along U.S. 180 to State Road 26, north along State Road 26 and 27 to State Road 90, east on State Road 90 to I-25.

[b] That portion of Luna County south of I-10.

[c] The portions of Luna, Sierra, and Doña Counties bounded on the south by the U.S./Mexico border; on the west by the western Luna County line north to I-10, I-10 east to U.S. 180 north to N.M. 26, N.M. 26 northwest to N.M. 27 at Nutt, N.M. 27 north to N.M. 152 at Hillsboro; on the north by N.M. 152 from Hillsboro to I-25; on the east by I-25 south to I-10, I-10 west to the Luna County line, and the eastern Luna County line south to the U.S./Mexico border.

[d] That part of the middle Rio Grande Valley bounded on the north by the south boundary of Isleta Pueblo; on the west by I-25, but including all portions of Bosque del Apache and Sevilleta NWR's lying west of I-25; on the south by the south boundary of Bosque del Apache NWR; on the east by Bosque NWR boundary from I-25 southeast then north to the refuge's northeast corner, then by a dirt road running southeast approximately 0.8 km from that corner to the western boundary of the White Sands Missile Range Extension Co-Use area (WSMREC), then by the WSMREC boundary north then east to the southeast corner of Sevilleta NWR, then by the eastern boundary of Sevilleta NWR north to U.S. 60, by U.S. 60 west to N.M. 47 (N.M. 304), and by N.M. 47 (N.M. 304) north to Isleta Pueblo.

[e] Socorro and Valencia Counties.

[f] Hunts combined.

classified as lessers, Canadians, or greaters. Adults for which the sex could not be determined or that did not have a critical measurement were not identified subspecifically. To estimate subspecific harvest of juveniles and appraise their differential vulnerability to hunting, we classified juvenile specimens subspecifically based on comparisons with measurements in

Table 3. Projected harvest, number of hunters, mean harvest per hunter, number of non-retrievables, and percent crippling loss of sandhill cranes in the middle Rio Grande Valley, New Mexico, winters 1986 to 1994.

Year	Harvest	No. of hunters	Harvest/ hunter ($\bar{x}$)	No. of cripples	Crippling loss (%)
1986	569	431	1.3	111	19.5
1987	985	498	1.9	166	16.8
1988	355	406	0.8	71	20.0
1989	700	366	1.9	202	28.8
1990	145	162	0.8	47	32.4
1991	209	252	0.8	21	10.0
1992	194	292	0.6	34	17.5
1993-94	459	158	2.9	66	15.0
1994-95	514	223	2.3	125	24.3
$\bar{x}$	458.2	309.7	1.4	94.0	20.4
SD	272.0	121.8	0.8	61.8	7.0

Table 4. Projected harvest, number of hunters, mean harvest per hunter, number of cripples, and percent crippling loss of sandhill cranes in southwest New Mexico, winters 1982 to 1994.

Year	Harvest	No. of hunters	Harvest/ hunter ($\bar{x}$)	No. of cripples	Crippling loss (%)
1982-83[a]	78	61	1.2		
1982-83[b]	35	98	0.3		
1983-84[a]	69	81	0.8		
1983-84[b]	33	91	0.3		
1984-85[a]	65	71	0.9		
1984-85[b]	36	41	0.8		
1989	212	140	1.5	57	26.8
1991	80	108	0.7	47	58.7
1992	73	80	0.9	32	43.8
1993	98	120	0.8	60	61.0
1994	64	44	1.4	20	31.2
1994-95	59	45	1.3	18	30.5
$\bar{x}$	100.2	108.8	0.9	39.0	41.9
SD	45.9	45.8	0.3	18.3	14.9

[a] Parts of Sierra, Doña Ana, and Luna Counties bounded by a line running south along I-25 from its junction with State Road 90 to intersection with I-10, along I-10 to U.S. 180, north along U.S. 180 to State Road 26, north along State Road 26 and 27 to State Road 90, east on State Road 90 to I-25.

[b] That portion of Luna County south of I-10.

Table 1. Comparable morphological data on juveniles from other studies were unavailable. We tested for differential vulnerability of juveniles to hunting by using a comparison based on the difference $\pi_1 - \pi_2$ methodology described by McPherson (1990). Age ratio data of RMP greaters and Mid-continent Population lessers occurring in winter flocks were from Drewien (1994) and Drewien et al. (1995).

Comparisons of Measurements of Subspecies

We did an analysis of variance of multiple comparisons (SAS Institute Inc. 1990) on 5 measurements (wing chord, posterior culmen, tarsus, tail, and mass) of males and females of the 3 subspecies to determine if these measurements differed between males and females of the same subspecies or among sexes of different subspecies. We also compared our measurements with those reported by Walkinshaw (1965), Johnson and Stewart (1973), Aldrich (1979), Hubbard (1986), and Lockman et al. (1987).

Other Information Collected from Crane Specimens

When possible, colors (Smithe 1975, 1981) of soft parts (tarsi, toes, bill, irides, and eye rings of juveniles) of specimens were recorded. Mass of each specimen was determined by Pesola balances with 50-g increments during 1982-90 and by electronic balances with accuracies of 1.0 g during 1991-95.

RESULTS

During 1982-83 to 1984-85, SWNM crane hunting seasons were conducted during October, November, December, and January and consisted of 3-day hunting seasons with 40 permits for each hunt. The daily bag and possession limits were 3 birds. SWNM crane hunting seasons were conducted during January in 1989-93 and in December in 1994-95. These seasons consisted of 2- to 3-day hunts with 100-350 permits available, a 2-3 daily bag limit, and a 3-9 possession limit. A summary of each SWNM season is provided in Table 2.

MRGV hunts during 1986-92 occurred during mid- to late October, and subsequent MRGV seasons occurred during October, December, and January. During 1986 and 1990, hunts consisted of a single season of 16 and 3 consecutive days, respectively. Subsequent hunting seasons consisted of 2-4 separate seasons consisting of 2-3 days. The number of permits during MRGV hunts varied from 75 to 750, daily bag limits varied from 2 to 3, and possession limits varied from 3 to 9 (Table 2).

In the MRGV the mean annual harvest during 9 seasons was 458.2 cranes and mean number of hunters was 309.7 (Table 3). Mean harvest per hunter was 1.4. Crippling losses averaged 93.6 cranes annually. The mean annual crippling loss was 20.4%. Similar data for SWNM hunts showed a mean annual harvest of 100.2 cranes and a mean of 108.8 hunters (Table 4). The mean harvest per hunter was 0.9. The

Table 5. Subspecific and sex composition of adult sandhill crane specimens examined at hunter check stations in the middle Rio Grande Valley, New Mexico, winters 1982 to 1994.

Year	Lessers				Canadians				Greaters			
	Males	Females	Total	%[a]	Males	Females	Total	%[a]	Males	Females	Total	%[a]
1983–84[b]	4	3	7	13.2	1	2	3	5.7	26	17	43	81.1
1984–85[b]	0	0	0	0.0	2	1	3	11.5	21	2	23	88.5
1986	33	30	63	26.0	3	11	14	5.8	78	87	165	68.2
1987	27	23	50	32.5	6	1	7	4.5	50	47	97	63.0
1988	9	13	22	35.5	2	2	4	6.4	22	14	36	58.1
1989	12	6	18	19.1	7	4	11	11.7	35	30	65	69.1
1990	13	9	22	36.7	4	1	5	9.0	19	14	33	55.0
1991	24	28	52	82.5	0	1	1	1.5	5	5	10	15.9
1992	20	28	48	40.7	3	1	4	3.4	32	34	66	55.9
1993–94	8	5	13	4.3	14	10	24	8.0	141	121	262	87.6
1994–95	18	9	27	7.6	23	18	41	11.6	143	142	285	80.7
$\bar{x}$	15.0	14.2	29.2	27.1	5.9	4.7	10.6	7.1	52.0	46.6	98.6	65.7
SD	9.6	11.6	20.7	23.1	6.9	5.6	12.0	3.5	48.3	48.3	96.4	20.6

[a] Percent of annual total.
[b] In 1983–84 and 1984–85, specimens were collected for research studies and were not taken during hunting seasons.

mean annual crippling loss was 39.0 cranes and mean crippling loss per season was 41.9%.

In the MRGV greaters comprised most of the harvest (65.7%), followed by lessers (27.1%) and Canadians (7.1%) (Table 5). Lessers were more abundant (66.6%) in SWNM, followed by greaters (20.3%) and Canadians (13.0%) (Table 6).

Sex and age of crane specimens are summarized in Tables 7 and 8. The harvest of adult males by subspecies in the MRGV (Table 9) was 55% for lessers (not different from 50%, P = 0.94), 54% for Canadians (not different from 50%, P = 0.49), and 57% for greaters (different from 50%, P = 0.001). In SWNM, males accounted for 56% of lessers (different from 50%, P = 0.04), 74% of Canadians (different from 50%, P = 0.002), and 51% of greaters (not different from 50%, P = 0.836). Mean harvest of juveniles (subspecies combined) was 13.4% in the MRGV and 14.2% in SWNM. In the MRGV, sex ratios (male:female) of lessers

Table 6. Subspecific and sex composition of adult sandhill crane specimens examined at hunter check stations in southwest New Mexico, winters 1982 to 1994.

Year	Lessers				Canadians				Greaters			
	Males	Females	Total	%[a]	Males	Females	Total	%[a]	Males	Females	Total	%[a]
1982–83	18	3	21	45.7	17	3	20	43.5	4	1	5	10.8
1983–84[b]	40	20	60	72.3	1	2	3	3.6	11	9	20	24.1
1984–85[b]	40	23	63	85.1	2	0	2	2.7	5	4	9	12.1
1989	19	20	39	60.9	3	5	8	12.5	12	15	27	26.6
1991	20	18	38	100.0	0	0	0	0.0	0	0	0	0.0
1992	13	9	22	64.7	3	1	4	11.7	2	6	8	23.5
1993	18	19	37	60.7	4	2	6	9.8	8	10	18	29.5
1994	7	14	21	41.2	9	4	13	25.5	12	5	17	33.3
1994–95	4	5	9	69.2	1	0	1	7.7	1	2	3	23.0
$\bar{x}$	19.8	14.5	34.4	66.6	4.4	1.8	6.3	13.0	6.1	5.7	9.8	20.3
SD	12.7	7.2	18.3	18.2	5.4	1.8	6.5	13.6	4.8	4.8	6.9	10.6

[a] Percent of annual total.
[b] In 1983–84 and 1984–85, specimens were collected for research studies and were not taken during hunting seasons.

Table 7. Sex and age of sandhill crane specimens examined at hunter check stations in the middle Rio Grande Valley, winters 1986–94.

	Adult			Juvenile			Sex undetermined			Combined total cranes			
Year	Male	Female	Total	Male	Female	Total	Adult	Juvenile	Total	Adult	Juvenile (%)[a]	Male	Female
1983–84[b]	32	22	54	1	2	3	1	3	4	55	6 (9.9)	33	24
1984–85[b]	25	13	38	0	1	1	6	0	6	44	1 (2.2)	25	14
1986	123	145	268	16	12	28	24	17	41	292	45 (13.3)	139	157
1987	87	81	168	12	3	15	7	16	23	175	31 (15.0)	99	84
1988	37	33	70	5	2	7	3	3	6	73	10 (12.0)	42	35
1989	55	51	106	4	3	9	10	3	13	116	12 (9.3)	59	54
1990	36	27	63	14	7	21	1	7	8	64	28 (30.4)	50	34
1991	30	36	66	12	10	22	2	5	7	68	27 (28.4)	42	46
1992	57	66	123	13	6	19	0	1	1	123	20 (13.9)	70	72
1993–94	170	153	323	6	13	19	10	3	13	333	22 (6.2)	176	166
1994–95	194	198	392	15	14	29	2	0	2	394	29 (6.8)	209	212
$\bar{x}$	76.9	75.0	153.5	8.9	6.6	15.7	6.0	5.2	11.2	149.2	20.2 (13.4)	85.9	81.6
SD	59.8	62.5	123.5	5.8	4.9	9.6	7.0	5.9	11.6	117.4	12.9 (8.8)	62.4	66.5

[a] Percent of annual total.
[b] Specimens collected in 1983–84 and 1984–85 were for research studies and were not taken during hunting seasons.

averaged 55:45, Canadians 54:46, and greaters 57:43 and, in SWNM, lessers 56:44, Canadians 74:26, and greaters 51:49 (Table 9). Sex ratios for juveniles (subspecies combined) were 57% males in the MRGV and 76% males in SWNM.

In the MRGV the subspecific composition consisted of 21.3% lessers, 7.5% Canadians, and 71.2% greaters. In the LRGV portion of Southwest Hunt Area 1 (Rio Grande floodplain), the subspecific composition was 48.1% lessers, 15.3% Canadians, and 36.6% greaters. The subspecific composition from the floodplain of the Rio Grande Valley (MRGV and LRGV combined) consisted of 24.6% lessers, 8.5% Canadians, and 66.9% greaters. In other SWNM hunting areas outside the LRGV, the subspecific composition was 87.6% lessers, 1.4% Canadians, and 10.9% greaters (Table 10).

We estimated that approximately 52.9% of the juveniles killed by hunters in the MRGV were lessers, 12.3% Canadians, and 34.7% greaters. Respective estimates for SWNM were 88.0% lessers, 1.5% Canadians, and 10.4% greaters. The mean harvest of 5.7% juvenile greaters we recorded in the MRGV varied from 2.7% in 1988 to 16.7% in 1991 (Table 11). In the MRGV the percent RMP harvest of

Table 8. Sex and age of sandhill crane specimens examined at hunter check stations in southwest New Mexico, winters 1982–94.

	Adult			Juvenile			Sex undetermined			Combined total cranes			
Year	Male	Female	Total	Male	Female	Total	Adult	Juvenile	Total	Adult	Juvenile (%)[a]	Male	Female
1982–83	92	15	107	17	3	20	0	0	0	107	20 (15.7)	19	18
1983–84	53	32	85	9	3	12	1	6	7	86	18 (17.3)	62	35
1984–85	51	31	82	6	1	7	6	6	12	88	13 (12.9)	57	32
1989	38	44	82	5	2	7	9	6	15	91	13 (12.5)	43	46
1991	21	19	40	9	5	14	0	0	0	40	14 (25.9)	30	24
1992	21	17	38	6	1	7	1	1	2	39	8 (17.1)	27	18
1993	31	32	63	3	1	4	0	0	0	63	4 (6.0)	34	33
1994	29	23	52	1	0	1	1	0	1	53	1 (1.9)	30	23
1994–95	6	7	13	1	2	3	0	0	0	13	3 (18.7)	7	9
$\bar{x}$	38.0	24.4	62.4	6.3	2.0	8.3	2.0	2.1	4.1	64.4	10.4 (14.2)	44.3	26.4
SD	25.1	11.3	29.4	5.0	1.5	6.0	3.2	2.9	5.8	30.7	6.7 (7.1)	29.3	11.1

[a] Percent of annual total.

Table 9. Sex ratios of sandhill crane specimens examined at hunter check stations in the middle Rio Grande Valley, winters 1986-94, and southwest New Mexico, winters 1982-94.

Hunt area	Lessers		Canadians		Greaters	
and year	*n*	♂♂:♀♀	*n*	♂♂:♀♀	*n*	♂♂:♀♀
Middle Rio Grande Valley						
1983-84[a]	7	57:43	3	33:67	43	61:39
1984-85[a]	0		3	67:33	23	91:9
1986	63	52:48	14	21:79	165	47:53
1987	50	54:46	7	86:14	97	52:48
1988	22	41:59	4	50:50	36	61:39
1989	18	68:32	11	64:36	65	54:46
1990	22	59:41	5	80:20	33	58:42
1991	52	46:54	1	0:100	10	50:50
1992	48	41:59	4	75:25	66	49:51
1993-94	13	61:39	24	58:42	262	54:46
1994-95	27	68:32	41	56:44	285	50:50
Southwest New Mexico						
1982-83	21	86:14	20	85:15	5	80:20
1983-84	60	67:33	3	33:67	20	55:45
1984-85	63	64:36	2	100:0	9	56:44
1989	39	49:51	8	60:40	27	44:56
1991	38	53:47	0		0	
1992	22	60:40	4	75:25	8	25:75
1993	37	49:51	6	68:32	18	44:56
1994	21	34:66	13	70:30	17	71:29
1994-95	9	44:56	1	100:0	3	33:67

[a] Specimens collected in 1983-84 and 1984-85 were for research studies and were not taken during hunting seasons.

juveniles was significantly higher in 1986 (14.9%, $P <$ 0.001) and 1993-94 (12.4%, $P < 0.001$) than the 5.1% (SD = 1.1) of juvenile greaters comprising RMP cranes reported from the MRGV and SWNM winter recruitment surveys conducted during 1986-94 (Drewien 1994, Drewien et al. 1995). The mean harvest of 36.9% juvenile lessers recorded from the MRGV ranged from 14.2% in 1989 to 48.8% in 1990 (Table 11). The percentage of juvenile lessers harvested in the MRGV during our study was significantly higher in 1986 (20.2%, $P = 0.026$), 1987 (25.3%, $P = 0.004$), 1990 (48.8%, $P < 0.001$), 1991 (36.6%, $P < 0.001$), 1992 (23.8%, $P = 0.010$), 1993-94 (33.3%, $P = 0.023$), and 1994-95 (25.0%, $P = 0.029$) than 11.2% (SE = 0.3) of juvenile lessers reported from western New Mexico during 1987-92 by Drewien et al. (1995). The mean percentage of juvenile lessers harvested in SWNM was 20.5% and ranged from 4.5% in 1994 to 26.7% in 1992 (Table 11). The percentage of juvenile lessers harvested in SWNM during our study was significantly higher in 1983-84 (18.9%, $P =$ 0.05) and 1992 (26.7%, $P = 0.028$) than the 11.2% reported by Drewien et al. (1995). For other years, percentages were either not significantly different or sample sizes were too small for analysis.

Soft part colors of adult cranes showed irides of varying shades of orange. Of 1,145 adults examined for a color match of the irides, 45.2% were spectrum orange, 29.4% were chrome orange, 21.2% were orange yellow, and 4.2% were flame scarlet. Juveniles had irides that were various shades of light brown to dark brown and most had conspicuous buff to buff-yellowish eye rings. Juveniles examined for color matches of irides revealed that approximately 75% were buff and 25% were burnt umber. Adults had dusky brown to blackish maxillas and mandibles with varying amounts of olive-brown and smoke gray-colored areas on the sides of the bill. Adults and juveniles typically had blackish-colored tarsi and toes. Juveniles typically had bills that were various shades of brown, including dusky brown, olive-brown, and raw umber.

An analysis of variance of multiple comparisons of the 15 subspecies-sex (6 combinations) and measurement (5 variables) combinations revealed that 13 of these combinations were significantly different ($P < 0.001$). Tail and wing chord measurements of male Canadians and female greaters were the only sex-subspecies and measurement combinations that did not differ significantly (Table 12).

Morphological data we collected are similar to those reported by Walkinshaw (1965), Johnson and Stewart (1973), Aldrich (1979), Hubbard (1986), and Lockman et al. (1987) (Table 13). Our mean values of wing chord measurements on female Canadians were significantly larger ($P < 0.05$) than similar measurements reported by Johnson and Stewart (1973) and Aldrich (1979). Our mean values of wing chord measurements on female greaters were significantly smaller ($P < 0.05$) than measurements reported by Lockman et al. (1987) and significantly larger ($P < 0.05$) than measurements reported by Johnson and Stewart (1973) and Hubbard (1986). Our mean values of wing chord measurements on male Canadians were significantly larger ($P < 0.05$) than measurements reported by Johnson and Stewart (1973) and Aldrich (1979). Our mean wing chord measurements on male greaters were significantly smaller ($P < 0.05$) than measurements reported by Lockman et al. (1987) and significantly larger ($P < 0.05$) than measurements reported by Johnson and Stewart (1973) and Hubbard (1986). Posterior culmen measurements on female lessers during our study were significantly smaller ($P < 0.05$) than measurements reported by Johnson and Stewart (1973); they were on female greaters significantly larger ($P < 0.05$) than measurements reported by Johnson and Stewart (1973) and significantly smaller ($P < 0.05$) than measurements reported by Aldrich (1979). Our mean measurements of posterior culmen on male lessers were

Table 10. Subspecific composition of sandhill cranes harvested in the Rio Grande Valley and southwest New Mexico, winters 1982–94.

Hunt area and county	Lessers		Canadians		Greaters	
	n	%	*n*	%	*n*	%
Middle Rio Grande Valley						
Soccoro	296	22.2	97	7.3	942	70.6
Valencia	31	15.3	19	9.4	152	75.2
Total and $\bar{x}$	327	21.3	116	7.5	1,094	71.2
Lower Rio Grande Valley[a]						
Sierra	82	45.6	30	16.7	68	37.8
Doña Ana	22	61.1	3	8.3	11	30.6
Total and $\bar{x}$	104	48.1	33	15.3	79	36.6
MRGV/LRGV combined	431	24.6	149	8.5	1,173	66.9
Southwest New Mexico[b]						
Sierra	5	50.0	0	0.0	5	50.0
Luna	179	89.5	3	1.5	18	9.0
Total and $\bar{x}$	184	87.6	3	1.4	23	10.9

[a] Areas within southwest hunting boundaries that were within the Rio Grande Valley floodplain.
[b] Areas within southwest hunting boundaries that were outside of the Rio Grande Valley floodplain.

significantly smaller ($P < 0.05$) than measurements reported by Johnson and Stewart (1973) and on male Canadians were significantly larger ($P < 0.05$) than measurements reported by Johnson and Stewart (1973) and Aldrich (1979). Our mean tarsus measurements were significantly larger ($P < 0.05$) for female Canadians than those reported by Aldrich (1979) and for female greaters than measurements reported by Hubbard (1986). Our mean tarsus measurements on male greaters were significantly larger ($P < 0.05$) than measurements reported by Hubbard (1986). Our mean mass measurements on lessers were significantly smaller ($P < 0.05$) than data reported by Johnson and Stewart (1973) and significantly larger ($P < 0.05$) for Canadians and greaters than data reported by Johnson and Stewart (1973).

Table 11. Estimated juvenile crane harvest by subspecies in the middle Rio Grande Valley, New Mexico, winters 1986 to 1994 and in southwest New Mexico, winters 1983 to 1994.

Hunt area and year	Percent of annual total		
	Lessers	Canadians	Greaters
Middle Rio Grande Valley			
1986	20.2	5.0	14.9
1987	25.3	36.3	8.5
1988	18.5	50.0	2.7
1989	14.2	21.4	5.7
1990	48.8	50.0	5.7
1991	36.6	50.0	16.7
1992	23.8	0	7.0
1993–94	33.3	4.0	12.4
1994–95	25.0	16.3	4.0
Southwest New Mexico			
1983–84	18.9	0	16.6
1984–85	16.0	0	10.0
1989	15.2	11.1	6.9
1991	20.8	0	0
1992	26.7	0	0
1993	9.6	0	0
1994	4.5	0	0
1994–95	27.2	0	0

DISCUSSION

A major concern in designing our hunting seasons was not to exceed annual harvest quotas of RMP greaters allocated to New Mexico. In all years except 1989, we achieved this goal. Achieving the yearly allotment during 1989 MRGV hunts required cancellation of January 1990 SWNM crane hunts. In 1993–94 MRGV hunts, the final 2 hunt segments were also canceled for similar reasons. In both instances, hunters were notified of these closures by letter and, during 1993–94 MRGV hunts, by mail and information made available on a toll-free telephone message. Hunter training and certification along with hunt designs appeared adequate to protect whooping cranes and none were known to be shot during the sandhill crane hunting seasons.

Subspecies distribution varied in different hunt areas. Greaters comprised a larger proportion of the harvest at northern winter sites in the MRGV, whereas lessers predominated in SWNM hunts. Drewien and Bizeau (1974) noted a

Table 12. Morphometry (mm) and mass (g) of sandhill crane specimens examined at hunter check stations in the Middle and Lower Rio Grande Valley and southwest New Mexico during winters 1982–94.

	Female						Male					
Subspecies	Wing chord	Posterior culmen	Tarsus	Anterior culmen	Tail	Mass	Wing chord	Posterior culmen	Tarsus	Anterior culmen	Tail	Mass
Lessers												
n	284	284	284	204	191	201	331	331	331	214	240	218
$\bar{x}$	448.4	68.8	176.2	57.9	162.7	2,954.5	472.8	73.2	186.0	61.4	171.2	3,434.1
SD	19.0	4.0	11.5	3.7	3.7	271.1	20.1	4.3	12.0	3.8	9.1	273.3
SE	1.1	0.2	0.7	0.7	0.3	19.1	1.1	0.2	0.7	0.3	0.6	18.5
min.	401	58.0	115	49.1	141	2,272	426	56.6	147	53.2	141	2,600
max.	545	80.0	206	70.4	209	3,827	535	84.1	214	76.1	211	4,250
Canadians												
n	65	65	65	50	44	50	89	89	89	80	72	79
$\bar{x}$	495.6	86.9	221.3	74.8	188.3	4,700.7	509.6	94.8	232.9	79.7	195.5	5,153.7
SD	19.5	2.0	7.9	2.5	10.3	366.4	12.6	2.8	9.0	3.1	10.5	385.8
SE	2.4	0.3	1.0	0.4	1.6	51.8	1.3	0.3	1.0	0.4	1.2	43.4
min.	382	81.1	210	69.3	162	3,800	434	87.6	216	70.9	165	4,155
max.	519	89.8	248	79.8	209	5,412	525	98.8	259	87.3	213	5,900
Greaters												
n	573	573	573	443	397	445	624	621	621	484	446	481
$\bar{x}$	513.9	97.7	227.5	82.0	193.5	4,928.3	537.6	103.6	239.3	86.8	203.8	5,554.3
SD	18.3	5.4	9.8	4.8	9.8	413.2	18.7	5.7	10.5	4.9	9.6	487.8
SE	0.7	0.2	0.4	0.2	0.5	19.6	0.8	0.2	0.4	0.2	0.5	22.2
min.	463	82.7	210	62.0	160	2,604	424	84.0	215	72.5	172	3,268
max.	597	121.0	271	103.9	228	6,350	590	132.5	280	105.2	250	6,850

similar distribution; greaters dominated on northern winter sites in the MRGV and lessers predominated on southern winter sites.

Crane hunts in the MRGV and SWNM can be administered to provide limited permit hunting commensurate with the status of RMP cranes. Generally, participating hunters have demonstrated a willingness to follow guidelines that protected whooping cranes and other non-hunted avian species. They have also become familiar with and supported a harvest allocation that was based on the status of RMP cranes. Permit requirements and mandatory check stations were accepted by most hunters. A recently adopted system of mandatory check in/out check stations has allowed for an accurate measure of harvest, hunter participation, and eliminated the need of projecting harvest estimates based on incomplete samples of questionnaires. This system has proven to be accurate, acceptable by most hunters, and more efficient for NMGF personnel.

Males consistently outnumbered females in hunter-harvested cranes; however, we have no data that supported findings by Tacha and Vohs (1984) that Mid-continent male cranes were 1.3 times more vulnerable than females to harvest. The sex ratio of 74% male Canadians in SWNM may have resulted from a relatively small sample size. We have no data that suggested Canadian males were more vulnerable to hunting than males of other subspecies. Our findings of 51–57% male greaters in the harvest are similar to 55% males in harvested greaters in Wyoming (Lockman et al. 1987).

Our data indicate that lesser and greater juveniles were more vulnerable to harvest than adults during most years. Greater vulnerability of juvenile sandhill cranes to hunting is not unexpected; however, the impacts of this differential vulnerability on the population are unknown and warrant further investigation.

Morphometric and mass data we collected on 3 migratory subspecies showed a number of significant differences when compared with similar data reported from other studies. We attribute most of these differences to our much larger sample sizes and associated larger variations resultings from large samples. We assume that in other studies morphometric data were collected with similar techniques, although most of these studies did not report specific methodology. Use of different techniques could contribute to some of the differences noted. For example, the diagonal measurement of the tarsus (Baldwin et al. 1931) in this study is different from the

Table 13. Comparison[a] of measurements of wing chord, posterior culmen, tarsus, and mass of sandhill cranes from various studies.

Measurement and subspecies	Female *n*	Female $\bar{x}$	Female SD	Female SE	Female min.	Female max.	Male *n*	Male $\bar{x}$	Male SD	Male SE	Male min.	Male max.	Reference[b]
Wing Chord (mm)													
Lessers	17	451.2			420	490	33	475.6			439	503	1
	17	445.7	12.0		421	464	31	467.0	15.5		419	502	2
	23	447.2	18.1	3.8	420	490	44	468.5	19.9	3.0			3
	35	450.2	17.1		414	487	57	469.8	17.6		420	510	4
	284	**448.4**	**19.0**	**1.1**	**401**	**545**	**331**	**472.8**	**20.1**	**1.1**	**426**	**535**	**6**
Canadians	3	472.6			456	495	7	507.3			480	524	1
	33	475.5*	14.3		434	505	51	503.0*	12.9		469	530	2
	4	474.3*	14.7	7.4	456	492	20	497.3*	12.8	4.0	482	525	3
	1	500.0					2	507.0			500	514	4
	65	**495.6**	**19.5**	**2.4**	**382**	**519**	**89**	**509.6**	**12.6**	**1.3**	**434**	**525**	**6**
Greaters	9	546.0			510	575	8	561.5			526	598	1
	10	487.3*	13.8		472	524	11	518.9*	9.4		504	531	2
	13	523.4	29.4	8.2	490	575	17	547	26.2		502	598	3
	19	500.9*	16.8		472	523	30	528.0*	13.6		497	556	4
	90	524.0*	20.1		485	575	105	545.0*	20.9		495	600	5
	573	**513.9**	**18.3**	**0.7**	**463**	**597**	**624**	**537.6**	**18.7**	**0.8**	**424**	**590**	**6**
Posterior culmen (mm)													
Lessers	17	67.4			63	74	33	71.4			65	76	1
	17	72.4*	4.0		63.7	78	31	76.5*	3.1		70.3	84.0	2
	14	69.9	4.7	1.3	62	76	20	71.5	4.6	1.0			3
	35	69.4	5.8		57	82	56	74.6	4.8		60	83	4
	284	**68.8**	**4.0**	**0.2**	**58.0**	**80.0**	**331**	**73.2**	**4.3**	**0.2**	**56.6**	**84.1**	**6**
Canadians	3	80.0			73	87	7	85			82	88	1
	33	86.7	3.3		82.8	94	51	92.1*	4.8		74	102	2
	3	82.7	8.4	4.8	73	88	6	89.5*	4.7	1.9	85	91	3
	1	87.0					2	92.5			87	98	4
	65	**86.9**	**2.0**	**0.3**	**81.1**	**89.8**	**89**	**94.8**	**2.8**	**0.3**	**87.6**	**98.0**	**6**
Greaters	1	97					8	107.2			101	116	1
	10	93.5*	2.5		90	98	10	100.7	4.5		95	107	2
	8	100.1*	2.1	0.7	97	103	9	107.1	6.9	2.3	100	120	3
	18	100.1	6.7		91	114	30	106.9*	4.7		100	117	4
	89	99	4.8		86	110	109	105	6.9		88	120	5
	573	**97.7**	**5.4**	**0.2**	**82.7**	**121.0**	**621**	**103.6**	**5.7**	**0.2**	**84**	**132.5**	**6**
Tarsus (mm)													
Lessers	17	182.0			162	198	33	186.6			156	210	1
	17	179.2	10.8		165	197	31	187.5	14.4		151	203	2
	23	180.7	12.0	2.5	156	200	44	187.8	12.1	1.8			3
	35	170.0	13.1		139	196	56	185.3	12.0		166	207	4
	284	**176.2**	**11.5**	**0.7**	**115**	**206**	**331**	**186.0**	**12.0**	**0.7**	**147**	**214**	**6**
Canadians	3	210.6			205	216	7	222			216	239	1
	33	217.0	7.6		203	232	51	230.6	9.5		211	249	2
	4	212.3*	2.2	1.1	210	215	11	226.8	10.8	3.3	203	240	3
	1	220.0					2	219.5			212	227	4
	65	**221.3**	**7.9**	**1.0**	**210**	**248**	**89**	**232.9**	**9.0**	**1.0**	**216**	**259**	**6**
Greaters	9	230.5			222	239	8	244.5			226	264	1
	10	228.3	6.3		219	235	11	236.5	8.3		227	253	2
	14	227.2	6.7	1.8	217	238	17	243.5	12.2	3.0	226	264	3
	19	220.4*	6.2		210	230	30	229.9*	7.6		217	242	4
	90	231	15.1		200	272	109	239	13.0		210	280	5
	573	**227.5**	**9.8**	**0.4**	**210**	**271**	**621**	**239.3**	**10.5**	**0.4**	**215**	**280**	**6**
Mass (g)													
Lessers	17	3,459.9*	250.1		2,866	3,829	31	3,945.6*	299.6		3,235	4,671	2
	201	**2,954.5**	**271.1**	**19.1**	**2,272**	**3,827**	**218**	**3,434.1**	**273.3**	**18.5**	**2,600**	**4,250**	**6**
Canadians	31	4,110.2*	251.9		3,676	4,895	49	4,794.8*	385.3		3,890	5,613	2
	50	**4,700.7**	**366.4**	**51.8**	**3,800**	**5,412**	**79**	**5,153.7**	**385.8**	**43.4**	**4,155**	**5,900**	**6**
Greaters	9	4,453.7*	427.1		3,645	5,052	11	4,885.6*	370.8		3,870	5,432	2
	88	4,845	338		3,975	5,675	101	5,430	432		4,425	6,600	5
	445	**4,928.3**	**413.2**	**19.6**	**2,604**	**6,350**	**481**	**5,554.3**	**487.8**	**22.2**	**3,268**	**6,850**	**6**

[a] *$P < 0.05$, significantly different from comparable measurements in this study.

[b] 1=Walkinshaw (1965), 2=Johnson and Stewart (1973), 3=Aldrich (1979), 4=Hubbard (1986), 5=Lockman et al. (1987), 6=**THIS STUDY**.

lateral measurement parallel to the tarsal axis (Johnsgard 1983:240).

From our past experience and information derived from crane hunts in the MRGV and SWNM, we recommend the following:

1. NMGF should continue to monitor the MRGV and SWNM hunts by establishing mandatory check in/out check stations where harvested cranes can be examined. Objectives of check stations should include an intensive effort to gather additional data on sex, age, and subspecific composition of harvested cranes. Investigations should include initiatives to determine effects of timing, duration, and hunting methods on the sex, age, and subspecific harvest of cranes in relation to their use of wintering areas in the MRGV and SWNM.
2. All crane hunting seasons within the New Mexico range of the RMP should be carefully reviewed and conducted in a manner consistent with guidelines outlined in the Pacific and Central Flyways Management Plan (Smith 1991). Changes in proportions of subspecies composition by hunt areas should be closely followed to ensure that excessive harvest of greaters does not occur.
3. NMGF should consider monitoring crane hunts in eastern New Mexico, including the Pecos Valley, primarily Chaves County, and major concentration areas in Curry and Roosevelt Counties.
4. NMGF should actively support and participate in Pacific and Central Flyway efforts to better understand the taxonomy, genetic composition, and distribution of all 3 migratory subspecies of cranes with particular emphasis on the Canadian subspecies, of which so little is known.

LITERATURE CITED

ALDRICH, J. W. 1979. Status of the Canadian sandhill crane. Pages 139-148 *in* J. C. Lewis, ed. Proc. 1978 crane workshop. Colorado State Univ. Printing Serv., Fort Collins.

BALDWIN, S. P., H. C. OBERHOLSER, and L. G. WORLEY. 1931. Measurements of birds. Sci. Publ. Cleveland Mus. Nat. Hist. 2:1-165.

BENNING, D. S., R. C. DREWIEN, D. H. JOHNSON, W. M. BROWN, and E. L. BOEKER. 1997. Spring population estimates of Rocky Mountain greater sandhill cranes in Colorado. Proc. North Am. Crane Workshop. 7:165-172.

DREWIEN, R. C. 1994. 1994 fall recruitment survey of the Rocky Mountain Population of greater sandhill cranes. Unpubl. Rep., Hornocker Wildl. Res. Inst., Univ. Idaho, Moscow. 3pp.

______, and E. G. BIZEAU. 1974. Status and distribution of greater sandhill cranes in the Rocky Mountains. J. Wildl. Manage. 38:720-742.

______, W. M. BROWN, and W. L. KENDALL. 1995. Recruitment in Rocky Mountain greater sandhill cranes and comparisons with other crane populations. J. Wildl. Manage. 59:339-356.

EDENS, R. H. 1992. Reducing conflicts between snow goose hunters and whooping cranes at Bosque del Apache National Wildlife Refuge. Pages 275-284 *in* D. A. Wood, ed. Proc. 1988 North Am. crane workshop. Florida Game Fresh Water Fish Comm. Nongame Wildl. Program Tech. Rep. 12.

HUBBARD, J. P. 1986. Populational composition of sandhill cranes occurring in New Mexico in 1982–1983. Unpubl. Rep., New Mexico Dep. Game and Fish, Santa Fe. 42pp.

JOHNSGARD, P. A. 1983. Cranes of the world. Indiana Univ. Press, Bloomington. 258pp.

JOHNSON, D. H., and R. E. STEWART. 1973. Racial composition of migrant populations of sandhill cranes in the northern plains states. Wilson Bull. 85:148-162.

LEWIS, J. C. 1979. Field identification of juvenile sandhill cranes. J. Wildl. Manage. 43:211-214.

______. 1992. The contingency plan for federal-state cooperative protection of whooping cranes. Pages 295-299 *in* D. A. Wood, ed. Proc. 1988 North Am. crane workshop. Florida Game Fresh Water Fish Comm. Nongame Wildl. Program Tech. Rep. 12.

LOCKMAN, D. C., L. SERDIUK, and R. C. DREWIEN. 1987. An experimental greater sandhill crane and Canada goose hunt in Wyoming. Pages 47-56 *in* J. C. Lewis, ed. Proc. 1985 crane workshop. Platte River Whooping Crane Maintenance Trust, Grand Island, Nebr.

MCPHERSON, G. 1990. Statistics in scientific investigation: its basis, application, and interpretation. Springer-Verlag, New York, N.Y. 666pp.

MILLER, R. S., and J. P. HATFIELD. 1974. Age ratios of sandhill cranes. J. Wildl. Manage. 38:224-242.

SAS INSTITUTE INC. 1990. SAS/STAT user's guide. Vol. 2. Version 6. Fourth ed. SAS Inst., Inc., Cary, N.C. 846pp.

SHARP, D. E. 1995. Status and harvests of the mid-continent and Rocky Mountain populations of sandhill cranes. Unpubl. Rep., U.S. Fish Wildl. Serv., Golden, Colo. 8pp.

SMITH, P., chairman. 1991. Management plan of the Pacific and Central flyways for the Rocky Mountain population of greater sandhill cranes. Unpubl. Rep., Joint Subcomm. Rocky Mountain Population Greater Sandhill Cranes, Pac. Flyway Study Comm. and Cent. Flyway Tech. Comm., U.S. Fish Wildl. Serv., Portland, Oreg. 55pp.

SMITHE, F. B. 1975. Naturalist's color guide. Part I, color guide. Am. Mus. Nat. Hist., New York, N.Y. [no pagination].

______. 1981. Naturalist's color guide. Part I, color guide. Am. Mus. Nat. Hist., New York, N.Y. [no pagination].

TACHA, T. C., and P. A. VOHS. 1984. Some population parameters of sandhill cranes from mid-continental North America. J. Wildl. Manage. 48: 89-98.

______, P. A. VOHS, and G. C. IVERSON. 1984. Migration routes of sandhill cranes from mid-continental North America. J. Wildl. Manage. 48: 1028-1033.

TAYLOR, J. P., and R. E. KIRBY. 1990. Experimental dispersal of wintering snow and Ross' geese. Wildl. Soc. Bull. 18:312-319.

WALKER, D. L., and S. D. SCHEMNITZ. 1987. Food habits of

sandhill cranes in relationship to agriculture in central and southwestern New Mexico. Pages 201-212 *in* J. C. Lewis, ed. Proc. 1985 crane workshop. Platte River Whooping Crane Maintenance Trust, Grand Island, Nebr.

WALKINSHAW, L. H. 1965. A new sandhill crane from central Canada. Can. Field-Nat. 79:181-184.

LONG BONE FRACTURE MANAGEMENT IN A SANDHILL CRANE: A CASE REPORT

PATRICE N. KLEIN,[1] Patuxent Wildlife Research Center, 11510 American Holly Drive, Laurel, MD 20708, USA
DOROTHY THOMPSON,[2] Patuxent Wildlife Research Center, 11510 American Holly Drive, Laurel, MD 20708, USA

Abstract: A 65-day-old, juvenile Mississippi sandhill crane (*Grus canadensis pulla*) sustained an oblique fracture of the right mid-diaphyseal femur. Internal fixation by use of an intramedullary pin and full cerclage wire was performed to stabilize the fracture site. Post-operative medical management of these fractures is often unsuccessful due to secondary complications associated with long-term restraint. This report describes the successful recovery of an endangered sandhill crane with a long bone fracture of the pelvic limb through a program of incremental physical therapy beginning 2 weeks post-femoral fracture repair. Sling support, assisted walking, and hydrotherapy were methods effective in preventing muscle atrophy during bone healing.

PROC. NORTH AM. CRANE WORKSHOP 7:232-236

Key words: Gruiformes, *Grus canadensis pulla*, Mississippi sandhill crane, long bone fracture, internal fixation, disuse atrophy, physical therapy.

A critical goal in the medical management of long bone fractures in cranes is the rehabilitation of skeletal muscles affected by disuse atrophy. This is particularly noted in pelvic limb fractures when the recovering bird may be non-weight bearing. Carpenter (1986) reported that captive-reared juvenile cranes appeared to be predisposed to leg injuries and pelvic limb developmental abnormalities. Modifications in captive management practices and nutrition have reduced the incidence of limb deformities, but traumatic injuries continue to present a husbandry and medical management challenge. Cranes sustaining long bone fractures of the legs are often placed into support structures during the post-operative period and become debilitated in the ensuing weeks of restraint and inactivity despite fracture healing (Curro et al. 1992, Olsen 1994). Consequently, the clinician is presented with a dilemma of a bird that cannot be returned to independent function. We therefore decided to initiate a program of physical therapy beginning 2 weeks after orthopedic surgical repair of the limb fracture to address the problems of skeletal muscle atrophy resulting from prolonged disuse.

The authors thank the Patuxent Wildlife Research Center captive propagation research group (CPRG) caretakers, volunteer staff, and G. H. Olsen for their support. Special thanks to K. Spencer for her assistance with illustrations.

CASE REPORT

A 2-month-old, male Mississippi sandhill crane sustained an oblique fracture of the right mid-diaphyseal femur in a handling accident. The fracture site was stabilized by internal fixation with an intramedullary pin and full cerclage wire (Martin and Ritchie 1994). The bird recovered from isoflurane anesthesia and was placed in a suspended support sling. Five days after surgery, the bird escaped overnight from the support sling and sustained oblique fractures of the left distal radius and ulna. Open reduction of these fractures and internal fixation of the distal ulna were accomplished by placement of an intramedullary pin and cerclage wire. A figure-8 support bandage secured the left wing. A new canvas sling suspended in a remodeled, adjustable support frame was constructed of 2.54- × 2.54-cm, vinyl-coated wire with reinforcing diagonal steel poles (Fig. 1). The support frame was 15.8 cm long × 9.5 cm wide × 14.2 cm high. The canvas sling was supported in the frame by metal dowels, running lengthwise, which could be adjusted to the appropriate height. A feeding tray similarly supported by adjustable dowels was placed at the open front end of the wire frame. Cloth ties woven through an edge of the canvas sling were stretched across the top of the frame and fastened to the opposite side to secure the bird in position.

One week later, the bird presented severe bilateral leg paresis, uncoordinated movements of the leg muscles, and moderate bilateral digital flexor muscle contraction. Generalized disuse atrophy in the pelvic limbs was apparent. The hocks were hyperextended and the digits were hyperflexed. Consequently, passive range of motion exercises (flexion and extension) of all leg joints from toes to hip were performed 3-4 times daily to stretch the contracted muscles. The bird could not support its body weight, nor ambulate, and required constant physical support. Twice daily, the bird and canvas sling were lifted from the support frame and, by holding the metal dowels, the bird was "hand-walked" for 30-minute intervals. Improved coordination but not muscle strength was noted in 7 days. Daily periods of graduated exercise and hydrotherapy were initiated to enable the bird to regain muscle strength and locomotor skills and to progress to

---

[1]Present address: The Humane Society of the U.S., 2100 L Street, N.W., Washington, DC 20037, USA.

[2]Present address: Wild Bird Rescue, 9058 Gorman Road, Laurel, MD 20708, USA.

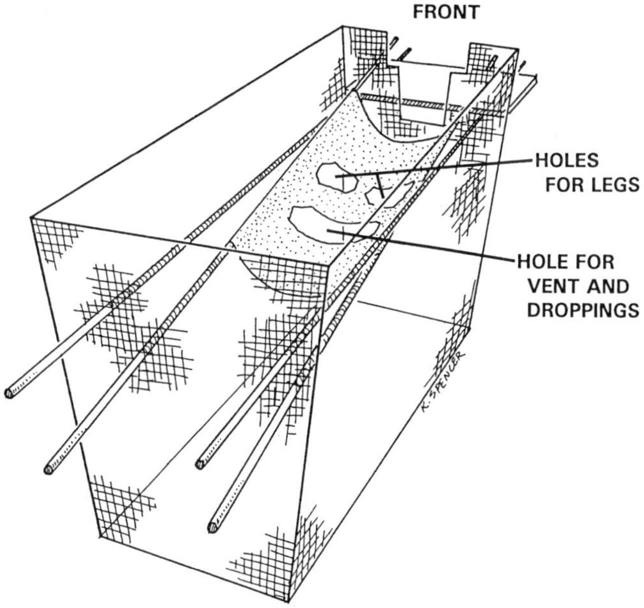

Fig. 1. Adjustable support frame with canvas sling used in long bone fracture rehabilitation in a sandhill crane. The sling is suspended by metal dowels running lengthwise through the frame. A feeding tray is positioned at the open end. The diagonal steel poles reinforcing the left and right sides of the frame are not pictured.

Table 1. Physical therapy protocol for long bone fracture rehabilitation in a sandhill crane.

Week in recovery	Physical therapy regimen
1	Passive range of motion exercises
2	Assisted walking with sling support
3	Sling-walking and hydrotherapy Brief hand-assisted walks (sling removed)
4	Assisted walking with nylon harness Hydrotherapy twice daily Increased assisted walks with hock-sitting
5	12 hours/day unrestricted exercise "Tail tugs" Hydrotherapy twice daily Support sling restraint (overnight only)
6	Unrestricted pen exercise/"tail tugs" Discontinued use of support sling Overnight monitoring to assist ambulation (Bird unable to rise without assistance)
7	Discontinued use of nylon harness Unrestricted pen exercise 24-hour monitoring without assistance
8	Bird returned to assigned pen with cohort Return to independent function

gradual elimination of all support devices.

Hydrotherapy sessions were added to the physical therapy regimen in the second week in recovery. These isotonic exercises were conducted in a solar-heated, 1-m-deep pool. Using the metal dowels to support the bird in the canvas sling, we "walked" the bird to the pool and then suspended it over the surface of the water with its legs submerged. The bird kicked its legs against the resistance of the water for 15–20 minutes, 2–3 times daily. Repeatedly, the bird was lowered to a standing position and supported to walk through the shallow pool.

In the fourth week, a nylon dog harness was fitted to the bird to act as a support harness during hand-assisted walking periods. The harness was long enough to accommodate the wing position and wide enough to adjust around the keel to the circumference of the body. Assisted walks with this support harness allowed the bird greater freedom of movement, including opportunities to begin hock-sitting and standing, using secure but minimized handling and support devices. Isotonic pool exercises were continued, the harness being used instead of the canvas sling for support. The duration of exercise sessions was increased by 15 minutes daily. Following each exercise period the bird was placed back in the sling and support frame.

Radiographs documented excellent callus formation at the femoral fracture site at 4 weeks, and the intramedullary pin was removed. Satisfactory bone healing of the left radius and ulna was present 4 weeks post-operation, permitting removal of the intramedullary pin. Muscle contraction in the wing was pronounced upon removal of the figure-8 bandage. Passive range of motion exercises (flexion and extension) at each joint was performed to promote wing extension.

At the fifth week, the bird had regained sufficient musculoskeletal strength and endurance to sustain 12 hours of daily walking and pool exercises with only partial assistance. The nylon harness was used as the principal support device, and the suspended sling was limited to overnight restraint. Isometric exercises were structured to stimulate muscle hypertrophy through muscle tension by repeated flexion and extension of various muscle groups. One example was "tail tugs." The tail feathers were gently grasped near the tail base as the bird attempted to walk away. The resistance applied was sufficient to force the bird to press its feet firmly against the ground while contracting the leg muscles. This 1- to 2-minute exercise was repeated 5 times during a 15-minute interval.

The suspended sling was discontinued in the sixth week. At this stage of recovery the bird was capable of independent walking and hock-sitting but could not rise from sternal recumbency without assisted support. Twenty-four-hour monitoring, unrestricted pen exercise, and the use of the nylon harness were continued through the seventh week until the bird was able to stand without assistance. Complete return to independent function was achieved in the eighth week, and the bird was returned to its assigned pen. Table 1 summarizes the type and duration of physical therapy used in the post-operative rehabilitation period.

DISCUSSION

Rehabilitation of skeletal muscles atrophied by disuse is a critical goal in the medical management of long bone fractures in cranes. Understanding avian myology apropos of structure and function of skeletal muscles is paramount to providing appropriate post-operative physical therapy (Vanden Berge 1975, Kendall and McCreary 1983:129-185, Orosz et al. 1992:62-97). In this case the muscles of the pelvic limb, particularly those of the hip and thigh, were severely weakened consequent to the femoral fracture. Identification of the specific muscle groups affected was a prerequisite to the development of an appropriate physical therapy program (Fig. 2).

The iliotibiales complex comprises the iliotibialis cranialis (sartorius), iliotibialis lateralis, and iliofibularis (biceps femoris). The first 2 muscles are flexors of the hip and extensors of the knee. Weakness of these muscles is associated with decreased strength of hip flexion and instablility of the knee joint. The iliofibularis is a flexor of the knee. Weakness results in hyperextension of the knee and lateral instability of the joint. The flexor cruris lateralis (semitendinosus) and f. cruris medialis (semimembranosus) which lie posterior to the iliofibularis extend the hip and flex the knee. The ambiens (pectineus) is distinctive to reptiles and birds. Its functional role as the "perching muscle" is to flex the digits in grasping. It also assists in extension of the knee and adduction of the leg. Weakness results in decreased strength for weight bearing.

The pubischiofemorales complex (adductor femoris) extends the hip and adducts the femur. Weakness results in hyperflexion and altered rotator action of the hip. The femorotibialis (quadriceps femoris) is the main extensor of the knee joint and is important in medial rotation of the tibiotarsus. Weakness impairs walking up inclines, stair-type climbing, and rising from a sitting position. The ischiofemoralis and obturator muscles extend the hip and act in lateral rotation of the femur during walking. Weakness causes medial rotation of the hip with a tendency towards knock-knees. The iliotrochanterici muscles (gluteus profundus and medius) flex the hip and rotate the femur with respect to the acetabulum. In general, these muscles act as shock absorbers during landing, bearing the strain of the load upon the other muscles.

Disuse atrophy of the muscles of the lower leg also was present. The fibularis (peroneus) brevis and tibialis cranialis flex the hock (tarsometatarsus). Weakness of these muscles could affect hock-sitting ability. The fibularis (peroneus) longus extends the hock. Muscle weakness decreases the ability to rise up from a sitting position. The gastrocnemius muscle consists of 2 or 3 heads (species dependent). It is a strong extensor of the hock and flexor of the digits. Postural function also has been associated with the gastrocnemius muscle. Therefore, muscle weakness results in hyperextension of the knee, inability to rise up on the digits, and postural deficits.

The digital flexor muscles (flexor perforans et perforatus digiti II–V) are grouped as intermediate, superficial and deep flexors respective to their levels of insertion. They extend the hock and flex their respective digits. Weakness results in hyperextension of the interphalangeal joints and reduces weight bearing by the hock. Contracture of the digital flexors results in flexion deformity of the distal phalanges and restricts flexion of the hock.

The pelvic limb muscles described were all affected by disuse atrophy of various degrees. The generalized paresis and incoordination first apparent in the juvenile Mississippi sandhill crane were manifestations of the acute fracture trauma and the secondary skeletal muscle atrophy associated with the post-operative restraint. These impairments were remedied through the use of the physical therapy techniques described in this report. Although labor and time intensive, the addition of a physical therapy regimen was considered to provide a greater chance for a successful outcome.

Successful recovery of long bone fractures in cranes is often hindered by both physical and psychological factors. Muscle atrophy, restraint intolerance pre-disposing to secondary trauma, food refusal, and stress related opportunistic infections complicate recovery efforts. Fortunately, this young crane tolerated the restraint of the support sling and maintained a good appetite and disposition. Hematology parameters measured throughout recovery remained within normal range. Serum SGOT (AST) and LDH values indicative of skeletal muscle damage were moderately elevated during the first 4 weeks of the post-operative period and gradually returned to normal. Other clinical chemistry parameters were within normal range.

The original, unsturdy support frame predisposed the crane to secondary trauma. The fractures of the radius and ulna did not hinder the general rehabilitation of the crane, but the prolonged figure-8 bandaging of the wing during fracture healing resulted in permanent tendon contracture and loss of full wing extension. Once substantive callus formation at the

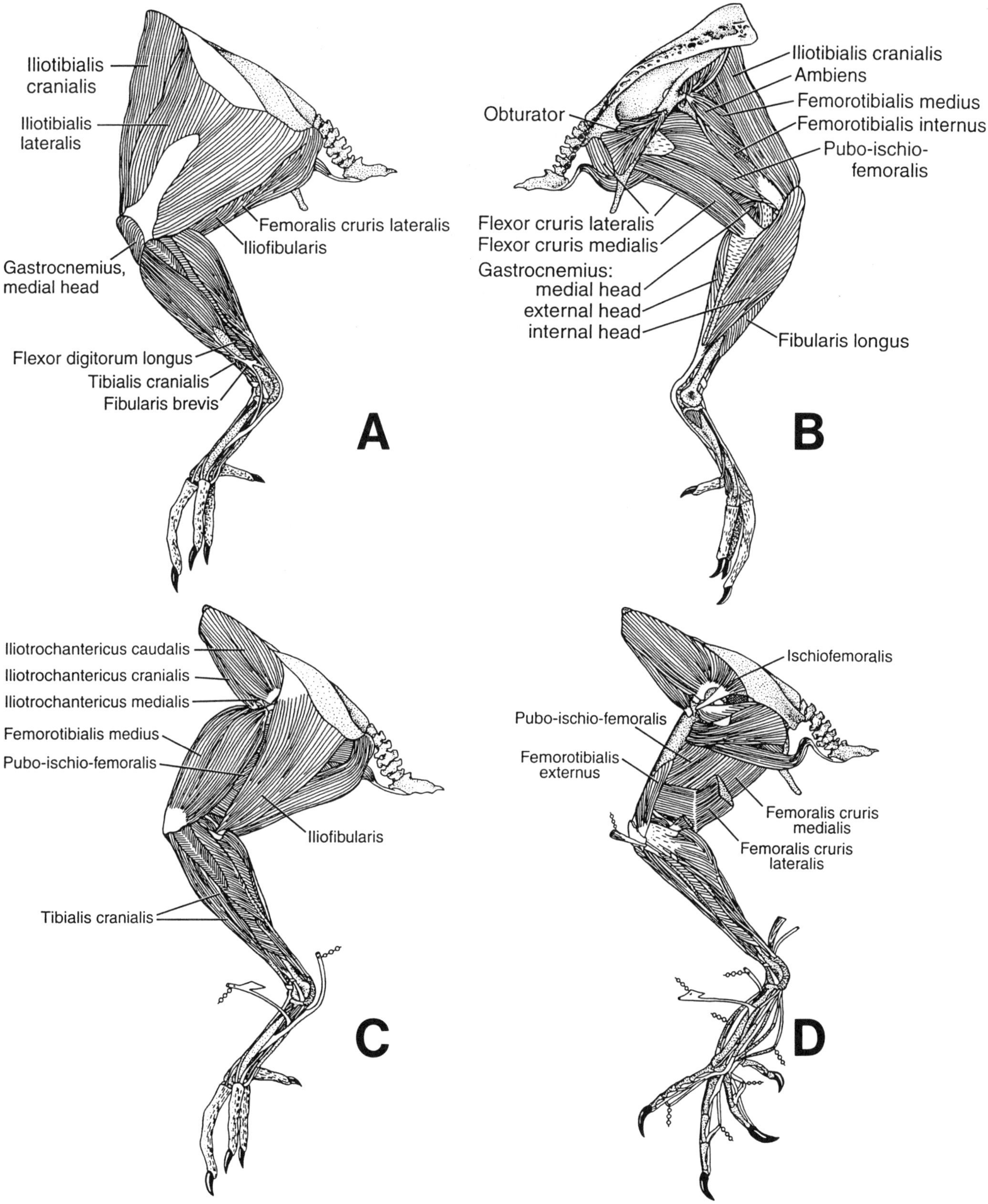

Fig. 2. Pelvic limb muscles of a typical bird (blue grouse [*Dendrogapus obscurus*]): (A) Superficial muscles of the left leg, lateral view; (B) Superficial muscles of the left leg, medial view; (C) Left leg, second layer of muscles, lateral view; (D) Left leg, third layer of muscles, lateral view. Redrawn from Vanden Berge (1975).

fracture site was documented, physical therapy for the wing fracture and more frequent bandage changes may have prevented this disability.

The successful rehabilitation of this bird may be partly attributed to the use of environmental and sensory stimuli to support its psychological well-being. Throughout the recovery period this juvenile crane was housed in a facility that maintained adult sandhill cranes in several adjacent pens. The vocalizations and behaviors of these adults provided appropriate imprinting models. Daily walking exercises took place in an outdoor yard where the young crane could probe the soil, forage, and observe the adult cranes in nearby pens. A standard pelleted feed diet was provided ad libitum, but curiosity items such as cracked corn and mealworms were often added.

One year later, the crane exhibited normal developmental physical and behavioral characteristics (T. G. Curro, J. Langenberg, and L. Deakin, School Vet. Med., Univ. Wisconsin, Madison, unpubl. data). The bird was well socialized in a community pen of age-matched cohorts, and demonstrated typical aggressive behavior towards penmates and caretakers. Although the physical therapy program succeeded in a captive management setting, the frequent human interaction during rehabilitation raises concerns of improper imprinting had this crane been a candidate for release to the wild. Notwithstanding, the use of an incremental physical therapy program can be the pivotal component to the effective medical management of long bone fractures in Gruiformes.

LITERATURE CITED

CARPENTER, J. W. 1986. Cranes (Order Gruiformes). Pages 316-326 *in* M. E. Fowler, ed. Zoo and wild animal medicine. Second ed. W. B. Saunders Co., Philadelphia, Pa.

CURRO, T. G., J. LANGENBERG, and J. PAUL-MURPHY. 1992. A review of lameness in long-legged birds. Proc. Annu. Conf. Assoc. Avian Vet. 1992:265-270.

KENDALL, F. P., and E. K. MCCREARY. 1983. Muscles: testing and function. Williams and Wilkins, Baltimore, Md. 451pp.

MARTIN, H., and B. W. RITCHIE. 1994. Orthopedic surgical techniques. Pages 1138-1160 *in* B. W. Ritchie, G. J. Harrison, and L. R. Harrison, eds. Avian medicine: principles and application. Wingers Publishing, Inc., Lake Worth, Fla.

OLSEN, G. H. 1994. Crane orthopedics: pediatrics and adults. Semin. Avian and Exotic Pet Med. 3:73-80.

OROSZ, S. E., P. K. ENSLEY, and C. J. HAYNES. 1992. Avian surgical anatomy: thoracic and pelvic limbs. W. B. Saunders Co., Philadelphia, Pa. 139pp.

VANDEN BERGE, J. C. 1975. Aves myology. Pages 1829-1843 *in* R. G. Getty, ed. Sisson and Grossman's the anatomy of the domestic animals. W. B. Saunders Co., Philadelphia, Pa.

METAL CONSUMPTION BY WHOOPING CRANES AND POSSIBLE ZINC TOXICOSIS

MARILYN G. SPALDING, Department of Pathobiology, College of Veterinary Medicine, University of Florida, Gainesville, FL 32611, USA
STEPHEN A. NESBITT, Florida Game and Fresh Water Fish Commission, 4005 S. Main St., Gainesville, FL 32601, USA
MARTIN J. FOLK, Florida Game and Fresh Water Fish Commission, 4825 Citrus Drive, St. Cloud, FL 34772, USA
LEE R. McDOWELL, Animal Science Department, University of Florida, Gainesville, FL 32611, USA
MARÍA S. SEPÚLVEDA, Department of Pathobiology, College of Veterinary Medicine, University of Florida, Gainesville, FL 32611, USA

Abstract: Fourteen captive-reared whooping cranes (*Grus americana*) were transferred to a pen for soft-release in Florida. Nine were killed by predators within 6 months. We found metal objects in the stomachs of 6 of these 9 either by radiography or at necropsy. All 6 had elevated serum zinc (Zn) concentrations ($\bar{x}$ = 10.3 ± 2.7 μg/ml, range 7.2–14 μg/ml) when compared to 3 without metal in their stomachs ($\bar{x}$ = 2.4 ± 0.92 μg/ml, range 1.6–3.4 μg/ml). We observed lethargy, abnormal posture, anemia, and elevated white blood cell counts in 2 cranes at or near the time of elevated serum Zn. The metal was surgically removed from these 2 and they appeared to recover. Metal debris from chain-link fence construction was the primary source of ingested objects.

PROC. NORTH AM. CRANE WORKSHOP 7:237-242

Key words: zinc, metal toxicosis, whooping cranes, *Grus americana*, sandhill cranes, *Grus canadensis*.

Metal ingestion is a common problem for captive cranes and can result in physical trauma to organs (Bush and Kennedy 1978), obstruction of the intestinal tract, or toxicosis from absorbed metals (Kennedy et al. 1977). Ingestion of metal has caused mortality from lead (Pb) toxicosis in wild sandhill cranes (*G. canadensis*) (Windingstad et al. 1984) and whooping cranes (Snyder et al. 1988). In this paper we report on cases of suspected Zn toxicosis associated with ingestion of metal objects by whooping cranes being released into Florida.

We thank the staffs of the National Wildlife Health Center (NWHC), Madison, Wisconsin; Patuxent Wildlife Research Center (PWRC), Laurel, Maryland; and the International Crane Foundation (ICF), Baraboo, Wisconsin, for providing some of the serum and liver Zn results. N. S. Wilkinson performed most of the Zn analyses. N. J. Thomas, J. C. Franson, and C. U. Meteyer of the NWHC also provided necropsy information from some of the carcasses. B. Wagner, J. A. Schmidt, S. T. Schwikert, J. M. Nicolich, and M. Wellington assisted with handling and made behavioral observations. J. A. Langenberg and G. H. Olsen provided considerable advice and background information. L. A. Young was responsible for care of the 2 cranes requiring surgical procedures at the Veterinary Medical Teaching Hospital (VMTH), University of Florida. Comments from D. J. Forrester and B. L. Homer improved the paper. Funding for this study was provided by the Florida Game and Fresh Water Fish Commission and the U.S. Fish and Wildlife Service. This is Florida Agricultural Experiment Stations Journal Series No. R-04946.

METHODS

An experimental release of whooping cranes into central Florida was described by Nesbitt et al. (1997). Fourteen captive raised cranes (6 from PWRC and 8 from ICF), 7–9 months of age, were held in a pre-release quarantine/acclimation pen in Osceola County in January 1993 and were allowed to fly free 1 month later. During quarantine they were examined upon arrival, when captured to change a brail from 1 wing to the other, 2 weeks after arrival, and upon release. Blood and feces were collected and they were re-vaccinated for eastern equine encephalitis. Hematology and blood chemistry parameters were compared to those of captive whooping cranes (Gee et al. 1981). They were observed daily: either directly from a blind near the release pen or by radiolocation with leg-mounted transmitters. Live birds were examined either in the field or at the University of Florida VMTH by using conventional radiographic techniques. Complete necropsy examinations were made on all birds found dead for determination of cause of death and collection of stomach contents. Blood was collected from the jugular vein into serum separator vacutainers or vacutainers with lithium heparin; serum or plasma was separated and frozen. In most instances serum was used to evaluate Zn concentrations; however, in a few cases plasma was used when serum was not available. Throughout the text serum and plasma are treated as equivalent. All of the serum and 2 of the liver samples were analyzed for Zn by air acetylene flame atomic absorption spectrophotometry on a Perkin-Elmer Model 5000 with AS-50 autosampler located at the University of Florida (Perkin-Elmer Corporation 1982). The remaining liver samples were analyzed for Pb and Zn for the NWHC by the Wisconsin Department of Agriculture, Trade and Consumer Protection Laboratory, Madison, Wisconsin. All serum values presented are wet weight basis, and liver values are listed in both wet and dry weight basis.

Comparisons between cranes that had metal in their stomachs and those that did not were made with the Mann-Whitney Test ($P < 0.05$) (Siegel 1956).

RESULTS

Two weeks after transfer to the release pen in Florida, 2 (#770 and #774) of the 14 cranes began to exhibit abnormal behavior (Table 1). They spent longer periods resting than other birds, were ataxic, assumed an abnormal posture, had unkempt feathers, and appeared emaciated. The back appeared hunched, probably due to a change in position at the base of the neck. They stood upright, moving the bill alongside the neck as if trying to preen but without grasping feathers.

Six weeks after arrival, the condition of crane #770 had deteriorated considerably and it was captured for examination. It could not fully extend the left wing. It had lost 0.9 kg body mass. An attempt at field radiography failed. The serum Zn concentration was 14 μg/ml; blood Pb was <2 mcg/dl. White blood cells were elevated (24,500/μl) with increases in numbers of heterophils, lymphocytes, and monocytes. Packed cell volume (35%) and hemoglobin (13 g/dl) were low; plasma uric acid was high (12.8 mg/dl). The bird was transported to the VMTH where 7 1.2- $\times$ 0.1-cm pieces of galvanized metal were surgically removed from the ventriculus. Acid digestion of this metal yielded 0.25% Zn. No abnormalities of the left wing were found by physical examination and radiology. The crane was returned to the release pen 4 days after surgery. Recovery was slow; use of the left wing gradually improved, and it first flew from the pen 1 month later. It was killed by a predator 4 months after surgery.

Two months after arrival in Florida, metal was detected in the stomach of crane #774 with a portable radiograph machine. The crane had lost 1.3 kg body mass and serum Zn was 4.4 μg/ml. In a blood sample collected 14 days after arrival, serum Zn was 8.2 μg/ml. White blood cells were elevated (57,000/μl) and most of the increase was due to heterophils (47,300/μl). Packed cell volume was low (27%). Alanine transaminase (ALT) was elevated at 176 IU/l. Surgery was performed 1 month later. A 5.4- $\times$ 0.2-cm piece of galvanized wire and 6 shiny pieces of 1.5- $\times$ 0.2-cm worn aluminum (Al)-like wire were removed from the ventriculus. Stones removed were covered with shiny material, perhaps from rubbing against the Al-like wire. The Al-like wire was only faintly detectable in radiographs due the poor ability of Al to deflect photons (Johns and Cunningham 1983:142). This bird was returned to the release pen the day following surgery. It flew from the pen the same day and spent the first night on the shore of Lake Marion. It remained away from other birds, its feathers were unkempt, and it was killed by a predator 2 months after surgery.

Four other birds (#'s 769, 778, 782, 783) had metal in their stomachs (Table 1) when they were killed by predators 1, 2, 4, and 5 months after arrival. Metal pieces found at necropsy included a pop-rivet, a stainless steel hog ring, 2 linked pieces of 7.2- and 6.3- $\times$ 0.2-cm galvanized wire, 12 pieces of Al-like wire, pieces of black plastic with imbedded stainless steel, and a piece of flat foil-like material, a piece of orange plastic, and pieces of green plasticized fiber. No behavioral abnormalities were observed in these birds.

The above 6 cranes with metal in their stomachs (Table 1) had $\geq$7 μg/ml serum Zn. The mean serum Zn calculated from the highest concentration for each of the birds (excluding the post-surgical high concentration in crane #770) was 10.3 $\pm$ 2.7 μg/ml serum Zn; the mean liver Zn was 51 $\pm$ 30 μg/g. Lead was not detected in the liver of 4 of these birds (#'s 769, 783, 774, 778). All 6 were killed by bobcats within 6 months of their arrival in Florida.

One crane (#773) was suspected to have ingested metal (Table 1) based upon a serum Zn of 11.6 μg/ml 2 weeks after arrival. However, this bird could not be recaptured and was killed by a predator 2 years after arrival. No metal was found in its stomach at death.

Seven cranes (#'s 772, 775, 776, 777, 779, 780, 784) exhibited no evidence of ingesting metal (Table 1). The mean of the highest serum Zn levels measured in these 7 birds was 2.7 $\pm$ 1.0 μg/ml. Liver Zn averaged 35 $\pm$ 12 μg/g in the 3 birds of this group that were killed by predators. Four of these birds survived for longer than 1 year; 3 were known to have been alive in January 1996.

Serum Zn concentrations of the 6 cranes with metal in stomach measured at 8–15 days prior to shipment to Florida were below 3 μg/ml ($\bar{x}$ = 2.0 $\pm$ 0.3) (Table 1). More of the cranes from PWRC were found with either metal in the stomach or elevated serum Zn (5/6, 83%) than cranes from ICF (2/8, 25%). Only 1 of the ICF cranes had ingested metal that resembled fencing wire. Serum from 4 wild Florida sandhill cranes (*G. c. pratensis*) from Osceola County, of similar age to the released whooping cranes, ranged from 0.49 to 2.39 μg/ml Zn (Table 1).

Upon arrival in Florida, cranes from ICF were generally heavier than cranes from PWRC. A direct comparison between the groups could not be made because weights were not available for all ICF cranes upon arrival and because 6 of the birds from ICF spent 2 weeks longer in quarantine in Florida. Two weeks prior to release (after 2–3 weeks in Florida), PWRC cranes averaged 5.3 $\pm$ 0.3 kg and ICF cranes averaged 6.0 $\pm$ 0.7 kg. On release (after 4–5 weeks in Florida) PWRC cranes averaged 5.0 $\pm$ 0.4 kg and ICF cranes averaged 5.4 $\pm$ 0.7 kg. PWRC cranes lost an

Table 1. Zinc concentrations (μg/g or ml) in serum and liver from whooping cranes released in Florida and from wild sandhill cranes of similar age from Osceola County, Florida.

Band no.	Source	Days in Florida	Serum	Liver Wet	Liver Dry	Comments
Whooping cranes with metal in stomach and clinical illness						
770	PWRC	-15	2.4			Preshipment sample
		14	4.7			
		27				Lethargic, head droop, bill against neck
		42				Appeared emaciated and unkempt, abnormal behavior
		46	14.0[a]			Emaciation, abnormal behavior
		53				First sign of recovery, flight efforts
		61	3.4			Moved to VMTH, radiographs, metal in ventriculus
		62				Surgery
		64	4.2			
		66	16.8			Released to pen
		80				Flew 50 m within pen
		97				First flew from pen
		106				Appeared normal
		171	6.1	49[b]	180	Predation
774	PWRC	-15	1.8			Preshipment sample
		14	8.2[a]			
		40				Lethargic, head droop, bill against neck
		42				Walk as if tranquilized, avoids other birds
		48				Head twisting behavior
		53	4.4			Radiographs- metal in ventriculus
		88				Moved to VMTH
		89	8.0			Surgery
		90				Released in pen, roosted alone on lake shore
		137				Solitary behavior, unkempt appearance
		149		51	192	Predation
Whooping cranes with metal in stomach						
782	PWRC	-8	2.3			Preshipment sample
		22	12.4[a]			
		164		34[b]	145	Predation
778	ICF	22	11.4[a]			
		123	6.6	27	90	Predation
769	PWRC	-15	1.7			Preshipment sample
		14	8.4[a]			
		38		110	420	Predation
783	ICF	13	7.2[a]			
		45		36	138	Predation
$\bar{x} \pm$ SE (n = 6)			10.3 ± 2.7	51 ± 30	194 ± 116	
Whooping crane suspected to have ingested metal based on serum Zn						
773	PWRC	-15	2.0			Preshipment sample
		14	11.6			
		26	9.6			
		852				Predation
Whooping cranes with no evidence of ingesting metal						
776	ICF	22	3.4[a]			
		40		48	204	Predation
772	PWRC	-15	2.0			Preshipment sample
		14	3.0[a]			
775	ICF	22	2.2[a]			
		47		26	90	Predation
779	ICF	22	2.0			
		160	2.6			
		237	4.1[a]			
777	ICF	22	1.6[a]			
		46		30	30	Predation
784	ICF	13	1.2			
		265	3.2[a]			
780	ICF	22	0.5			
		161	1.5[a]			
$\bar{x} \pm$ SE (n = 7)			2.7 ± 1.0	35 ± 12	137 ± 49	
Florida sandhill cranes						
G92-267			2.0[a]			8 months of age
G92-277			0.4[a]			8 months of age
G92-270			2.4[a]			9 months of age
G92-256			1.6[a]			1 year old
$\bar{x} \pm$SE (n = 4)			1.6 ± 0.9			

[a] Values used to calculate means.
[b] Wet weight not available, estimated by multiplying dry weight by 0.27 (mean conversion factor for other birds in this sample).

estimated 0.6 kg during Florida quarantine while ICF birds lost 1.3 kg.

DISCUSSION

Six whooping cranes that ingested metal objects had elevated (>7 μg/ml, Mann Whitney Test, $P = 0.001$) serum Zn concentrations and 2 of these exhibited signs of weight loss, abnormal behavior, and weakness while in a pre-release quarantine/acclimation pen. Crane #770 appeared weakest 46 days after arrival in Florida when its serum Zn was 14 μg/ml and appeared to recover somewhat before surgery when its serum Zn was 3.4 μg/ml. The highest serum Zn concentration (16.8 μg/ml) occurred in crane #770 4 days after surgery. It is suspected that this concentration might have been caused by Zn mobilization following stress and reduced food intake associated with surgery. Crane #774 was not handled when it was most lethargic (between 40 and 50 days after arrival). Thus, the peak serum Zn for this and other asymptomatic birds may not have been detected.

Muscular atrophy and weakness were observed in diving ducks that had ingested galvanized wire and pennies with serum Zn concentrations of 13 to 17 μg/ml and liver Zn concentrations 242 to 548 μg/g (Zdziarski et al. 1994). Morris et al. (1986) reported a blue and gold macaw (*Ara ararauna*) that was ataxic with 15.5 μg/ml plasma Zn.

Liver Zn concentrations in birds are usually less than 210 μg/g dry weight (Eisler 1993). Only 1 of the whooping cranes in this study that had ingested metal (#769) had a higher liver Zn concentration (420 μg/g dry weight). No difference was found in liver Zn when cranes that had metal in their stomach were compared with those that did not at the time of death. This was expected, however, because no cranes were exhibiting signs of toxicosis at the time of death. Whooping crane livers collected opportunistically (Lewis et al. 1992; J. Lewis, pers. commun.) from Aransas and Rocky Mountain flocks had 40/174, 44/–, 67/206, 75/–, –/234, and 100/346 μg/g wet weight/dry weight Zn. Liver Zn was 1,910 μg/g in a chachalaca (*Ortalis cinereiceps*) that had consumed a penny (Droual et al. 1991). Peach-faced lovebirds (*Agapornis roweicollis*) housed in a new galvanized wire cage died with metallic flakes in their gizzards; liver Zn concentrations were 75–156 μg/g dry weight (Reece et al. 1986).

Howard (1992) suggested the pancreas may be a more relevant tissue to test for Zn toxicosis. Experimental oral dosing of chickens with Zn resulted in exocrine pancreatic structural and functional changes (Lu and Combs 1988). Pancreatic changes, primarily acinar atrophy and ductular proliferation, were noticed in psittacines ingesting galvanized wire (Reece et al. 1986) and in diving ducks ingesting pennies and fence clips (Zdziarski et al. 1994). Zinc toxicosis was associated with erosion of the gizzard in a chachalaca (Droual et al. 1991) and with necrotizing ventriculitis in diving ducks (Zdziarski et al. 1994). We found no such lesions in the whooping cranes; however, clinical signs of toxicosis were not observed near the time of death in cranes killed by predators.

Surgical removal of metal was chosen in the 2 clinical cases due to the number and large size of the metal objects present. It was believed that the large stones in the ventriculis might present a problem with endoscopic removal. Other options for treatment in such circumstances include fluoroscopy or endoscopy guided removal with forceps, chelation therapy, or the use of bulk cathartics (reviewed by Dumonceaux and Harrison 1994). Chelating agents may have improved the recovery rate; however, this therapy was not selected because frequent handling would have delayed return of the bird to the wild.

A cause for the elevated white blood cell count in the 2 clinically affected cranes was never determined. Although handling stress may have contributed, it cannot account for the magnitude of the elevation. Many of the birds killed by predators, including these 2, had lesions that suggested infection with *Eimeria* sp. (disseminated visceral coccidiosis). Experimentally dosed psittacines (Howard 1992) had dose-related uric acid and ALT activity increases and PCV and hemoglobin decreases; some birds had monocytosis. Similar but inconsistent findings were found in these cranes; however, the lack of changes in individual birds may have been due to the time of blood collection relative to the course of toxicosis.

Weight loss was expected and observed in all cranes during the Florida quarantine period. Weight loss could not be correlated with metal ingestion due to the discrepancy in initial weight between the cranes from the 2 breeding facilities. Cranes from ICF were generally heavier upon arrival and lost more weight during the quarantine period. This was likely a result of their better nutritional condition (body fat and muscle mass) when compared to birds from PWRC. The 2 clinically ill birds lost more weight than their PWRC cohorts. Howard (1992) reported weight loss in psittacines experimentally dosed with Zn and galvanizing Zn. The poorer nutritional condition of the PWRC cranes is a possible driving force for their greater propensity to pick up metal objects.

Most of the wire pieces removed from the birds were of the type found in the area of newly constructed chain-link fencing. The construction contract for the fencing in Florida specified that no metal fragments were to be dropped at the construction site and that all metal was to be removed upon completion. More than a dozen pieces were found with a metal-detector the day before the shipment of the cranes.

Serum collected from the 6 PWRC birds prior to transportation had low Zn, suggesting that much of the metal was either acquired at the breeding facility near the time of transfer or in Florida.

Stones, except for small pieces of limestone, are generally absent from the sandy and organic soils in Osceola County (Brown et al. 1990). Florida sandhill cranes usually have small limestone pebbles and sometimes small quartz-like stones (possibly from roadsides) in their stomachs. The stomachs of all of the whooping cranes examined at death contained many large (about 1 cm diameter) quartz or hard stones that were probably acquired in captivity because they were unlike those found in central Florida soils. The 2 birds that underwent surgery to remove metal (and most stones) had either limestone pebbles or very few (4) stones in their stomachs when re-examined at death. Whooping cranes may pick up metal pieces in their search for large gastroliths. Bush and Kennedy (1978) noticed a greater propensity for the larger sarus cranes (*G. antigone*) to ingest foreign objects than the smaller sandhill cranes. Foreign objects and stones larger than 0.7 cm across are very rarely observed in the stomachs of dead sandhill cranes in Florida. No difference was apparent between the gastric stones in birds from ICF and PWRC that would explain the greater propensity for metal consumption by PWRC birds. We could find no information on gastroliths of wild whooping cranes.

Survival of the birds ingesting metal ($\bar{x}$ = 115 ± 59 days in Florida, range 38–171 days) was not significantly lower than in those with no evidence of metal ingestion ($\bar{x}$ = 436 ± 367 days, range 40–730 days [presently surviving birds were calculated at 730 days]). Survival of the 2 clinically ill birds was undoubtedly artificially increased, because these birds remained within the release pen while they were ill and were not subject to predation. Although there was no direct evidence that Zn toxicosis contributed to mortality, compromised health could predispose birds to predation. Long-term sublethal effects of birds recovering from toxicosis should also be considered. The bird suspected of ingesting metal (#773) survived for more than 2 years; however, it never associated with other birds and showed no evidence of pair formation at the time that its release cohorts did.

Three steps were taken to resolve the problem of metal ingestion in subsequent releases, and instances of metal ingestion have been reduced significantly in the 38 birds released in Florida since the first release reported here. (1) Metal detectors were used extensively (3–4 passes through the pen) prior to the introduction of any cranes. Because of proximity to the fence, metal detectors were not effective near the bottom of the fence, the area most used by birds pacing the fence and probing. We installed a 75-cm-wide concrete skirting at the base of the fence so that a metal detector could be used effectively up to the edge of the skirting. (2) Commercially purchased stones of an appropriate size (about 1 cm diameter) were placed in the release pen to serve as gastroliths and thereby diminish the bird's attraction to other hard objects. (3) Radiographic and serologic examination for evidence of metal ingestion, and surgical removal of metals present, were added to the 2-week preshipment quarantine protocol. Metal objects were found in the stomachs of 20% of 38 cranes examined at breeding facilities prior to shipment to Florida during 1994–95. Surgery 2 weeks before shipment allows time for the complete recovery of the cranes prior to the stress of transportation. Another reason for resolving metal ingestion problems prior to shipment was that the prolonged recovery period of the birds that became ill and required surgery in Florida appeared to delay the time of departure from the pen for other birds in the release cohort.

Ingestion of metal, plastic, and glass objects continues to be a potential hazard for whooping cranes in Florida. Twelve of 26 cranes (46%) released since January 1993 that were killed by predators contained foreign objects. In 2 cases the metal was known to be there from preshipment radiographs, 4 others contained metal, and 6 contained other foreign objects (plastic bases of shotgun shells, pieces of plastic, worn pieces of glass). Thus far they have not appeared to present a health problem. Evidence presented here suggests that whooping cranes ingest foreign objects because appropriate gastroliths are not readily available in Florida, but this does not explain such behavior in captive cranes where stones are available.

LITERATURE CITED

BROWN, R. B., E. L. STONE, and V. W. CARLISLE. 1990. Soils. Pages 35-69 *in* R. L. Myers and J. J. Ewel, eds. Ecosystems of Florida. Univ. Central Florida Press, Orlando.

BUSH, M., and S. KENNEDY. 1978. Ventriculostomy for removal of foreign bodies from sarus cranes. J. Am. Vet. Med. Assoc. 173:1107-1110.

DROUAL, R., C. U. METEYER, and F. C. GALEY. 1991. Zinc toxicosis due to ingestion of a penny in a gray-headed chachalaca (*Ortalis cinereiceps*). Avian Dis. 35:1007-1011.

DUMONCEAUX, G., and G. J. HARRISON. 1994. Toxins. Pages 1030-1052 *in* B. W. Ritchie, G. J. Harrison, and L. R. Harrison, eds. Avian medicine: principles and application. Wingers Publishing, Inc., Lake Worth, Fla.

EISLER, R. 1993. Zinc hazards to fish, wildlife, and invertebrates: a synoptic review. U.S. Fish Wildl. Serv. Biol. Rep. 10. 106pp.

GEE, G. F., J. W. CARPENTER, and G. L. HENSLER. 1981. Species differences in hematological values of captive cranes, geese, raptors, and quail. J. Wildl. Manage. 45:463-483.

HOWARD, B. R. 1992. Health risks of housing small psittacines in galvanized wire mesh cages. J. Am. Vet. Med. Assoc. 200:1667-1674.

JOHNS, H. E., and J. R. CUNNINGHAM. 1983. The physics of radiology. Fourth ed. Charles C. Thomas, Springfield, Ill. 522pp.

KENNEDY, S., J. P. CRISLER, E. SMITH, and M. BUSH. 1977. Lead poisoning in sandhill cranes. J. Am. Vet. Med. Assoc. 171:955-958.

LEWIS, J. C., R. C. DREWIEN, E. KUYT, and C. J. SANCHEZ. 1992. Contaminants in habitat, tissues, and eggs of whooping cranes. Proc. North Am. Crane Workshop 6:159-165.

LU, J., and G. R. COMBS. 1988. Effect of excess dietary zinc on pancreatic exocrine function in the chick. J. Nutr. 118:681-689.

MORRIS, P. J., J. JENSEN, and F. APPLEHAUS. 1986. Lead and zinc toxicosis in a blue and gold macaw (*Ara ararauna*) caused by ingestion of hardware cloth. Annu. Proc. Am. Assoc. Zoo Vet. 1985:13-17.

NESBITT, S. A., M. J. FOLK, M. G. SPALDING, J. A. SCHMIDT, S. T. SCHWIKERT, J. M. NICOLICH, M. WELLINGTON, and J. C. LEWIS. 1997. An experimental release of whooping cranes in Florida–the first three years. Proc. North Am. Crane Workshop 7:79-85.

PERKIN-ELMER CORPORATION. 1982. Analytical methods for atomic absorption spectrophometry. Perkin-Elmer, Norwalk, Conn. 227pp.

REECE, R. L., D. B. BICKSON, and P. J. BURROWES. 1986. Zinc toxicity (new wire disease) in aviary birds. Aust. Vet. J. 63:199.

SIEGEL, S. 1956. Nonparametric statistics for the behavioral sciences. McGraw-Hill Inc., New York, N.Y. 312pp.

SNYDER, S. B., M. J. RICHARD, J. P. THILSTED, R. C. DREWIEN, and J. C. LEWIS. 1988. Lead poisoning in a whooping crane. Pages 207-210 *in* D. A. Wood, ed. Proc. 1988 North Am. crane workshop. Florida Game Fresh Water Fish Comm. Nongame Wildl. Program Tech. Rep. 12.

WINDINGSTAD, R. M., S. M. KERR, L. N. LOCKE, and J. J. HURT. 1984. Lead poisoning of sandhill cranes (*Grus canadensis*). Prairie Nat. 16:21-24.

ZDZIARSKI, J. M., M. MATTIX, R. M. BUSH, and R. J. MONTALI. 1994. Zinc toxicosis in diving ducks. J. Zoo and Wildl. Med. 25:438-445.

WHOOPING CRANE MORTALITY AT PATUXENT WILDLIFE RESEARCH CENTER, 1982-95

GLENN H. OLSEN, Patuxent Wildlife Research Center, 11510 American Holly Drive, Laurel, MD 20708, USA
JOANNA A. TAYLOR,[1] Patuxent Wildlife Research Center, 11510 American Holly Drive, Laurel, MD 20708, USA
GEORGE F. GEE, Patuxent Wildlife Research Center, 11510 American Holly Drive, Laurel, MD 20708, USA

Abstract: Whooping cranes (*Grus americana*) have been reared at Patuxent Wildlife Research Center since 1966. During 1982-95 there were 103 mortalities caused by infectious and parasitic diseases (46%), trauma (21%), anatomic abnormalities (17%), miscellaneous conditions (12%), and open or no diagnoses (5%). The implications that disease may have on new whooping crane flocks in Florida and Canada are discussed, based on these mortality factors in captivity.

PROC. NORTH AM. CRANE WORKSHOP 7:243-248

Key words: congenital deformities, crane, disease, *Grus americana*, infectious disease, mortality, parasites, scoliosis, whooping crane.

The endangered species research program was established in 1965 (Erickson 1968) at the Patuxent Wildlife Research Center (Patuxent), Laurel, Maryland. Captive propagation forms an integral part of gene pool preservation, research, and reintroduction mission of the program. Researchers develop effective husbandry, medical, and reproductive techniques to build self-sustaining captive populations of critically endangered native species. Eventually, captive-bred animals are released to augment and restore wild populations. The most intensive effort in captive propagation research has focused on the whooping crane. The first whooping crane (an injured young-of-the-year) was transferred from Canada to the U.S. Bureau of Sport Fisheries and Wildlife in 1964 and moved to Patuxent in 1966. Patuxent received 6 eggs removed from nests in Wood Buffalo National Park, Northwest Territories, Canada in 1967 and 173 eggs in subsequent years.

The first successful breeding of whooping cranes at Patuxent occurred in 1975 (Ellis et al. 1992). Although a steadily increasing number of eggs has been produced annually, mortality factors, including disseminated visceral coccidiosis (Carpenter et al. 1980), eastern equine encephalitis virus (Dein et al. 1986), mycotoxins (Olsen et al. 1995), and delayed sexual maturity have kept recruitment rates low. Also, 73 eggs (61 fertile) produced at Patuxent were used in an unsuccessful reintroduction project at Grays Lake National Wildlife Refuge in Idaho (Ellis et al 1992). These factors slowed the growth of the captive flock. In 1989, 22 of 58 captive whooping cranes were transferred to the International Crane Foundation in Baraboo, Wisconsin, to minimize the impact a stochastic disaster would have on the captive breeding program and gene pool. From 1992 through 1995, 11 whooping cranes were transferred to the Devonian Science Center of the Calgary Zoo, Alberta, Canada, to establish a third captive flock. Other than papers documenting specific diseases, there has been 1 summary report of captive whooping crane mortality covering the years 1966-81 (Carpenter and Derrickson 1982).

Mortality information from captive flocks provides an understanding of the impact of disease on life span, reproduction, and population demographics. Mortality information can also be examined to estimate its impact on proposed releases. Whooping cranes are being released in Florida to establish a non-migratory flock, and plans are underway for a release in the Prairie Provinces of Canada. We discuss possible implications of mortality on releases in these 2 locations, as well as captive mortality.

We thank M. T. Childress and J. L. Kodak for help with preparation and editing of this manuscript, J. M. Nicolich for her help in preparing the information used in this study and for presenting the paper at the workshop, and J. Lewis for his editorial comments. The necropsies were performed by the authors and veterinarians at Patuxent, the National Wildlife Health Center (NWHC), Madison, Wisconsin, and the Maryland Department of Agriculture, College Park, Maryland. These veterinarians were J. W. Carpenter, F. J. Dein, C. Driscoll, D. Green, C. Franson, P. Klein, and N. Thomas.

METHODS

Patuxent has reared whooping cranes by using various techniques. For the first 11 years (1967-77), whooping crane eggs were artificially incubated for 28 days and then transferred to a mechanical hatcher after pipping. Crane chicks were reared indoors by caretaking staff, then moved to outdoor pens as they grew older. Between 1978 and 1989, some whooping crane eggs were incubated by cochin chickens and sandhill cranes to improve embryo survival.

[1]Present address: Okefenokee National Wildlife Refuge, Route 2, Box 3330, Folkston, GA 31537, USA.

Hatching took place in artificial hatchers or under sandhill cranes. Chicks were reared as before, by caretakers or outdoors by surrogate sandhill crane parents. Beginning in 1990, whooping crane chicks were no longer reared by sandhill cranes to avoid cross-species imprinting. However, since 1990 some whooping crane chicks have been successfully reared by whooping crane pairs. Due to the low number of experienced whooping crane pairs, most chicks continue to be reared by staff, most recently by individuals wearing shrouds and masks to camouflage their human form. Sandhill cranes continue to act as surrogate incubators.

Adult whooping cranes are housed in outdoor pens of various sizes. Most of these are covered with nylon flight netting. Each pen has a gravity-flow poultry feeder and constantly flowing water provided in stainless steel drinking cups (Van Es waters) (Hinshaw 1952). Breeding pairs are visually isolated from each other by opaque green nylon tennis netting and separated by empty adjacent pens. Since the early 1970's, all cranes have been fed starter, maintainer, or breeder diets, depending upon age and breeding status (Carpenter 1977).

Cause of death for each bird was determined from a combination of clinical findings, gross pathological changes observed at necropsy, and microbiological, parasitological, serological, and histopathological results. Gross necropsies were performed at Patuxent, although cases were occasionally referred to the NWHC. Radiographs were taken when trauma, foreign body, or skeletal deformity was suspected. Specimens from liver, lungs, kidneys, spleen, heart, brain, gastrointestinal tract, and other appropriate organs were obtained for microbiological testing. Samples were processed at Maryland Medical Laboratory in Baltimore, Maryland, Maryland State Board of Agriculture, Animal Health Department Laboratory in College Park, Maryland, or the NWHC.

We conducted parasitological analysis (both direct smear and flotation techniques) for isolation and examination of intestinal parasites. Parasites were preserved in 70% isopropyl alcohol for later identification.

Representative tissue samples collected at necropsy were preserved in 10% buffered formalin for histopathological study and sent to the NWHC or the Maryland State Board of Agriculture. When appropriate, additional tests were performed, including analysis of tissues for suspected chemical residues, and food and water for contaminants and toxins.

RESULTS

During the 14 years of this study, 103 whooping cranes died and were examined (Table 1). Nearly one-half of these cranes (46%) died from infectious or parasitic diseases. Fewer died from trauma (21%), anatomic abnormalities (17%), or from miscellaneous causes (12%). Cause of death was unknown in 5 cases. The causes of death were similar to those reported earlier (Carpenter and Derrickson 1982).

Whooping cranes were assigned to 3 age categories based on earlier studies: chicks (<2 months), juveniles (≥2 and <12 months), and adults (≥12 months) (Carpenter and Derrickson 1982). Most deaths occurred in chicks <2 months of age (62%, n = 63) (Table 1). Ten chicks that died during the hatching process are included in the 63 deaths. Four of the chicks that died during hatching died from scoliosis, 3 from omphalitis, and 1 each from malpositioning and delayed hatching. No definitive cause of death was discovered in 1 death.

There were 15 juvenile (≥2 and <12 months) crane mortalities, and 25 subadult/adult (≥1 yr) mortalities between 1982 and 1995. Of these 25, 12 were males and 13 were females. The subadult/adult cranes ranged from 1 to 23 years of age at death.

The largest concentration of deaths occurred in spring and summer (April n = 6, May n = 39, June n = 17, and July n = 12, August n = 9) and was associated with chick hatching and early growth. The mortality rates in the autumn (September n = 7, October n = 3, and November n = 5) were elevated due to eastern equine encephalitis in 1984 (7 birds), and mycotoxins in 1987 (3 birds). Mortality was lowest during the cold months of the year (December n = 2, January n = 2, February n = 1, and March n = 0).

DISCUSSION

The primary causes of death for captive whooping cranes were infectious and parasitic diseases. Several of these diseases are of importance to the future of the captive propagation program and subsequent release efforts. Pneumonia and associated respiratory diseases were the most common infectious diseases (9%). Aspergillosis was also a common sequel to respiratory infections (7% of the cases). Infectious diseases caused 31.7% of deaths during 1966–81 (Carpenter and Derrickson 1982) and 32% of deaths in this study. However, there has been a shift in causative agents. *Aspergillus* spp. was not reported in the earlier study but casued 7% of deaths in this study. Factors contributing to this rise in incidence were improved diagnostic tests and the increased use of powerful antibiotics, which will control bacterial infections but may make the cranes more susceptible to secondary fungal infections. Although one-third of all infectious diseases were diagnosed as either pneumonia or aspergillosis, young chicks accounted for most of these deaths. Respiratory disease is an important mortality factor in any wild flock, but the extent of such mortality has been and will remain unknown. When there is whooping

Table 1. Causes of mortality in whooping cranes by age class at Patuxent Wildlife Research Center, Laurel, Maryland, 1982–95.

Cause of mortality	Age class[a] (*n*) Chick	Juvenile	Adult	All ages *n*	%
Trauma	**10**	**5**	**7**	**22**	**21**
Aggression	4	0	2	6	6
Spinal trauma	0	2	1	3	3
Fractures	3	1	1	5	5
Miscellaneous	3	2	3	8	7
Congenital abnormalities	**15**	**2**	**0**	**17**	**17**
Scoliosis	9	0	0	9	9
Heart defects	0	1	0	1	1
Leg deformities	6	1	0	7	6
Infectious or parasitic disease	**33**	**6**	**8**	**47**	**46**
Pneumonia/respiratory disease	8	1	0	9	9
Omphalitis/rupture	7	0	0	7	7
Disseminated visceral coccidiosis	1	0	0	1	1
Peritonitis/intestinal rupture	9	1	0	10	9
Acanthocephalans	4	0	0	4	4
Eastern equine encephalitis	0	0	7	7	7
Aspergillosis	4	2	1	7	7
Miscellaneous infections	0	2	0	2	2
Miscellaneous mortality factors	**3**	**1**	**0**	**12**	**12**
Mycotoxins	0	0	3	3	3
Nutritional deficiency	0	0	1	1	1
Drug reaction	0	0	3	3	3
Capture myopathy	1	0	1	2	2
Malposition/delayed hatching	2	0	0	2	2
Miscellaneous	0	1	0	1	1
No diagnosis/open	**2**	**1**	**2**	**5**	**5**
Totals	**63**	**15**	**25**	**103**	**100**

[a] Chicks (<2 months); juveniles (≥2 and <12 months); adults (≥12 months).

crane reproduction at the Florida release site, we should be able to document chick mortality and determine if respiratory disease is a major factor affecting chick survival.

During 1966–81, 4 deaths (10%) were attributed to disseminated visceral coccidia (DVC). Anti-coccidial medication was incorporated into the pelleted feed after identification of this disease in the late 1970's. Between 1982 and 1995, only 1 whooping crane death resulted from DVC, a parent-reared chick <2 months of age that may not have received adequate amounts of medicated pelleted feed to prevent the infection. The 2 coccidia protozoans, *Eimeria gruis* and *E. riechenowi*, found in whooping cranes are found in both Florida and Canada, and adults can act as carriers. The disease is primarily of concern in chicks; both the enteric and visceral forms can be severe. Symptoms of the enteric form include diarrhea, weight loss, bloody stools, and electrolyte and fluid imbalances. Small white nodules (2–4 mm diameter) in various internal organs, including heart, liver, lungs, and kidney, and in the soft tissues of the mouth, are observed in the visceral form. Coccidiosis is unlikely to be a major problem in new wild populations; nevertheless, occasional chick mortality may be encountered. Although *Eimeria* organisms occur in Canada, harsh winters should limit their spread, especially to chicks.

Eastern equine encephalitis (EEE), an arbovirus spread by mosquitoes (primarily *Culiseta melanura*), can lead to lethargy, incoordination, and acute death in whooping cranes (Dein et al. 1986). Pathology is rather nonspecific: ascites, hepatomegaly, splenomegaly, and congestion or mild hemorrhage of the gastrointestinal tract. This agent does not cause disease in most native birds, but it does in whooping cranes and is potentially significant to the Florida release because the arbovirus and mosquitoes occur in the release area. At Patuxent, 5 of the 7 deaths occurred in adult females. Any outbreak could have serious consequences for a reintroduction program, affecting young and older birds of breeding age. EEE has not been reported in the Prairie Provinces of Canada. However, a similar disease (western equine encephalitis) does occur, but this entity has never been documented as causing disease or death in whooping cranes.

Acanthocephalans and the associated category of peritonitis/intestinal rupture in chicks are important causes of mortality (14% total, Table 1). Acanthocephalans are small parasites (2–3 mm long) that lose torpor and detach from the intestinal wall when the host dies, causing difficulty in locating the actual parasite (Greiner and Ritchie 1994). The cause of death is usually attributed to intestinal rupture and peritonitis. The parasite life cycle is thought to involve invertebrate hosts that cranes are exposed to in the field. Most deaths occur in birds <1 year of age. Adult acanthocephalans are controlled by benzimidazole antihelmintics such as oxfendazole, oxibendazole, and thiabendazole (Macwhirter 1994). Thiabendazole is used to control nematodes in whooping cranes in the non-breeding season, but its use in combination with monensin medicated feed during the breeding season should be investigated. There is nothing readily apparent in the captive situation that enhances the incidence of acanthocephalan infestation. On the contrary, if invertebrates do act as intermediate hosts, the pelleted diet reduces the number of invertebrates a crane may be capturing and eating. In 1966–81 only 4.8% of deaths were due to acanthocephalans (Carpenter and Derrickson 1982). However, adding the 2.4% of mortalities caused by peritonitis would give a mortality rate of 7.2%, still half of what we found in 1982–95. Because this is a disease found in young cranes, the increase may be associated with the increasing numbers of chicks being raised in recent years. Regardless of the cause of the increase, this mortality factor requires further research to learn more about control measures.

Trauma as observed in captivity would generally not occur in the wild. Trauma was the direct cause of 21% of whooping crane mortalities; however, nearly every instance was an artifact of captive propagation. Aggression was primarily of 2 types. The first, and most common, was 1 or more cranes attacking and killing a penmate. When a pair forms in a community pen (large pen enclosures housing multiple unpaired adult cranes), the 2 may become territorial and attack other cranes. Also, the adult crane may direct aggression against a chick. Inexperienced parents have been known to kill the chick they are raising. Chicks that wander and enter an occupied pen are at an even higher risk of being attacked. Trauma/aggression accounted for 6% of deaths in this study and 7.3% during 1966–81 (Carpenter and Derrickson 1982). This indicates trauma continues to be a factor in the captive propagation program that has not been resolved.

Fractures and associated complications, especially leg fractures, are often fatal (Olsen 1994). Fractures are usually associated with handling or other aspects of the captive situation. In wild cranes, collisions with powerlines and fences often result in fractures and death. Indeed, several whooping cranes received by Patuxent from the wild flock came as injured adult birds. However, leg fractures are not seen as often in the wild as in the captive flock. In the period 1966–81, there was only 1 fatality (2%) associated with a fracture (Carpenter and Derrickson 1982). We reported 6 cases, indicating a rise in this type of mortality, possibly associated with increased efforts to rear wild, non-imprinted cranes for release.

Spinal trauma occurs when a crane runs or flies into a fence, luxating or fracturing 1 or more vertebrae. If this injury is not immediately fatal, the paralyzed crane has to be euthanized. Recently, this type of mortality has increased as juveniles are raised to be wilder (to enhance survival after release), resulting in increased injury after birds are startled by occasional disturbances.

Mycotoxin-related mortality can occur in wild sandhill cranes (Windingstad et al. 1989) and is a potential threat to released whooping cranes. Mycotoxins are produced by certain aflatoxin or tricothecene fungi. Clinical signs are variable (Olsen et al. 1995) but can include inappetence, weight loss, oral ulceration, and paralysis of wings or neck. Clinical pathology may include elevated uric acid levels. There are no significant gross pathological lesions, and toxin detection can be difficult. Mycotoxin distribution is sporadic and unpredictable, and concentrations often are related to abnormal weather cycles, such as excess moisture. There are no reports of the disease in cranes in Canada, though it is likely that mycotoxins occur, especially where abundant waste grain is a common food source for cranes. Mycotoxins can be a potential problem for any release, but the sporadic nature of the toxin production indicates any problem would be stochastic.

Anatomic abnormalities resulting in fatalities at hatching and of young chicks occur in captivity and may occur in wild cranes. Scoliosis resulted in the death of 9 whooping cranes. We have seen very few instances of scoliosis in the 3 subspecies of sandhill cranes raised at Patuxent (Olsen and Gee 1997). Scoliosis was first diagnosed in 1983 and is a lethal congenital factor in the gene pool of the whooping crane. Therefore, scoliosis will be a likely mortality factor in other whooping crane populations.

Several types of leg deformities are common in the captive setting, including hock rotation associated with valgus deformity, slipped gastrocnemius tendons, and deformed feet. The exact congenital nature of these conditions is not well understood, may not affect the wild populations, and requires further investigation. Anatomic abnormalities resulted in 19.5% of mortalities from 1966 through 1981 (Carpenter and Derrickson 1982) and had a similar incidence (17%) during the years 1982–95 of this study.

Several diseases known to occur in the Canadian Prairie Provinces (Wobeser 1985) are not seen in the captive flock. Avian cholera caused by the bacterial organism *Pasteurella multocida* occurs in lesser snow (*Chen hyperborea*) and Ross's geese (*C. rossii*) in the Kindersley area of Saskatchewan during April and early May. Other waterfowl and sandhill cranes occasionally succumb to avian cholera. The disease is peracute; most birds do not appear clinically ill before death. Some birds may exhibit symptoms such as weakness, depression, inability to fly, and convulsions prior to death. Avian cholera causes scattered hemorrhagic lesions on the viscera, and white or yellow 1- to 2-mm-diameter necrotic foci on the liver. Diagnosis is confirmed by bacterial culture. The outbreaks in Saskatchewan are not as explosive as others seen in waterfowl wintering areas; but because the outbreaks occur in spring with occasional sandhill crane mortality, released whooping cranes may encounter some losses to avian cholera.

Salmonellosis caused by bacteria of the genus *Salmonella* (most commonly *S. typhimurium*) occurs in the Canadian prairies. House sparrows (*Passer domesticus*) around Saskatoon, Saskatchewan, have died from this disease (Wobeser 1985). Generally, sick birds appear depressed, are dehydrated, and have diarrhea. A postmortem examination reveals caseous degeneration of the crop, liver, spleen, kidney, or brain. Yolk sac or intestinal inflammation is seen in young birds. Culture and strain identification is important because virulence varies with strain. Sandhill cranes in Mississippi have been affected by salmonellosis (Langenberg et al. 1994). An occasional non-virulent strain has been diagnosed in captive whooping cranes after routine cultures were taken, but no mortality has been seen. All cranes destined for release are tested for salmonellosis. This is a potential, but unlikely, problem for released whooping cranes in either Florida or Canada. Most cases would involve isolated individuals.

Avian tuberculosis, a chronic, debilitating disease caused by the bacteria *Mycobacterium avium*, has been reported in wild birds in Saskatchewan (Wobeser 1985). A sick bird appears weak and has an elevated white blood cell count with elevated monocyte levels. White caseous nodules are found in the liver, spleen, and gastrointestinal tract on postmortem examination. The disease does occur in whooping cranes in Idaho and Texas (Snyder and Richard 1994) and therefore would be of concern for a release program. However, avian tuberculosis concerns do not preclude the use of the present release sites in Florida or future releases in Canada.

Newcastle disease is caused by a paramyxovirus and there are 4 strains of the virus. The most virulent (viscerotropic velogenic Newcastle disease strain [VVND]) causes high morbidity and mortality in poultry, psittacines, raptors, and waterfowl. There are no pathognomonic signs, and young birds develop a peracute disease resulting in high mortality. Symptoms include yellowish or hemorrhagic diarrhea, coughing, sneezing, dyspnea, incoordination, or ataxia. Gross pathologic findings include enlarged liver and spleen, airsacculitis, and generalized inflammatory response of visceral organs. In the past, North America had been considered free of VVND, but in 1990 about 9,000 birds (7,000 double-crested cormorants [*Phalacrocorax auritus*], 2,000 gulls [*Larus* spp.], and a small number of white pelicans [*Pelecanus erythrorhynchos*]) died in the Canadian Prairie Provinces (Wobeser et al. 1993). This disease must be monitored because any occurrence in the area of the proposed whooping crane reintroduction would pose a threat to the released birds.

In summary, some mortality factors seen in the captive flock at Patuxent, such as infectious and parasitic diseases and congenital abnormalities, will have similar impacts on reintroduction programs. Other conditions are artifacts of the captive propagation situation. Some mortality factors in the captive breeding flock require further investigation to facilitate better understanding and reduction of mortality.

LITERATURE CITED

CARPENTER, J. W. 1977. Propagation and management of endangered species at the Patuxent Wildlife Research Center. Proc. Annu. Conf. Assoc. Avian Vet. 1977:23-33.

______, and S. R. DERRICKSON. 1982. Whooping crane mortality at the Patuxent Wildlife Research Center, 1966–1981. Pages 175-179 *in* J. C. Lewis, ed. Proc. 1981 crane workshop. Natl. Audubon Soc., Tavernier, Fla.

______, T. R. SPRAKER, and M. N. NOVILLA. 1980. Disseminated visceral coccidiosis in whooping cranes. J. Am. Vet. Med. Assoc. 177:815-848.

DEIN, J. F., J. W. CARPENTER, G. C. CLARK, R. J. MONTALI, C. L. CRABBS, T. F. TSAI, and D. E. DOCHERTY. 1986. Mortality of captive whooping cranes caused by eastern equine encephalitis virus. J. Am. Vet. Med. Assoc. 189:1006-1010.

ELLIS, D. H., J. C. LEWIS, G. F. GEE, and D. H. SMITH. 1992. Population recovery of the whooping crane with emphasis on reintroduction efforts: past and future. Proc. North Am. Crane Workshop 6: 142-150.

ERICKSON, R. C. 1968. A federal program for endangered wildlife. Trans. North Am. Wildl. Nat. Resour. Conf. 33:418-433.

GREINER, E. C., and B. W. RITCHIE. 1994. Parasites. Pages 1007-1029 *in* B. W. Ritchie, G. J. Harrison, and L. R.

Harrison, eds. Avian medicine: principles and applications. Wingers Publishing, Inc., Lake Worth, Fla.

HINSHAW, W. R. 1952. Principles of disease prevention. Pages 70-92 *in* H. E. Biester and Z. H. Schworte, eds. Diseases of poultry. Fourth ed. Iowa State Univ. Press, Ames. 1103pp.

LANGENBERG, J., R. WINDINGSTAD, and G. OLSEN. 1992. Summary of diseases and disease issue recommendations. Pages 88-103 *in* U. S. Seal and S. Hereford, eds. Mississippi sandhill crane (*Grus canadensis pulla*) population and habitat viability assessment. IUCN/SSC Conserv. Breeding Specialist Group, Apple Valley, Minn.

MACWHIRTER, P. 1994. Passeriformes. Pages 1172-1199 *in* B. W. Ritchie, G. J. Harrison, and L. R. Harrison, eds. Avian medicine: principles and applications. Wingers Publishing, Inc., Lake Worth, Fla.

OLSEN, G. H. 1994. Crane orthopedics: pediatrics and adults. Semin. Avian and Exotic Pet Med. 3:73-80.

_____, J. W. CARPENTER, G. F. GEE, N. J. THOMAS, and F. J. DEIN. 1995. Mycotoxin caused disease in captive whooping cranes (*Grus americana*) and sandhill cranes (*Grus canadensis*). J. Zoo and Wildl. Med. 26:569-576.

_____, and G. F. GEE. 1997. Causes of Mississippi sandhill crane mortality in captivity, 1984–95. Proc. North Am. Crane Workshop 7:249-252.

SNYDER, S. B., and M. J. RICHARD. 1994. Treatment of avian tuberculosis in a whooping crane (*Grus americana*). Annu. Proc. Am. Assoc. Zoo Vet. 1994:167-169.

WINDINGSTAD, R. M., R. J. COLE, P. E. NELSON, T. J. ROFFE, R. R. GEORGE, and J. W. DORNER. 1989. *Fusarium* mycotoxins from peanuts suspected as a cause of sandhill crane mortality. J. Wildl. Dis. 25:38-46.

WOBESER, G. A. 1985. Handbook of diseases of Saskatchewan wildlife. Saskatchewan Parks and Renewable Resour., Regina. 65pp.

_____, F. A. LEIGHTON, R. NORMAN, D. J. MYERS, D. ONDERKA, M. J. PYBUS, J. L. NEUFELD, G. A. FOX, and D. J. ALEXANDER. 1993. Newcastle disease in wild water birds in western Canada, 1990. Can. Vet. J. 34:353-359.

CAUSES OF MISSISSIPPI SANDHILL CRANE MORTALITY IN CAPTIVITY, 1984–95

GLENN H. OLSEN, Patuxent Wildlife Research Center, 11510 American Holly Drive, Laurel, MD 20708, USA
GEORGE F. GEE, Patuxent Wildlife Research Center, 11510 American Holly Drive, Laurel, MD 20708, USA

Abstract: During 1984–95, 111 deaths were documented in the captive flock of Mississippi sandhill cranes (*Grus canadensis pulla*) housed at the Patuxent Wildlife Research Center. Trauma was the leading cause of death (37%), followed by infectious/parasitic diseases (25%), anatomic abnormalities (15%), and miscellaneous (8%). No positive diagnosis of cause of death was found in 19% of the necropsies. Chicks <2 months old suffered 76% of captive deaths. Trauma, the greatest cause of deaths of captive juveniles and adults, is likely limited to collisions in the wild. Infectious/parasitic diseases and anatomic abnormalities could affect wild chick survival at similar rates to those of captive chicks.

PROC. NORTH AM. CRANE WORKSHOP 7:249-252

Key words: disseminated visceral coccidiosis, *Grus canadensis pulla*, heart defects, Mississippi sandhill crane, mortality, necropsy, trauma.

The Patuxent Wildlife Research Center (Patuxent) held the only captive propagation colony of endangered Mississippi sandhill cranes (MSCH) until 1995. In the early 1970's the population was at a low of approximately 38 birds (Anonymous 1973). The captive flock was started in 1966 from 4 eggs collected in the wild.

Mortality continues to be a significant factor affecting the status of this crane in captivity and the wild. A summary of mortality factors in the captive propagation program at Patuxent (Carpenter et al. 1976) covered all 3 subspecies of sandhill cranes resident at Patuxent (greater sandhill crane [*G. c. tabida*], Florida sandhill crane [*G. c. pratensis*], and Mississippi sandhill crane) during the years 1966–75 but did not separate the subspecies for cause of death. At that time the Mississippi sandhill crane propagation program was in its infancy, and only 11 had been necropsied. A Population and Habitat Viability Assessment Workshop reported mortality of Mississippi sandhill cranes in the wild from 1981 to 1992 and in captivity (Langenberg et al. 1992).

In 1994 the U.S. Fish and Wildlife Service decided to end propagation of this subspecies at Patuxent and divide the captive flock between White Oak Conservation Center in Yulee, Florida, and the Audubon Institute in New Orleans, Louisiana. Therefore, a final examination of the mortality factors affecting this crane flock at Patuxent is needed to aid in managing the 2 new flocks and as a comparison with mortality factors in the wild flock.

We thank L. Lammons, M. T. Childress, and J. L. Kodak for helping prepare and edit the manuscript. We thank J. C. Lewis and D. W. Stahlecker for editorial comments. As with any epidemiological study involving multiple cases over a number of years, necropsies were performed by the authors and others at Patuxent, the National Wildlife Health Center, Madison, Wisconsin, and the Maryland Department of Agriculture, College Park, Maryland. We acknowledge the following veterinarians for their contributions to the Mississippi sandhill crane program: J. W. Carpenter, F. J. Dein, C. Driscoll, D. Green, C. Franson, P. Klein, and N. Thomas.

METHODS

The captive propagation program at Patuxent cares for 300–350 cranes annually, primarily sandhill cranes, but 25% of the flock is whooping cranes (*G. americana*). Adult cranes are housed in pairs in outdoor pens and provided with fresh water and a diet of specially formulated pellets (Olsen and Carpenter 1996). Juveniles (≥2 and <12 months) and some unpaired adults live in community pens containing 2–12 unmated birds. Chicks (<2 months) are housed with surrogate sandhill crane parents or are raised in individual pens by costumed caretakers. Since 1989, Patuxent caretakers have worn gray costumes to disguise their human shape. Crane chicks raised by costumed caretakers are also exposed to adult cranes in adjacent pens.

Diagnoses of cause of mortality were based on medical history, clinical signs, radiology, necropsy results, and laboratory findings (microbiological, parasitological, toxicological, clinical pathological, and histopathological) as delineated in previous studies (Carpenter et al. 1976, Carpenter and Derrickson 1982).

RESULTS

From 1984 through August 1995, 111 Mississippi sandhill cranes died in captivity from various causes (Table 1). Most deaths (76%) occurred among chicks (from hatching to fledging at age 2 months); only 20% were juveniles, and 14% were subadults or adults (≥12 months old). Of the 5 major categories of mortality listed, trauma was the most common cause of death (34%), followed by infectious/parasitic diseases (25%), no definitive diagnosis (19%), anatomic abnormalities (15%), and miscellaneous conditions (8%) (Table 1).

Among chicks, the causes of death were infectious/

Table 1. Causes of mortality in Mississippi sandhill cranes by age class at Patuxent Wildlife Research Center, Laurel, Maryland, 1984–95.

Cause of mortality	Age class[a] (*n*)			All ages	
	Chick	Juvenile	Adult	*n*	%
Trauma	**19**	**8**	**11**	**38**	**34**
Aggression	14	2	6	22	20
Predation	2	2	2	6	5
Spinal trauma	0	0	2	2	2
Fractures	0	4	1	5	4
Miscellaneous	3	0	0	3	3
Congenital abnormalities	**16**	**0**	**0**	**16**	**14**
Scoliosis	3	0	0	3	3
Heart defects	8	0	0	8	7
Miscellaneous	5	0	0	5	4
Infectious or parasitic disease	**23**	**4**	**0**	**27**	**25**
Pneumonia/respiratory disease	8	2	0	10	9
Colibacillosis	1	0	0	1	1
Yolk sac infection or rupture	12	0	0	12	11
Cloacal prolapse or septicemia	1	0	0	1	1
Disseminated visceral coccidiosis	1	2	0	3	3
Miscellaneous mortality factors	**4**	**3**	**2**	**9**	**8**
Maggot infestation	0	1	0	1	1
Intestinal adhesions	1	0	0	1	1
Drowning	0	1	0	1	1
Cloacal impaction	1	0	0	1	1
Capture myopathy	0	0	2	2	2
Stunting syndrome	0	1	0	1	1
Enteritis	2	0	0	2	2
No diagnosis/open	**12**	**7**	**2**	**21**	**19**
Totals	**76**	**22**	**13**	**111**	**100**

[a] Chicks (<2 months); juveniles (≥2 and <12 months); adults (≥12 months).

parasitic diseases (30%), trauma (25%), anatomic abnormalities (21%), miscellaneous conditions (8%), and no definitive diagnosis (16%). For juvenile and adult cranes, trauma was the greatest cause of mortality (juvenile 37%, adult 73%). No mortalities were attributed to anatomic abnormalities in cranes older than 2 months. Likewise, infectious/parasitic diseases were not a mortality factor in subadults and adults. Most deaths were seasonal, occurring in May (15%), June (37%), and July (22%), as expected given the high mortality rate in chicks. There were no deaths in January, and only 1 death each in February, March, and April during the study. Deaths during August through December ranged from 2% to 8% of total mortalities.

DISCUSSION

Trauma was the most common cause of mortality, and the highest ranking mortality factor in cranes >2 months old. Trauma is often associated with pair formation in subadult and adult cranes housed in community pens. When a pair forms, they become aggressive, often defending the food or water resources from penmates. Another form of aggression occurs when a crane jumps or flies into a nearby pen occupied by a pair. The trespasser is usually severely attacked by the pair. Trauma to chicks being reared by parent cranes can result when one of the adult cranes accidently injures the chick, often by stepping on the chick while defending the pen from intruders. An inexperienced parent crane will occasionally attack and kill a chick. Of the different forms of trauma/aggression that occur in the captive flock, only that directed against chicks may occur in the wild.

Predation accounted for 5% of mortality. The predators were primarily great horned owls (*Bubo virginianus*). In the wild flock, predation caused 30.8% of the mortality seen in released birds (Langenberg et al. 1992) and also contributed to mortality in chicks. Five percent of the mortality in the captive flock has been a result of fractures of long bones and ensuing complications. This problem characterizes several crane species raised in captivity (Olsen 1994) and may not be as serious a mortality factor in the wild. Cervical or spinal trauma represented 3% of mortality in captivity, resulting in the deaths of 3 MSHC in 12 years. These injuries often occur when birds fly into obstacles such as fences or feed sheds. It is likely that this type of mortality is also found in the wild flock, as it occurs in other sandhill crane populations, when individual cranes fly into fences or powerlines (Morkill and Anderson 1991, Faanes and Johnson 1992, Fannin 1992, Brown and Drewien 1995).

Congenital anatomic abnormalities accounted for 21% of deaths in chicks. Chicks with severe abnormalities were often euthanized if their chance of survival was low, rather than waiting for natural death to occur. If a similar mortality rate occurred in the wild flock, we would expect 1 of 5 wild-hatched chick deaths to be due to anatomic abnormalities.

Cardiac murmurs accounted for 50% of the deaths listed under anatomic abnormalities. Lesions seen at necropsy were either a ventricular septal defect or an atrial septal defect. Preliminary studies of the genetics of these chicks show a possible recessive genetic factor in certain members of the captive population. Currently, there is no evidence that such a factor exists in the wild population, but we assume such a recessive gene does exist. Not all Mississippi sandhill cranes diagnosed with serious heart defects die as chicks. Some were reared successfully to fledging and released in Mississippi. Eleven cranes with heart murmurs were released in 1989-91. Of these, 7 (64%) died, compared to a 17% mortality rate for the 53 cranes released without murmurs during the same period (Langenberg et al. 1992). Clearly, those birds with serious identifiable murmurs are at high risk. Since 1992 such cranes have not been included in the release program. However, when the wild population becomes self-sustaining, there may be some mortality of juveniles or young adults where a heart defect is a contributing factor.

All identifiable anatomic abnormalities may occur due to congenital factors inherent in the population, but many may also be attributed to contaminants, improper incubation, or other factors (Olsen et al. 1990, Olsen and Clubb 1996). However, because of the controlled situation at Patuxent, contaminant exposure in the captive flock is unlikely. Likewise, incubation technique is not suspect because natural and artificial incubation is used, and no link has been identified between incubation technique and anatomic abnormalities in any of the crane species found at Patuxent.

Infectious diseases are also more common in chicks. Omphalitis/ruptured yolk sac/yolk peritonitis caused one-half of all chick mortality. No one causative agent was common to all cases, although *Escherichia coli* was the most often isolated bacteria. Infectious diseases may occur with the same frequency in the wild population, but this is unknown at the present time.

Pneumonia/respiratory disease-related mortality occurs each year, especially among crane chicks; 8 of the 10 deaths from infectious disease were in chicks, the other 2 deaths occurred in juveniles. No single identifiable etiological agent linked all cases. However, cranes with chronic respiratory problems often develop aspergillosis. Because respiratory infections occur in chicks raised by costumed caretakers and by surrogate crane parents, and because no common etiological agent was identified, we believe this disease would occur with similar frequency in the wild population.

Disseminated visceral coccidiosis (DVC) has caused the death of a number of crane chicks of all species/subspecies in captivity at Patuxent in the past. However, only 3 deaths of Mississippi sandhill cranes in the 12 years covered by this study were attributed to DVC. Two of these deaths occurred in 1988, when the coccidiostat amprollium failed to prevent the disease for the first time. Subsequently, Patuxent switched to the coccidiostat monensin, which has since been effective in controlling DVC. Further investigation into methods for controlling coccidiosis in captive cranes continues. DVC as a cause of mortality in wild Mississippi sandhill cranes is not documented and is probably rare in chicks.

MANAGEMENT IMPLICATIONS

The results of necropsies from captive Mississippi sandhill cranes dwelling at Patuxent are useful to monitor the status of the captive flock as it is divided and moved to new locations. Any increases in mortality factors may indicate problems for the captive propagation program. In addition, by continuing to monitor mortality in the captive flock and comparing these losses to those of the wild flock, we have the opportunity to identify health issues that warrant further investigation.

LITERATURE CITED

ANONYMOUS. 1973. Threatened wildlife in the United States. Endangered Species and Int. Activities Fish Wildl. Res. Publ. 114. 289pp.

BROWN, W. M., and R. C. DREWIEN. 1995. Evaluation of two power line markers to reduce crane and waterfowl collision mortality. Wildl. Soc. Bull. 23:217-227.

CARPENTER, J. W., and S. R. DERRICKSON. 1982. Whooping crane mortality at the Patuxent Wildlife Research Center, 1966-1981. Pages 175-179 *in* J. C. Lewis, ed. Proc. 1981 crane workshop. Natl. Audubon Soc., Tavernier, Fla.

______, L. N. LOCKE, and J. C. MILLER. 1976. Mortality in captive sandhill cranes at the Patuxent Wildlife Research Center, 1966–1975. Pages 268-283 *in* J. C. Lewis, ed. Proc. [1975] int. crane workshop. Oklahoma State Univ. Publ. Printing, Stillwater.

FAANES, C. A., and D. H. JOHNSON. 1992. Cranes and powerlines: an analysis of the issue. Pages 197-202 *in* D. A. Wood, ed. Proc. 1988 North Am. crane workshop. Florida Game Fresh Water Fish Comm. Nongame Wildl. Program Tech. Rep. 12.

FANNIN, T. E. 1992. Contaminant residues in sandhill cranes killed upon striking powerlines in central Nebraska. Proc. North Am. Crane Workshop 6:166-170.

LANGENBERG, J., R. WINDINGSTAD, and G. OLSEN. 1992. Summary of diseases and disease issue recommendations. Pages 88-103 *in* U. S. Seal and S. Hereford, eds. Mississippi sandhill crane (*Grus canadensis pulla*) population and habitat viability assessment workshop. IUCN/SSC Conserv. Breeding Specialist Group, Apple Valley, Minn.

MORKILL, A. E., and S. H. ANDERSON. 1991. Effectiveness of marking powerlines to reduce sandhill crane collisions. Wildl. Soc. Bull. 19:442-449.

OLSEN, G. H., and J. W. CARPENTER. 1996. Cranes. Pages 973-991 *in* R. Altman, K. Quesenbury, G. M. Dorrestein, and S. L. Clubb, eds. Avian medicine and surgery. W. B. Saunders Co., Philadelphia, Pa.

______, and S. CLUBB. 1996. Embryology, incubation, and hatching. Pages 54-71 *in* R. Altman, K. Quesenbury, G. M. Dorrestein, and S. L. Clubb, eds. Avian medicine and surgery. W. B. Saunders Co., Philadelphia, Pa.

______, J. M. NICOLICH, and D. J. HOFFMAN. 1990. A review of some causes of death in avian embryos. Proc. Annu. Conf. Assoc. Avian Vet. 1990:106-111.

AVIAN TUBERCULOSIS IN A WHOOPING CRANE: TREATMENT AND OUTCOME

S. BRET SNYDER, Albuquerque Biological Park, Rio Grande Zoo, 903 10th Street S.W., Albuquerque, NM 87102, USA
MICHAEL J. RICHARD, Albuquerque Biological Park, Rio Grande Zoo, 903 10th Street S.W., Albuquerque, NM 87102, USA
CAROL U. METEYER, National Wildlife Health Center, 6006 Schroeder Road, Madison, WI 53711, USA

Abstract: A whooping crane (*Grus americana*) confirmed as suffering from *Mycobacterium avium* infection was treated for 1 year with daily doses of rifampin (45 mg/kg) and ethambutol (30 mg/kg) and 2 doses of *M. vaccae* antigen. Remission of disease occurred during therapy; however, recrudescence to active infection was suspected by 10 months after the antitubercular drugs were discontinued when the crane exhibited weight loss and had thickening of bowel wall as seen on radiographs. A second therapeutic regimen using azithromycin was then initiated (40 mg/kg fed daily) and was accompanied by a second remission within 6 weeks. After 16 weeks of azithromycin therapy, ethambutol (30 mg/kg daily) was added to the azithromycin to reduce the probability of emergent drug resistance. Three weeks later the crane developed severe hepatic dysfunction as suggested by blood chemistry values. This contributed to the crane's collapse and eventual death. All tissues cultured were negative for *M. avium* infection. A severe hepatopathy and chronic fibrosing cardiomyopathy may have resulted from the drug combinations. This case suggests azithromycin as a promising therapeutic agent in treatment of avian tuberculosis and warrants further investigation.

PROC. NORTH AM. CRANE WORKSHOP 7:253-255

Key words: avian tuberculosis, azithromycin, ethambutol, *Grus americana*, *Mycobacterium avium*, *Mycobacterium vaccae*, rifampin, treatment, whooping crane.

Treatment of avian tuberculosis is seldom attempted because of the long time interval for assessing outcome of treatment and the potential for suppression of disease rather than cure. This may result in recrudescence, further transmission of infection, and increased zoonotic exposure (Bush et al. 1978, Farer 1978, Beehler 1990). However, endangered species or otherwise valuable birds may justify special consideration. Recent reports suggest treatment can be successful in some cases of avian mycobacteriosis (Rosskopf and Woerpel 1991, VanDerHeyden 1994) although these diagnoses were not confirmed as *M. avium* infection by bacterial isolation. Drugs used in treatment of psittacine species were various combinations of isoniazid, rifampin, ethambutol, streptomycin, clofazimine, and cyclocerine. This report follows the course of treatment and outcome over 29 months of a whooping crane infected in the wild with *M. avium*.

We thank J. P. Thilsted, New Mexico Veterinary Diagnostic Services (NMVDS), New Mexico Department of Agriculture, for biopsy histopathology and R. B. Ferguson, Tuberculosis and Mycology Section, Scientific Laboratory Division, New Mexico Department of Health, for isolation and identification of *M. avium* complex. D. G. Teubert, Senior Research Specialist, Department of Pathobiological Sciences, School of Veterinary Medicine, University of Wisconsin-Madison, verified the isolate as *M. avium* by DNA probe. R. L. Cromie, The Wildfowl and Wetlands Trust, United Kingdom, provided *M. vaccae* antigen, and S. B. Kadin and D. J. Williams, Pfizer Laboratories, Inc., Groton, Connecticut, donated azithromycin. Helpful comments were given by J. C. Lewis, U.S. Fish and Wildlife Service (USFWS), and by other reviewers of the manuscript.

CASE REPORT

A debilitated 5-year-old male whooping crane from the experimental Rocky Mountain flock (Drewien and Bizeau 1978) was captured at the Monte Vista National Wildlife Refuge, Colorado, and transported to the Rio Grande Zoo, Albuquerque, New Mexico, on 5 November 1992 by University of Idaho biologists. Details of the initial case presentation, diagnostic workup, and patient management have been reported (Snyder and Richard 1994). At initial examination the bird was thin (weight 5.78 kg) and had a large, easily palpable, mid-coelomic mass and a chronic cloacal prolapse. A firm 2-cm tissue mass was palpable within the prolapsed tissue.

Initial diagnostic tests included blood samples for hematology and serum biochemistry, feces for ova and parasites, whole body radiographs, and biopsy and culture of the tissue mass in the wall of the prolapsed cloaca. Radiographs and biopsy were done under isoflurane anesthesia. The prolapse was reduced and held in place with a purse-string suture.

Radiographs confirmed a large, mid-coelomic soft tissue density and splenomegaly. The initial white blood cell count (WBC) was estimated to be elevated (8-10 WBC/oil immersion field). Dissection of the cloacal mass revealed pockets of brownish, inspissated exudate within a firm fibrous-like stroma. Histologically, there were numerous coalescing granulomas with necrotic centers surrounded by zones of macrophages. An acid fast strain showed rod-shaped, acid-fast bacteria in the centers of many of the granulomas. The preliminary diagnosis was avian tuberculosis, and a decision was made to attempt treatment based on reports of success in some psittacine species (Rosskopf and Woerpel 1991, VanDerHeyden 1994). Mycobacteriosis or other granulomatous diseases were differential diagnoses. *M. avium* complex was confirmed in 3 weeks by isolation from the biopsy tissue (organism on file, NMVDS). The isolate was later verified as *M. avium* by use of a specific DNA probe for this organism.

First Treatment Regimen

Antitubercular treatment was initiated on 12 November 1992 when the bird began to self feed, and was continued for 1 year. Rifampin (Rifadin, Marion Merrell Dow, Inc.) and ethambutol (Myambutol, Lederle) were started at 30 mg/kg daily. The measured portion of powder and tablet were loaded into gelatin capsules and placed into the body cavity of a mouse or herring, which the bird ate. The dosage of rifampin was increased to 45 mg/kg daily (on 10 December 1992) for the remainder of the treatment year. Isoniazid (Isoniazid, USP, Rugby Laboratories) was unsuccessfully added as a third antitubercular drug at a dose of 30 mg/kg daily on 2 occasions early in treatment. On both attempts the bird became anorectic within 3 days but resumed eating when isoniazid was discontinued.

The purse-string suture at the cloaca was removed in mid-January when pericloacal swelling had regressed sufficiently. *M. vaccae* antigen (Cromie 1991) was started at that time. Two 0.05-ml doses of the vaccine were given intradermally 8 weeks apart in the thick skin on top of the head.

Monitoring of the patient throughout treatment was by physical examinations, radiography, hematology, serum biochemistry, and feces collection for demonstration of acid-fast bacilli by staining and by mycobacterial culturing. These data were collected at various intervals from monthly to quarterly throughout treatment. The mid-coelomic mass and spleen were markedly reduced in size by 9 March 1993, 4 months after treatment with antitubercular drugs was begun. At this time the cloacal mass was palpable at less than 1 cm. Nine months after treatment was started (6 August 1993), all tests were within normal parameters, including radiographs. The cloacal mass was no longer palpable. The WBC (by eosinophil Unopette method) had decreased from 32,787/μl on 16 December 1992 to 6,206/μl on 6 August 1993.

The bird continued to appear normal on 10 June 1994, 6 months after administration of antitubercular drugs was discontinued, although it had experienced a weight loss of 0.22 kg from 25 February 1994. This weight loss was possibly an early manifestation of recrudescence. Sequential blood values during treatment and observation have been reported (Snyder and Richard 1994).

Second Treatment Regimen

By 15 September 1994 there was additional weight loss of 0.28 kg and an increase in the WBC from 10,304/μl in June 1994 to 20,280/μl. Radiographs taken 23 September 1994 showed prominent loops of intestine in the pelvic region which were interpreted as uniform thickened bowel wall. No other abnormal findings were evident; the bird's appetite was normal and fecal cultures were negative for mycobacteria.

Although *M. avium* was not again isolated from the bird by fecal culture, we considered recrudescence of infection as the cause of the recurrent wasting illness and bowel wall thickening. Treatment with azithromycin (Pfizer Laboratories, Inc.) was initiated 26 October 1994 at 20 mg/kg given in food daily. After 1 week the dose was increased to 40 mg/kg and was continued 16 weeks until 15 February 1995 with no ill effect. We decided to initially give azithromycin alone to test its potential usefulness in treatment. The bird made a dramatic improvement; weight gain was 0.57 kg over the first 6 weeks on azithromycin (26 October to 8 December 1994) and the WBC dropped to $<10{,}000/\mu$l. The bird again appeared in full remission.

Third Treatment Regimen

The drug manufacturer recommended using azithromycin in combination with other antibiotics to reduce the probability of emergence of drug resistant *M. avium*. Thus, after 16 weeks of azithromycin therapy, a second antibiotic, ethambutol (30 mg/kg daily), was added (starting 15 February 1995). After 3 weeks of treatment with this combination (8 March 1995), the bird was found collapsed and unable to stand; both antibiotics were discontinued. Body weight had dropped 1.5 kg (to 4.69 kg), WBC was 10,442/μl, and serum aspartate aminotransferase (AST) was >760 mg/dl. The serum AST had been 310 mg/dl in June 1994 and 321 mg/dl in September 1994. Tube feeding and intensive supportive care were initiated and continued for 3 weeks as the bird's condition worsened until euthanasia was performed on 4 April 1995. The carcass was submitted to National Wildlife Health Center, Madison, Wisconsin, for complete necropsy.

Feces were collected for acid-fast bacilli (AFB) stain and mycobacterial culture on 17 occasions between 12 November 1992 and 10 March 1995 (29 months). AFB stains were negative on all samples. Cultures were negative for mycobacteria in 9 of the samples; *M. fortuitum* was isolated from 1 sample (28 January 1994), *M. terrae* was isolated from 2 samples (23 August 1994 and 23 November 1994), and 5 samples were overgrown by contaminant bacteria.

Necropsy Findings

A complete necropsy was performed including gross and histopathology, virology, parasitology, bacteriology, and TB culture. Gross necropsy revealed significant lesions involving the heart and liver. There was serosanguinous fluid within the pericardial sac, a mildly cloudy pericardium, and a pale area on the left ventricular wall. The liver was large with rounded borders and was orange/red and mottled. Bile was red-tinged and watery. There was no evidence of the coelomic or cloacal masses found at initial presentation and no tubercular-type lesions were seen in the gastrointestinal tract, liver, or spleen. Histologic lesions were a severe chronic active fibrosing cardiomyopathy and a severe hepatopathy that was suggestive of a toxic reaction. Acid-fast stains failed to reveal evidence of acid-fast bacteria, and no mycobacteria were recovered from cultures of spleen, liver, and ceca.

DISCUSSION

Avian tuberculosis, confirmed by isolation of *M. avium* from granulomatous tissue, was resolved in this whooping crane before its death from an apparent adverse drug reaction (azithromycin + ethambutol). The first drug regimen (rifampin + ethambutol + *M. vaccae*) induced remission of disease, but 10 months after the antitubercular drugs were discontinued, the recurrence of wasting illness suggested recrudescence of infection. Treatment with azithromycin brought about a rapid and pronounced second remission. It is unknown if the final combination (azithromycin + ethambutol) contributed to the control of infection or if the apparent cure had already been effected. Although the results of necropsy suggest that mycobacterial infection was resolved, a sufficient time interval to assess the long-term outcome was not possible.

Azithromycin, a macrolide antibiotic, has shown in vitro and in vivo activity against *M. avium* complex infection in persons with AIDS (Young et al. 1991, Cynamon and Klemens 1994). The results of this clinical trial suggest further investigation of azithromycin as a promising therapeutic agent for *M. avium* infection in birds.

The etiology of the hepatopathy and cardiomyopathy found at necropsy cannot be proven, but the toxic nature of the changes and their temporal relationship to drug therapy (azithromycin + ethambutol) suggest an adverse reaction to the drugs. A recent report describes fulminant hepatic failure in humans undergoing antitubercular therapy with rifampin and/or isoniazid (Mitchell et al. 1995). A similar event likely occurred in this bird when given azithromycin and ethambutol.

Fecal AFB strains and cultures were not reliable for either diagnosis or surveillance of infection in this whooping crane. These tests were negative before and during treatment as well as during recrudescence. Accurate diagnosis required biopsy of affected tissue.

LITERATURE CITED

BEEHLER, B. A. 1990. Management of *Mycobacterium avium* in a mixed species aviary. Annu. Proc. Am. Assoc. Zoo Vet. 1990:125-129.

BUSH, M., R. J. MONTALI, E. E. SMITH, and W. S. PERANTINO. 1978. Clinical experience with tuberculosis in exotic birds. Pages 199-204 *in* R. J. Montali, ed. Mycobacterial infections in zoo animals. Smithsonian Inst. Press, Washington, D.C.

CROMIE, R. L. 1991. Development of a vaccine for avian tuberculosis in wildfowl. Ph.D. Thesis, Univ. London, London, U.K. 393pp.

CYNAMON, M. H., and S. P. KLEMENS. 1994. Chemotherapeutic agents for mycobacterial infections. Pages 250-251 *in* L. N. Friedman, ed. Tuberculosis: current concepts and treatment. CRC Press, Inc., Boca Raton, Fla.

DREWIEN, R. C., and E. G. BIZEAU. 1978. Cross-fostering whooping cranes to sandhill crane foster parents. Pages 201-222 *in* S. A. Temple, ed. Endangered birds: management techniques for preserving threatened species. Univ. Wisconsin Press, Madison.

FARER, L. S. 1978. Mycobacterial infections in zoo animals: health implications for humans. Pages 223-226 *in* R. J. Montali, ed. Mycobacterial infections of zoo animals. Smithsonian Inst. Press, Washington, D.C.

MITCHELL, I., J. WENDON, S. FITT, and R. WILLIAMS. 1995. Anti-tuberculous therapy in acute liver failure. Lancet 345:555.

ROSSCOPF, W. J., and R. W. WOERPEL. 1991. Successful treatment of avian tuberculosis in pet psittacines. Proc. Annu. Conf. Assoc. Avian Vet. 1991:238-251.

SNYDER, S. B., and M. J. RICHARD. 1994. Treatment of avian tuberculosis in a whooping crane. Annu. Proc. Am. Assoc. Zoo Vet. 1994:167-170.

VANDERHEYDEN, N. 1994. Update on avian mycobacteriosis. Proc. Annu. Conf. Assoc. Avian Vet. 1994:53-61.

YOUNG, S. L., L. WIVIOTL, M. WU, P. KOLONOSKI, R. BOLAN, and C. B. INDERLIED. 1991. Azithromycin for treatment of *Mycobacterium avium-intracellulare* complex infection in patients with AIDS. Lancet 338:1107.

IDENTIFICATION OF POTENTIAL REINTRODUCTION SITES FOR WHOOPING CRANES IN THE PRAIRIE PROVINCES OF CANADA

DALE G. HJERTAAS, Saskatchewan Environment and Resource Management, 3211 Albert Street, Regina, SK S4S 5W6, Canada
BRIAN W. JOHNS, Canadian Wildlife Service, 115 Perimeter Road, Saskatoon, SK S7N 0X4, Canada
WENDY J. LYON, Canadian Plains Research Centre, University of Regina, Regina, SK S4S 0A2, Canada

Action 2.3.3.2.2.1 of the national recovery plan for the whooping crane (*Grus americana*) calls for the establishment of an independent flock of whooping cranes in Canada, separate from the Wood Buffalo-Aransas flock, that attains a minimum of 25 breeding pairs by 2020 A.D. Potential reintroduction sites include the Prairie Provinces and possibly parts of western Ontario (Edwards et al. 1994). The first step in completing this action is to identify potential reintroduction sites in the 3 Prairie Provinces. We initially screened areas suggested to us by government and Ducks Unlimited biologists. The most important screening criteria were location outside the main Wood Buffalo-Aransas migratory pathway; habitat quality, quantity, and security; and significant mortality and disturbance factors. The screening eliminated many basins as reintroduction sites because of small size or other restrictions. Elimination did not necessarily mean that these sites would be unsuitable breeding habitat for whooping cranes but simply that we did not consider them to be ideal sites for a reintroduction.

Screening of 33 sites yielded 3 complexes that were further evaluated in the field. These were the Saskatchewan River Delta; The Overflowing River Region including Leaf Lake and Plummer's Marsh; and the Yorkton Region including the Yorkton South marshes, Rokeby Marsh, and Horseshoe Lake Heritage Marsh (Fig. 1). Detailed field evaluations were conducted at Bloodsucker Marsh in the Saskatchewan River Delta, Plummer's Marsh, and Rokeby Marsh. On each of these basins we examined:

1. Water: quantity for territorial breeding space, quality, permanency, and management.
2. Vegetation: general type and distribution for nesting, foraging, and roosting habitat.
3. Food: abundance and diversity of aquatic macroinvertebrates and, secondarily, minnows, amphibians, and some vegetative sources.
4. Site security: land ownership and use, disturbance factors, and mortality factors.
5. Sandhill crane (*G. canadensis*) presence.
6. Whooping crane records: historic and current.

Most results were restricted to the primary study basins in each complex and did not necessarily reflect conditions present on other marshes within each complex. The detailed results were reported in Lyon et al. (1995*a,b*). Each of the areas we studied had advantages and disadvantages as a potential release site.

Fig. 1. Canada's Prairie Provinces showing the historic breeding range of the whooping crane as mapped by Allen (1952), the principal migration corridor from Wood Buffalo National Park (Johns 1992), and locations of the 3 areas evaluated for possible reintroduction: (A) Saskatchewan River Delta, (B) Overflowing River Region, (C) Yorkton Region.

Plummer's Marsh is a secure wilderness site. Invertebrate populations were high on the small portion of the lake we sampled, and there were shallow open water areas and an extensive wet meadow zone where cranes should be able to forage. Islands and extensive clumps of bulrush (*Scirpus* spp.) should offer good nest habitat. Disturbance would be minimal and only predation by gray wolves (*Canis lupus*), lynx (*Lynx canadensis*), and black bear (*Ursus americanus*) was identified as a possible important mortality factor. However, access for working on and around the marsh is very difficult. The bottom is very mucky, making walking almost impossible. Access around the lake would be only by airboat or helicopter. Sandhill cranes nest on the bogs between Leaf Lake and Plummer's Marsh. We tracked 2 birds by satellite in 1994. One wintered close to Aransas, the other near the U.S.-Mexico border along the Gulf Coast.

The Saskatchewan River Delta offers a huge expanse of productive wetlands. Our sampling at Bloodsucker Marsh showed the highest invertebrate populations of the 3 study areas. Because of a series of channels and airboat ramps, accessibility is greater than at Plummer's Marsh, but access remains limited to airboat or helicopter. Disturbance factors are minimal, and again predation was the only important mortality factor identified. The amount of foraging and roosting habitat in shallow open water was a concern because of a tendency for dense emergents, especially common reed (*Phragmites australis*), to form dense stands extending from shore into about 80 cm of water. However, sedge (*Carex* spp.) meadows, bogs, and some shallow zones are present, and, given the vast expanses of marshes, 37,000 ha in the Cumberland South complex alone, there is undoubtedly a significant amount of foraging habitat. Sandhill cranes were rare in the Cumberland marshes, although they can be found in adjacent bog areas. There are no confirmed records of breeding whooping cranes from the Cumberland Delta.

The Yorkton Region has good access to all basins and would be the easiest area in which to conduct a release. However, it is the smallest wetland area and has the greatest number of mortality and disturbance factors. These include 3 electrical transmission lines, as well as highways and occupied farmsteads near the marshes. Most uplands are privately owned, and many are used for agricultural purposes. The Yorkton South marshes are typically surrounded by grassland used for hay or pasture and by cultivated fields. Horseshoe Lake Heritage Marsh includes 2,300 km^2 controlled by Ducks Unlimited. Approximately 4,100 ha of marsh in the remainder of the Yorkton complex are also controlled by Ducks Unlimited through wetland easements. Ducks Unlimited also controls through ownership, leases, and management agreements 2200 ha of upland in this complex, but the majority of the uplands are controlled by local farmers. Rokeby Marsh had the lowest numbers of invertebrates of the 3 areas studied, possibly an effect of the heavy and extensive cattail (*Typha* spp.) stands in this marsh. Additional study would be necessary to assess invertebrate populations in the other marshes of the complex. The Yorkton region is a parkland site and thus subject to periodic drought impacts. Although the entire complexes have never become totally dry, individual marsh units have. The parkland is within the core of the breeding range described by Allen (1952), and several historic breeding records are identified in the area including a record from the Horseshoe Lake Heritage Marsh. The Yorkton area was visited by whooping cranes in 1994 and 1995. Thus some contact between a population at this site and the Wood Buffalo-Aransas population is inevitable. A small sandhill crane population survived the settlement period at Rokeby Marsh and gradually expanded through parts of the Yorkton South complex. This population currently consists of about 20 pairs.

The type of release now being considered, with captive-reared cranes and ultralight aircraft (Clegg et al. 1997, Lishman et al. 1997), requires accessibility to facilitate manipulation, tracking, and retrieval of the reintroduced cranes. These actions will be most feasible at the Yorkton sites. All of the parkland sites contain people and associated hazards. Disturbance factors such as those found near Yorkton will be present at any reintroduction site within the traditional parkland breeding areas. The Recovery Team must decide whether these risks and risks associated with contact with the existing whooping crane flock are acceptable.

LITERATURE CITED

ALLEN, R. P. 1952. The whooping crane. Natl. Audubon Soc. Res. Rep. 3. 246pp.

CLEGG, K. R., J. C. LEWIS, and D. H. ELLIS. 1997. Use of ultralight aircraft for introducing migratory crane populations. Proc. North Am. Crane Workshop 7:105-113.

EDWARDS, R., S. BRECHTEL, R. BROMLEY, D. HJERTAAS, B. JOHNS. E. KUYT, J. LEWIS, N. MANNERS, R. STARDOM, and G. TARRY. 1994. National recovery plan for the whooping crane. Rep. 6. Recovery Nat. Endangered Wildl. Comm., Ottawa, Ont. 39pp.

JOHNS, B. W. 1992. Preliminary identification of whooping crane staging areas in Prairie Canada. Pages 61-66 *in* D. A. Wood, ed. Proc. 1988 North Am. crane workshop. Florida Game Fresh Water Fish Comm. Nongame Wildl. Program Tech. Rep. 12.

LISHMAN, W. A., T. L. TEETS, J. W. DUFF, W. J. L. SLADEN, G. G. SHIRE, K. M. GOOLSBY, W. A. BEZNER KERR, and R. P. URBANEK. 1997. A reintroduction technique for migratory birds: leading Canada geese and isolation-reared sandhill cranes with ultralight aircraft. Proc. North Am. Crane Workshop 7:96-104.

LYON, W. J., D. HJERTAAS, and B. W. JOHNS. 1995*a*. Development of screening criteria and identification of potential reintroduction sites for whooping cranes in Prairie Canada. Unpubl. Rep., Can. Plains Res. Cent., Regina, Sask. 26pp.

______, ______, and ______. 1995*b*. Evaluation of three potential reintroduction sites for whooping cranes in Canada. Unpubl. Rep., Can. Plains Res. Cent., Regina, Sask. 115pp.

PROC. NORTH AM. CRANE WORKSHOP 7:256-257

Key words: *Grus americana*, Prairie Provinces, Canada, reintroduction, whooping crane.

SEMEN COLLECTION AND FERTILITY IN NATURALLY FERTILE SANDHILL CRANES

GUOJUN CHEN,[1] Patuxent Wildlife Research Center, 11510 American Holly Drive, Laurel, MD 20708, USA
GEORGE F. GEE, Patuxent Wildlife Research Center, 11510 American Holly Drive, Laurel, MD 20708, USA
JANE M. NICOLICH, Patuxent Wildlife Research Center, 11510 American Holly Drive, Laurel, MD 20708, USA
JOANNA A. TAYLOR,[2] Patuxent Wildlife Research Center, 11510 American Holly Drive, Laurel, MD 20708, USA

Abstract: Aviculturists often ask if semen collection will interfere with fertility in naturally fertile pairs of cranes. We used 12 naturally fertile Florida sandhill crane (*Grus canadensis pratensis*) pairs for this study, 6 control and 6 experimental. All pairs had produced fertile eggs in previous years and were in out-of-doors pens scattered throughout different pen complexes, within auditory range but physically isolated. Semen was collected on Tuesday mornings and Friday afternoons from 26 February 1993 to 4 June 1993. We used standard artificial insemination methods to collect and to evaluate the semen and spermatozoa. Semen collection did not affect semen quality or quantity. Semen volume, sperm density, sperm motility, sperm morphology, sperm live, sperm number per collection, and male response to semen collection exhibited significant daily variation ($P < 0.05$). Although semen collection began 13 days before the first egg in the experimental group, we observed no differences in the date of first egg laid or in fertility between experimental and control groups. Also, we observed no differences in the interval between clutches or in the percentage of broken eggs between experimental and control groups. Sires consistently producing better semen samples produced fewer fertile eggs than sires producing poorer semen samples ($r = 0.60$).

PROC. NORTH AM. CRANE WORKSHOP 7:258

Key words: Florida sandhill crane, *Grus canadensis pratensis*, semen collection, semen, spermatozoa quality and quantity, egg production, fertility.

---

[1]Present address: Heilongjiang Research Institute of Wildlife, Haping Road, Harbin, 150040, P. R. China.
[2]Present address: Okefenokee National Wildlife Refuge, Route 2, Box 3330, Folkston, GA 31537, USA.

RESULTS OF EIGHT YEARS OF PREDATOR CONTROL TO ENHANCE SANDHILL CRANE PRODUCTION ON MALHEUR NATIONAL WILDLIFE REFUGE, OREGON

GARY L. IVEY, Malheur National Wildlife Refuge, HC 72 Box 245, Princeton, OR 97721, USA

Abstract: Because of a 21% decline in breeding pairs of greater sandhill cranes (*Grus canadensis tabida*) on Malheur National Wildlife Refuge from 1971 to 1985, a predator control program was initiated in 1986 to enhance production. The primary cause for the decline was low recruitment of young due to high predation by ravens (*Corvus corax*), raccoons (*Procyon lotor*), and coyotes (*Canis latrans*). On average, predators destroyed 46% of all crane nests and 90% of the prefledged colts. Mink (*Mustela vison*) were added to the program in 1993 after a study showed them to be an important predator of colts. During the 8 years of the predator control program, crane nest success averaged 68%, compared to 47% before the control program. Predators destroyed an average of 21.5% of crane nests during the control program, compared to 46% during years when no predator control was practiced. Average colt survival was 15.1% during the predator control years vs. 9.8% during years without predator control, and average productivity was 15.9 young per 100 pairs vs. 11.0. The best validation of success of the program was the recovery of the breeding population, which increased from 168 breeding pairs in 1989 to 238 in 1994.

PROC. NORTH AM. CRANE WORKSHOP 7:258

Key words: Greater sandhill crane, *Grus canadensis tabida*, predators, predator control, Oregon.

FOOD AND ENERGY INTAKE RATES OF WINTERING WHOOPING CRANES FORAGING ON TWO SELECTED FOOD ITEMS

FELIPE CHAVEZ-RAMIREZ, Department of Wildlife and Fisheries Sciences, Texas A&M University, College Station, TX 77843-2258, USA
JAY T. NELSON, Department of Wildlife and Fisheries Sciences, Texas A&M University, College Station, TX 77843-2258, USA
R. DOUGLAS SLACK, Department of Wildlife and Fisheries Sciences, Texas A&M University, College Station, TX 77843-2258, USA

Abstract: Number, mass, and potential energy of food items consumed by wintering whooping cranes (*Grus americana*) foraging on blue crab (*Callinectes sapidus*) and wolfberry (*Lycium carolinianum*) fruit were estimated and compared between 2 consecutive winter periods. Mass consumed was determined through field observations by quantifying the number of items taken per unit time and the average, or modal, weight of each food item determined from collections. Gross energy intake was determined by calculating the product of mass consumed times gross energy (kJ/g) of each food item. Potential energy retained was calculated by determining metabolizable energy coefficients for each food item determined in studies of captive whooping cranes. Energy retained varied with food item and between different winter seasons. Calculations of energy retained per hour from foraging on wolfberry fruit resulted in 118.4 and 196.8 kJ during winters 1992–93 and 1993–94, respectively. Energy retained per hour from feeding on blue crab was estimated at 843 and 396.8 kJ during the first and second season, respectively. The decrease in blue crab intake rates apparently resulted from a significant decrease in blue crab abundance from 1 year to the next, whereas the increase in wolfberry fruit appeared to be an adjustment to compensate for the decreased crab availability in salt marsh areas. Comparisons of overall energetic intake to energy expenditure (based on time activity budgets) during the 2 winters showed that whooping cranes were consistently in positive energy balance during winter 1992–93 but in energetic deficit during October–December of winter 1993–94.

PROC. NORTH AM. CRANE WORKSHOP 7:259

Key words: Whooping crane, *Grus americana*, foraging, energy balance, winter, food items, blue crab, wolfberry.

CORRIGENDUM

STAHLECKER, D. W. 1992. Crane migration in northern New Mexico. Pages 1–12 *in* D. A. Wood, ed. Proc. 1988 North Am. crane workshop. Florida Game Fresh Water Fish Comm. Nongame Wildl. Program Tech. Rep. 12.

These tables were not included in the above article. They are published here so that interested readers will have access to them while reading the results and discussion presented in that paper. It is now well documented that sandhill cranes (*Grus canadensis*) from outside the Rocky Mountain Population stage in spring and autumn in the San Luis Valley of southern Colorado (Benning et al. 1997). Therefore, the high autumn counts reported in the 1988 paper and discussed in relation to Table 3 could also have included cranes from other populations.

Table 1. Weather data and crane counts from days of major flights of cranes (>500 cranes) in northern New Mexico during autumns, 1984–87.

	No. of cranes counted				Weather[a]					
					Winds (hr)[b]			Sun[c]		
Date	Jemez River	Rio Grande	Afternoon flight	Totals	1100	1400	1700	AM	PM	Frontal position[d]
Roosting										
11 Nov 1984	340		220	560	-1	-2	-1	1	2	-1
12 Nov 1984	320	1,045	282	1,657	-1	-2	-2	1	3	-1
15 Nov 1984	820		117	937	-2	-2	-1	2	3	-1
18 Nov 1984	420		250	670	-1	1	1	3	3	-1
12 Oct 1985	350	180	0	530	1	-1	-1	3	3	-1
26 Oct 1985	950	5	405	1,360	2	-1	1	0	2	-1
16 Nov 1985	2,140	303	10	2,453	0	-1	-1	1	3	-1
23 Oct 1986	750	40		790	0	-2	-1	3	0	1
4 Nov 1986	750	330	50	1130	2	-1	-1	0	0	-1
7 Nov 1986	106	1683	27	1,816	3	2	-2	2	2	-1
9 Nov 1986	2,380	5,027	150	7,557	1	-2	1	3	3	-1
16 Nov 1986	225	298	15	538	-1	1	2	2	0	-1
17 Nov 1986	152	542		798	-1	-1	-2	2	0	-1
20 Nov 1986	1,122	62	54	1,238	-1	-1	-2	3	3	-1
17 Oct 1987	363		300	663	0	-2	2	3	3	0
26 Oct 1987	1,846	998	410	3,254	2	-2	0	1	3	-1
31 Oct 1987	420	407		827	-1	-1	-2	1	3	-1
3 Nov 1987	2,131	671	131	2,933	1	1	1	0	3	-1
Overflights										
10 Oct 1984[e]	400	50	2,865	3,315	0	-2	0	3	1	-1
14 Oct 1985[f]	0		395+	395+	2	1	2	3	3	1
28 Oct 1985	107		861+	968+	1	-1	-1	3	3	-1
2 Nov 1985	348		396	744	0	-1	2	0	1	1
6 Nov 1985	0		40+	40+	2	1	3	3	3	1
19 Nov 1985	0		1,014+	1,014+	2	2	2	3	1	1
25 Oct 1986	1		3,600	3,601	1	2	-1	3	3	0
10 Nov 1986	1,776	592	2,782	5,115	1	-2	2	3	3	1
23 Nov 1986	42		4,454	4,496	1	1	-1	2	0	1
4 Nov 1987	1,730	227	3,957	5,914	1	1	-2	3	2	1
8 Nov 1987	125		7,270	7,395	-1	2	2	0	2	1
9 Nov 1987	300	15	586	901	1	-1	-2	3	3	0
16 Nov 1987	0		1,055	1,055	0	-1	2	3	3	1
18 Nov 1987	37		1,078	1,115	0	1	1	3	3	1

[a] Weather: - = Unfavorable migration conditions.

[b] Winds: (1100) at Alamosa, Colorado; (1400) at White Rock, New Mexico; (1700) at Albuquerque, New Mexico. Wind speed and direction data were categorized as follows: 0 = calm; 1 = 0.5–2.6 m/sec; 2 = 3.1–7.7 m/sec; 3 = >7.7 m/sec; - = unfavorable direction (100–260°).

[c] Sun (percent cloud cover): 0 = 90–100% ; 1 = 60–80%; 2 = 30–50%; 3 = 0–20%. AM = average of 0800 and 1100 readings for Alamosa, Colorado; PM = average of 1400 and 1700 readings for Albuquerque, New Mexico.

[d] Frontal Position: -1 = approaching or stationary front, northwest of Alamosa, Colorado; 0 = no frontal activity; 1 = cold front that had passed Albuquerque, New Mexico.

[e] 1984 afternoon flight total was the count at White Rock, New Mexico, minus known roost counts (usually just Jemez River). Only on 11 Nov 1984 was coverage of roosts complete.

[f] + = a 1985 afternoon overflight of cranes where count was not complete.

Table 2. Stepwise regression analysis of the effect of weather on crane migration in northern New Mexico, 1984–87. A variable had to exceed the 0.15 significance level for entry into the model.

			r^2 attributable to variables								
			Wind				Sun				
	No. of cranes	n	1100	1400	1700	$\bar{x}$	AM	PM	$\bar{x}$	Front	TOTAL
Roosting	>100	49			4.3					5.7	10.0
	>500	32								10.0	10.0
	>1,000	18								16.4	15.4
	>2,000	10		30.7				20.3			51.0
Overflight	>100	49		4.7						14.2	18.9
	>500	32								20.3	20.3
	>1,000	18	8.0						8.6	40.3	56.9
	>2,000	10	56.6							27.5	84.1
Ratio	>100	49		4.3						17.9	22.2
	>500	32								28.0	28.0
	>1,000	18								64.4	64.4
	>2,000	10								37.4	37.4

Table 3. Downgraded estimates of the autumn 1986 sandhill crane migration over northcentral New Mexico with several potential rates of estimation error.

Estimates	Count type	Count	Estimate	Cooperator reports	Total	Grand total
Reported	Roost	9,992	5,770	1,485	17,247	
	Overflight	3,399	4,124	4,283	11,806	29,053
Assume estimate 25% high	Roost	9,992	4,328	1,114	15,433	
	Overflight	3,399	3,093	3,212	9,704	25,137
Assume estimate 33% high	Roost	9,992	3,866	995	14,853	
	Overflight	3,399	2,763	2,869	9,031	23,884
Assume estimate 50 % high	Roost	9,992	2,885	743	13,620	
	Overflight	3,399	2,062	2,142	7,603	21,223
Assume estimate 75% high	Roost	9,992	1,442	371	11,805	
	Overflight	3,399	1,031	1,071	5,501	17,306

LITERATURE CITED

BENNING, D. S., R. C. DREWIEN, D. H. JOHNSON, W. M. BROWN, and E. L. BOEKER. 1997. Spring population estimates of Rocky Mountain greater sandhill cranes in Colorado. Proc. North Am. Crane Workshop 7:165-172.

RESOLUTIONS PASSED BY THE NORTH AMERICAN CRANE WORKING GROUP

SEVENTH NORTH AMERICAN CRANE WORKSHOP, BILOXI, MISSISSIPPI, U.S.A.

10 – 13 JANUARY 1996

RECOGNITION OF THE CONTRIBUTIONS OF TOM TACHA TO CRANE RESEARCH

Participants of the Seventh North American Crane Workshop wish to recognize Tom Tacha's many contributions to the understanding of Mid-continent sandhill crane behavior, ecology, and taxonomy. Tom published 30 papers on cranes between 1979 and 1995 and was chief editor of *Migratory Shore and Upland Game Bird Management in North America*, published shortly before his death. Tom's Ph.D. work and other morphological and behavioral studies strongly influenced our notions of the boundaries among, and validity of, the 3 migratory sandhill crane subspecies. He set new standards for quantitative analysis of crane data on subjects ranging from time budgets and mate choices to body size and fat reserves. Tom's varied behavioral work included a *Wildlife Monograph* about crane social organization. Workshop participants mourn the loss of Tom and extend our sympathy to family and friends.

HOST RESOLUTION

Participants of the Seventh North American Crane Workshop hereby express their thanks and appreciation to the Mississippi Sandhill Crane National Wildlife Refuge for hosting this workshop, to the staff of the Biloxi Beach Resort Inn, to the various sponsors (particularly the Office of Endangered Species, U.S. Fish and Wildlife Service, Jackson, Mississippi), to the field trip leaders, and to the committees and individuals who made the workshop such a success. The paper sessions were excellent, and the spirit of friendship and dedication to crane conservation among the participants was revitalizing. Special thanks go to North American Crane Working Group Vice President Scott Hereford for his tireless efforts in planning and organizing the workshop, to the session leaders who edited the technical papers, to Wendy Brown for her contributions as NACWG President and newsletter editor from 1992 to 1996, to the 5 other board members for their efforts, and to Richard Urbanek and Dale Stahlecker for the gargantuan task of editing these proceedings.